KT-484-677

electronics

Second Edition

David Crecraft
David Gorham

NORWICH CITY COLLEGE LIBRARY
Stock No 198176
Class 621.381 CRE
Cat. M Proc. 3WKL

The Open University

© The Open University 2003

The right of D. I. Crecraft and D. A. Gorham to be identified as authors of this work has been asserted by them in accordance with the Copyright, Designs and Patents Act 1988.

All rights reserved. No part of this publication may be reproduced or transmitted in any form or by any means, electronic or mechanical, including photocopy, recording or any information storage and retrieval system, without permission in writing from the publisher or under licence from the Copyright Licensing Agency Limited, of 90 Tottenham Court Road, London W1T 4LP.

Any person who commits any unauthorised act in relation to this publication may be liable to criminal prosecution and civil claims for damages.

First published in 1993 by:
Chapman & Hall in association with The Open University

Reprinted in 2000 by:
Stanley Thornes (Publishers) Ltd

Second edition published in 2003 by:
Nelson Thornes Ltd (in association with The Open University)
Delta Place
27 Bath Road
CHELTENHAM
GL53 7TH
United Kingdom

03 04 05 06 07 / 10 9 8 7 6 5 4 3 2 1

A catalogue record for this book is available from the British Library

ISBN 0 7487 7036 4

Page make-up by Alden Bookset Ltd, Exeter

Printed in Great Britain by Ashford Colour Press

Contents

Preface

This book is intended as a teaching text in electronics. It therefore contains a structure and is written in a style that are designed to help you understand the basic principles of the subject and to design a range of electronic circuits. At the same time its content list, its 'signposts', its section-numbering and figure-numbering, its section titles and its full index are intended to enable those who prefer to study by dipping into the subject, as their interest takes them, to learn equally effectively. All chapters have their own aims and objectives, and summaries.

Notes and asides are printed in the margin, so that the main text can flow without interruption.

The book was originally adapted from the Open University course *Analogue and Digital Electronics* (T202) by some of the authors of that course and in this edition has been developed to suit university courses everywhere. It lays the foundations of electronics for those who wish to take the subject further. But it also reaches a level in the subject at which it is possible for you to design a range of high-quality circuits. The emphasis is on developing a clear understanding of the principles behind the design of circuits, rather than on presenting a catalogue or file of well-tried circuits from which you could take your pick. It is, as a consequence, not a comprehensive reference book on electronics.

Following the well-tried and extremely successful Open University style, the texts of most chapters contain questions for you to answer before proceeding further. There are two kinds of question included in the text. Firstly there are worked examples. These are simple questions arising out of the previous paragraphs that you should be able to answer without much difficulty. They are intended to make your study of the book a little less passive, and to emphasize key points. The answers follow on immediately from the questions. Secondly there are 'self-assessment questions' which are intended to be a little more challenging, and to enable you to test whether you are achieving the aims or the objectives of the chapter concerned. The expected answers to these questions are at the end of the chapter in which they appear. Neither kind of question is much concerned with testing your powers of recall. Although every chapter introduces terminology which is likely to be new to you, and expects you to remember these words for later use, neither the worked examples nor the self-assessment questions test you on your grasp of the meanings of these words.

The terms introduced in each chapter are listed at the beginning of the chapter under the heading of 'General Objectives' and they are printed in **bold** when they are first introduced and explained. You should try to make sure you have understood their meanings sufficiently for you to use them correctly yourself if required. This is a somewhat lesser aim than being able fully to explain them to others, since in many cases the full meaning of a term (such as 'transistor', 'transformer', 'integrated circuit') embraces aspects of the topic concerned that go well beyond the scope of the book. That is, there are types of transistor or integrated circuit, and aspects of transformers, for example, which are not discussed in this book. The book does, however, deal with the principles underlying the meanings of all the terms it introduces.

Before studying this book, we expect you to have studied basic electricity; that is direct current (d.c.) and voltage, batteries, resistance and power. This book revises these concepts briefly before taking the subject further.

For analogue circuit simulation on a PC, use a simple derivative of SPICE. For digital circuits, there are many packages available.

All but basic mathematics is explained in the text as and when it is needed. However, since one of the aims of the book is to enable you to do back-of-envelope calculations, rather than very detailed analyses, the book avoids advanced mathematical analyses. This can be left to computers and the many versions of circuit simulation software that are now available. Suggestions are made in the text where computer simulation of circuit performance can help you understand how a circuit works, and can easily be implemented using the kind of software that can be run on microcomputers. These points are flagged by a computing 'icon' in the margin.

DAVID CRECRAFT
DAVID GORHAM

Acknowledgements

The first edition was based on the 'Analogue and Digital Electronics' course at the Open University, which involved the following people.

AUTHORS

Dr David Crecraft *Joint Course Team Chairman*
Dr David Gorham
Roger Loxton
Dr Mike Meade *Joint Course Team Chairman*
Dr Phil Picton
Dr Ed da Silva
Prof John Sparkes
Mirabelle Walker

COURSE SUPPORT TEAM

Steve Best *Graphic Artist*
Dr Keith Cavanagh *Editor*
Ian Every *Academic Computing Service*
John Harne *Academic Computing Service*
Ruth Hall *Designer*
Dr Mavis Hamilton *Course Manager*
Roger Harris *Course Manager*
Andy Reilly *Editor*
Karen Shipp *Academic Computing Service*

Symbols used in this book

Unless otherwise specified the **subscripts** used are as follows:
b, B = base; c, C = collector; e, E = emitter; be, BE, etc. = base-emitter, etc.; of a bipolar transistor.
d, D = drain; g, G = gate; s, S = source; dg, DG, etc. = drain-gate, etc.; of a MOSFET or JFET.
i = input; o = output; in equivalent circuits.
n = n-region; p = p-region; of a semiconductor.
h = high; l = low in digital circuits.

Doubled subscripts, such as V_{CC}, refer to supply voltages, in this case to collectors. A bar over a Boolean logic symbol indicates its complement.

Symbol	***Representing***	***Unit (SI)***
$\mathbf{A}_v A_v$	Voltage gain (open loop) of an amplifier or other network	dimensionless **phasor** or ratio
A_0	D.C. (zero-frequency) voltage gain	dimensionless
a.c.	Adjective applied to alternating voltages and currents	
B	Bandwidth	hertz
B	Susceptance: the quadrature part of $\mathbf{Y}$	siemens
B	Magnetic flux density	tesla
C	Capacitance	farad
CMRR	Common-mode rejection ratio	dimensionless
C_{PD}	Power dissipation capacitance	farad
d.c.	Adjective applied to non-alternating voltages and currents	
e	electromotive force (e.m.f.), instantaneous value	volt
E	Source electromotive force (e.m.f.)	volt
E	Electric field strength	$V\ m^{-1}$
f	Frequency	hertz
f_P	Full power bandwidth	hertz
$\mathbf{G}$	Closed-loop gain of feedback amplifier	dimensionless phasor
G	D.C. (zero-frequency) value of G	dimensionless
G	Conductance; reciprocal of resistance	siemens
GB	Gain–bandwidth product	hertz
$\mathbf{g}$, g	Small-signal conductance	siemens
g_m	Mutual conductance or transconductance	amp per volt
h_{fe}	Small-signal common-emitter current gain (dI_C/dI_B) of a bipolar transistor	dimensionless
I	Current; d.c. value; or amplitude or r.m.s. value of a.c. current	ampere
I_B	Input bias current of an amplifier	ampere
I_D	Diode current	ampere
I_F	Maximum forward current for a diode	ampere
I_S	Diode saturation current	ampere
I_{SC}	Short-circuit current	ampere
I_{IH}, I_{IL}, I_{OH}, I_{OL}	Input and output currents for high and low logic levels	ampere
I_{IO}	Input offset current of a differential amplifier	ampere

(continued)

Symbol	*Representing*	*Unit (SI)*
I_n	Equivalent input noise current of an amplifier (r.m.s.)	ampere
i	Current, instantaneous value	ampere
i_{sc}	Short-circuit small-signal current	ampere
L	Inductance	henry
NM	Noise margin	volt
N_p, N_s	Primary, secondary turns in a transformer	dimensionless
P	Power	watt
Q	Quality factor of a resonant circuit	dimensionless
Q	Charge	coulomb
Q_n	Label for a logic output after n clock pulses	
q	Instantaneous value of charge	coulomb
R	D.C. resistance of a resistor	ohm
R_L	Load resistance connected to a circuit	ohm
$R_{L(total)}$	Total load on the current source of a transistor	ohm
R_{in}, R_{out}	Input, output resistance of a circuit	ohm
r	Slope resistance, small-signal resistance	ohm
S	Slew rate	$V\ s^{-1}$
T	Temperature	kelvin
t	Time	second
t_{pd}, t_{su}, t_h	Propagation delay time, set-up time, hold time of digital circuits	second
V	Voltage: d.c. value or amplitude or r.m.s. value of a.c. voltage	volt
VA	Early voltage in a bipolar transistor	volt
V_D	D.C. voltage applied to a diode	volt
V_F	Forward voltage drop for a diode	volt
V_{IH}, V_{IL}, V_{OH}, V_{OL}	Input and output voltages for high and low logic levels	volt
V_{IO}	Input offset voltage of a differential amplifier	volt
V_n	Equivalent input noise voltage of an amplifier (r.m.s.)	volt
V_{oc}	Open-circuit voltage	volt
V_P	Pinch-off voltage of a JFET	volt
V_{ref}	Reference voltage	volt
V_R	Peak reverse voltage for a diode	volt
V_S	Supply voltage or source voltage	volt
V_T	Threshold voltage of a MOSFET	volt
v	Instantaneous value of voltage	volt
X	Reactance: the imaginary or quadrature part of Z	ohm
Y	Admittance; reciprocal of impedance	siemens
Z	Impedance	ohm
Z_L	Load impedance	ohm

	Constant quantities	
c	Speed of light and other electromagnetic waves	$300\ Mm\ s^{-1}$
K	Constant ≈ 40 at $T = 293$ K	V^{-1}
K	In a digital context $= 2^{10} = 1024$	
M	In a digital context $= 2^{20} = 1\ 048\ 576$	
ε_0	Permittivity of free space: 8.854	$pF\ m^{-1}$
μ_0	Permeability of free space: 1.257	$\mu H\ m^{-1}$
π	Circumference/diameter of circle: 3.1416	

	Symbols using Greek characters	
α (alpha)	D.C. current ratio of a bipolar transistor (I_C/I_D)	dimensionless
β (beta)	D.C. current ratio of a bipolar transistor (I_C/I_B)	dimensionless
β	Feedback ratio: the fraction of the output fed back to the input	dimensionless
δ (delta)	A tiny increment	
Δ (capital delta)	A finite increment	
ε (epsilon)	Electric permittivity ($\varepsilon = \varepsilon_0 \varepsilon_r$)	F m^{-1}
ε_r	Relative permittivity	dimensionless
θ (theta)	Phase angle	radian
θ_{JC}, θ_{CS}, θ_{SA}	Thermal resistance: junction–case, case–sink, sink–ambient	°C W^{-1}
λ	Channel length modulation factor (MOSFETs)	V^{-1}
μ (mu)	Magnetic permeability ($\mu = \mu_0 \mu_r$)	H m^{-1}
μ_r	Relative permeability	dimensionless
ρ (rho)	Resistivity	Ω m^{-1}
σ (sigma)	Conductivity	S m^{-1}
Σ (capital sigma)	Sum (of a mathematical series)	
τ (tau)	Time constant	second
ϕ (phi)	Phase angle	radian
Φ (capital phi)	Magnetic flux	weber
ω (omega)	Angular frequency ($\omega = 2\pi f$)	radian s^{-1}

Some SI (Systéme Internationale) units and their abbreviations

Variable	*SI unit*	*Abbreviation*	*Notes*
Current	ampere	A	Commonly abbreviated to 'amp'
Charge	coulomb	C	
Power ratio	bel	B	Not strictly an SI unit
	decibel	dB	Not strictly an SI unit
Capacitance	farad	F	
Inductance	henry	H	
Frequency	hertz	Hz	
Energy	joule	J	
Temperature	kelvin	K	Never use capital K for 'kilo'!*
Length	metre	m	
Angle	radian	rad	
	degree	°	Not strictly an SI unit
Time	second	s	
Conductance	siemens	S	Never use capital S for 'seconds'!
Voltage (e.m.f., potential)	volt	V	
Power	watt	W	
Resistance	ohm	Ω	

*Note that a capital K is used as the multiplier 1024 ($= 2^{10}$) in digital quantities, such as KB (kilo byte). This of course is not an SI usage.

SI prefixes (multipliers) used with SI units

Prefix	*Abbreviation*	*Value*
femto	f	10^{-15}
pico	p	10^{-12}
nano	n	10^{-9}
micro	μ	10^{-6}
milli	m	10^{-3}
kilo	k	10^{3}
mega	M	10^{6}
giga	G	10^{9}

D.C. circuits and methods of circuit analysis

1

AIMS

The aims of this chapter are

1. To revise some of the basic concepts of direct-current (d.c.) electrical circuits.
2. To explain measurements of voltage and current.
3. To introduce some methods for the analysis of d.c. resistive circuits, and to develop familiarity with, and confidence in, their use.

When you have completed your study of this chapter you should have achieved the following]objectives.

GENERAL OBJECTIVES

1. Understand the meaning of, and use correctly, the following terms:

- accuracy (of a meter)
- ammeter
- ampere
- charge
- conductance
- conductor
- current divider
- electrical energy
- electrical power
- electrical resistance
- electromotive force (e.m.f.)
- equivalent circuit
- equivalent resistance
- insulator
- internal resistance
- linear resistor
- non-linear resistor
- ohm
- open circuit voltage
- parallel connection of resistors
- potential difference
- potential divider
- power dissipation
- resistance
- sensitivity (of meter)
- series connection of resistors
- series resistors
- shunt resistor
- siemens
- temperature coefficient of resistance
- variable potential divider (pot)
- volt
- voltage divider
- voltmeter

2. Understand, and be able to explain to someone else, what is meant by the terms 'circuit analysis' and 'network analysis'.

3. Understand the meaning of and use correctly the following terms:

- branch
- common connection
- duality
- earth (ground) connection
- ideal current source
- ideal voltage source
- internal resistance
- Kirchhoff's current law
- Kirchhoff's voltage law
- linear device
- loop
- mesh
- nodal analysis
- node
- Norton equivalent circuit
- principle of superposition
- reference node
- Thévenin equivalent circuit

SPECIFIC OBJECTIVES

1. State the formulae relating, and perform simple calculations involving, the electrical quantities: charge, current, voltage, resistance, conductance, power and energy.
2. State and use the formulae for the equivalent resistance and/or equivalent conductance of resistors in series and in parallel.
3. Use the voltage and current divider rules to calculate voltages and currents in simple circuits.
4. Explain how a basic meter movement may be converted to a multimeter capable of measuring current, voltage and resistance over a wide range of values.
5. Calculate the effect on a simple circuit of connecting an ammeter into it or connecting a voltmeter across a component of it.
6. Give a formal statement of Kirchhoff's current and voltage laws and use them to develop the current law and voltage law equations for a circuit.
7. Explain the use of a Thévenin equivalent circuit and a Norton equivalent circuit to represent a voltage source.
8. Use Thévenin's and Norton's theorems to calculate the equivalent circuit of a network.
9. Use the principle of superposition to calculate the Thévenin and Norton equivalent circuits for a circuit containing multiple voltage and/or current sources.
10. Describe the sequence of steps which constitutes the method of nodal analysis.
11. Use the method of nodal analysis to develop the set of simultaneous equations which allow evaluation of the node voltages.
12. Use a CAD package to analyse a range of d.c. resistive circuits of varying complexity.

1.1 INTRODUCTION

In writing this first chapter, we have assumed that you, dear reader, already know the basic facts about voltage (e.m.f.), current, power and resistance in d.c. circuits. We start Section 1.2 with resistance, by way of revision, and go on to non-linear resistance and the concept of internal resistance of an e.m.f. source.

We have brought forward a summary of Section 1.2 so that you can see at a glance the concepts upon which the rest of this chapter, and this book, are built. This should provide a useful revision at this point, and an *aide-memoire* in your further reading.

Methods of circuit analysis are described here, and illustrated, for d.c. circuits only. You will find that a thorough understanding of these methods for d.c. will make your subsequent use of them for a.c. circuits fairly painless.

We have also assumed familiarity with some of the standard prefixes and their symbols to indicate fractions or multiples of basic units. Examples are:

micro-(μ) $\times 10^{-6}$
milli-(m) $\times 10^{-3}$
kilo-(k) $\times 10^{3}$
mega-(M) $\times 10^{6}$

Hence,

1 microampere (1 μA) $= 10^{-6}$ A
1 kilojoule (1 kJ) $= 10^{3}$ J
1 megohm (1 MΩ) $= 10^{6}$ Ω
1 millivolt (1 mV) $= 10^{-3}$ V
etc.

Note that abbreviations do *not* pluralize; e.g. 10 cm, *not* 10 cms: this means 10 centimetre seconds!

All the upper-case symbols for *units* are abbreviations for the names of famous scientists and engineers, but when these units are written in full they have *lower-case* initial letters as shown above.

1.2 ELECTRIC CURRENT, ELECTROMOTIVE FORCE, POTENTIAL DIFFERENCE, ENERGY AND POWER, RESISTANCE AND CONDUCTANCE

1.2.1 Summary of Section 1.2

1. An electric current consists of the movement or drift of electrons along a conductor. The unit of current is the ampere (A), which numerically is equal to a flow rate of electrons of 6.24×10^{18} electrons per second through any cross-section of the conductor. An electric current exists when electrons are caused to move around a closed conducting path by a source of electromotive force (e.m.f.).

2. A d.c. electromotive force can be generated by chemical reactions in a battery, by using mechanical power in a generator or by converting light energy in a solar cell. The unit of e.m.f. is the volt (V), and 1 volt is that e.m.f. which, when causing a current of 1 ampere to flow, dissipates energy at a rate of 1 joule per second, that is, 1 watt.
3. The ratio of the potential difference in a circuit to the current flowing around the circuit is called the resistance of the circuit. The resistance of a device is equal to the voltage across the device divided by the current flowing through it. The unit of resistance is the ohm (Ω). A resistor of 1 ohm is one in which a current of 1 ampere flows when there is a potential difference of 1 volt across it. Devices that obey Ohm's law (or do so very nearly) are called **resistors**.
4. Voltage, current and resistance are related to each other by the relationships

   $$R = V/I, \quad V = IR, \quad I = V/R$$

5. The ratio of current to potential difference (i.e. the reciprocal of resistance) is called the conductance. The unit of conductance is the siemens (S).
6. Voltage, current and conductance are related to each other by the relationships

   $$G = I/V, \quad V = I/G, \quad I = VG$$

7. Electrical power (that is, the rate at which electrical energy is consumed) is measured in watts (W), where 1 watt is equal to 1 joule per second. The power (in watts) dissipated in a resistor is equal to the product of the voltage drop and the current:

   $$P = VI = I^2R = V^2/R$$

   In resistors, electrical energy is dissipated as heat.
8. Electrical resistance tends to change with temperature. For metallic conductors the resistance increases with increasing temperature. In semiconductors it usually decreases with increasing temperature.
9. All e.m.f. sources possess internal resistance which reduces the terminal voltage when current flows from the source. The e.m.f. source can be represented by an equivalent circuit consisting of an ideal (i.e. resistance-less) e.m.f. source in series with the internal resistance.

1.2.2 Resistance and conductance

The magnitude of the current flowing in a circuit depends on the magnitude of the e.m.f. source, and on the device or devices through which the current is flowing. For example, the rear window demister on a car has a current of 10 A flowing through it when connected to the car battery of e.m.f. 12 V. The ratio of voltage to current is called the **resistance** of the demister, R. So

$$R = V/I$$

The unit in which we measure resistance is the **ohm** (symbol Ω). A resistance of 1 Ω will have a current of 1 A flowing through it when the potential difference across it is 1 V. Therefore the resistance of the demister is

$$R = 12\ \text{V}/10\ \text{A} = 1.2\ \Omega$$

If the 12 V battery were replaced with a 6 V battery, the current in the demister would be only 5 A. In this case halving the e.m.f. causes the current to be halved too. In fact the magnitude of the current is proportional to the magnitude of the e.m.f. source, so resistance is independent of the current flowing.

For many of the electrical devices we shall meet, the current is not proportional to the voltage, so that the resistance varies as the current varies. However, devices that have a more or less constant value of resistance over as wide as possible a range of operating conditions are called **resistors**. A voltage/current graph for a resistor would ideally be a straight line, as shown in Fig. 1.1, where the slope of the line is the resistance. Practical resistors have nearly constant resistance and hence an almost linear voltage/current graph. The physicist Ohm was only familiar with materials having a constant resistance when he stated his law that 'the current flowing in an electrical conductor is proportional to the voltage across it at constant temperature'. Devices which obey Ohm's law, like resistors, are often called **ohmic devices**.

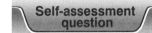

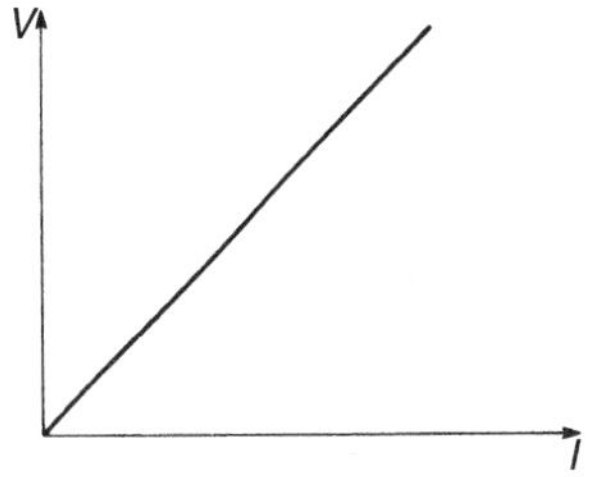

Fig. 1.1 The graph of voltage against current of a resistor at constant temperature.

The equation $R = V/I$ is most frequently written in the form

$$V = IR$$

and is commonly, though erroneously, given the title 'Ohm's law'. In fact all the equation says is that the voltage across a device is equal to the product of current and resistance; it says nothing about the constant nature of R which Ohm's Law states.

Self-assessment question

1 (a) If a current of 3 A flows through a light bulb when the potential difference across it is 12 V, what is the resistance of the bulb?
(b) If an e.m.f. source causes a current of 10 mA (10×10^{-3} A) to flow in a circuit of resistance 300 Ω, what is the voltage of the e.m.f. source?
(c) A resistor of resistance 0.1 Ω has a current of 200 mA flowing through it. What is the potential difference across the resistor?

Self-assessment question

2 Using the relationship $V = IR$ and the expression for the power ($P = V \times I$) in a circuit, find alternative expressions for P which could be used (a) to find the power dissipated as heat in a resistor R given the current I flowing through it and (b) to find the power dissipated as heat in the resistor given the voltage V across it.

An alternative way of relating current and voltage in a circuit or device is to use the ratio I/V instead of V/I. The ratio I/V is called the **conductance** of the circuit, symbol G:

$$G = I/V$$

The unit of conductance is the **siemens** (symbol S). A resistor whose conductance is 1 S, through which a current of 1 A is flowing, has a potential difference across it of 1 V.

Clearly, comparing the definition of the ohm and the siemens, the relationship between them is simply

$$G = 1/R \quad \text{or} \quad R = 1/G$$

Which you choose to use is an entirely personal choice. Traditionally, resistance has been used more commonly than conductance, but as you will see later in this chapter, and in later chapters, there are occasions when conductance is a more convenient property than resistance to use in calculations.

Self-assessment question

3 (a) If a current of 0.5 A flows through a circuit when the voltage of the e.m.f. source is 200 V, what is the conductance of the circuit?
(b) A current of 5 A flows through a resistor whose conductance is 10 S. What is the potential difference across it?
(c) A resistor whose conductance is 0.3 S has a potential difference across it of 2.4 V. What is the current flowing through it?
(d) What is the conductance of the demister described earlier in this section?

Conductors are made from the materials in which a small potential difference can cause a large current flow. They therefore have very low resistance and high conductance. Copper is the most commonly encountered conducting material; other examples of good electrical conductors are aluminium, silver, gold and platinum. Notice that all these materials are pure metals. Part of the definition of a metal is that it has large numbers of mobile electrons, and is therefore a good conducting material. Not all conducting materials are necessarily metals, but all metals are conductors.

Insulators are made from materials in which only a very small current flows, even when a very large potential difference exists across them. They have very high resistance and very low conductance. Some examples of common insulating materials are most plastics, rubber, glass and ceramic materials. Air is also an insulating material.

Some materials have neither good insulating properties nor good conducting properties. Examples of such materials are carbon, some metallic alloys and silicon. Silicon is one of the range of materials known as **semiconductors**, which are used in the manufacture of transistors and integrated circuits (both of which you will meet later in this book).

All materials change their characteristics, including resistance, with changes in temperature. Most conductors experience an increase in resistance when their temperature rises. They are said to have a positive **temperature coefficient of resistance**. Semiconductors and many insulators experience a decrease in resistance with increase in temperature and are said to have a negative temperature coefficient.

Current flowing in a device causes heat to be generated, so that whenever a current increases so does the rate of heat generation, with an inevitable rise in temperature. Since this temperature rise is associated with a change in resistance, how can a resistor have a constant resistance over a range of currents? The answer is that resistors are designed so that (a) the temperature rise is not large (provided the power dissipated in them is within a specified limit) and (b) the temperature coefficient of resistance is small. Under these conditions the resistor can usually be considered to have a constant resistance as the current through it varies.

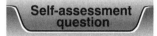

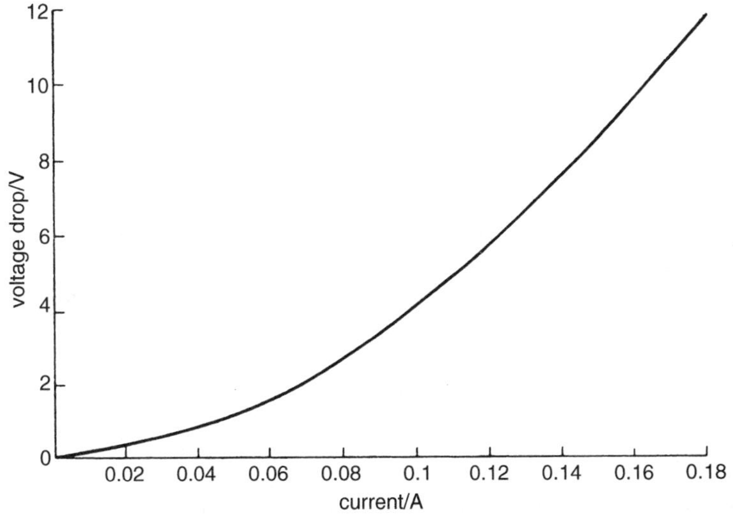

Fig. 1.2 Voltage drop against current for an electric bulb.

A device such as a torch bulb, or a domestic lighting filament bulb, is designed to get very hot when supplied with its rated voltage, so its resistance changes quite dramatically from its cold value to its value under normal operating conditions. The resistance of a torch bulb filament can change from 8 Ω at 20° C to about 33 Ω at 1500° C, the temperature at which it operates. The resistance of the filament of a 240 V 60 W domestic light bulb changes from 70 Ω at 20° to about 1000 Ω at 2000° C. Figure 1.2 is a graph of the voltage/current relationship of a 12 V 2 W tungsten filament bulb, clearly showing that the ratio *V*/*I* is not constant.

Some components have a voltage/current graph which is not a straight line even at constant temperature. Figure 1.3 is the graph of current against voltage for a junction diode (described later in the book). Two examples of junction diodes are shown in Fig. 1.4. It is not possible to designate a single value of resistance for such a device, because the resistance varies with the current flow.

Even though the value of resistance for many devices is not a constant, the d.c. resistance at any chosen current is still equal to the ratio of the voltage drop to the current flowing, and the conductance is still the inverse of the resistance.

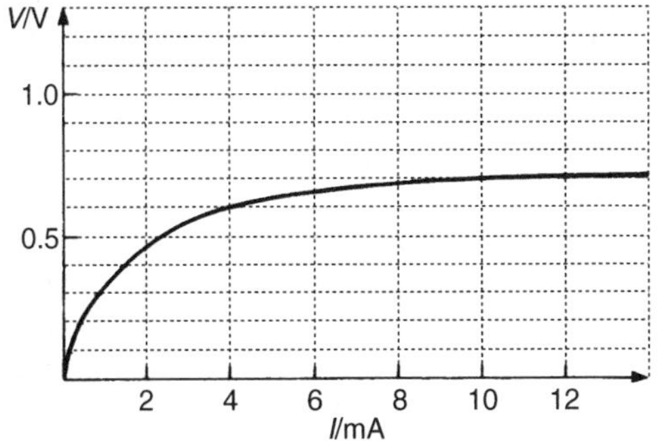

Fig. 1.3 A graph of voltage against current for a junction diode at constant temperature.

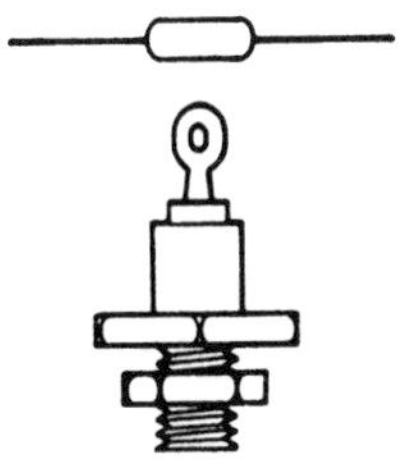

Fig. 1.4 Examples of junction diodes.

Self-assessment question

4 (a) Calculate the resistance of the light bulb of Fig. 1.2 at (i) 0.1 A, (ii) 8 V. What is the power dissipated in the bulb in each case?
(b) Calculate the conductance of the junction diode of Fig. 1.3 at (i) a current of 4 mA, (ii) a current of 10 mA.

For non-ohmic devices like the junction diode, we often wish to refer to the value of the ratio of a *change* in voltage, ΔV, to the resulting *change* in current, ΔI. This ratio is called the **incremental resistance**, or the **small-signal resistance**. As you can see in Fig. 1.5, the incremental resistance $\Delta V/\Delta I$ is the slope of the line AB. This is clearly different from the resistance at A, which is V_1/I_1 and is the slope of the line 0A. For very small changes in voltage and current, the slope of the line AB becomes the slope of the curve itself, and the incremental resistance becomes $r = \mathrm{d}V/\mathrm{d}I$, the differential coefficient of V with respect to I. With ohmic resistors, like the demister, the resistance $R = V/I$ is the same as the incremental resistance, $r = \mathrm{d}V/\mathrm{d}I$. Ohmic resistors are alternatively called **linear resistors**, and non-ohmic resistors are usually called **non-linear resistors**.

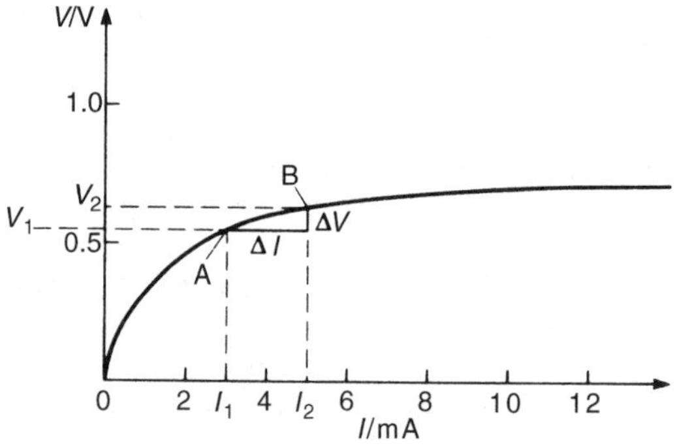

Fig. 1.5 The incremental resistance of the diode is $\Delta V/\Delta I$.

1.2.3 Internal resistance and output voltage of an e.m.f. source

With any practical e.m.f. source, the voltage that can be measured at the terminals of the device is reduced when current is taken from the device. An ideal e.m.f. source would have a voltage which was independent of the current. This departure of real devices from the ideal model of an e.m.f. source must be taken into account when performing circuit calculations. To take account of this voltage change, we say that the e.m.f. source has an **internal resistance**.

Figure 1.6 shows two ways of representing a battery. In Fig. 1.6a the usual symbol for the battery is drawn, together with a note of its internal resistance. In Fig. 1.6b, the two quantities V_S and R_S are separated; the e.m.f. generated is again represented by the symbol for a battery, but the internal resistance of the battery is separately represented. So in Fig. 1.6b the symbol for the source represents an 'ideal' (i.e. resistanceless) source of e.m.f. V_S. Its internal resistance is then put alongside it. Separating the e.m.f. and internal resistance in this way is not, of course, a true representation of where, in the battery, the two phenomena occur. They are certainly intermixed in practice. But from

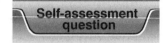

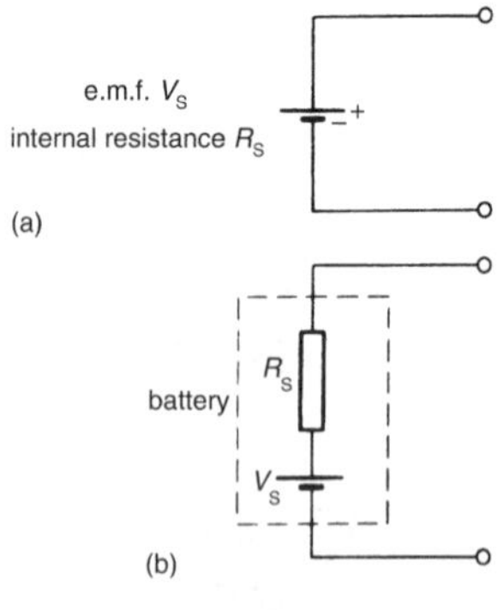

Fig. 1.6 (a) A battery has internal resistance R. (b) An equivalent circuit for the battery.

the point of view of calculating the terminal voltage it represents the battery quite accurately. The two symbols enclosed in the box in Fig. 1.6b are called the **equivalent circuit** of the battery, since they represent in a convenient form all that need be known about the battery to describe its performance in the circuit.

When an external load resistor R is connected to the battery, as in Fig. 1.7, a current I flows around the circuit. A potential difference V_R will exist across the internal resistance R_S, its magnitude being the product of the current and resistance, namely IR_S. The terminal voltage of the battery (that is, the voltage which can be measured across the terminals) will then be

$$V_B = V_S - IR_S$$

So the terminal voltage falls as the current increases, which is the observed effect that the equivalent circuit is designed to represent.

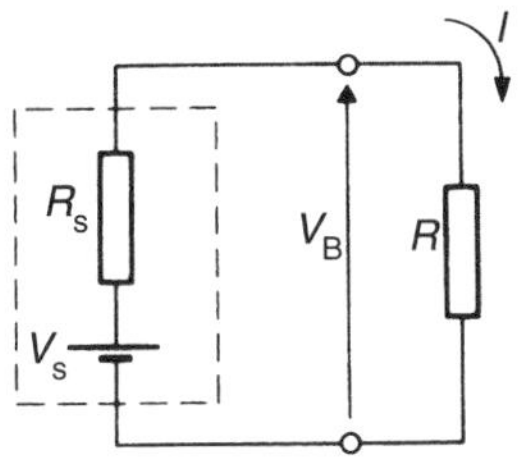

Fig. 1.7 The equivalent circuit of the battery with a resistor completing the circuit.

Self-assessment question

5 The terminal voltage V_B is also the potential drop across the load resistor R, and is therefore equal to IR. Using this result and the expression for V_B derived above, derive an expression which shows that if R is large compared with R_S, the change in voltage is small.

Self-assessment question

6 (a) If a 1.5 V torch battery has an internal resistance of 0.7 Ω, calculate the change in terminal voltage of the battery as the current taken is increased from zero to 250 mA. What value of load resistance will cause this current to flow?

(b) A car battery has a terminal voltage which varies from 13.2 V to 11.5 V as the current taken changes from zero to 70 A. What is the internal resistance of the battery? What is the load resistance when the current is 70 A?

1.3 RESISTORS IN SERIES AND IN PARALLEL

1.3.1 Resistors in series

Figure 1.8 shows three resistors in series, connected to an e.m.f. source. Notice the symbol for the voltage source, which we use when we do not wish to be specific about the method of e.m.f. generation. The arrow shows the polarity of the source; by convention *the arrow head is at the positive end.* The same convention is used for the potential difference across each resistor.

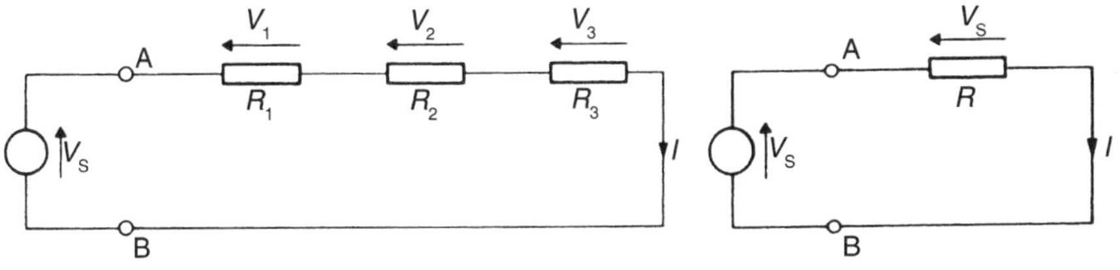

Equivalent resistance = R.

Fig. 1.8 Series connection of resistors.

The potential at the left-hand end of each resistor is more positive than that at the right-hand end because the current is flowing from left to right. Hence the directions of the arrows representing the polarity of V_1, V_2 and V_3.

The same current I flows through each component in the circuit, so we can evaluate each potential difference using Ohm's relationship:

$$V_1 = IR_1, \quad V_2 = IR_2, \quad V_3 = IR_3$$

The total potential difference across all three resistors is $V_1 + V_2 + V_3$, and this must be equal to the applied e.m.f. V_S, since there are no other components in the circuit across which potential can be dropped. (We make the assumption that the internal resistance of the source, and the resistances of the connecting cables, can be neglected because they are very much smaller than the resistance values of the resistors.) Hence

$$\begin{aligned} V_S &= V_1 + V_2 + V_3 \\ &= IR_1 + IR_2 + IR_3 \\ &= I(R_1 + R_2 + R_3) \end{aligned}$$

In the equivalent circuit the equivalent resistance R has a value such that the same current I flows, and

$$V_S = IR$$

Hence,

$$R = R_1 + R_2 + R_3$$

The result we have achieved is that *the resistance equivalent to the three resistors connected in series is the sum of the individual resistances*. The same argument can be applied to any number of resistors in series, giving the general result that the *equivalent resistance of n resistors in series connection is equal to the sum of the n individual resistances*.

When performing calculations to find the equivalent resistance of a *series* connection of resistors, always bear in mind that the equivalent resistance should turn out to be larger than the largest of the individual resistor values. This is extremely useful if you are using a calculator, when mis-keying of the input data can give rise to erroneous results. A quick mental estimate of the expected answer is invaluable in spotting such errors.

Self-assessment question

7 Three resistors R_1, R_2 and R_3 of resistance values 100 Ω, 200 Ω and 500 Ω, respectively, are connected in series across a 10 V e.m.f. source of negligible internal resistance. Calculate (a) the equivalent resistance, (b) the current flowing in the circuit, (c) the potential difference across each resistor, (d) the power dissipated in each resistor and (e) the power taken from the e.m.f. source.

1.3.2 Resistors in parallel

Figure 1.9 shows three resistors in parallel, connected across an e.m.f. source. This time the current through each resistor is different, but the potential

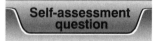

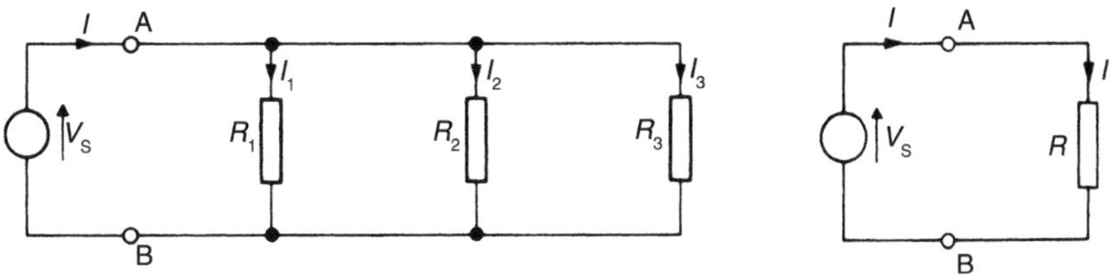

Equivalent resistance = R.

Fig. 1.9 Parallel connection of resistors.

difference across each resistor must be the same because the e.m.f. source V_S is connected directly across each resistor.

Using the relationship $I = V/R = VG$, we can write an expression for the current in each resistor:

$$I_1 = V_S G_1, \quad I_2 = V_S G_2, \quad I_3 = V_S G_3$$

The e.m.f. source supplies all the current in the circuit, so the current I is the sum of the three currents I_1, I_2 and I_3. In the equivalent circuit,

$$I = V_S G$$

Hence,

$$V_S G = V_S G_1 + V_S G_2 + V_S G_3$$

So the equivalent conductance is

$$G = G_1 + G_2 + G_3$$

This result says that *the equivalent conductance of the three resistors in parallel is the sum of the individual conductances*. Again, the same argument can be applied to any number of resistors (and their conductances) in parallel, giving the general result that *the equivalent conductance of n resistors in parallel is equal to the sum of the n individual conductances*.

The equivalent conductance of a parallel connection of resistors should always be greater than the largest individual conductance, and so the equivalent resistance should always be smaller than the smallest individual resistor value.

For the particular case of two resistors in parallel, there is a useful expression for the equivalent resistance. Here,

$$\begin{aligned} G &= G_1 + G_2 \\ &= 1/R_1 + 1/R_2 \\ &= \frac{R_1 + R_2}{R_1 R_2} \end{aligned}$$

Hence,

$$R = \frac{1}{G} = \frac{R_1 R_2}{R_1 + R_2}$$

So 'the equivalent resistance of two resistors in parallel is equal to the product of their resistance values, divided by their sum'.

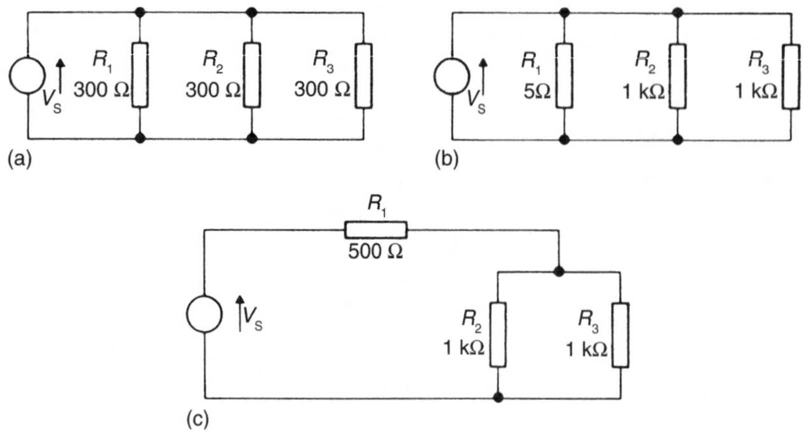

Fig. 1.10 Circuits of Self-assessment question 8.

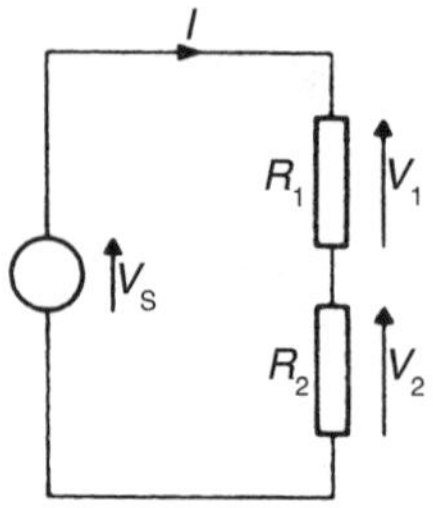

Fig. 1.11 The voltage divider circuit.

Self-assessment question

8 For each of the circuits of Fig. 1.10, estimate (*without* performing calculations) the equivalent resistance of the circuit [*Hint*: for (c), look at the parallel resistors first, then include the series resistor R_1.]

1.3.3 The voltage divider

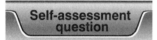

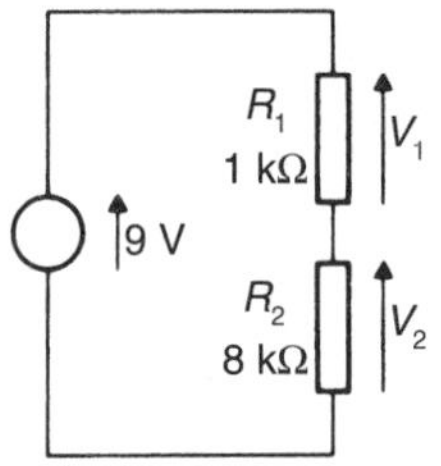

Fig. 1.12 An example voltage divider.

Figure 1.11 shows a particular case of series connection where only two resistors R_1 and R_2 are connected across the source of e.m.f. V_S. The ratio V_1/V_2 is

$$\frac{V_1}{V_2} = \frac{IR_1}{IR_2} = \frac{R_1}{R_2}$$

This expression illustrates the *voltage divider rule*, which, in words, says the voltage across two resistors in series divides between them in the ratio of their resistances. So, for example, in the circuit of Fig. 1.12, the voltage divides in the ratio 1:8 and V_1 is $\frac{1}{9}$ of the applied e.m.f. while V_2 is $\frac{8}{9}$ of it. Hence V_1 is 1 V and V_2 is 8 V.

The voltage divider circuit is particularly useful when we need to obtain a potential difference which is a fraction of the voltage of an applied e.m.f. source. For example, the circuit of Fig. 1.13 has a potential difference V_{OUT} between terminals C and D which is 1/1000 of the voltage V_{IN} between points A and B.

Self-assessment question

9 Suppose you have a hand torch that has a 0.5 W bulb and is designed to take a 3 V battery. The battery is flat, you cannot get another; all you have available is a 12 V car battery. You decide to use a series connection of a resistor and the bulb across the car battery in order to continue to get light from the torch bulb. What value of series resistor should you use to ensure that the bulb is working at its correct power rating? what power is being dissipated in the resistor? When you connect up the circuit, the bulb operates as expected, but takes a noticeably longer time to reach full brightness than is normal. Explain this phenomenon. (*Hint*: remember that the cold resistance of a bulb is much less than its operating resistance.)

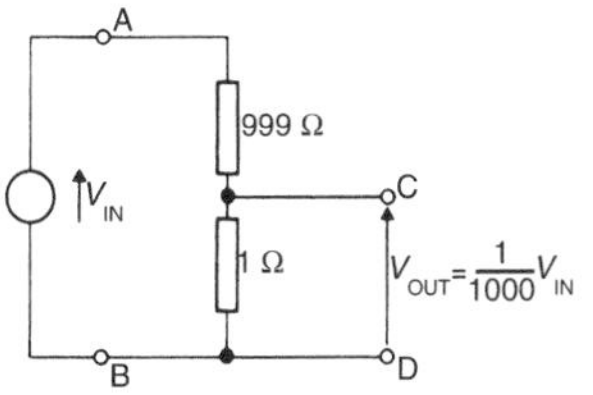

Fig. 1.13 A 1000 voltage divider.

1.3.4 The current divider

Figure 1.14 shows a particular parallel connection of two resistors across an e.m.f. source.

Self-assessment question

10 (a) Show that, for the circuit of Fig. 1.14, the current I divides between the two resistors R_1 and R_2 in proportion to their conductances.
(b) What will be the values of I_1 and I_2 if $V_S = 5$ V, $R_1 = 2\ \Omega$ and $R_2 = 18\ \Omega$?

1.3.5 The variable potential divider (or 'pot')

Figure 1.15 shows several different types of variable potential divider or 'pot'.* These devices have three terminals, unlike the resistors considered up to now which only have two. As shown in Fig. 1.16, two of the terminals are connected to either end of a resistor. The third which is called the **wiper connection** makes electrical contact with the resistor along its length. Figure 1.17 shows this in more detail for a pot with a wire-wound resistor. As the shaft of the pot is turned, the wiper connection is moved along the length of the resistor.

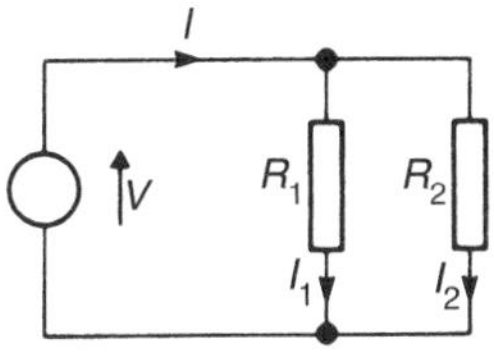

Fig. 1.14 The current-divider circuit.

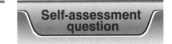

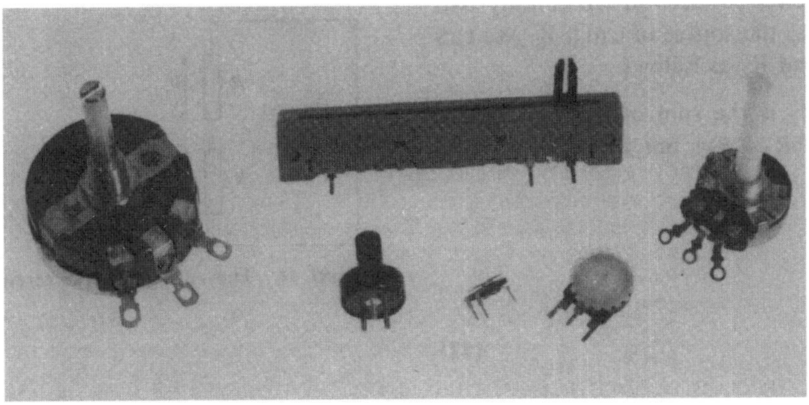

Fig. 1.15 Several types of electric-potential divider or 'pot'.

The pot circuit can be considered to be part of the voltage divider circuit of Fig. 1.11 in which the ratio of R_1 to R_2 can be varied by rotating the shaft of the pot. If a pot is used in a circuit such as Fig. 1.18 the output voltage can be varied by adjusting the position of the wiper contact. (In the straight-slider type pot of Fig. 1.15 the resistor is arranged in a straight line.)

* Sometimes referred to as a potentiometer.

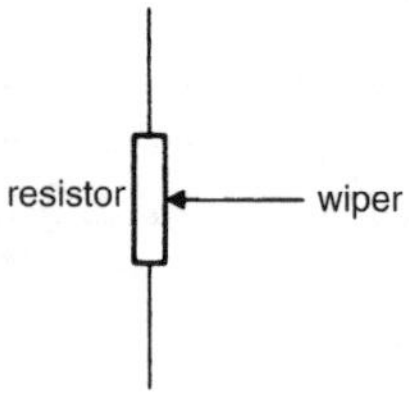

Fig. 1.16 Electrical symbol for a pot.

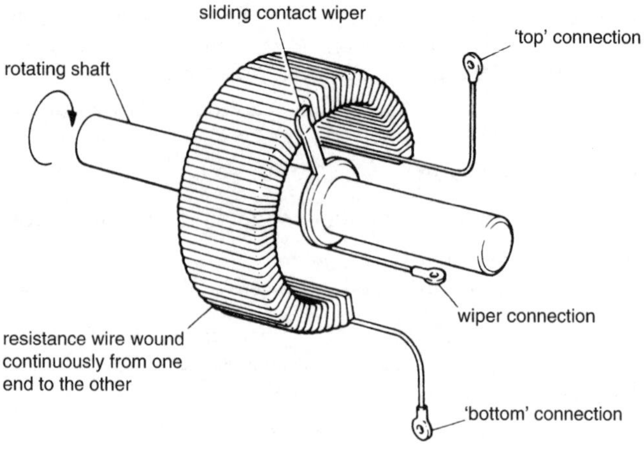

Fig. 1.17 A pot shown in more detail.

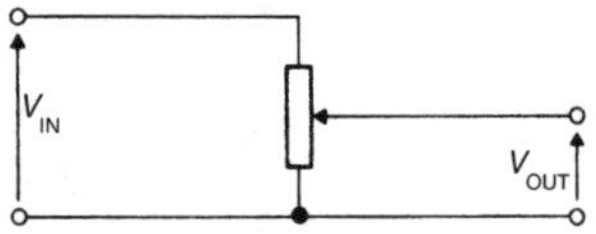

Fig. 1.18 A pot being used to vary the output voltage.

Pots are made with resistors whose resistance varies, in some cases linearly and in other cases logarithmically, with length, with many different types of resistor and values of resistance. Pots are therefore specified with their resistance value as being 'lin' or 'log' and with the maximum power that they can dissipate.

Self-assessment question

11 For the circuit shown in Fig. 1.19a:

(a) Calculate the output voltage for a wiper position three-quarters of the way along the track towards B; and
(b) Calculate the power dissipated in the pot.
(c) For the circuit shown in Fig. 1.19b the pot wiper is positioned three quarters of the way along the resistor towards B and a load resistance of 10 Ω is connected between the pot wiper and the common connection as shown. What will be the output voltage V_{OUT} across the load resistor?
(d) Without performing calculations, try to estimate the wiper's position where the change in output voltage, due to the connection of the load resistor, would be greatest.

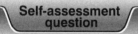

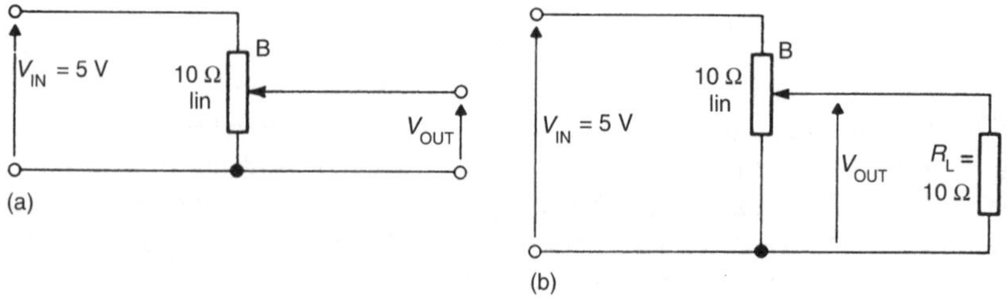

Fig. 1.19 Circuits of Self-assessment question 11.

1.3.6 Summary of Section 1.3

1. A **series connection** of resistors across an e.m.f. source results in the current being the same in each resistor, while the applied e.m.f. voltage is divided between the resistors.
2. A **parallel connection** of resistors across an e.m.f. source results in the voltage being the same across each resistor, while the source current is divided between the resistors.
3. Series and parallel resistor networks can be represented by an **equivalent circuit** consisting of an **equivalent resistance** (or equivalent conductance) connected across the e.m.f. source. The equivalent resistance is one which gives rise to the same current flow through the voltage source as does the network.
4. For a series connection of resistors, the equivalent resistance is equal to the sum of the individual resistance values.
5. For a parallel connection of resistors, the equivalent conductance is equal to the sum of the individual conductances. The equivalent resistance is the reciprocal of the equivalent conductance.
6. Two resistors in series across an e.m.f. source form a **voltage divider**, the applied voltage being divided between the resistors in proportion to their **resistances.**
7. Two resistors in parallel across an e.m.f. source form a **current divider**, the source current dividing among the resistors in proportion to their **conductances.**
8. A variable **potential divider** (or 'pot') is a resistor with a moving contact called the wiper. As the wiper moves, the output voltage of the pot, measured between the wiper and one end of the pot, varies from zero to the value of the voltage supplied across the ends of the whole resistive track.

1.4 THE MEASUREMENT OF CURRENT, VOLTAGE AND RESISTANCE

1.4.1 Using ammeters and voltmeters

To measure the current flowing in a circuit, the circuit usually has to be broken into and an ammeter inserted in the path of the unknown current, as in Fig. 1.20. To measure the voltage across a circuit component, the voltmeter is connected across the component as in Fig. 1.21.

In using a meter to measure either current or voltage it is important to be able to decide whether connecting the meter into the circuit will have any

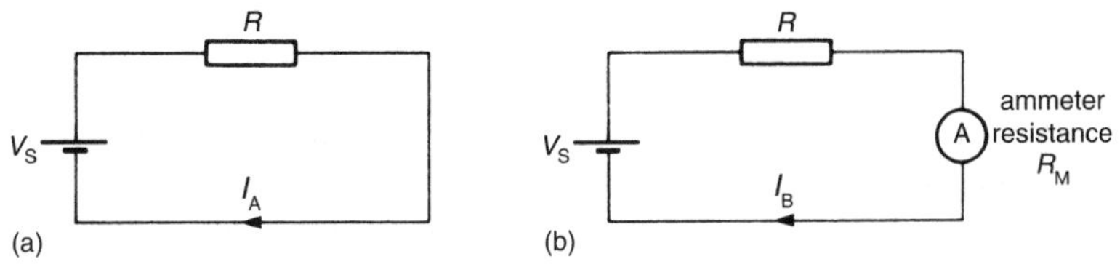

Fig. 1.20 Measuring current.

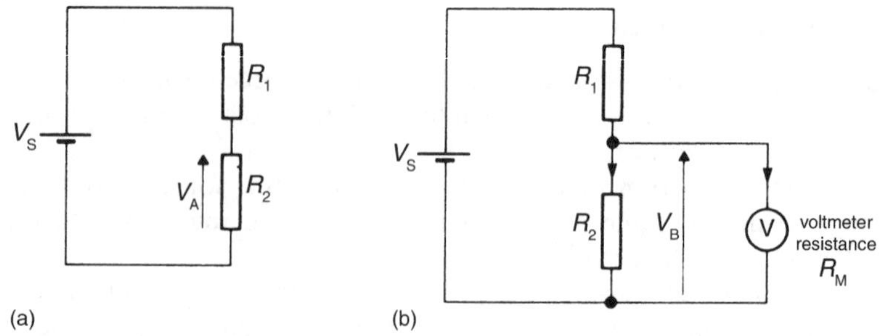

Fig. 1.21 Measuring voltage.

significant effect on the quantity you are trying to measure. In the case of current measurement, if the insertion of the meter increases, to any significant extent, the total circuit resistance, then the current flowing through the meter will be less than the current which would flow without the meter. In the case of voltage measurement, it is the current which flows through the meter which may lower the value of the voltage you are trying to measure.

Ideally then, a current meter should possess zero resistance and a voltmeter should take no current from the circuit, so it should have infinite resistance. In practice meters do have finite resistance, although a good current meter has low resistance while a good voltmeter has high resistance. What is important when using a meter is to be able to estimate whether the meter is having any significant effect on the quantity being measured. This is left to a later section of this chapter, where the Thévenin and Norton equivalent circuits make analysis much easier.

1.4.2 The multimeter

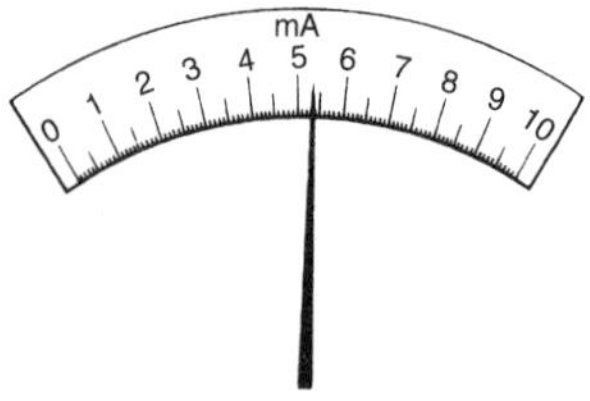

Fig. 1.22 An analogue meter display.

Fig. 1.23 A digital meter display.

A multimeter is a general-purpose electrical measuring instrument which is capable of measuring current and voltage (both a.c. and d.c., although only d.c. concerns us here) and resistance. It normally has several ranges for each of these quantities. Each range has a different full-scale value of the measured quantity, that is, the value of the current, voltage or resistance which causes the meter to reach its maximum indication. Switches are provided to select the required function and usually to select the required full-scale value, although many modern multimeters are auto-ranging in that they automatically select the most appropriate range for the measured value.

There are two common types of multimeter, the moving-coil multimeter, and the digital multimeter. The moving-coil multimeter displays the value of the measured quantity as the mechanical deflection of a pointer across a graduated scale, as in Fig. 1.22. It is described as an analogue meter because the movements of the pointers are a copy, or analogue, of the changes occurring in the measured quantity. In Fig. 1.22, the meter is indicating a measured current value of 5.3 mA on a 10 mA full-scale deflection range.

A digital multimeter displays the measured value as a collection of denary digits, as in Fig. 1.23, where the meter is shown indicating a measured voltage value of 5.35 V on a 10 V full-scale range (actually the maximum reading is only 9.99 V).

1.4.3 The moving-coil multimeter

The moving-coil multimeter is based on the moving-coil meter movement, in which a coil of wire—to which a pointer is attached—is supported on a pivot in the magnetic field of a permanent magnet, as shown in Fig. 1.24. If a current is passed through such a coil, the current and the magnetic field interact to produce a force which acts against a return spring, or taut band, to deflect the pointer. This pointer then indicates the magnitude (and direction) of the current causing the deflection.

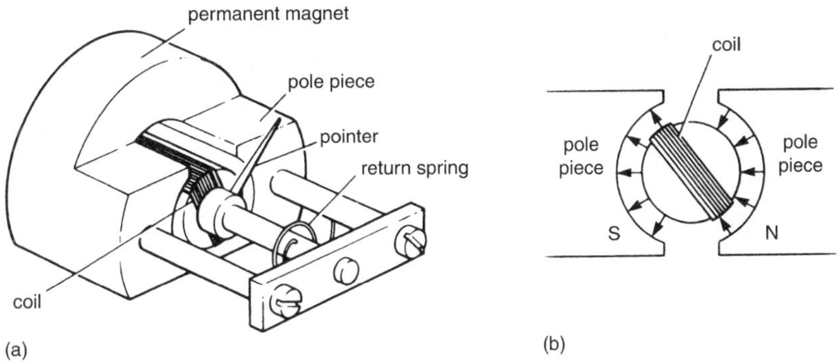

Fig. 1.24 (a) The construction of a moving-coil current meter. (b) A cross-section through the coil showing the form of the magnetic field.

Such meter movements can measure currents up to a specified maximum values; say, 50 μA full-scale deflection (f.s.d.). There is also, in effect, a minimum value corresponding to the current at which the deflection is too small to give an accurate reading. If, for example, the meter has a scale marking each microampere (i.e. 50 divisions on the scale) and a reading can be expected to be accurate to half a division, for whatever reason, then a current of 50 μA can be measured to an accuracy of $\pm 1\%$; i.e. half a division, up or down, in 50 divisions. A current of 5 μA on the same meter will give a deflection of 5 divisions on the scale but is still only correct to ± 0.5 divisions. So in this instance the accuracy is only $\pm 10\%$.

This idea of 'accuracy', namely the range of readings on a meter scale within which the true value lies, is different from 'sensitivity'. Sensitivity is a measure of how small a value of the quantity to be measured will be indicated by the meter at full-scale deflection. A meter with a f.s.d. of 5 μA is 10 times more sensitive than one with a f.s.d. of 50 μA. Accuracy is often expressed as a percentage of the quantity being measured. Thus, in the above example, an accuracy of ± 0.5 μA can alternatively be expressed as $\pm 1\%$ accuracy at 50 μA or as $\pm 10\%$ accuracy at 5 μA.

A current that causes a deflection of one-tenth full scale is about the least that can normally be measured with acceptable accuracy. So a 50 μA f.s.d. meter can usefully measure currents between 5 and 50 μA.

A typical moving-coil meter movement might, for example, have a resistance of 2500 Ω and a full-scale deflection current of 50 μA.

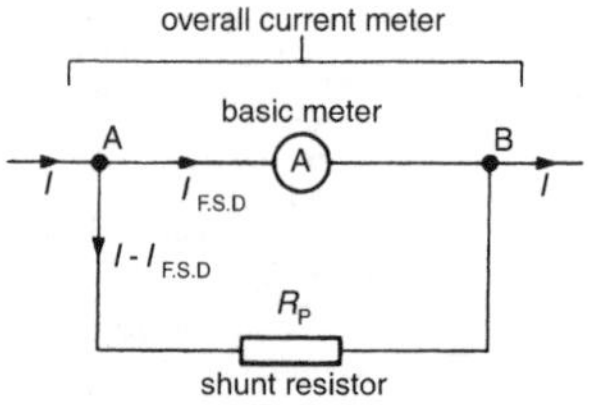

Fig. 1.25 Adding a parallel (or shunt) resistor to convert a basic ammeter into a less sensitive current meter.

Current measurement

Figure 1.25 shows the circuit used for the measurement of currents higher than the full-scale deflection current of the movement. Some of the current is allowed to bypass the coil by flowing through the parallel (or shunt) resistor R_P. For example, if R_P is 1/99 of the coil resistance, then 99 times as much current will flow through R_P as flows through the coil, so that a current of 5 mA (100 times 50 μA) is required through the overall meter to give full-scale deflection of the pointer.

Self-assessment question

12 What value of parallel resistance R_P is required with a 50 μA, 2500 Ω meter movement if the full-scale deflection current of the overall meter is to be 5 A? What will be the overall meter resistance?

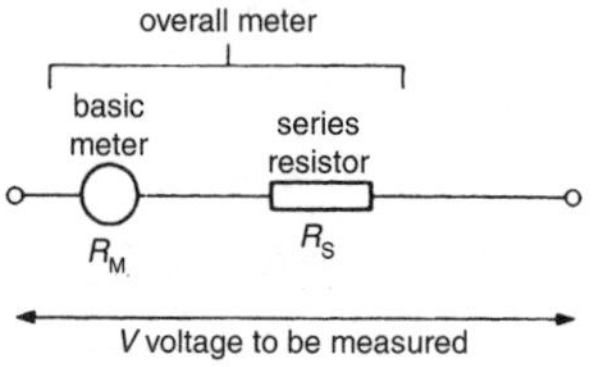

Fig. 1.26 Adding a series resistor R_S to convert a 'basic' meter into a voltmeter; the overall resistance of the voltmeter is $R_V = R_S + R_M$.

Voltage ranges

Figure 1.26 shows the circuit used for the measurement of voltage. A resistor R_S is connected in series with the meter movement so that full-scale deflection current flows in the coil when the required full-scale deflection voltage is applied across the overall meter. For example, if R_S is 97.5 kΩ, so that $(R_S + R_m)$ is 100 kΩ, then the full-scale deflection voltage of the meter is 100 kΩ × 50 μA = 5 V.

Self-assessment question

13 (i) What value of series resistance R_S is required with the 50 μA, 2500 Ω meter movement if the full-scale deflection voltage of the overall meter is to be (a) 100 V and (b) 25 V? What will be the overall meter resistance in each case?

(ii) For the three voltmeter ranges 5 V, 25 V and 100 V f.s.d., calculate the ratio of the overall meter resistance to the full-scale deflection voltage.

If you performed the calculations in Self-assessment question 13 correctly, you will have found that the ratio of the overall meter resistance to the full-scale deflection voltage is the same for all three voltmeter ranges, that is 20 000 Ω V^{-1}. This figure is another way of expressing the sensitivity of a multi-range voltmeter; it is called the 'ohms-per-volt' figure. The higher this figure, the better the voltmeter in terms of its effect on the voltage you are trying to measure. It is commonly used as a 'figure of merit' for moving-coil voltmeters. (The fact that the ohms-per-volt figure is the same for all voltage ranges should come as no surprise—it is simply the reciprocal of the full-scale deflection current of the basic meter movement!)

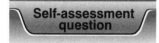

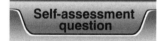

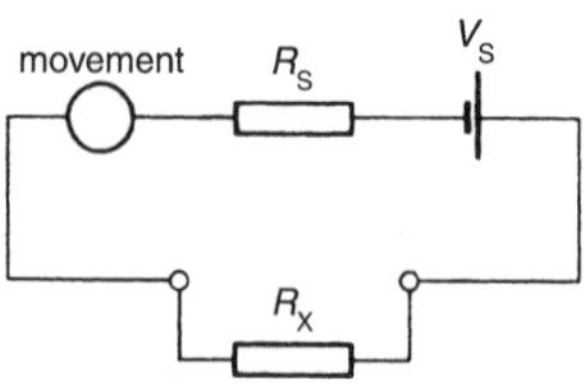

Fig. 1.27 The measurement of resistance using a moving-coil meter movement.

Resistance ranges

Figure 1.27 shows the circuit used in a moving-coil multimeter for the measurement of resistance. A battery supplies an e.m.f. V_S which causes current to flow through the meter movement, R_S and the unknown resistor R_x. R_S and V_S are chosen so that when $R_x = 0$ (that is, the overall meter terminals are joined by a zero-resistance link (a short circuit), the current flowing through the meter movement is I_{FSD}.

For example, if the basic meter is the 50 μA, 2.5 kΩ meter movement used in earlier examples, and the e.m.f. source is a 1.5 V battery, then

$$1.5\,\text{V} = (R_m + R_S) \times 50\,\mu\text{A}$$

So

$$(R_m + R_S) = \frac{1.5\,\text{V}}{50\,\mu\text{A}} = 30\,\text{k}\Omega$$

If the external resistor R_x is now made equal to the internal resistance $(R_m + R_s)$, the total resistance in the circuit is doubled, and so the current flowing is halved. For our example then, the half-way mark on the scale represents $R_x = 30$ kΩ.

Self-assessment question

14 Calculate the value of R_x corresponding to (a) one-quarter full-scale deflection, (b) one-eighth full-scale deflection, (c) three-quarters full-scale deflection.

The results of Self-assessment question 14, together with the half-scale value of 30 kΩ, have been drawn onto a meter scale in Fig. 1.28 to show how non-linear such a resistance scale turns out to be. Clearly, zero deflection is obtained only when the meter terminals are open-circuit, corresponding to an infinitely large value of R_x.

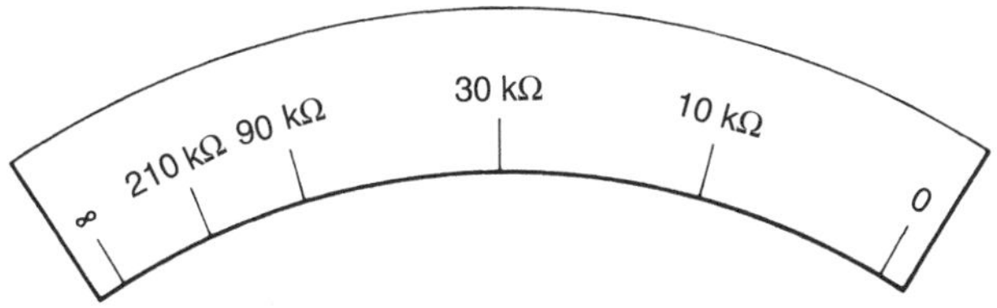

Fig. 1.28 A typical ohms range showing the non-linear scale.

1.4.4 The digital multimeter

The basic measuring device in a digital multimeter (the equivalent of the moving-coil meter movement) is an electronic circuit which converts an input d.c. voltage into a digital waveform and presents its value on a denary digital display. (Digital waveforms are discussed later in the book.). Figure 1.29 shows the arrangement. The input voltage is converted into a digital waveform in an analogue-to-digital (A-D) converter. The input voltage to the converter which produces a full-scale display might typically be 10 V. The A-D converter

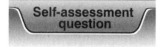

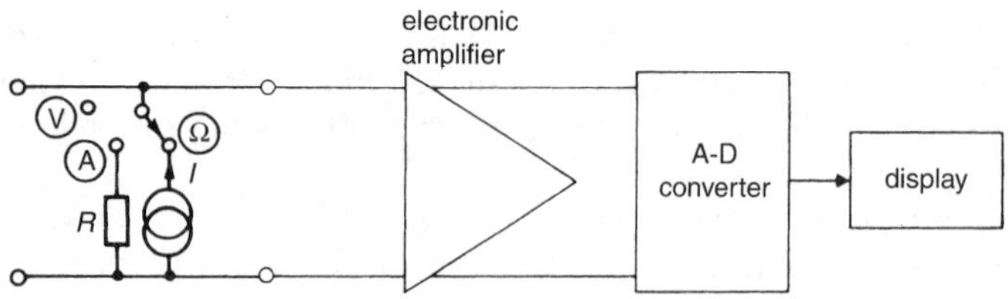

Fig. 1.29 The structure of a digital multimeter.

samples the input voltage at regular intervals and updates the display after each sample. Between the samples, the display is held constant at the value of the last sample taken. A-D converters are described later in the course.

The A-D converter is preceded by an electronic amplifier, as shown in Fig. 1.29. Amplifiers have the following properties:

(i) Very little current flows into the amplifier input terminals; that is, it has a very high *input resistance*.
(ii) The ratio of the output voltage of the amplifier to the voltage applied to its input can be varied over a very wide range, for example from much less than 1 up to a thousand or more.

Figure 1.29 shows the three basic methods used in the digital multimeter to measure (a) voltage, (b) current and (c) resistance. With the switch set to **V** the amplifier **voltage gain** (the ratio of the output voltage of the amplifier to its input voltage) is varied to provide different full-scale deflection voltage ranges. For example, if the full-scale input of the A-D converter is 10 V, to create a voltmeter having a full-scale range of 1 V requires an amplifier gain of 10; to create a full-scale range of 10 mV requires an amplifier gain of 1000. For all voltage ranges, the input resistance of the multimeter is the high input resistance of the amplifier, which, in a typical mid-cost-range instrument, might be 100 MΩ.

With the switch set to **A**, the current to be measured flows through a low-resistance resistor, while the voltage across the resistor is measured by the voltmeter. The amplifier is given its highest possible voltage gain, so that the input voltage required for full-scale indication is as small as possible. This means that the input resistance of the current meter, for any given current range, is as low as possible. Different resistor values are used for each required current range so that the input voltage to the amplifier at full-scale indication is the same whatever the current range.

Self-assessment question

15 Suppose the A-D converter input for full-scale indication is 10 V, and that the maximum voltage gain of the amplifier is 500. What value of resistor will be required to give full-scale deflection when the current to be measured is (a) 1 mA, (b) 100 mA?

With the switch set to Ω the circuit can measure resistance. An electronic circuit inside the multimeter generates a constant current which flows through the resistor under test. The voltage drop which then exists across the unknown resistor is measured using the digital voltmeter. Different resistance ranges are obtained by changing the magnitude of the constant current. The advantage of using a constant current (as against a constant voltage as is used in the moving-coil multimeter) is that the voltage drop across the unknown resistor is proportional to its resistance, so giving a linear resistance range. This is convenient when converting the voltage drop, via the A-D converter, into a read-out.

1.4.5 Summary of Section 1.4

1. Ideally, ammeters would possess zero resistance and voltmeters would possess infinite resistance. In practice, an ammeter has a finite, low resistance while a voltmeter has a finite, high resistance.
2. Connecting an ammeter into a circuit affects the current you are trying to measure. The resistance of the ammeter reduces the current flowing to some extent.
3. Connecting a voltmeter across a component of a circuit affects the voltage you are trying to measure. The current taken by the voltmeter reduces the measured voltage to some extent.
4. A multimeter is an electrical measuring instrument capable of measuring current, voltage and resistance over a wide range of values.
5. The two commonly available types of multimeter are (a) the moving-coil multimeter, based around a moving-coil meter movement and indicating the value of the measured quantity as the deflection of a pointer across a graduated scale and (b) the digital multimeter, based around an A-D converter and denary digital display.
6. The *sensitivity* of a meter is the current and/or voltage which the meter measures at full-scale indication. The smaller this value, the greater the sensitivity.
7. The *accuracy* of a meter indication is the extent of the range of values around the indicated value within which the true value of the quantity being measured lies (e.g. $\pm 5\ \mu A$, ± 20 mV, $\pm 2\%$, and so on).
8. The moving-coil meter is a high-sensitivity current meter. It can be made to measure large currents by connecting shunt resistors in parallel with the movement. It can be made to measure voltages by connecting resistors in series with the movement. (The ohms-per-volt figure for a voltmeter is the total meter resistance divided by the full-scale deflection voltage. It is a *figure of merit* for a moving-coil voltmeter.) Resistance measurement is achieved by applying a known voltage to the unknown resistor and measuring the current flow. This results in a non-linear resistance scale.
9. An A-D converter and display can form a low-sensitivity voltmeter. An electronic amplifier preceding the converter allows the sensitivity to be increased. Different voltage gains of the amplifier are used to establish the different voltage ranges. Current is measured by passing the current through a low-resistance resistor and measuring the voltage drop across it. Resistance is measured by passing a constant current through the unknown resistor and measuring the voltage drop across it. This results in a linear resistance scale, convenient for conversion to the read-out.

1.5 KIRCHHOFF'S LAWS

1.5.1 Introduction

The rest of this chapter describes some useful methods of circuit analysis. The treatment is confined to d.c. circuits, but the same methods are used for a.c. circuit analysis throughout the rest of this book. We start with two simple, and intuitively obvious, laws known as Kirchhoff's laws.

1.5.2 Kirchhoff's current law

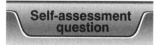

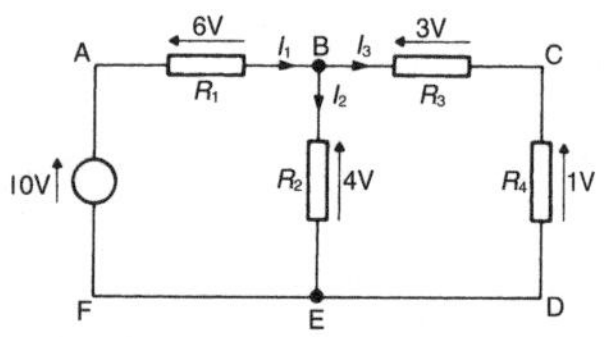

Fig. 1.30 A circuit used to illustrate Kirchhoff's laws.

Consider Fig. 1.30. Points such as A, B and C, where two or more circuit components are connected together (with connections of negligible resistance) are called **nodes**. Kirchhoff's current law is a formalization of the implications for circuit analysis of the fact that current cannot accumulate at a circuit node, so that the current flowing towards the node must be equal to the current flowing away from it. This leads to the formal statement which is Kirchhoff's current law (sometimes called his first law).

At any node of a network, at every instant of time, the sum of the currents into the node is equal to the sum of the currents out of the node.

In the circuit of Fig. 1.30, the application of this law to node B gives the equation $I_1 = I_2 + I_3$.

An alternative, but equivalent, form of Kirchhoff's current law can be obtained by considering currents directed into a node as positive in sense, while currents directed out of a node are considered negative in sense. In this case, Kirchhoff's current law can be stated as

At any node of a network, at every instant of time, the algebraic sum of the currents at the node is zero.

For node B of Fig. 1.30, this formulation of the law gives the equation $I_1 - I_2 - I_3 = 0$ which is clearly an alternative form of the previous equation.

Self-assessment question

16 For the circuit of Fig. 1.31, write down the Kirchhoff's current law equation for each of the nodes A, B, C, D and E.

1.5.3 Kirchhoff's voltage law

This law is a formalization of a result which we have used intuitively up to now, that the voltage drops around a circuit add up to the voltage of the e.m.f. source. The law states that

The algebraic sum of the voltages across all the components around any loop of a circuit is zero.

Because the law uses the term 'algebraic sum' it is clear that we expect voltage to be both positive and negative in a circuit. Which voltages are positive and

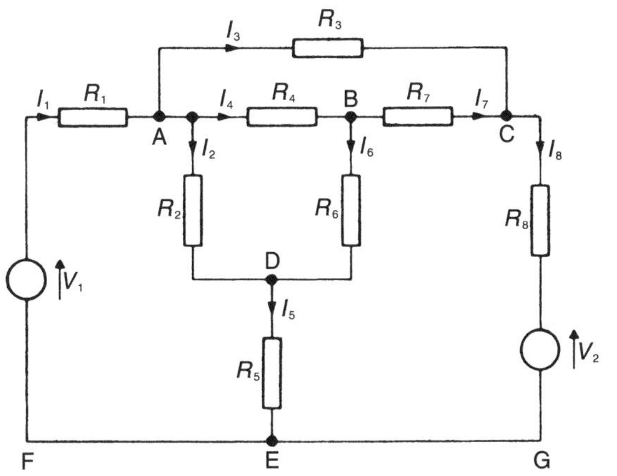

Fig. 1.31 A complex circuit to illustrate the use of Kirchhoff's voltage law.

which are negative can be seen from Fig. 1.30. As explained previously, the arrow next to the source shows the direction in which the source drives current, and the arrowhead indicates the more positive voltage. In a resistor, the voltage *drop* is proportional to the current flowing, and so the voltage arrow must be in the opposite direction to the direction of current flow (the voltage gets *less* positive within the resistor in the direction of current flow).

Kirchhoff's voltage law (also called his second law) applies to the loops in Fig 1.30 as follows. Going clockwise around loop ABEFA, with clockwise-pointing voltage arrows regarded as positive and anti-clockwise ones as negative, we get

AB	−6 V
BE	−4 V
FA	+10 V
Algebraic sum	0 V

The algebraic sum is clearly zero, in accordance with Kirchhoff's voltage law.

Going round the loop anticlockwise, i.e. around the loop AFEBA, we would get $-10+4+6=0$. For the loop ABCDEFA, the result is $-6-3-1+10=0$; and for loop BCDEB, the result is $-3-1+4=0$.

Kirchhoff's voltage law can be used in circuit analysis to produce equations which enable us to evaluate an unknown voltage or current. You can see the type of equation that can be obtained by applying the voltage law to the loops of the circuit of Fig. 1.31, and remembering the relationship that the voltage drop in a resistor is the product of the current and resistance. Applying the law to loop FADEF gives $V_1 - I_1R_1 - I_2R_2 - I_5R_5 = 0$. For loop ABDA, the law yields $-I_4R_4 - I_6R_6 + I_2R_2 = 0$.

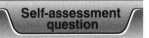

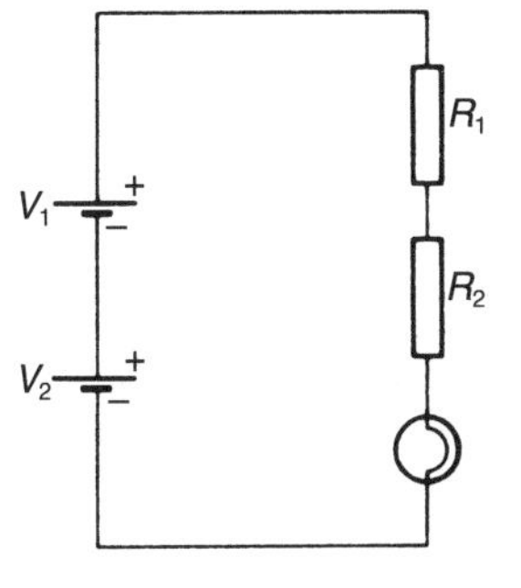

Fig. 1.32 Circuit for Self-assessment question 17. $V_1 = 10$ V, $V_2 = 5$ V, $V_{R1} = 1$ V and $V_{BULB} = 2.5$ V.

Self-assessment question

17 Using Kirchhoff's voltage law, find the voltage across R_2 in the circuit shown in Fig. 1.32.

1.6 THÉVENIN AND NORTON EQUIVALENT CIRCUITS

1.6.1 Voltage and current sources

Section 1.2.3 introduces the idea of the equivalent circuit of a battery, which represents the battery as an ideal e.m.f. source in series with a resistor, called the **internal resistance**. The e.m.f. source V_S has an e.m.f. equal to the open-circuit terminal voltage of the battery, that is the terminal voltage measured by a voltmeter drawing negligible current, and with no other load connected. The internal resistance R_S has a value such that, when load current I is drawn, the drop in terminal voltage is correctly predicted by the product IR_S. That is,

$$V_B = V_S - IR_S$$

This expression predicts a *linear* fall in terminal voltage as current increases. In other words, a graph of terminal voltage against current would be a straight line. The equivalent circuit is based on an assumption that this *linearity* is obtained, which means that both the source e.m.f. V_S and the internal resistance R_S are independent of current.

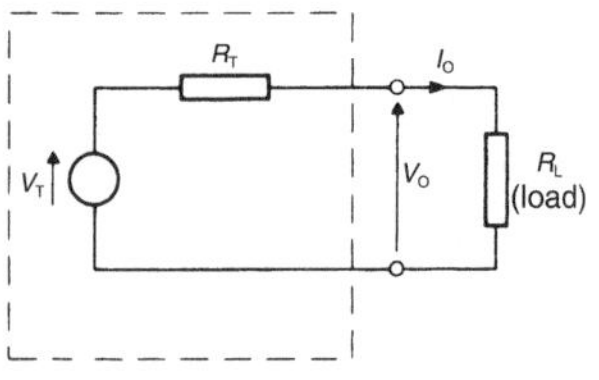

Fig. 1.33 The Thévenin equivalent circuit with load resistor.

The equivalent circuit of the battery is an example of a **Thévenin equivalent circuit**. This is shown in Fig. 1.33. It turns out that *all* d.c. circuits that have an electrical output can be thought of in terms of a Thévenin equivalent, that is an equivalent circuit comprising an ideal e.m.f. source and a resistor. However, not all such circuits are linear. Some devices—batteries, for example—while behaving linearly over a wide range of currents, stop behaving linearly at higher load currents.

The idea of the linearity of a system or component is extremely important, and crops up frequently in this book.

A voltage source such as a battery has a relatively small internal resistance compared with its normal range of load resistances, so its terminal voltage is little different from its source e.m.f. (its open-circuit voltage). Thus a battery is a good approximation to an *ideal* voltage source for a wide range of high-resistance loads.

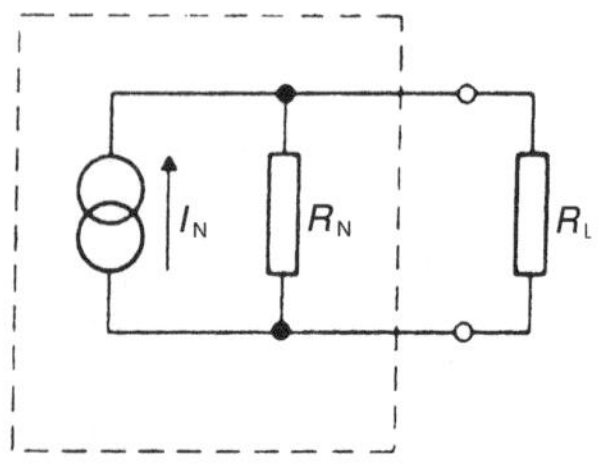

Fig. 1.34 The Norton equivalent circuit with load resistor.

An alternative equivalent circuit for a d.c. circuit with an electrical output is the **Norton equivalent circuit**, shown in Fig. 1.34. Here, the voltage source is replaced by an ideal **current source** I_N, and the internal resistance R_N appears in parallel with it. With the correct choice of I_N and R_N, this equivalent circuit models correctly the behaviour of the real circuit it represents as far as the load is concerned.

An ideal current source is a source whose *current* remains unchanged, whatever the load may be. There is, in reality, no such thing as an ideal current source, just as there is not an ideal voltage source. A battery

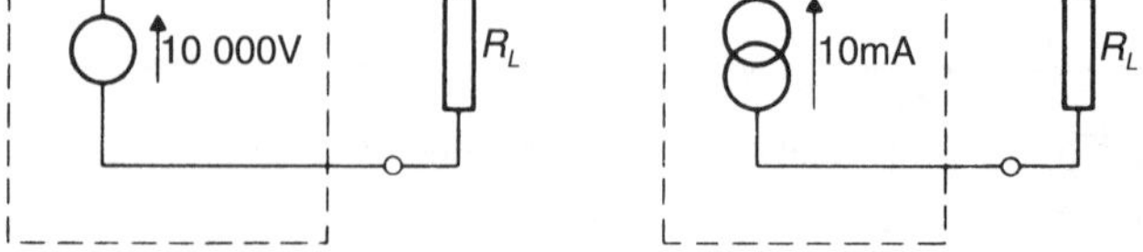

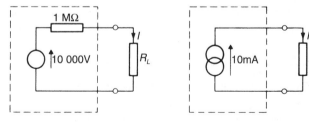

Fig. 1.35 A practical realization of a near-ideal current source for values of R_L between 0 and 10 kΩ.

approximates to an ideal voltage source for a range of high resistance load resistors. A close approximation to an ideal current source can be achieved by connecting a resistor having high resistance in series with a high-voltage source, for example a 1 MΩ resistor in series with a 10 kV d.c. e.m.f. source as in Fig. 1.35. The current flowing in the load resistor will be equal to 10 mA when $R_L = 0$, will be within 0.1% of 10 mA for values of R_L between 0 and 1 kΩ, and will be within 1% of 10 mA for values of R_L between 0 and 10 kΩ. However, such a current source would be extremely expensive and could be dangerous.

The closest approach to a cheap and practical current source is achieved by an electronic circuit designed to produce a constant output current whatever the load (within, of course, its specified range of operation).

The theoretical concept of a source of constant current is very useful in the analysis of many circuits. As you will see later in this book, a transistor produces an output current whose value is dependent on an input voltage to the transistor, but which is almost independent of the load resistor connected to it. Analysis of circuits containing transistors depends heavily on the use of a constant current source as a component of the equivalent circuit representing the transistor.

The ideal voltage source has zero internal resistance because, however the current through it changes, the change in the voltage across it is zero, so the ratio (change of voltage)/(change of current) is zero. This is strictly what we mean when we talk about internal resistance, that is, R_T is measured in terms of voltage and current *changes*. What then is the internal resistance of an ideal current source? Using the same approach, that the internal resistance is the ratio of change of voltage across the device to the change in current through it, we see that the internal resistance must be infinitely large. The voltage across the source changes but the current, in an ideal source, does not. Hence the change in the current is zero, and any number, when divided by zero, gives an infinitely large quotient.

Thus **the ideal constant current source has an infinite source resistance**. There is a 'duality' between the concepts of an ideal voltage source and an ideal current source: the voltage source has constant voltage and zero source resistance, the current source has constant current and zero source *conductance*.

If the Thévenin and Norton equivalent circuits are both to represent a circuit accurately, we would expect there to be fixed relationships between V_T, R_T (the source e.m.f. and internal resistance of the Thévenin equivalent) and I_N, R_N (the source current and internal resistance of the Norton equivalent). We can easily deduce the relationships by considering both circuits under two different load conditions, the case of a load resistor R_L of infinitely large resistance (the open-circuit condition) and the case when R_L has zero resistance (the short-circuit condition). For the Thévenin equivalent circuit, the results for these two conditions are:

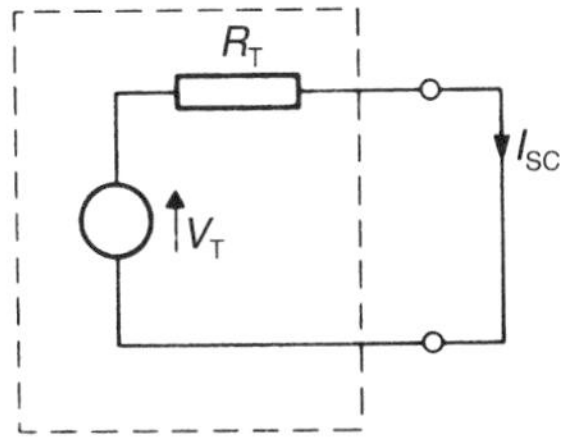

Fig. 1.36 The Thévenin equivalent circuit under short-circuit conditions.

(a) For the open-circuit condition the output voltage V_{OC} is, by definition, V_T.
(b) From Fig. 1.36, the short-circuit current I_{SC} is V_T/R_T.

Figures 1.37a and b show these two same conditions for the Norton equivalent circuit.

(a) In Fig. 1.37a, the current I_N can only flow through the resistance R_N, so

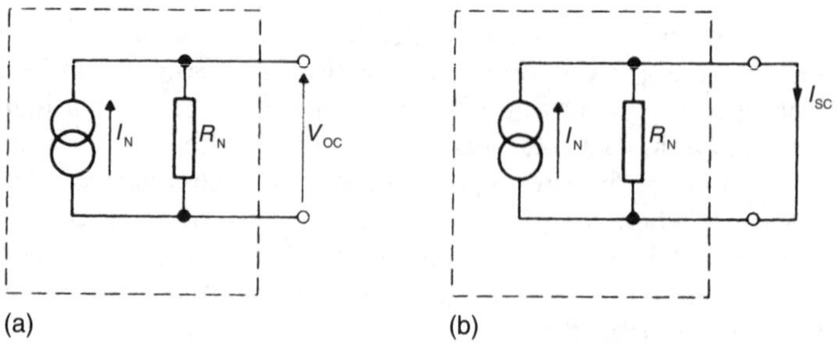

Fig. 1.37 The Norton equivalent circuit under (a) open-circuit conditions and (b) short-circuit conditions.

$$V_{OC} = I_N R_N.$$

(b) In Fig 1.37b, because of the short-circuit, the voltage across R_N must be zero, and the current through it also zero. Hence the current I_{SC} must be equal to the current source I_N. Hence $I_N = I_{SC}$.

Equating the values of V_{OC} and I_{SC} from the two equivalent circuits, we get

$$\text{For } V_{OC}: \quad V_T = I_N R_N$$
$$\text{For } I_{SC}: \quad V_T/R_T = I_N$$

Comparing these two equations, it is clear that R_N must equal R_T, that is, the parallel resistance in the Norton equivalent circuit is equal to the series resistance in the Thévenin equivalent circuit. We can see also that the constant current generator in the Norton equivalent circuit is equal to the ratio V_T/R_T of the Thévenin equivalent circuit.

The use of equivalent circuits of this sort must be treated with some caution. They are meant to represent only the current/voltage relationship of the output of a circuit. To show that the equivalent circuit representation has limitations, imagine that the Thévenin and Norton equivalent circuits both represent the same battery. When the battery is open-circuited, no current flows in the Thévenin circuit but the whole of current I_N apparently flows through R_N in the Norton circuit. This seems to imply that R_N will dissipate $I_N^2 R_N$ watts of power and will therefore get hot, whereas it will stay cool in the Thévenin circuit. Also the battery in the Norton circuit looks as if it will go flat faster than the battery in the Thévenin when left open-circuited. But the two circuits represent the same battery and the actual behaviour of the battery cannot depend merely on how you choose to think about it.

As explained earlier, the equivalent circuits represent only the output of a device or circuit, they tell us nothing about what goes on inside. We cannot therefore necessarily use the circuits to deduce anything about energy dissipation inside the battery or about the rate at which the battery will run down when left open-circuited.

1.6.2 Thévenin's theorem

The component values for a Thévenin equivalent circuit of a combination of sources and resistances can be obtained using a rule usually known as *Thévenin's theorem*:

> As far as any load connected across its output terminals is concerned, a linear circuit consisting of voltage sources, current sources and resistances is equivalent to an ideal voltage source V_T in series with a resistance R_T. The value of the voltage source is equal to the open-circuit voltage of the linear circuit. The resistance is equal to the resistance which would be measured between the output terminals if the load were removed and all sources were replaced by their internal resistances.

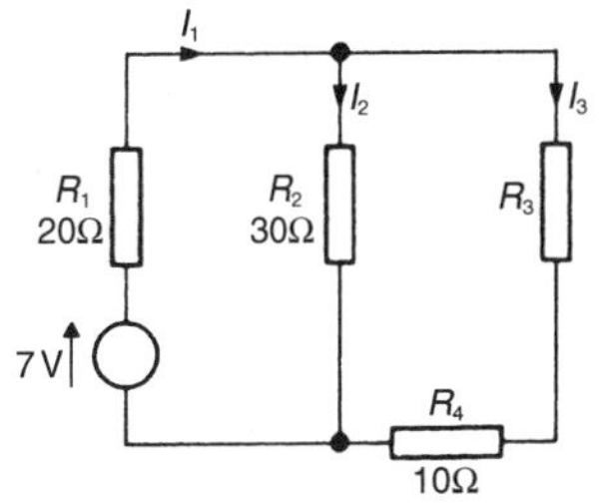

Fig. 1.38 An example circuit for the application of Thévenin's theorem.

Knowing how to use the theorem is much more important than being able to state it. As an example, suppose we wished to find the current in R_3 in the circuit of Fig. 1.38, for a range of values of resistor R_3. One method is to consider R_3 as being a load on the remainder of the circuit. Then by calculating a Thévenin equivalent circuit of the remainder we could find a simple expression for I_3 in terms of R_3.

Figure 1.39a shows how we are going to consider the circuit: as a network together with the load R_3 on the network. We need to calculate the component values V_T and R_T of the Thévenin equivalent circuit which will replace the network, as shown in Fig. 1.39b. Figure 1.40 shows the network in its open-circuit condition, that is, with the load R_3 removed, and from this configuration the open-circuit voltage V_{OC} (which by the theorem is the voltage V_T) can be found. In Fig. 1.40, no current flows in the 10 Ω resistor, so no voltage drop occurs across it. The other two resistors form a voltage divider across the 7 V source, so V_T can be calculated using the potential divider rule:

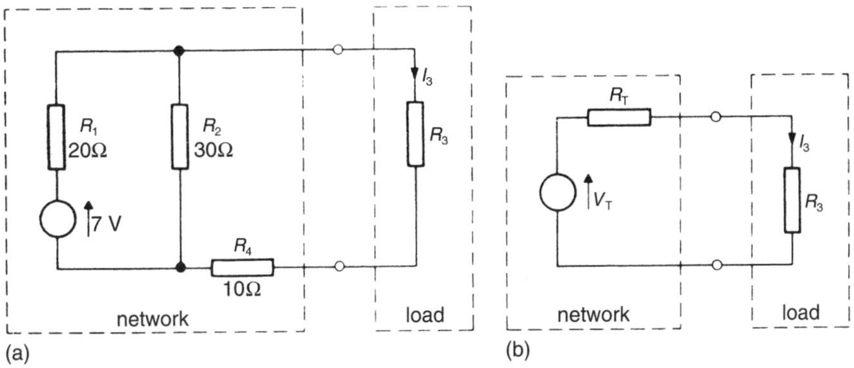

Fig. 1.39 (a) The conceptual split of Fig. 1.38 into a network and a load. (b) The Thévenin equivalent circuit replaces the network.

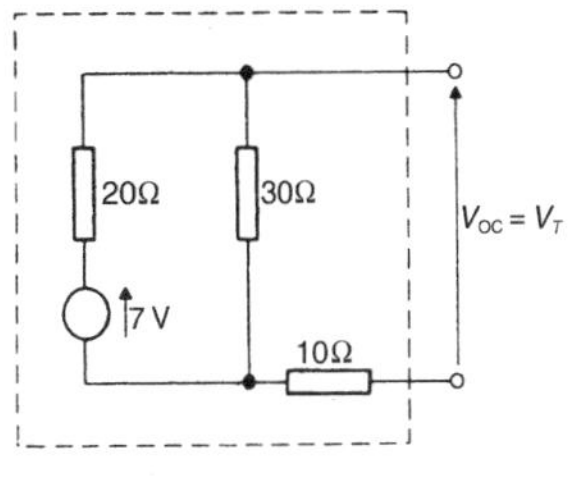

Fig. 1.40 The circuit for the calculation of V_T.

$$V_T = 7\,\text{V} \times \frac{30}{20+30} = \frac{21}{5}\,\text{V} = 4.2\,\text{V}$$

Figure 1.41 shows the network with the ideal e.m.f. source replaced by its internal resistance (which you know is zero). R_T is the resistance between the output terminals of the network when the load is removed. In this case it is clear by examination that R_T is the sum of the 10 Ω resistor and the parallel combination of the 20 Ω and 30 Ω resistors. Hence:

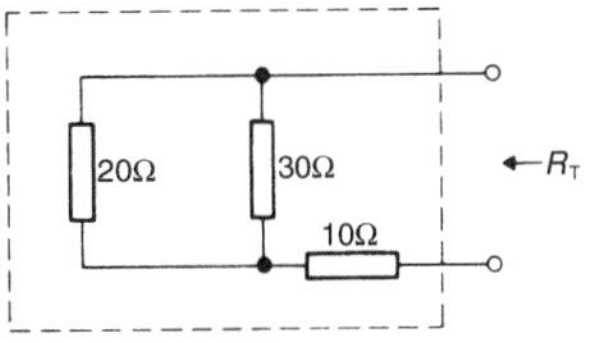

Fig. 1.41 The circuit for the calculation of R_T.

$$R_T = 10\,\Omega + \frac{20 \times 30}{20+30}\,\Omega = 22\,\Omega$$

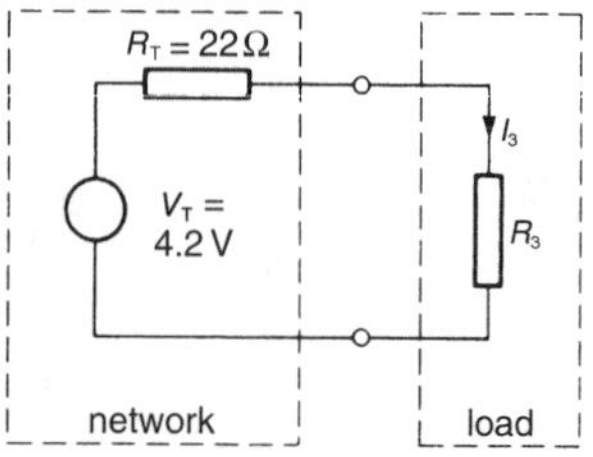

Fig. 1.42 Fig. 1.39b redrawn with the numerical values of V_T and R_T.

We can now put these values into the Thévenin equivalent circuit of Fig. 1.39b and obtain the circuit of Fig. 1.42. This circuit is the equivalent of the circuit of Fig. 1.38, as far as R_3 is concerned, so the value of I_3, calculated from Fig. 1.42, will be the value of I_3, in Fig. 1.38. From Fig. 1.42,

$$I_3 = 4.2\,\text{V}/(22\,\Omega + R_3)$$

Now, for any value of R_3, I_3 can be calculated without recalculating any of the voltages or currents in the network.

Self-assessment question

18 For the circuit of Fig. 1.43, use Thévenin's theorem to find the current through the 50 Ω resistor R_4. (Remember, an ideal current source has infinite resistance.)

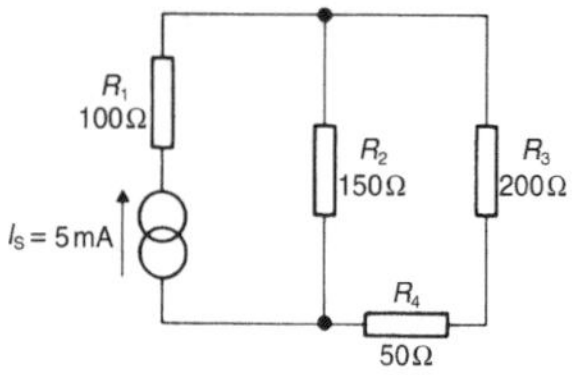

Fig. 1.43 Circuit of Self-assessment question 18.

If the circuit you wish to analyse exists physically, and providing it behaves linearly over the whole range of load resistances which can be applied to it, finding the Thévenin equivalent circuit is simply a matter of taking measurements on the circuit. One method is to measure the output voltage when the terminals are open-circuit, and the current which flows in a short-circuit placed across the output terminals (Fig. 1.44). The voltage is, by definition, the Thévenin equivalent circuit voltage V_T. Comparing Fig. 1.44b with its Thévenin's equivalent, Fig. 1.45, you can see that $I_{SC} = V_T/R_T$ and so $R_S = V_T/I_{SC}$. The two measurements therefore provide the values of both V_T and R_T.

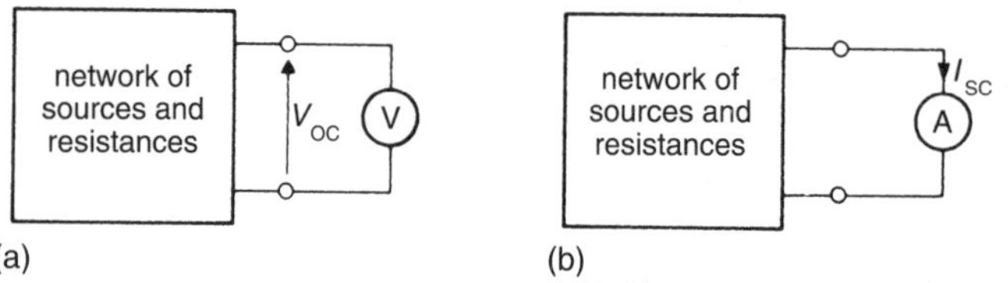

Fig. 1.44 The measurement of V_{OC} and I_{SC} to determine the Thévenin equivalent circuit of a network. The voltmeter and ammeter are assumed to be perfect.

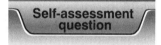

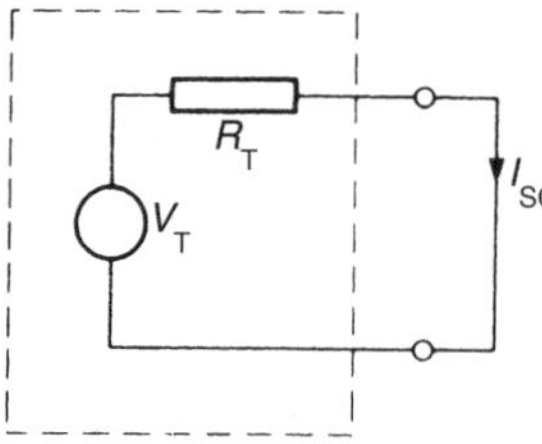

Fig. 1.45 A short-circuit Thévenin equivalent circuit.

Many circuits, while behaving linearly with a wide range of load resistances, do not behave linearly under short-circuit conditions, and in such cases the short-circuit current measurement is not a true measure of the behaviour of the circuit over its linear region. An alternative method of finding R_T is to apply a variable load to the network and to adjust its resistance value until the output voltage of the network is exactly $\frac{1}{2}V_T$. In this condition, the value of R_T is equal to the value of R_L, so if R_L is known, R_T is also known.

1.6.3 Norton's theorem

From what we have said up to now about the Norton equivalent circuit, and from the statements of Thévenin's theorem, you should be able to state Norton's theorem for yourself. Try doing so before reading the theorem below.

As far as any load connected across its output terminals is concerned, a linear circuit consisting of voltage sources, current sources and resistances is equivalent to an ideal current source I_N in parallel with a resistance R_N. The value of the current source is equal to the short-circuit current of the linear circuit. The value of the resistance is equal to the resistance which could be measured between the output terminals if the load were removed and all sources were replaced by their internal resistances.

The definition seems to suggest that you need to calculate or measure the short-circuit current in order to obtain the Norton equivalent circuit, and in some cases this is the most convenient method. Alternatively, remember that R_N in the Norton equivalent circuit is the same as R_T in the Thévenin equivalent circuit, and that $V_T = I_N R_N$, so you still need only calculate or measure the open-circuit voltage V_T and the resistance R_N in order to create the Norton equivalent circuit. In any case, all you need is two of the three parameters: the third is a function of the other two.

When using the Norton equivalent circuit to evaluate the current in a load resistor (Fig. 1.46), the current divider rule allows I_L to be found in terms of I_{SC}. The current divider rule says that the current divides between two parallel resistors in proportion to their conductances; thus

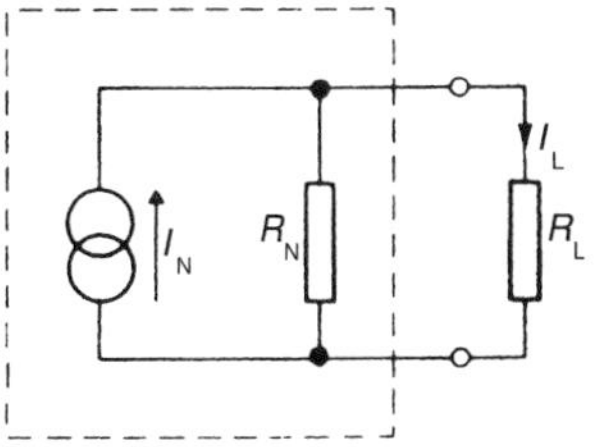

Fig. 1.46 A Norton equivalent circuit with load R_L.

$$\frac{I_L}{I_N} = \frac{1/R_L}{1/R_N + 1/R_L} = \frac{R_N}{R_N + R_L}$$

So

$$I_L = \frac{I_N R_N}{R_N + R_L}$$

Hence, knowing I_N and R_N, I_L can be found for any value of R_L.

Self-assessment question

19 Use Norton's theorem to find the value of the current in R_5 in the circuit of Fig. 1.47. (*Hint*: you can find I_N either by calculation of V_{OC} and R_N, or by calculation of I_{sc}, the short-circuit current, and R_N. There is something about the values of the resistors which makes the calculation of I_N easier than would at first appear.)

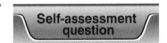

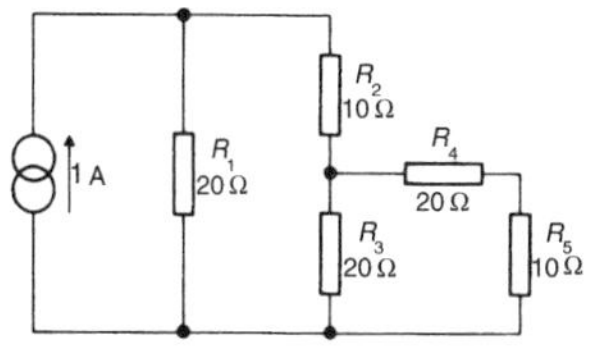

Fig. 1.47 Circuit of Self-assessment question 19.

1.6.4 Calculating the effects of voltmeters and ammeters

In Section 1.4 we discussed the effects of connecting a voltmeter across part of a circuit, or inserting an ammeter into a circuit.

In the case of the voltmeter, its internal resistance draws a little current from the circuit and lowers the voltage across the part of the circuit where it is connected. So the measured voltage is *less* than the voltage which exists without the meter inserted.

In the case of the ammeter, its internal resistance drops a little voltage and lowers the current flowing through the part of the circuit where it is connected. So the measured current is *less* than the current which exists without the meter inserted.

Thévenin's and Norton's theorems are very useful for calculating the errors caused by connecting meters to circuits.

Measuring voltage

Suppose we intend to measure the voltage across two points in a circuit, by connecting a voltmeter across them. In order to calculate the error in the measurement caused by the meter resistance, we first consider the Thévenin equivalent of the circuit *at the points where the meter is to be connected*. We want to know the open-circuit voltage at this point, which is the source voltage V_S of the Thévenin equivalent. The equivalent source resistance R_S is the resistance looking into the circuit, with all sources replaced by their equivalent source resistances.

When a voltmeter of internal resistance R_M is connected, it draws a current I and the voltage across it becomes

$$V_M = V_S - IR_S$$

The error in the required voltage is $V_S - V_M = IR_S$. If we express this error as a fraction of the measured voltage V_M, it becomes

$$\begin{aligned}(V_S - V_M)/V_M &= IR_S/IR_M \\ &= R_S/R_M\end{aligned}$$

Thus the error in the measured voltage, as a proportion of the measured voltage, is the ratio of the equivalent source resistance of the circuit at the terminals where the voltage is measured, divided by the voltmeter's internal resistance.

Measuring current

Suppose we now intend to measure the current flowing in a part of a circuit by breaking the circuit at that point and inserting an ammeter. We first consider the Norton equivalent circuit, between the two points where the meter is to be connected. We want to know the short-circuit current at this point, which is the source current I_S of the Norton equivalent. As before, the source resistance (or conductance) is the resistance (or conductance) R_S (or G_S) looking into the circuit, with all sources replaced by their equivalent source resistances (or conductances).

When an ammeter of internal resistance R_M is connected, the current through it becomes

$$I_M = I_S - V/R_S$$

where V is the voltage across the meter.

The error in the measured current is V/R_S. As a fraction of the measured current, this is

$$\begin{aligned}(I_S - I_M)/I_M &= (V/R_S)/(V/R_M) \\ &= R_M/R_S\end{aligned}$$

Thus the error in the measured current, as a proportion of the measured current, is the ratio of the ammeter's internal resistance divided by the equivalent source resistance of the circuit at the terminals where the current is measured.

1.6.5 Circuits with multiple sources—the superposition principle

Many circuits you may wish to analyse contain several sources rather than just one. When it comes to the use of the Thévenin or Norton theorem to produce an equivalent circuit for a multiple-source network, we can make use of the **principle of superposition.**

> The principle of superposition is that, in a linear network, the contribution of each source to the output voltage or current can be worked out independently of all other sources, and the various contributions then added together to give the net output voltage or current.

As an example, consider the circuit of Fig. 1.48, which has a voltage source V_S of source resistance R_1, in parallel with a current source I_S of source resistance R_2, feeding a load R_L. The problem is to find an expression for the current in R_L using Thévenin's theorem. The source resistance R_T of the network is easy to find. Replace the ideal voltage source by zero resistance (a short-circuit) and the ideal current source by infinite resistance (an open-circuit) giving R_T as R_1 in parallel with R_2. Hence

$$R_T = \frac{R_1 R_2}{R_1 + R_2}$$

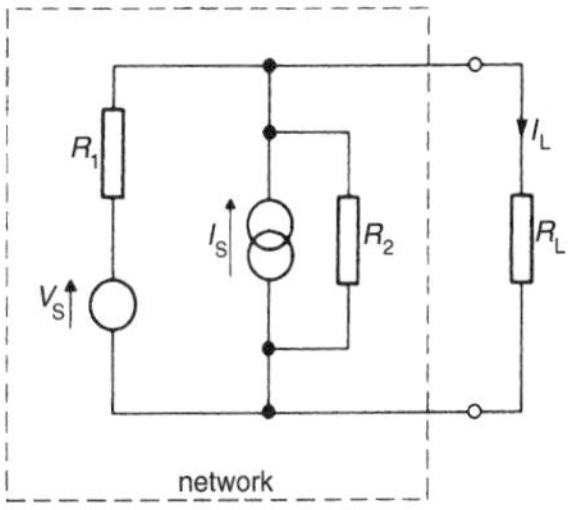

Fig. 1.48 A circuit to demonstrate the use of the superposition theorem.

The Thévenin voltage, the output voltage of Fig. 1.49a, will, by the superposition theorem, be the sum of the contribution from the voltage source V_{OC1} (Fig. 1.49b) and the contribution from the current source V_{OC2} (Fig. 1.49c). Notice from Fig. 1.49 that to find the contribution of V_S alone we must replace the ideal current source with its internal resistance (infinite resistance or open-circuit), while to find the contribution of the current source alone we must

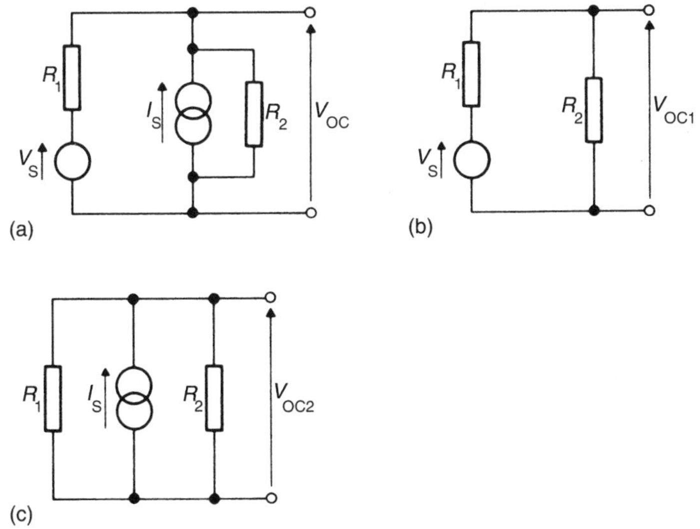

Fig. 1.49 Application of the superposition theorem to the calculation of the open-circuit output voltage of the network. V_{OC} in (a) will be the sum of V_{OC1} from (b) and V_{OC2} from (c).

replace the ideal voltage source with its internal resistance (zero ohms or a short-circuit). From Fig. 1.49b,

$$V_{\mathrm{OC1}} = V_{\mathrm{S}}\frac{R_2}{R_1 + R_2}$$

From Fig. 1.49c, since R_1 and R_2 are in parallel across I_{S},

$$V_{\mathrm{OC2}} = I_{\mathrm{S}}\frac{R_1 R_2}{R_1 + R_2}$$

Hence,

$$V_{\mathrm{T}} = \frac{(V_{\mathrm{S}} + I_{\mathrm{S}}R_1)R_2}{(R_1 + R_2)}$$

The load current I_{L} can now be calculated from the equivalent circuit of Fig. 1.50 as

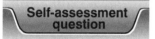

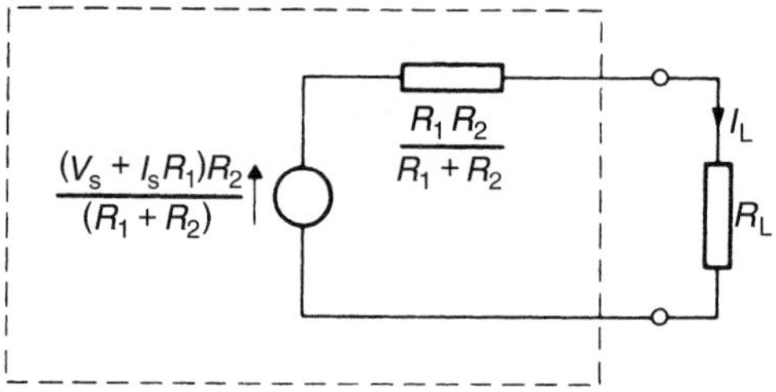

Fig. 1.50 The Thévenin equivalent circuit of the network of Fig. 1.48.

$$I_{\mathrm{L}} = \frac{(V_{\mathrm{S}} + I_{\mathrm{S}}R_1)R_2/(R_1 + R_2)}{[R_1 R_2/(R_1 + R_2)] + R_{\mathrm{L}}}$$

$$= \frac{(V_{\mathrm{S}} + I_{\mathrm{S}}R_1)R_2}{R_1 R_2 + R_{\mathrm{L}}(R_1 + R_2)}$$

which is the required expression for I_{L}.

Self-assessment question

20 Calculate, using the principle of superposition, the short-circuit current of the network of Fig. 1.48. Hence construct the Norton equivalent circuit and find the expression for I_{L}.

In general, when applying the principle of superposition to a network containing multiple sources, the contribution of each source acting alone is found by replacing *all* other ideal sources by their internal resistances (zero resistance for an ideal voltage source, infinite resistance for an ideal current source).

1.6.6 Summary of Section 1.6

Any linear circuit containing voltage sources, current sources and resistances can be represented by a Thévenin equivalent circuit or a Norton equivalent circuit. The Thévenin equivalent circuit consists of an ideal voltage source V_T in series with a resistor R_T. The Norton equivalent circuit consists of an ideal current source I_N in parallel with a resistor R_N.

V_T is the open-circuit voltage of the circuit, I_N is the short-circuit current of the circuit and $R_T = R_N = V_T/I_N$.

R_T (or R_N) can be calculated by removing the load on the circuit, replacing each ideal source by its internal resistance and calculating the resistance across the output terminals. To form either the Thévenin equivalent circuit or the Norton equivalent circuit, any two of the three quantities V_T, I_N and R_T (or R_N) must be calculated, the third being found from the other two.

These two equivalent circuits are particularly useful for analysing a circuit in terms of the current flowing in any branch of the circuit, and deducing how that current will vary if the branch resistance varies.

To apply Thévenin's or Norton's theorem to a network containing multiple sources, the principle of superposition must be used. This principle is that the total output current or voltage of a linear network can be found by finding the contribution to the output of each source acting alone, and adding together the individual contributions. To find the contribution from any one source, all other sources must be replaced with their internal resistances.

1.7 NODAL ANALYSIS

1.7.1 Introduction

The use of the Thévenin or Norton equivalent circuits allows us to evaluate unknown voltages and currents in a wide range of circuits. Unfortunately, however, there are some circuits where the use of these equivalent circuits is very difficult or impossible. Consider Fig. 1.51a. This circuit is amenable to analysis by the 'trick' of substituting Norton current generators and shunt resistors for the two voltage sources and series resistors. The trick for Fig. 1.51b is to substitute a Thévenin equivalent for V_S, R_1 and R_2, and another Thévenin equivalent for V_S, R_3 and R_4. Finding all the voltages and currents is then fairly straightforward.

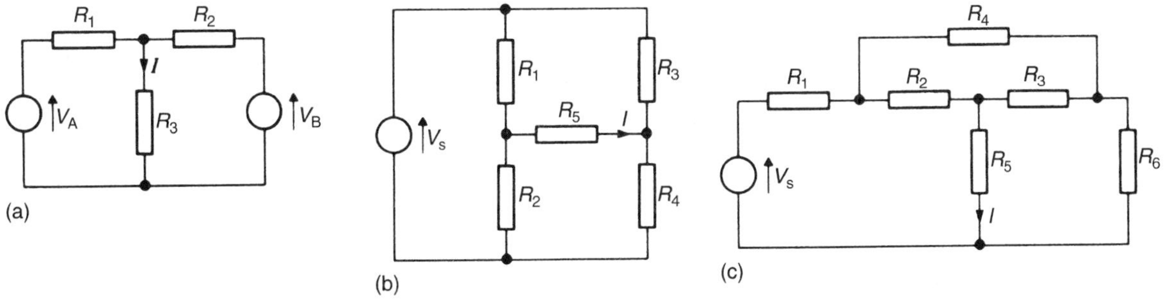

Fig. 1.51 Three example circuits where analysis requires the use of Kirchhoff law equations.

With the circuit of Fig. 1.51c, however, these techniques are difficult to apply, and much experience is needed to guess which trick is the best. In such cases, and with more complicated circuits, we need to apply Kirchhoff's laws to obtain a set of simultaneous equations which must be solved to analyse the circuit. The number of simultaneous equations that can be obtained rises rapidly with the complexity of the circuit, and their subsequent solution can become difficult and error-prone.

Clearly, it is necessary to have a methodical approach to the formation of the equations so that their solution is both possible and as straightforward as possible. There are two methods of formalizing this process of circuit analysis, and they are called **nodal analysis** and **mesh analysis**.

Nodal analysis uses Kirchhoff's current law at the nodes of a circuit. Mesh analysis uses Kirchhoff's voltage law on the meshes of a circuit. There are networks where mesh analysis is more difficult to apply than nodal analysis, and most computer-aided design packages for network analysis use nodal analysis, so this is the method we have chosen to describe in this section.

1.7.2 An outline of the method

Nodal analysis produces equations in terms of the voltages at each of the nodes of a network. To achieve this the method consists of six distinct steps.

1. One of the nodes of the circuit is chosen as a **reference node**, that is, a node which will be used as a reference point for the specification of all the other node voltages. So it will be regarded as having zero voltage. The choice of reference node is arbitrary, though it is often best to choose one side of a voltage source as the reference node. (Note that, in d.c. circuits, if there are two or more voltage sources they nearly always have one terminal in common, which you can make the reference node.)
2. Label all the other nodes voltages V_1, V_2, V_3, etc. Again the node number is arbitrary.
3. If any of the nodes have their voltages fixed by a source of e.m.f., label them V_{S1}, V_{S2}, etc. These voltages are not variable so their values do not have to be calculated. This re-labelling will eliminate some of the voltages you have already labelled.
4. At each other node (i.e. not the reference node and not the known nodes) apply Kirchhoff's current law. That is, equations expressing the fact that the current into a node equals the current leaving it. For each equation, express each current in the form $(V_X - V_Y)/R$ where V_X and V_Y are the voltages at either end of the resistor R (unless of course the value of the current is known).
5. The result of step 4 is a set of simultaneous equations which can be solved to obtain the unknown node voltages.
6. For a full analysis, the node voltages can be used to obtain a value for each branch current.

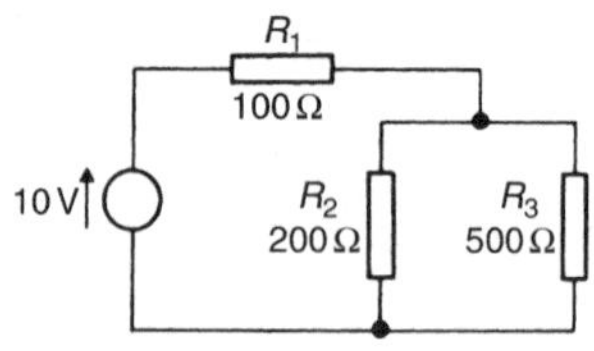

Fig. 1.52 A simple network.

We can use the relatively simple circuit of Fig. 1.52 to explain and illustrate the method. As the analysis proceeds, you may well think that the solution is more complex than it need be, and that you can obtain the same answers much more simply. For this particular example you would be correct. However, the method of nodal analysis is capable of providing a solution however complex the network may be, and where other techniques might not be applicable.

1.7.3 Choosing a reference node

Because we are really interested in finding voltage drops across all components of the circuit, we can arbitrarily choose any node as a reference node, and consider the voltage at that node to be zero. The method then allows us to calculate the voltage at any other node relative to the reference node, that is, the voltage differences between each other node and the reference node.

Figure 1.53 is a repeat of Fig. 1.52. The reference node is chosen arbitrarily at the junctions of R_2, R_3 and one side of the e.m.f. source, and indicated with the triangle symbol included on the figure. In Fig. 1.53 the node voltages V_1 and V_2 are labelled, and V_1 equals the supply voltage, 10 V. Currents are allocated to each of the branches of the network.

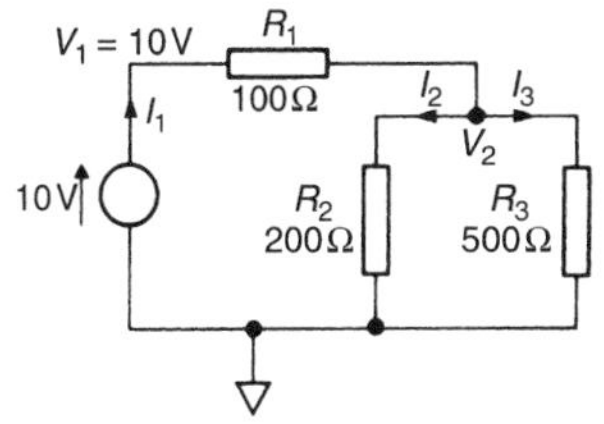

Fig. 1.53 The simple network with reference node identified and node voltage and branch currents marked.

1.7.4 Obtaining the node voltage equations

We must now apply Kirchhoff's current law at the one significant node of the circuit, the one with node voltage V_2. This gives

$$I_1 = I_2 + I_3$$

We must now write each current in the form $(V_X - V_Y)/R$ where V_X and V_Y are the voltages at either end of the resistor R. Remembering that the reference node voltage is considered zero, this gives:

$$I_1 = (10 - V_2)/R_1$$
$$I_2 = (V_2 - 0)/R_2$$
$$I_3 = (V_2 - 0)/R_3$$

Substituting these values into the current equation gives

$$\frac{(10 - V_2)}{R_1} = \frac{V_2}{R_2} + \frac{V_2}{R_3}$$

or

$$V_2\left(\frac{1}{R_1} + \frac{1}{R_2} + \frac{1}{R_3}\right) = \frac{10}{R_1}\ \text{V}$$

For this simple network there is only one node voltage to be found, and hence only one equation.

1.7.5 Completing the analysis

Using the resistance values of each resistor in the network, we can calculate the conductances $1/R_1$, $1/R_2$, etc. and substitute these values into the above equation to find V_2:

$$V_2(0.01\ \text{S} + 0.005\ \text{S} + 0.002\ \text{S}) = 10\ \text{V} \times 0.01\ \text{S}$$

$$V_2 = \frac{100}{17}\text{V} = 5.88\ \text{V}$$

If a full analysis of the circuit is required, then we must find the value of each branch current, as well as the value of the node voltage. To obtain the branch currents, we can use the expressions for each current obtained in the previous section. So,

$$I_1 = \frac{10 - V_2}{R_1} = \frac{10 - 5.88}{100}\text{A} = 41.2\ \text{mA}$$

$$I_2 = V_2/R_2 = 5.88/200 = 29.4\ \text{mA}$$

$$I_3 = V_2/R_4 = 5.88/500 = 11.8\ \text{mA}$$

This completes the analysis, since we now know the values of all the node voltages and all the branch currents.

Self-assessment question

21 Use nodal analysis to obtain the node voltages and branch currents of the network of Fig. 1.54.

Self-assessment question

22 Use nodal analysis to obtain the simultaneous equations in the node voltages for the network of Fig. 1.55.

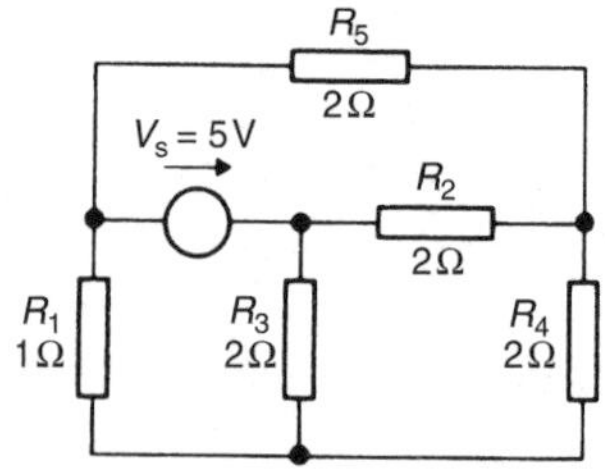

Fig. 1.54 Circuit of Self-assessment question 21.

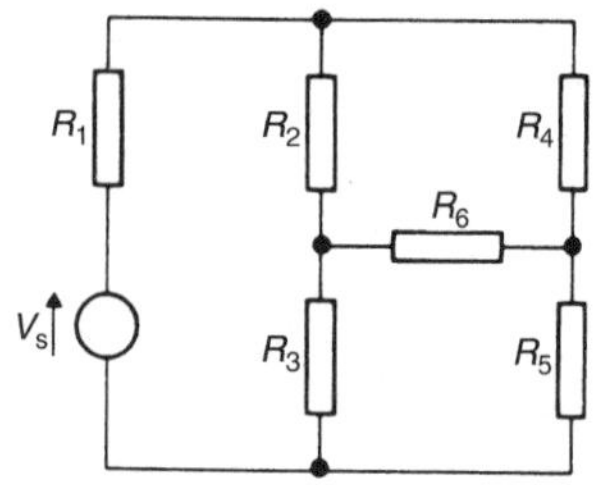

Fig. 1.55 Circuit of Self-assessment question 22.

1.7.6 A circuit with multiple e.m.f. sources

The nodal analysis of a circuit containing more than one e.m.f. source, such as the circuit of Fig. 1.51a, follows exactly the same steps as are used for a circuit with a single e.m.f. source. Figure 1.56 is a repeat of Fig. 1.51a with numerical values for the resistors and voltage sources, with a reference node chosen and with the node voltages and branch currents identified. In this case the problem is to find the current I_3.

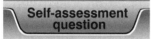

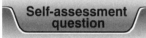

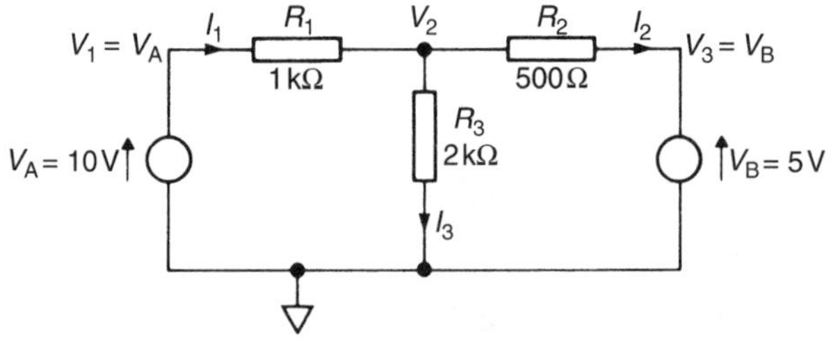

Fig. 1.56 Fig. 1.51(a) redrawn to include component values, reference node, node voltages and branch currents.

Only one current law equation is required, that at node 2. It is

$$I_1 = I_2 + I_3$$

The expressions for the branch currents in terms of the node voltages are

$$I_1 = \frac{V_A - V_2}{R_1}, \quad I_2 = \frac{V_2 - V_B}{R_2}, \quad I_3 = \frac{V_2}{R_3}$$

Substituting these values in the current law equation gives

$$\frac{V_A - V_2}{R_1} = \frac{V_2 - V_B}{R_2} + \frac{V_2}{R_3}$$

or

$$V_2\left(\frac{1}{R_1} + \frac{1}{R_2} + \frac{1}{R_3}\right) = \frac{V_A}{R_1} + \frac{V_B}{R_2}$$

This equation allows the unknown node voltage V_2 to be found, and hence the required current I_3.

Self-assessment question

23 For the network of Fig. 1.56, use the equation above to find the value of V_2 and hence find I_3.

1.7.7 Summary of Section 1.7

Nodal analysis is a method of circuit analysis which is applicable to any circuit configuration. The method allows the calculation of all the unknown node voltages in a circuit and hence the determination of any required branch current. The steps involved in nodal analysis are:

1. Choose a reference node.
2. Label the other node voltages.
3. If any of the nodes have fixed, known, voltages from e.m.f. sources, label them with their fixed value.
4. At each unknown node, apply Kirchhoff's current law. For each current equation, express each current in the form $(V_x - V_y)/R$ or $(V_x - V_y)G$ (unless the value of a current is known).
5. Solve the resulting set of simultaneous equations to obtain the unknown node voltages.
6. If required, use the node voltages to calculate the branch currents. The method is equally applicable to circuits with multiple e.m.f. sources. As the circuit complexity increases (in particular as the number of nodes and meshes increases) the number of simultaneous equations to be solved to obtain the node voltages increases. This increases the work involved in evaluating the node voltages. A computer program could be used to solve the set of simultaneous equations, but a more appropriate solution is to use a circuit analysis software package both to create and solve the node voltage equations.

1.8 ANALYSING CIRCUITS USING A COMPUTER-AIDED DESIGN (CAD) PACKAGE

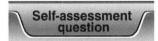

The method of nodal analysis (and indeed the method of mesh analysis which we have not covered) allows you to analyse quite complex circuits provided you have either a computer program to solve simultaneous equations or a lot of mathematical skill. Even so, you still have to perform quite a lot of work to set up the simultaneous equations which lead to the solution.

If you were a circuit designer trying to examine what might happen in the circuit if you changed one or more components, then the analysis might have to be performed many times over for the same circuit configuration. To make life easier for circuit designers, computer-aided design (CAD) packages are available which, when given the circuit configuration and component values, create the equations necessary for a full analysis of the circuit, as well as solving them to evaluate node voltages and branch currents.

We assume that you have access to such a package, and a computer. This is the appropriate time for you to learn how to 'drive' the software package. When you have learnt how to use it you should analyse the circuits contained in the following figures: Fig. 1.53, Fig. 1.54, Fig. 1.55 (with $R_1 = 30\ \Omega$, $R_2 = R_3 = 100\ \Omega$, $R_4 = 200\ \Omega$, $R_5 = 250\ \Omega$, $R_6 = 500\ \Omega$, and $V_S = 5$ V), Fig. 1.56. You will find that you have analysed some of these circuits in Self-assessment questions or at least have formulated the simultaneous equations. For those you have already analysed, you should check the answers provided by the CAD package against the answers obtained previously.

If possible, do the CAD work now.

Answers to self-assessment questions

1. (a) The resistance R of the bulb $= V/I = 12\text{ V}/3\text{ A} = 4\ \Omega$.

(b) Using the equation $V = IR$,

$$V = 10 \times 10^{-3}\text{A} \times 300\ \Omega = 3\text{ V}$$

(c) Again, since $V = IR$,

$$V = 200 \times 10^{-3}\text{A} \times 0.1\ \Omega = 20 \times 10^{-3}\text{V} = 20\text{ mV}$$

2. (a) To find the power dissipated in a resistor, given I and R, substitute $V = IR$ in the expression $P = VI$ giving

$$P = (IR) \times I = I^2R$$

(b) To find the power, given V and R, substitute $I = V/R$ into $P = VI$, giving

$$P = V^2/R$$

3. (a) Using the equation $G = I/V$,

$$G = 0.5\text{ A}/200\text{ V} = 2.5 \times 10^{-3}\text{ S} = 2.5\text{ mS}$$

(b) Using the equation $V = I/G$,

$$V = 5\text{ A}/10\text{ S} = 0.5\text{ V}$$

(c) Using the equation $I = VG$,

$$I = 2.4\text{ V} \times 0.3\text{ S} = 0.72\text{ A}$$

(d) For the demister, $V = 12$ V and $I = 10$ A, so, using $G = I/V$,

$$G = 10\text{ A}/12\text{ V} = 0.83\text{ S}$$

4. (a) (i) From the graph (Fig. 1.2), when $I = 0.1$ A, $V = 4.2$ V, so

$$R = V/I = 4.2\text{ V}/0.1\text{ A} = 42\ \Omega$$

(ii) When $V = 8$ V, $I = 0.144$ A, $R = V/I = 55.6\ \Omega$

The power dissipated in the bulb when $I = 0.1$ A can be found using $P = VI$, so

$$P = 4.2\text{ V} \times 0.1\text{ A} = 0.42\text{ W}$$

When $V = 8$ V,

$$P = 8\text{ V} \times 0.144\text{ A} = 1.15\text{ W}$$

(b) (i) From Fig. 1.3, when $I = 4$ mA, $V = 0.65$ V, so

$$G = I/V = 4 \times 10^{-3}\text{ A}/0.65\text{ V} = 6.02 \times 10^{-3}\text{ S} = 6.02\text{ mS}$$

(ii) When $I = 10$ mA, $V = 0.7$ V, so

$$G = I/V = 10 \times 10^{-3}\text{ A}/0.7\text{ V} = 14.3\text{ mS}$$

5. From the equation $V_B = IR_L$ we can write $I = V_B/R_L$. Substituting this value of I in the equation for V_B, we get

$$V_B = V_S - \frac{V_B}{R_L} R_S$$

Rearranging,

$$V_S - V_B = V_B \frac{R_S}{R_L}$$

So, if R_L is large compared with R_S, the quotient R_S/R_L is small and the change in terminal voltage ($V_S - V_B$) is also small compared with V_B.

Alternatively, rearranging the above equation differently,

$$V_B + V_B \frac{R_S}{R_L} = V_S$$

so

$$V_B = \frac{V_S}{1 + R_S/R_L}$$

Therefore, if $R_S \ll R_L$, $V_B \approx V_S$.

6. (a) The fall in terminal voltage of the battery is equal to the voltage drop across the internal battery resistance.
Using $V = IR$,
Voltage drop $= 250 \times 10^{-3}$ A $\times$ 0.7 $\Omega = 175 \times 10^{-3}$ V $= 0.175$ V
Hence, at a current of 250 mA.
Terminal voltage $=$ (1.5 V $-$ 0.175 V) $=$ 1.325 V
This is also the potential drop across the load resistor, so using $R = V/I$

$$R = 1.325 \text{ V}/250 \times 10^{-3} \text{ A} = 5.3\ \Omega$$

(b) The fall in terminal voltage is (13.2 – 11.5) V $=$ 1.7 V when the current taken is 70 A. So Internal resistance $= V/I =$ 1.7 V/70 A $=$ 0.024 Ω.

The load resistance is equal to the terminal voltage divided by the current, so load resistance $=$ 11.5 V/70 A $=$ 0.164 Ω

7. (a) Using the equation $R = R_1 + R_2 + R_3$, equivalent resistance R is

$$R = 100\ \Omega + 200\ \Omega + 500\ \Omega = 800\ \Omega$$

(b) The current flowing in the circuit is given by $I = V/R$, so

$$I = 10 \text{ V}/800\ \Omega = 12.5 \text{ mA}$$

(c) Using the expression $V = IR$ for each resistor in turn:

For the 100 Ω resistor,

$$V = 12.5 \times 10^{-3} \text{ A} \times 100\ \Omega = 1.25 \text{ V}.$$

For the 200 Ω resistor,

$$V = 12.5 \times 10^{-3} \text{ A} \times 200\ \Omega = 2.5 \text{ V}.$$

For the 500 Ω resistor,

$$V = 12.5 \times 10^{-3} \text{ A} \times 500\ \Omega = 6.25 \text{ V}.$$

(d) Using the expression $P = I^2R$ for the power dissipated as heat in a resistor:

For R_1, $P = (12.5 \times 10^{-3} \text{A})^2 \times 100\ \Omega = 16$ mW.

For R_2, $P = (12.5 \times 10^{-3} \text{A})^2 \times 200\ \Omega = 31$ mW.

For R_3, $P = (12.5 \times 10^{-3} \text{A})^2 \times 500\ \Omega = 78$ mW.

(e) The power taken from the e.m.f. source is

$$P = V \times I = 10 \text{ V} \times 12.5 \times 10^{-3} \text{ A} = 125 \text{ mW}.$$

This is also the sum of the powers dissipated in the three resistors

8. (a) The equivalent resistance of three equal resistors in parallel must be 1/3 of the value of each (since the equivalent conductance will be three times the conductance of each), so $R = 100\ \Omega$.
(b) The resistance of the 1 kΩ resistors are so much larger than that of the 5 Ω resistor that the equivalent resistance is hardly less than 5 Ω. [The error introduced by calling it 5 Ω, and neglecting the parallel resistance of 500 Ω (1000 Ω in parallel with 1000 Ω), is only 1%.]
(c) In Fig. 1.10c, the two 1000 Ω resistors in parallel will have an effective resistance of 500 Ω. This, in series with R_1, also 500 Ω, gives an equivalent resistance of 1000 Ω.

9. The bulb is designed to work at 3 V, 0.5 W. Its normal operating current is therefore given by the equation $I = P/V$. Hence,

$$I = 0.5 \text{ W}/3 \text{ V} = 0.167 \text{ A} = 167 \text{ mA}$$

To apply a voltage of 3 V to the bulb at a current of 167 mA requires a voltage drop of 9 V in a series resistance, R. So

$$R = 9 \text{ V}/0.167 \text{ A} = 54\ \Omega$$

Power dissipated in $R = V^2/R =$ (9 V)^2/54 Ω $=$ 1.5 W.

The bulb, when cold, has a much lower resistance than its operating value. In normal operation, the only resistance across the 3 V battery at switch-on is this lower resistance, so the initial current is much larger than the final operating current, giving rapid heating of the bulb filament. In the new circuit, although the bulb resistance is low at switch-on, the current has to pass through the 54 Ω resistor, which does not change its value. The initial surge of current at switch-on is therefore only slightly larger than the final operating current, so the filament heating occurs more slowly.

10. (a) In Fig. 1.14, let $G_1 = 1/R_1$ and $G_2 = 1/R_2$. Now $I_1 = VG_1$ and $I_2 = VG_2$. So $I_1/I_2 = VG_1/VG_2 = G_1/G_2$. Hence the ratio of the two currents is equal to the ratio of their conductances. Notice that since $I = V(G_1 + G_2)$,

$$\frac{I_1}{I} = \frac{G_1}{G_1 + G_2} \text{ so } I_1 = \frac{I \times G_1}{G_1 + G_2} \text{ and similarly } I_2 = \frac{I \times G_2}{G_1 + G_2}$$

(b) If $R_1 = 2\ \Omega$ and $R_2 = 18\ \Omega$ then $G_1 = 0.5$ S and $G_2 = 0.055$ S. So $I_1 = VG_1 = 5\text{ V} \times 0.5\text{ S} = 2.5$ A and $I_2 = VG_2 = 5\text{ V} \times 0.055\text{ S} = 0.28$ A.

11. (a) In Fig. 1.19a, when the wiper is 3/4 of the way towards B,

$$V_{OUT} = \frac{5\text{ V} \times 7.5\ \Omega}{10\ \Omega} = 3.75\text{ V}$$

(b) The power dissipated in the pot $= V_S^2/R = (5\text{ V})^2/10\ \Omega = 2.5$ W.

(c) In Fig. 1.19b, the pot, together with the resistor R_L will form a potential divider across the 5 V supply. The lower part of the potential divider is the parallel resistance of the 10 Ω load and the lower 3/4 of the pot resistance, namely 7.5 Ω. The upper resistor of the potential divider is the other 1/4 of the pot resistance. The resistance of the lower part of the potential divider is

$$\frac{7.5 \times 10}{7.5 + 10}\Omega = 4.29\ \Omega$$

The resistance of the upper part is 2.5 Ω. Hence

$$V_{OUT} = 5\text{ V} \times \frac{4.29\ \Omega}{2.5\ \Omega + 4.29\ \Omega} = 3.16\text{ V}$$

The presence of the 10 Ω load resistor has lowered the output voltage by 0.69 V at the 3/4 position of the wiper.

(d) Your reasoning should go something like this: With the slider at the end B of the pot, the load is connected directly to the 5 V supply, and so $V_{OUT} = 5$ V. The presence of R_L causes no error in this position. With the slider at the other end of the pot, the output voltage is zero, with or without R_L, so again there is no error. The maximum change in V_{OUT} is therefore likely to be around the central position of the wiper. [Exact calculation shows that the change in output voltage caused by attaching a load resistor across the output of a pot, expressed as a *percentage* of the unloaded output, occurs exactly at the centre of the pot. However, the maximum *voltage* change will occur about 2/3 of the way towards B.]

12. The ratio of the overall current to the meter current is 5 A/50 μA = 100 000. Hence the ratio of the shunt resistance to the coil resistance is

$$\frac{1}{99999}$$

Hence, shunt resistance ≈ 2500 Ω/100 000 = 0.025 Ω. The overall meter resistance is also 0.025 Ω.

Note that in practice such a low shunt resistance is difficult to achieve reliably, so to measure large currents a resistance is usually put in series with the meter (e.g. 10 kΩ in this case) thus necessitating a larger shunt resistance (e.g. 1.25 Ω for the meter with 10 kΩ in series with it).

13. (i) (a) The overall meter resistance = 100 V/50 μA = 2 MΩ.
Hence the series resistance = 2 MΩ − 2.5 kΩ = 1.998 MΩ.
(b) The overall meter resistance = 25 V/50 μA = 500 kΩ.
Hence, series resistance = 500 kΩ − 2.5 kΩ = 497.5 kΩ.

(ii) For the 5 V range, overall meter resistance = 100 kΩ. Required ratio = 100 kΩ/5 V = 20 000 Ω V^{-1}.
For the 25 V range, the ratio ≈ 500 kΩ/25 V = 20 000 Ω V^{-1}.
For the 100 V range, the ratio ≈ 2 MΩ/100 V = 20 000 Ω V^{-1}.

14. (a) For 1/4 f.s.d. the total resistance must be 4 times $(R_m + R_S)$ so $R_X = 3 \times 30\text{ k}\Omega = 90\text{ k}\Omega$. That is the point which is 1/4 f.s.d. must be calibrated as 90 kΩ.

(b) For 1/8 f.s.d. $R_X = 7 \times 30\text{ k}\Omega = 210\text{ k}\Omega$.

(c) For 3/4 f.s.d. the total resistance must be (4/3) $(R_m + R_S)$, so

$$R_X = \frac{1}{3} \times 30\text{ k}\Omega = 10\text{ k}\Omega.$$

15. (a) The input voltage to the amplifier plus A-D converter which will give full scale deflection is 10 V/voltage gain = 10 V/500 = 20 mV. The resistor needed to produce a voltage drop of 20 mV when a current of 1 mA is flowing through it is 20 mV/1 mA = 20 Ω.

(b) Similarly the resistance required to give full-scale deflection when 100 mA is flowing is 20 mV/100 mA = 0.2 Ω.

16. For node A, $I_1 = I_2 + I_3 + I_4$.
For node B, $I_4 = I_6 + I_7$.
For node C, $I_3 + I_7 = I_8$.
For node D, $I_2 + I_6 = I_5$.
For node E, $I_5 + I_8 = I_1$.

17. Applying Kirchhoff's voltage law to the circuit of Fig. 1.32, clockwise round the circuit starting with V_2:

$$5\text{ V} + 10\text{ V} - 1\text{ V} - V_{R2} - 2.5\text{ V} = 0$$

Hence $V_{R2} = 11.5$ V.

18. Figure 1.57 is the circuit of Fig. 1.43 redrawn as a network with R_4 as the load on the network. The open-circuit voltage V_{OC} is the voltage across the network output terminals with R_4 removed. Since there is then no current in R_3 (and hence no voltage drop across it), V_{OC} will be equal to the voltage across R_2. Hence

$$V_{OC} = V_T = I_S \times R_2 = 5\text{ mA} \times 150\ \Omega = 0.75\text{ V}.$$

The source resistance R_T can be found by replacing the current source by its internal resistance (an open-circuit), when R_T is clearly R_3 in series with R_2. Hence, $R_T = 200\ \Omega + 150\ \Omega = 350\ \Omega$

From the Thévenin equivalent circuit of the network,

$$I = \frac{0.75\text{ V}}{(350 + 50)\ \Omega} = 1.88\text{ mA}$$

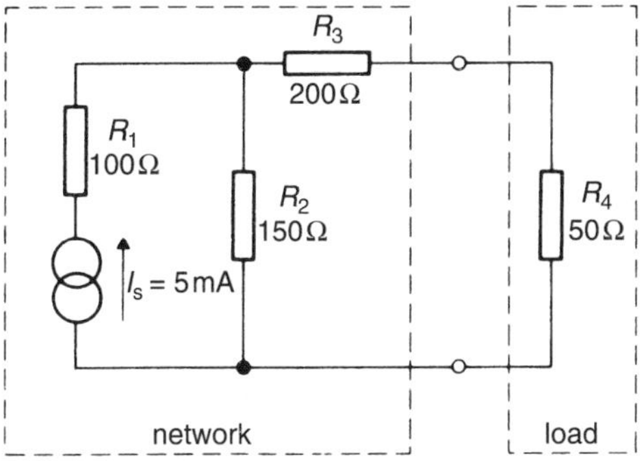

Fig. 1.57

19. You can calculate I_N by two methods, either by finding V_{OC} and R_N, or by finding I_{SC}, the short-circuit current. Either method is a full solution of the problem.

(a) Figure 1.58a shows the circuit for the calculation of V_{OC}. There is no voltage drop across R_4, so V_{OC} is equal to the voltage drop across R_3. The current from the source divides between R_1 and the branch containing R_2 and R_3 in the ratio of their conductances, so three-fifths of the current flows through R_1, while two-fifths flows through R_2 and R_3. The voltage drop across R_3 is therefore $20\ \Omega \times 0.4\text{ A} = 8$ V and hence $V_{OC} = 8$ V. Figure 1.58b shows the circuit for the calculation of R_N. The current source has been replaced with an open-circuit. R_N is made up of R_1 and R_2 in series across R_3, the whole of that being in series with R_4. Hence,

$$R_N = R_4 + \frac{R_3(R_2 + R_1)}{R_3 + (R_2 + R_1)} = 20\ \Omega + \frac{20 \times 30}{50}\ \Omega = 32\ \Omega$$

Using $I_N = V_{OC}/R_N$,

$I_N = 8\text{ V}/32\ \Omega = 0.25$ A

Figure 1.58c is the Norton equivalent circuit loaded with the resistor R_5. From the figure, the current in R_5 is given by

$$I = I_{SC} \times \frac{G_5}{G_N + G_5} = 0.25 \times \frac{0.1}{0.031 + 0.1}\text{ A} = 0.19\text{ A}$$

(b) Figure 1.58d shows the circuit for the calculation of I_{SC} (which is equal to I_N). It looks at first sight as though the calculation of I_{SC} requires a nodal analysis of the network, but observation of the resistor values reveals a short-cut to the required value of I_{SC}. Because R_3 and R_4 are equal, any current passing through R_2 will divide equally

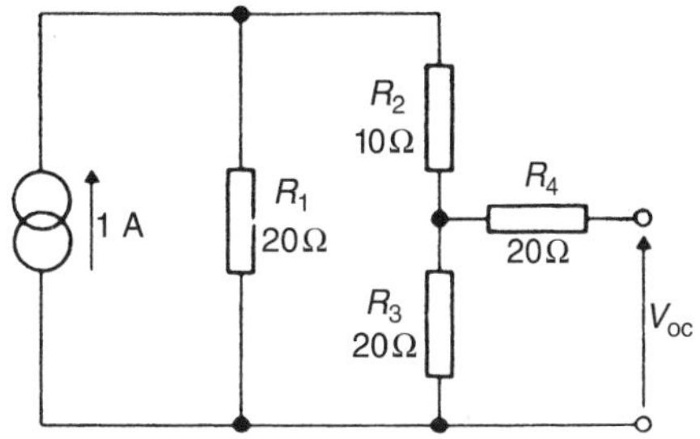

Fig. 1.58(a)

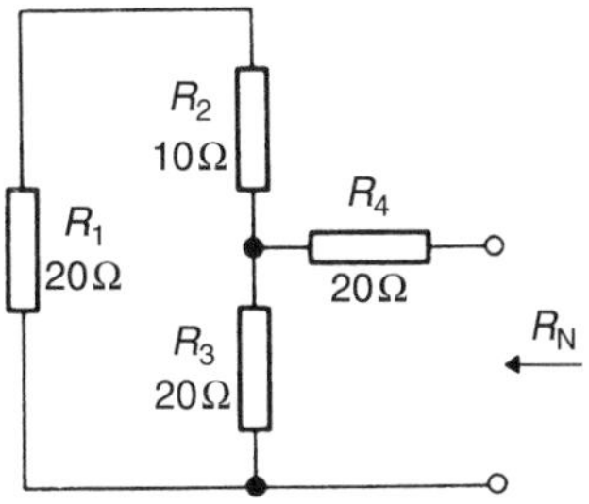

Fig. 1.58(b)

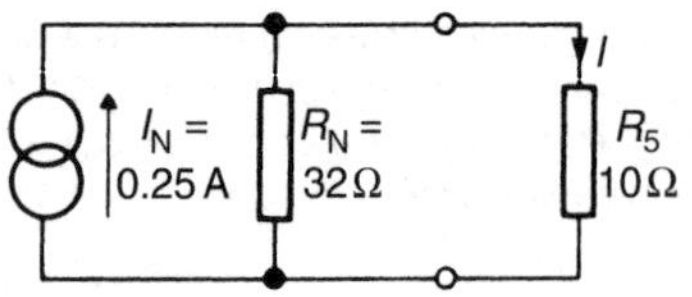

Fig. 1.58(c)

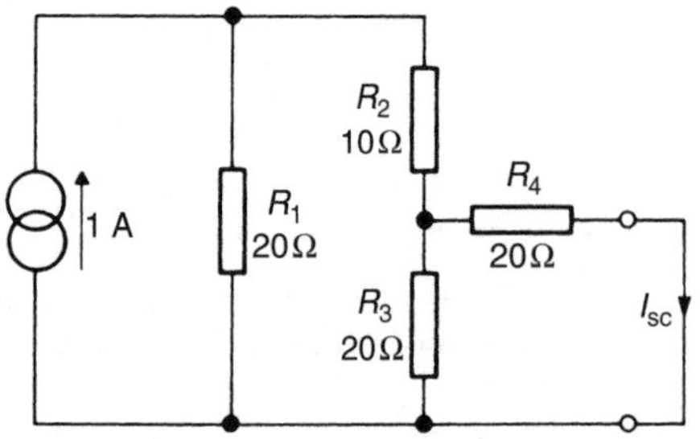

Fig. 1.58(d)

between the two, so I_{SC} is half the current through R_2. R_3 and R_4 in parallel have an equivalent resistance of 10 Ω (20 Ω in parallel with 20 Ω) which, in series with R_2, makes 20 Ω. So the current from the source divides equally between R_1 and R_2. The current in R_2 is therefore $\frac{1}{2}$A, and the current I_{SC} is $\frac{1}{4}$A. Hence, $I_N = I_{SC} = 0.25$ A. This is the same value as was found using the alternative method in (a).

20. Figure 1.59a shows the required short-circuit current I_{SC}. By the principle of superposition it will be the sum of I_{SC1} from Fig. 1.59b and I_{SC2} from Fig. 1.59c. From Fig. 1.59b, $I_{SC1} = V_S/R_1$. From Fig. 1.59c, $I_{SC2} = I_S$ (no current can flow in R_1 or R_2 since there is no voltage across them). Hence,

$$I_{SC} = I_S + V_S/R_1$$

R_N has already been found; it is $R_1R_2/(R_1 + R_2)$. The Norton equivalent circuit is therefore that shown in Fig. 1.59d with the load R_L attached. By the current divider rule,

$$I_L = \frac{I_N G_L}{G_N + G_L}$$

$$= \frac{(I_S + V_S/R_1)/R_L}{(R_1 + R_2)/R_1R_2 + (1/R_L)}$$

$$= \frac{(I_S + V_S/R_1)R_1R_2}{R_L(R_1 + R_2) + R_1R_2}$$

$$= \frac{(I_SR_1 + V_S)R_2}{R_1R_2 + R_L(R_1 + R_2)}$$

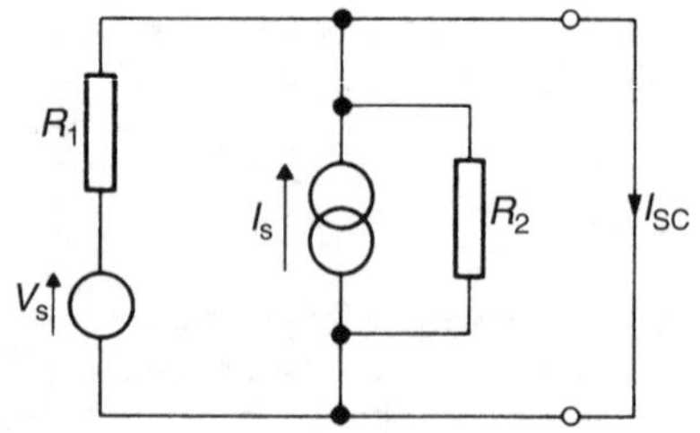

Fig. 1.59(a)

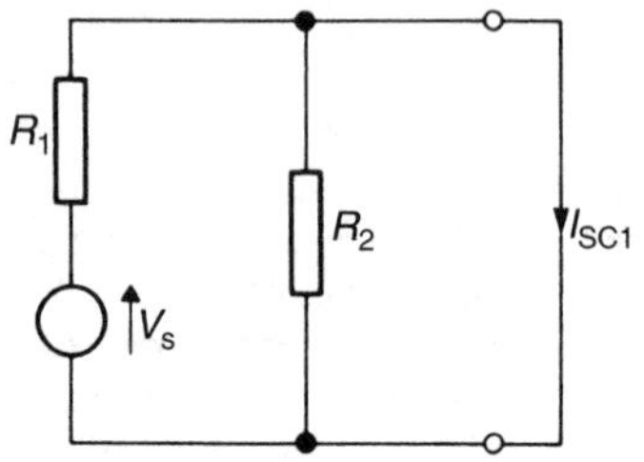

Fig. 1.59(b)

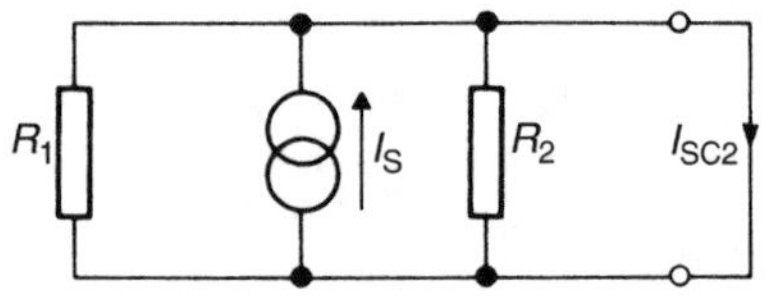

Fig. 1.59(c)

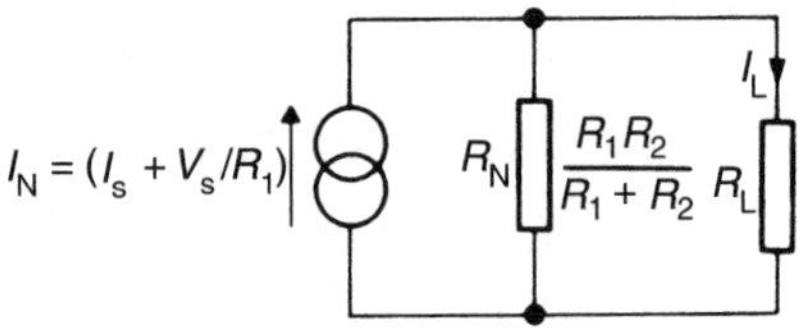

Fig. 1.59(d)

which is the same result as was obtained in Section 1.6.5. using Thévenin's theorem.

21. Figure 1.60 is Fig. 1.54 redrawn to show the reference node, and with node voltages and branch currents identified. Applying Kirchhoff's current law:

At node 2, $I_2 + I_5 = I_4$;

At node 3, $I_3 + I_4 = I_1$.

Writing the currents in terms of node voltages,

$$I_1 = \frac{V_3}{R_1}, \quad I_2 = \frac{V_S - V_2}{R_2}, \quad I_3 = \frac{V_S - V_3}{R_3},$$

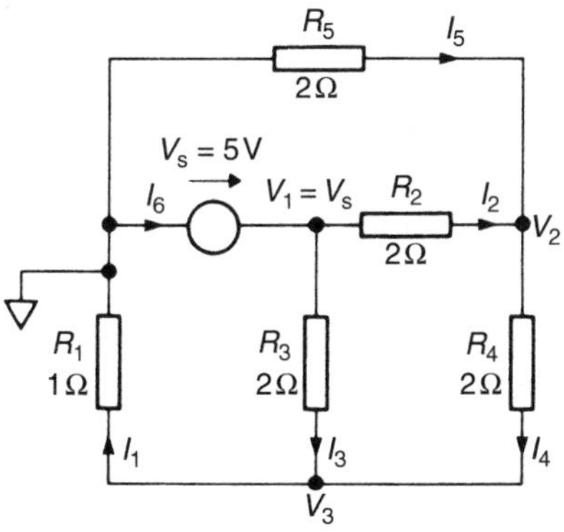

Fig. 1.60

$$I_4 = \frac{V_2 - V_3}{R_4}, \quad I_5 = \frac{0 - V_2}{R_5} = -\frac{V_2}{R_5}$$

So, for node 2,

$$\frac{V_S - V_2}{R_2} - \frac{V_2}{R_5} = \frac{V_2 - V_3}{R_4}$$

$$V_2\left(\frac{1}{R_2} + \frac{1}{R_4} + \frac{1}{R_5}\right) - \frac{V_3}{R_4} = \frac{V_S}{R_2}$$

substituting numerical values,

$$V_2(0.5 + 0.5 + 0.5) - 0.5V_3 = 0.5 \times 5\text{ V}$$

$$1.5V_2 - 0.5V_3 = 2.5\text{ V} \qquad \text{(A)}$$

For node 3,

$$\frac{V_S - V_3}{R_3} + \frac{V_2 - V_3}{R_4} = \frac{V_3}{R_1}$$

$$\frac{V_2}{R_4} - V_3\left(\frac{1}{R_1} + \frac{1}{R_3} + \frac{1}{R_4}\right) = -\frac{V_S}{R_3} \qquad \text{(B)}$$

$$0.5V_2 - 2V_3 = -2.5\text{ V}$$

From (B), $V_2 = 4V_3 - 5$ V. Substitute in (A):

$$1.5(4V_3 - 5\text{ V}) - 0.5V_3 = 2.5\ V$$

$$(6 - 0.5)V_3 = 2.5\text{ V} + 7.5\text{ V}$$

$$V_3 = \frac{10}{5.5}\text{V} = \frac{20}{11}\text{V} = 1.82\text{ V}$$

Hence,

$$V_2 = 4V_3 - 5\text{ V}$$

$$= \left(\frac{80}{11} - 5\right)\text{V} = \frac{25}{11}\text{V} = 2.27\text{ V}$$

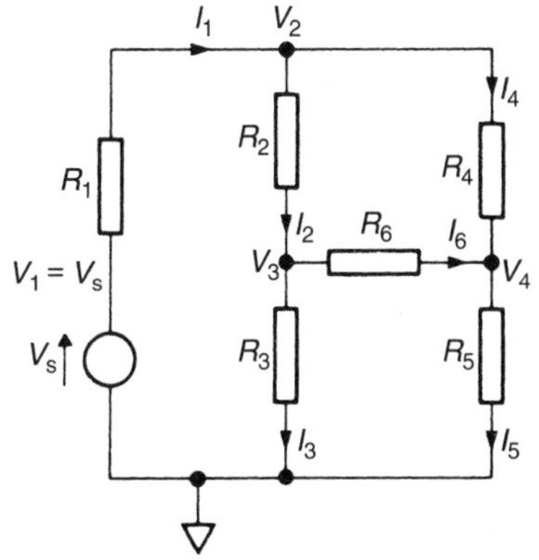

Fig. 1.61

Using these values of V_2 and V_3, the currents are:

$$I_1 = \frac{V_3}{R_1} = \frac{1.82\text{ V}}{1\ \Omega} = 1.82\text{ A}$$

$$I_2 = \frac{V_S - V_2}{R_2} = \frac{(5 - 2.27)\text{ V}}{2\ \Omega} = 1.37\text{ A}$$

$$I_3 = \frac{V_S - V_3}{R_3} = \frac{(5 - 1.82)\text{ V}}{2\ \Omega} = 1.59\text{ A}$$

$$I_4 = \frac{V_2 - V_3}{R_4} = \frac{2.27\text{ V} - 1.82\text{ V}}{2\ \Omega} = 0.23\text{ A}$$

$$I_5 = -\frac{V_2}{R_5} = -\frac{2.27\text{ V}}{2\ \Omega} = -1.14\text{ A}$$

$$I_6 = I_2 + I_3 = (1.37 + 1.59)\text{ A} = 2.96\text{ A}$$

22. Figure 1.61 is Fig. 1.55 re-drawn for nodal analysis. Applying Kirchhoff's current law;

At node 2, $I_1 = I_2 + I_4$;

At node 3, $I_2 = I_3 + I_6$;

At node 4, $I_6 + I_4 = I_5$.

Evaluating the currents in terms of the node voltages:

$$I_1 = \frac{V_S - V_2}{R_1}, \quad I_2 = \frac{V_2 - V_3}{R_2}, \quad I_3 = \frac{V_3}{R_3}$$

$$I_4 = \frac{V_2 - V_4}{R_4}, \quad I_5 = \frac{V_4}{R_5}, \quad I_6 = \frac{V_3 - V_4}{R_6}$$

So, at node 2,

$$\frac{V_S - V_2}{R_1} = \frac{V_2 - V_3}{R_2} + \frac{V_2 - V_4}{R_4}$$

$$V_2\left(\frac{1}{R_1} + \frac{1}{R_2} + \frac{1}{R_4}\right) - \frac{V_3}{R_2} - \frac{V_4}{R_4} = \frac{V_S}{R_1} \qquad \text{(A)}$$

At node 3,

$$\frac{V_2 - V_3}{R_2} = \frac{V_3}{R_3} + \frac{V_3 - V_4}{R_6}$$

$$\frac{V_2}{R_2} - V_3\left(\frac{1}{R_2} + \frac{1}{R_3} + \frac{1}{R_6}\right) + \frac{V_4}{R_6} = 0 \qquad \text{(B)}$$

At node 4,

$$\frac{V_3 - V_4}{R_6} + \frac{V_2 - V_4}{R_4} = \frac{V_4}{R_5}$$

$$\frac{V_2}{R_4} + \frac{V_3}{R_6} - V_4\left(\frac{1}{R_4} + \frac{1}{R_5} + \frac{1}{R_6} = 0\right) \qquad \text{(C)}$$

Equations (A), (B) and (C) are the required simultaneous equations in the three node voltages V_2, V_3 and V_4.

23. The equation is

$$V_2\left(\frac{1}{R_1} + \frac{1}{R_2} + \frac{1}{R_3}\right) = \frac{V_A}{R_1} + \frac{V_B}{R_2}$$

Substituting the conductances G_1, G_2, G_3 in mS and the values of the source voltages:

$$V_2(1 + 2 + 0.5) = (10 \times 1) + (5 \times 2)$$

$$V_2 = \frac{20}{3.5} = 5.7\ \text{V}$$

Now

$I_3 = V_2/R_3 = 5.7\ \text{V}/2\ \text{k}\Omega = 2.85\ \text{mA}$.

Signals, waveforms and a.c. components

2

AIMS

1. To distinguish between analogue and digital signal waveforms. To distinguish between periodic waveforms and non-periodic (signal and noise) waveforms.
2. To explain the reason for the study of sinusoidal waveforms, and to define their properties.
3. To explain the ways in which voltage, current and power are expressed for periodic waveforms (including sinusoids) and for non-periodic waveforms.
4. To introduce the concept of the frequency spectrum and Fourier analysis, for both periodic and non-periodic waveforms.
5. To explain the properties of capacitors and inductors, and their response to sinusoidal and step waveforms.
6. To explain the properties of transformers.

GENERAL OBJECTIVES

After reading this chapter you should understand the meanings of the following terms:

- a.c. voltage
- amplitude
- analogue signal waveform
- angular frequency
- bandwidth
- capacitance, capacitor
- cosine wave
- dielectric
- electromagnetic induction
- filter: low-pass, high-pass
- Fourier analysis, Fourier coefficient, Fourier component, Fourier series
- frequency
- frequency distribution
- fundamental frequency
- Gaussian distribution
- Harmonic
- inductance, inductor
- input resistance
- magnetic field, flux
- magnetic flux density
- magnetizing current
- mean-square value
- noise
- non-periodic waveform

- normal distribution
- period
- periodic waveform
- permeability
- permittivity
- phase response, phase shift
- r.m.s. (root mean square)
- reactance: capacitive, inductive
- self-induction
- signal
- sine wave, sinusoid
- sinusoidal waveform
- spectrum: amplitude; continuous; line; phase;power density
- square wave
- step response
- time constant
- transformer: step-down; step up
- waveform

SPECIFIC OBJECTIVES

After completing your study of this chapter, you should be able to:

1. Distinguish between analogue and digital waveforms.
2. Calculate the frequency, and the angular frequency, of a sinusoidal waveform from its period, and vice versa.
3. Calculate the phase shift of a sinusoidal waveform which is shifted in time with respect to some reference.
4. Explain the meanings of the terms mean-square voltage and r.m.s. voltage of periodic waveforms and of signals and noise (non-periodic) waveforms.
5. Calculate the mean-square voltage and r.m.s. voltage of simple periodic waveforms.
6. Describe methods of measuring these quantities.
7. State properties of the frequency spectra of periodic waveforms (line spectra, Fourier series).
8. Calculate the r.m.s. voltage of a periodic waveform from its spectrum.
9. Explain the terms power density and power density spectrum used for non-periodic waveforms.
10. Calculate the mean-square voltage of a signal or noise waveform, over a band of frequencies, from its power density spectrum.
11. Define capacitance and inductance, and calculate the reactance of a capacitor or inductor at a given frequency.
12. Sketch the frequency response of simple *CR* and *LR* low-pass and high-pass networks (filters).
13. Describe the response of simple *CR* and *LR* low-pass and high-pass networks to voltage steps, and write down expressions for their exponential response.

14. Describe the behaviour of a transformer in terms of input and output currents and voltages, and relate these currents and voltages to the number of turns of the primary and secondary windings.

2.1 ELECTRICAL WAVEFORMS

The word 'waveform' is used to refer to the graph of the voltage, or current, of a varying quantity against time. Consider the simple circuit shown in Fig. 2.1a. If the switch is closed, current will almost instantaneously flow through the lamp, causing it to shine. If the switch is opened and closed repeatedly at 1-second intervals, the graph of current will be as shown in Fig. 2.1b. This waveform is called a **square wave**. It is of course also the graph of the voltage across the lamp and approximately the graph of the brightness of the lamp. If the switch is opened and closed irregularly, according to some code, as in Fig. 2.1c the circuit can be used to transmit information from the sender to

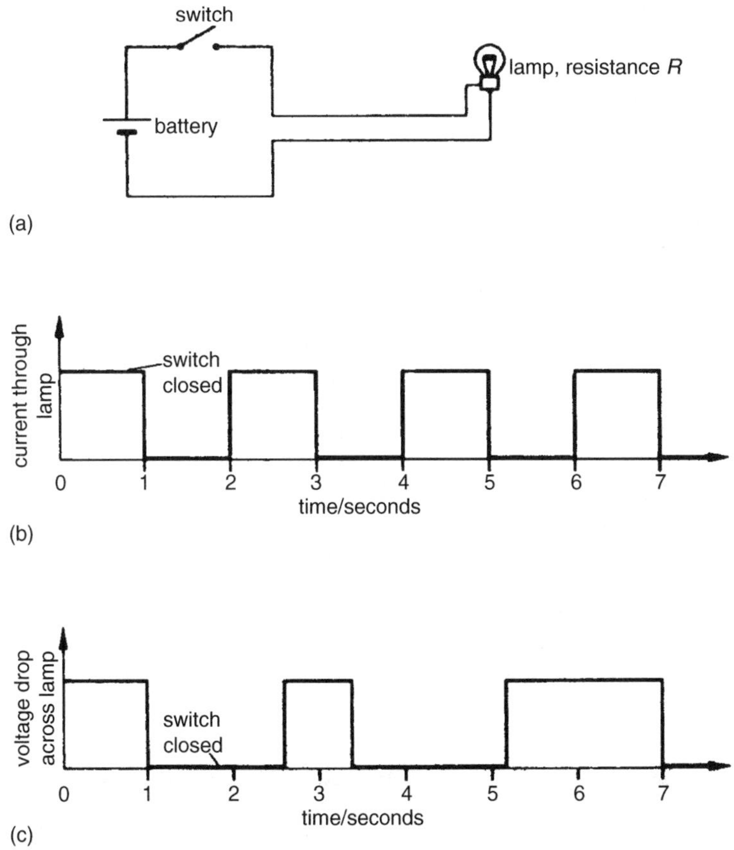

Fig. 2.1 (a) A simple communication circuit. (b) The 'squarewave' current waveform through the lamp when the switch is operated at 1-second intervals. (c) A digital signal produced by operating the switch according to a code in order to transmit a message.

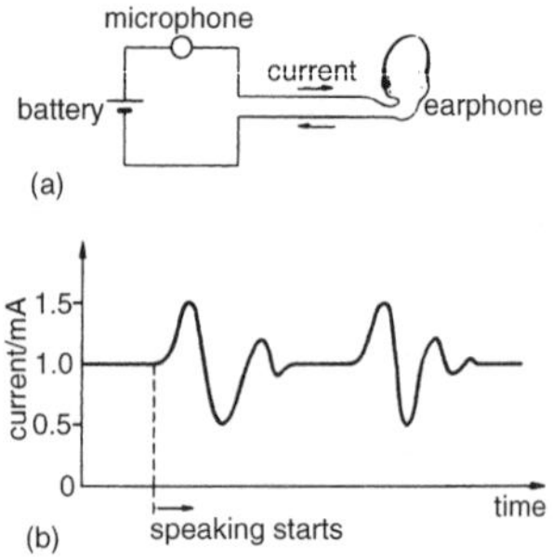

Fig. 2.2 (a) A communication circuit with microphone, battery and earphone. (b) The current in the circuit before and after speaking starts. The current varies around the mean value, in this case, of 1 mA.

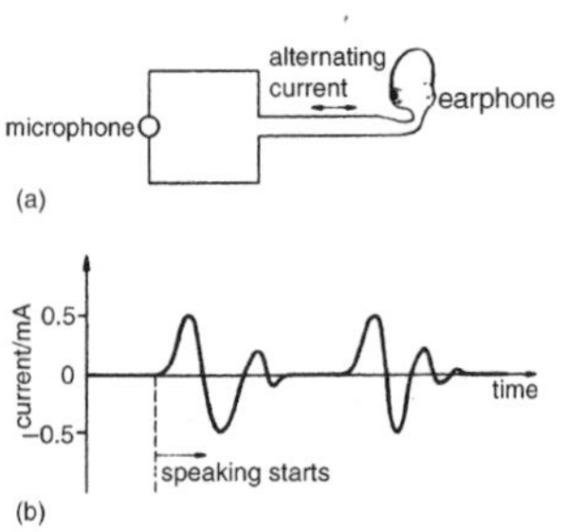

Fig. 2.3 (a) A communication circuit with electromagnetic microphone and earphone. (b) The current in the circuit before and after speaking starts. The current varies around a zero mean.

people watching the lamp—provided they know the code. So the waveform of Fig. 2.1c can be called a **signal**. This type of on-off waveform is an example of a **digital waveform**.

A voltage applied in one part of the circuit of Fig. 2.1 appears across the lamp almost instantaneously, even though the electrons comprising the current move quite slowly. This is because all the electrons begin to drift around the circuit almost simultaneously. Similarly, if the voltage applied to a circuit is varied continuously, by means of a signal generator or microphone for example, the current in the whole circuit will vary in a similar way, whether the current is alternating (i.e. changing direction) or just varying in magnitude.

Figures 2.2b and 2.3b show waveforms of the kinds of currents which microphones can cause to flow. Consider first the circuit of Fig. 2.2a. Before you speak into this kind of microphone a steady d.c. current of perhaps 1 mA flows in the circuit caused by the battery, as shown by the waveform of Fig. 2.2b. As soon as you speak this current is changed. The microphone modifies the current, causing it to *vary* around the average value (1 mA in this case). It varies between about 0.5 mA and 1.5 mA in this illustration. *It is usual to regard this kind of current variation as the sum of a d.c. current and an a.c. current waveform.*

The microphone in the circuit of Fig. 2.3a does not need a battery and so the quiescent current is zero as shown in the graph of Fig. 2.3b. The microphone makes use of the electromagnetic effect to create a voltage whenever the diaphragm (similar to an eardrum) in the microphone is disturbed by the sound waves impinging on it. For the same speech input as the other microphone received, this one might produce a similar waveform as shown in Fig. 2.3b. The resulting current in this case is an *a.c. current* and alternates between perhaps 0.5 mA in one direction and 0.5 mA in the opposite direction. Note that, in this graph, current in one direction is called a positive current whilst in the other direction it is called a negative current. The positive and negative parts of the waveform tend to cancel out, producing an average current of zero.

As regards the communication of information, the waveforms of Fig. 2.2b and 2.3b are identical; the average current of 1 mA in Fig. 2.2b might as well not be there as far as their information-carrying capacity is concerned. Indeed, if the two currents represented by these two waveforms were each to flow through an earphone you would not hear much difference between them because only the *changes* in current produce audible sounds. Additional d.c. currents may be necessary to make certain devices work properly, but they are not usually used to convey messages.

The waveform of Fig. 2.3b is called an *analogue waveform*, because it is an electrical analogue of the sound waveform which it represents.

2.2 SINUSOIDAL WAVEFORMS AND FREQUENCY

The waveforms shown in Figs. 2.4a and 2.4b are examples of **sinusoidal waveforms**. Figure 2.4a is a **sine wave** because it is a graph of $y = A \sin \theta$. Figure 2.4b is a **cosine wave** because it is a graph of $y = A \cos \theta$. The difference between the waveforms is the value of y at $\theta = 0$. With the

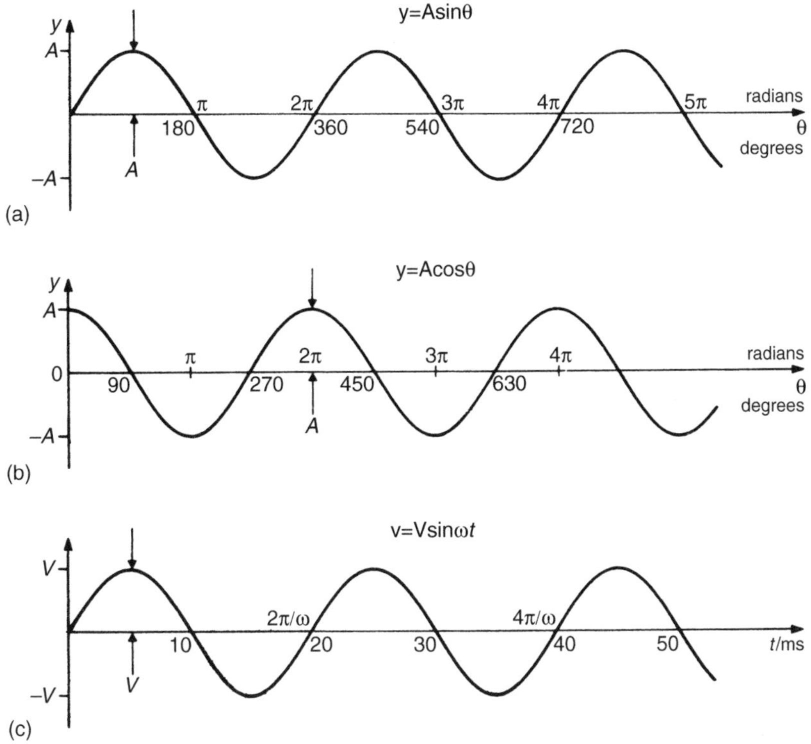

Fig. 2.4 Sinusoidal waveforms. (a) A sine wave. (b) A cosine wave. (c) A sinusoidal voltage waveform, $v=V\sin\omega t$, in which the instantaneous value of the voltage is plotted against time.

sine wave, $y=0$ when $\theta=0$, but with the cosine wave, $y=A$ when $\theta=0$. Both sinusoidal waveforms have the same shape and are both usually called **sinusoids** for short.

In electronics y usually stands for either a voltage or a current. Suppose y stands for a sinusoidal voltage; then A is the maximum value of the voltage, called the **amplitude** of the waveform, and is usually symbolized by a capital V, often with a pair of subscripts to identify it. So in general the equation can be written as $v=V\sin\theta$. However, in an electrical waveform, θ increases uniformly with time and so is usually expressed as a constant multiplied by time t. That is, $\theta=\omega t$, where ω is called the **angular frequency** of the sinusoid and is the rate of increase of the angle θ per second. In Fig. 2.4a and 2.4b, θ is expressed in both degrees and radians, so ω could in principle be expressed in either degrees per second or radians per second. (Note that there are 2π radians in 360°.) In practice, however, ω *is always expressed in radians per second*. The equation for the sinusoid is then $v=V\sin\omega t$. Figure 2.4c is a graph of this equation. It is a repeat of Fig. 2.4a re-drawn with a time axis. Note that a complete *cycle* of the waveform occurs in a time of $2\pi/\omega$.

The **frequency**, f, of a sinusoid is the number of cycles of the wave that occur per second. The unit of frequency is the hertz, or Hz for short. (It is named after Heinrich Hertz, 1857–94, a German physicist who discovered radio waves in 1885.) For example, in Fig. 2.4c one cycle of the waveform occupies a period of 20 milliseconds, so the frequency of this sinusoid is

$$\frac{1}{20\text{ ms}} = \frac{1000}{20\text{ s}} = 50\text{ Hz}$$

This is the frequency of the a.c. mains electricity power supply in Europe. (In North America the frequency of the a.c. supply is 60 Hz.)

Worked example

If there are 2π radians in one cycle, what is the relationship between frequency and angular frequency?

The frequency f is the number of cycles per second, and there are 2π radians per cycle, so the number of radians per second, which is ω, is $\omega = 2\pi f$. (The angular frequency of a 60 Hz supply is $2\pi \times 60 = 377$ radian/s.)

Figure 2.5 shows some more sinusoidal voltage waveforms. Although they differ in various ways they have the characteristic sinusoidal shape.

Self-assessment question

1 What are the frequencies and amplitudes of the sinusoids in Fig. 2.5? Are they sine waves or cosine waves?

The waveforms in Fig. 2.6 are not sinusoidal. They are **periodic waveforms**; that is, they possess a shape which is repeated over and over at a constant repetition rate. The duration of the repeating pattern is called the **period** of the waveform. The sinusoid is one example of a periodic waveform. The waveform of Fig. 2.6a might be that produced by a steady note played on a flute; that of Fig. 2.6b might be the voltage produced by a regular heart beat; and that of Fig. 2.6c is the waveform applied to the deflection plates of an oscilloscope; each time the voltage increases steadily, the luminous spot is drawn across the screen, then as the voltage abruptly falls the spot rapidly flies back again.

Self-assessment question

2 Would you say that the term 'signal' could be used when referring to each of the waveforms in Fig. 2.6?

Most electronic circuits are not built in order to handle periodic non-sinusoidal waveforms, such as those in Fig. 2.6; they are designed and built to deal with *signals*, so why is it worth discussing simple waveforms like sine waves? One reason is that circuits that can handle waveforms such as those in Fig. 2.6 can usually handle signals too; and furthermore, as explained in Section 2.4, such waveforms are, *in effect*, no more than several different sinusoidal waveforms added together. This is true of other simple waveforms too, but a sinusoidal waveform of voltage or current is unique in that when it is applied continuously to a *linear* circuit the currents and voltages anywhere in the circuit are always sinusoidal at that frequency. They will differ in other respects in different parts of the circuit, but not in frequency. This makes the analysis in terms of sinusoids much simpler than other forms of analysis.

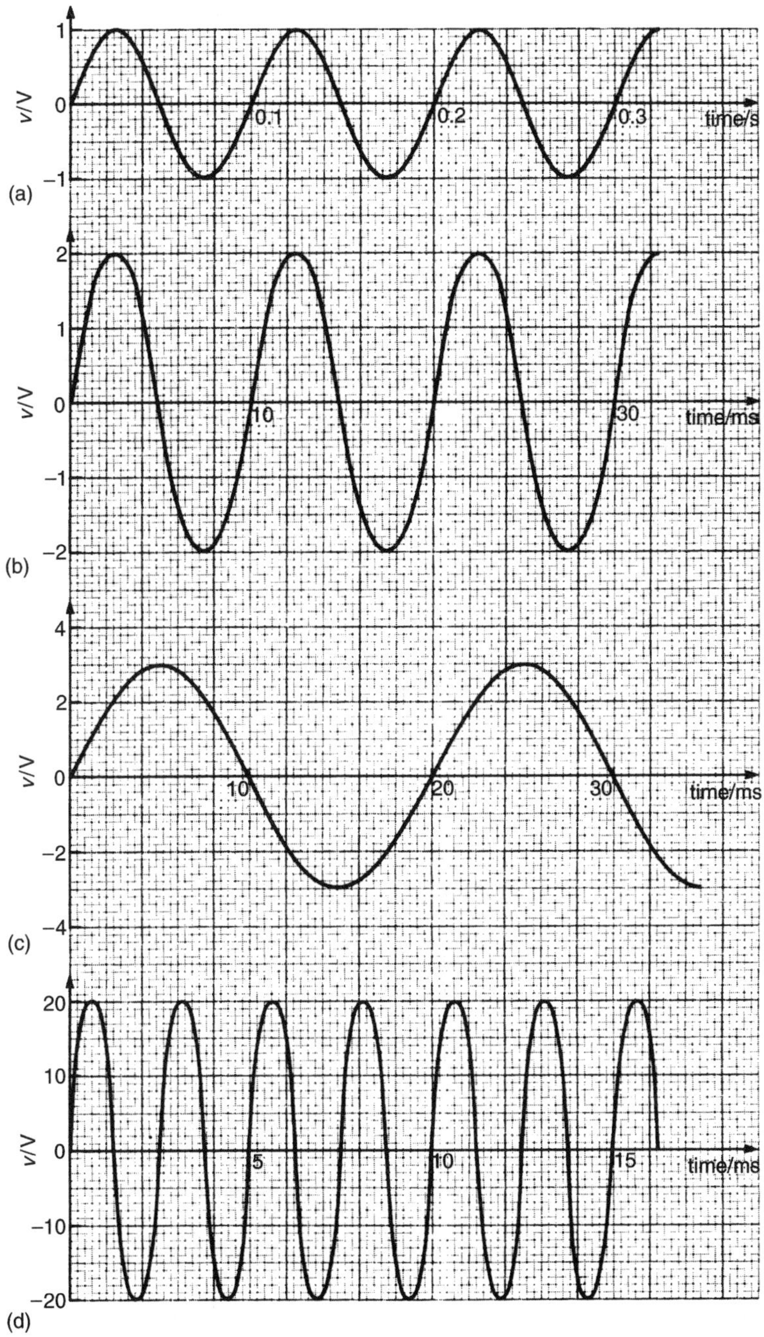

Fig. 2.5 Sinusoidal voltage waveforms of different frequency and amplitude.

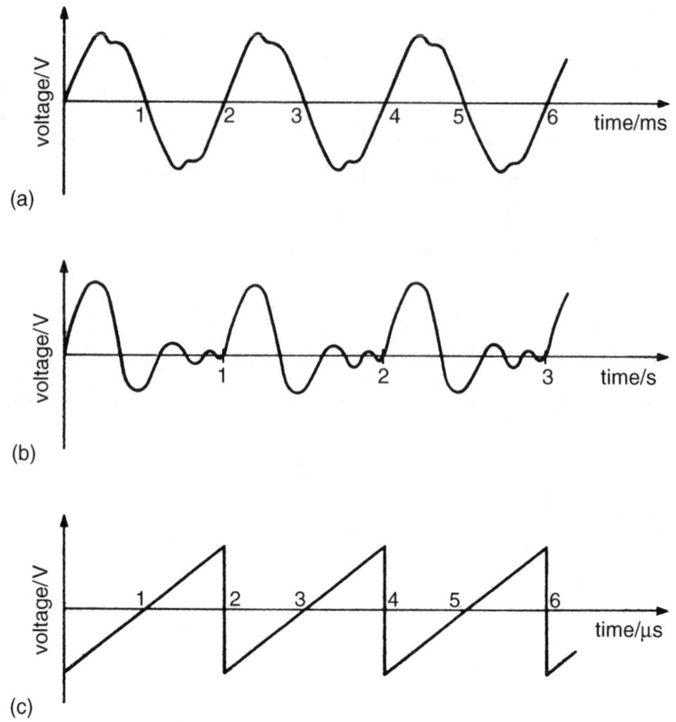

Fig. 2.6 Examples of periodic, non-sinusoidal waveforms.

Worked example

(Revision) What is a linear resistor?

A linear resistor is one that obeys Ohm's law; its resistance is independent of the magnitude of the current flowing through it. So the current flowing through it is always linearly related to the voltage across it.

A somewhat similar definition of linearity can be stated for circuit components other than resistors, as explained later in this chapter.

There are three properties of sinusoids that are of importance in electronics. The first two are the *amplitude* and the *frequency*. The third is the **phase**. The waveforms of Fig. 2.4a and 2.4b, for example, differ in phase. This concept of phase is a little more difficult to grasp than amplitude or frequency. To begin with *phase* and *phase difference* are not quite the same. First let's consider the meaning of phase.

All the waveforms in Fig. 2.5 show $v=0$ when $t=0$. Furthermore, the value of v initially increases from zero (rather than decreases) as time increases. This means that all the waveforms are *sine waves with zero phase*. The phase of a sinusoid depends on the value of v at a particular instant in time usually regarded as $t=0$.

If the waveform of voltage or current versus t has a non-zero value at $t=0$ it can be described as a sine wave which has a phase angle ϕ. The equation of the graph of such a current waveform would then be $i=I_m \sin(\omega t+\phi)$.

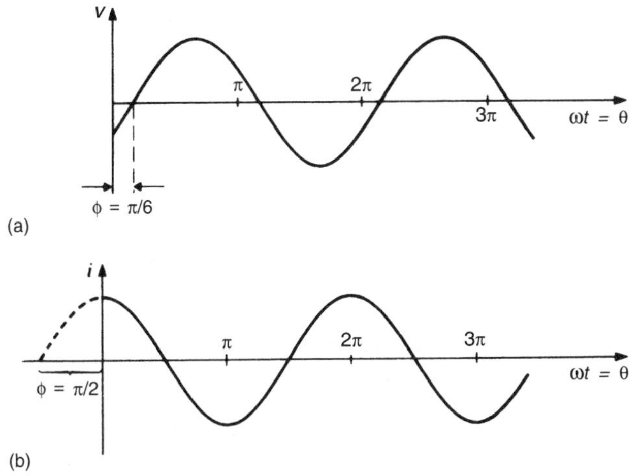

Fig. 2.7 (a) A sinusoid that lags in phase by $\pi/6$ radians with respect to a sine wave. (b) A sinusoid that leads in phase by $\pi/2$ radians with respect to a sine wave.

If the waveform is in advance of, or leading, a sine wave—that is $i=0$ a little before $t=0$—the phase is said to be positive. If the waveform is lagging, or delayed with respect to a sine wave—that is $i=0$ a little after $t=0$—then the phase is said to be negative.

Figure 2.7 shows how the value of ϕ can be determined. In Fig. 2.7 two sinusoidal waveforms are plotted against ωt, which is the same as plotting the waveform against the angle θ. As you can see, the waveform of curve (a) appears to lag behind a sine wave by the angle ϕ, which in this case is $\pi/6$. (The voltage does not rise to 0 V until $\omega \mathrm{t}=\pi/6$, that is a time $t=\pi/(6\omega)$ *after* a zero-phase sine wave would.) The peaks of the waveform are to the right of the corresponding peaks of a sine wave. A lagging phase is represented mathematically by a minus sign, so the equation for curve (a) is $i=I \sin(\omega \mathrm{t}-\pi/6)$.

Similarly, graph 2.7b is of a sinusoid that is *in advance of* or *leading* sin ωt by the angle ϕ, where $\phi=\pi/2$. So the equation of curve (b) is $v=V \sin(\omega \mathrm{t}+\pi/2)$.

Worked example

What alternative equation also describes curve (b) in Fig. 2.7?

Curve (b) is also a graph of $v=V \cos \omega t$. Therefore $\cos \omega t$ and $\sin(\omega t+\pi/2)$ are the same. In other words, $\cos \omega t$ is $\pi/2$ radians in advance of $\sin \omega t$, or $\sin \omega t$ lags behind $\cos \omega t$ by $\pi/2$ radians.

The point about *phase*, as distinct from *phase difference*, is that you can use *phase* to refer to waveforms of *different* frequencies but you must refer to a particular instant. Thus you can say that all the waveforms in Fig. 2.5 have the same phase at $t=0$ even though they have different frequencies. If any instant other than that shown in Fig. 2.5 were regarded as the reference instant, all the waveforms would be said to have different phases.

Phase difference, on the other hand, refers to a comparison between *two sinusoids of the same frequency*, like those of Figs. 2.4a and 2.4b. Because they have the same frequency, the phase difference between the two sinusoids is the same at all instants of time. Thus there is a constant phase difference of $\pi/2$ between Figs. 2.4a and 2.4b. The sine wave lags behind the cosine wave by $\pi/2$ or, if you prefer, the cosine wave leads the sine wave by $\pi/2$. It is meaningless to talk about the *phase difference* between the waveforms of Fig. 2.5.

To obtain the phase difference between two graphs of sinusoids, given the time difference, you simply express the time difference as a fraction of the period of one cycle and then multiply by the number of radians (or degrees) in one cycle:

$$\text{Phase difference, } \phi = \frac{\text{time difference } (\Delta t)}{\text{period of one cycle}} \times 2\pi \text{ radians}$$

For example, the *time* difference between the waveform of Fig. 2.8b and the reference waveform of Fig. 2.8a is marked as Δt, where $\Delta t = -0.25$ ms (the minus sign again indicates delay). Since the period of one cycle is 2 ms,

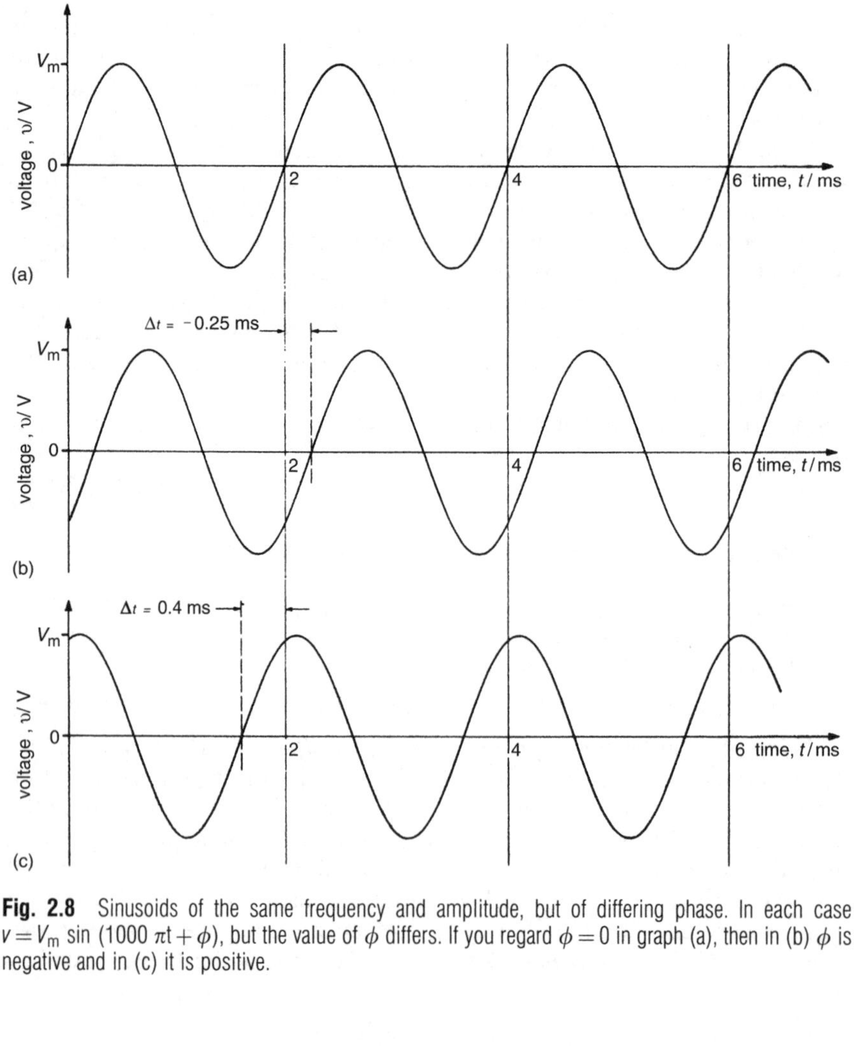

Fig. 2.8 Sinusoids of the same frequency and amplitude, but of differing phase. In each case $v = V_m \sin(1000\,\pi t + \phi)$, but the value of ϕ differs. If you regard $\phi = 0$ in graph (a), then in (b) ϕ is negative and in (c) it is positive.

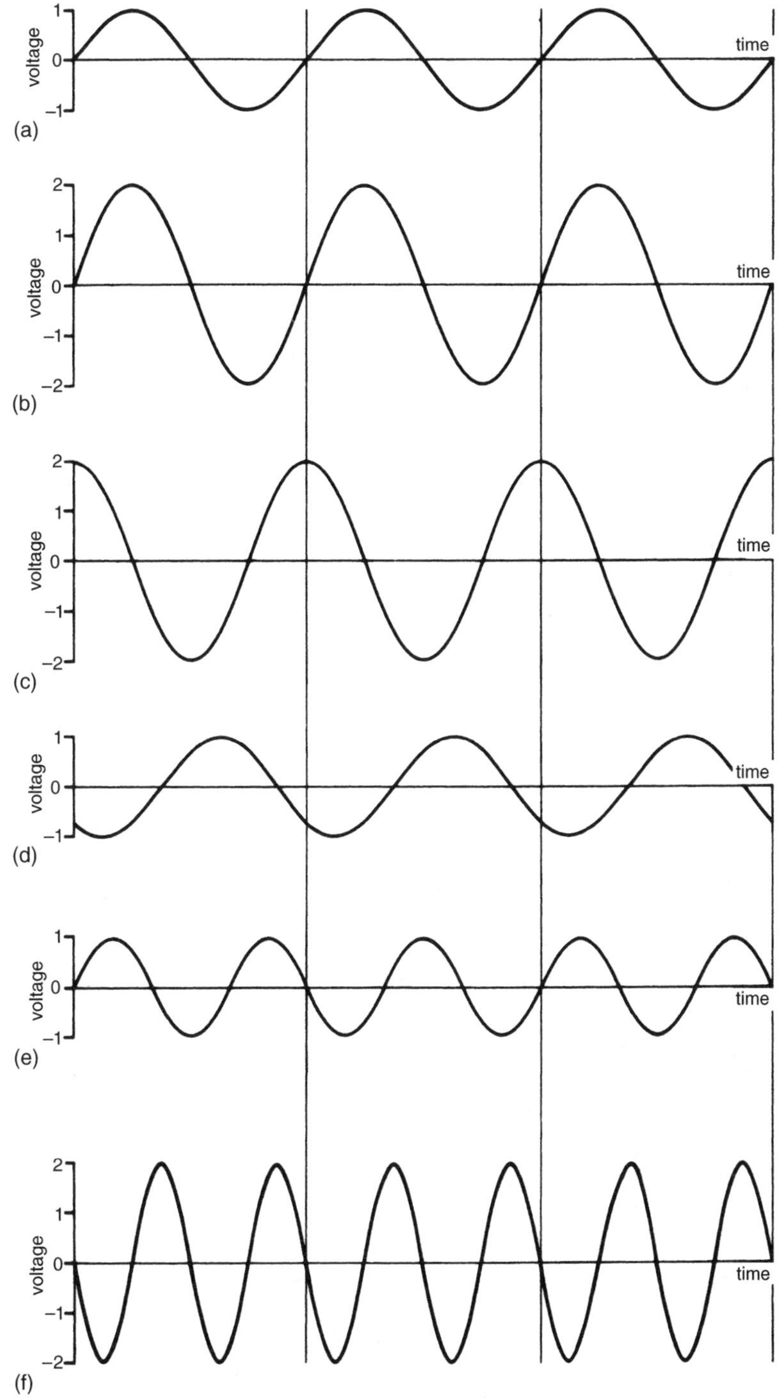

Fig. 2.9 Various sinusoidal waveforms. The time scale for each waveform is the same.

this time difference is 0.25 ms/2 ms or 1/8 of one cycle of the sinusoid. Hence the *phase* difference ϕ between the two waveforms is

$$\phi = -\frac{1}{8} \times 2\pi = \pi/4 \text{ rad}$$

or

$$\phi = -\frac{1}{8} \times 360^\circ = 45^\circ$$

Self-assessment question

3 (a) What is the phase of the waveform in Fig. 2.8c relative to that of the waveform in Fig. 2.8a?
(b) What is the phase difference between waveforms (b) and (c)?

Sinusoids of differing phase, but of the same frequency, arise in circuits because certain components, such as capacitors and inductors, give rise to phase differences between the current through them and the voltage across them. How this happens is explained in Section 2.5.

As you will probably have realized already, it is possible to express a phase difference either as a lead of ϕ or equivalently as a lag of $(2\pi - \phi)$ since a waveform can be shifted in time in either direction in order to make it coincide with another waveform of the same frequency. In practice it is usual to look for the smaller phase angle. So a sine wave is usually said to lag behind a cosine wave by $\pi/2$ rather than lead it by $3\pi/2$.

Self-assessment question

4 It is possible to delay a waveform by means of a transmission line. If a transmission line delays both a 2 MHz sinusoid and a 5 MHz sinusoid by 0.4 μs, by what phase angles do these two sinusoids lag behind the input waveform?

Self-assessment question

5 Figure 2.9 shows six sinusoidal waveforms. Take waveform (a) as a reference. Each of the remaining five differs from it in one of the ways, A to E, stated in the list below. For each waveform (b) to (f) select the statement below which best describes how it differs from waveform (a).

A It differs in phase only.
B It differs in amplitude only.
C It differs in frequency only.
D It differs in amplitude and phase.
E It differs in amplitude and frequency.

2.3 VOLTAGE, R.M.S. AND POWER

2.3.1 Periodic waveforms

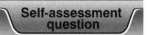

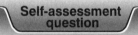

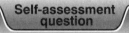

Examples of simple periodic waveforms are sinusoids and squarewaves. They have repeatedly recurring shapes. The interval over which the shape repeats is called the **period**. In the case of non-sinusoids, such as squarewaves, the rate of repetition is called the **fundamental frequency**.

Average value

The average value of any periodic waveform can be found by taking n equally spaced samples of the waveform over one period. Then

$$\text{Average value} = \bar{v} \approx \frac{v_1 + v_2 + v_3 + \cdots + v_n}{n}$$

Notice the use of the 'bar' notation to signify the average value $\bar{v}$ of a waveform. This method gives an approximate result because the number of samples is finite. To obtain the true average of the waveform a continuous averaging process must be used. This can be done by integrating the waveform with respect to time over one period, and dividing by the period T:

$$\bar{v} = \frac{1}{T}\int_0^T v\,\mathrm{d}t$$

This can be interpreted as follows: To find the average value $\bar{v}$ of a periodic waveform, find the area contained between the waveform and the time axis over one period, and then divide by the period. (The area below the time axis should be taken as negative.)

Figure 2.10b shows a periodic waveform of considerable practical importance. This is obtained by 'rectifying' the sinusoid of Fig. 2.10a to obtain a periodic sequence of half sinusoids.

Worked example

What are the period and fundamental frequency of the rectified sinusoid compared with the original?

The rectified sinusoid has half the period. Its fundamental frequency is twice the frequency of the sine wave.

To obtain the average value of the rectified sinusoid we can represent one period of its waveform by a half-period of the original sinusoid. Thus,

$$v = V_a \sin \omega t \qquad \text{for values of } \omega t \text{ from } 0 \text{ to } \pi$$

In Fig. 2.10c, this is redrawn as

$$v = V_a \sin \phi \qquad \text{for values of } \theta \text{ from } 0 \text{ to } \pi, \text{where } \theta \text{ is the angle } \omega t$$

This substitution of $V_a \sin \theta$, averaged over the angular interval π, simplifies the analysis.

The average value is given by forming an integral over one period and dividing by the period:

$$\bar{v} = \frac{1}{\pi}\int_0^{\pi} V_a \sin\theta \, \mathrm{d}\theta = \frac{V_a}{\pi}[-\cos\theta]_0^{\pi}$$
$$= \frac{V_a}{\pi}[1+1] = \frac{2}{\pi}V_a$$

Rectifying a sinusoidal waveform provides an important way of measuring its voltage, as you will see shortly.

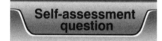

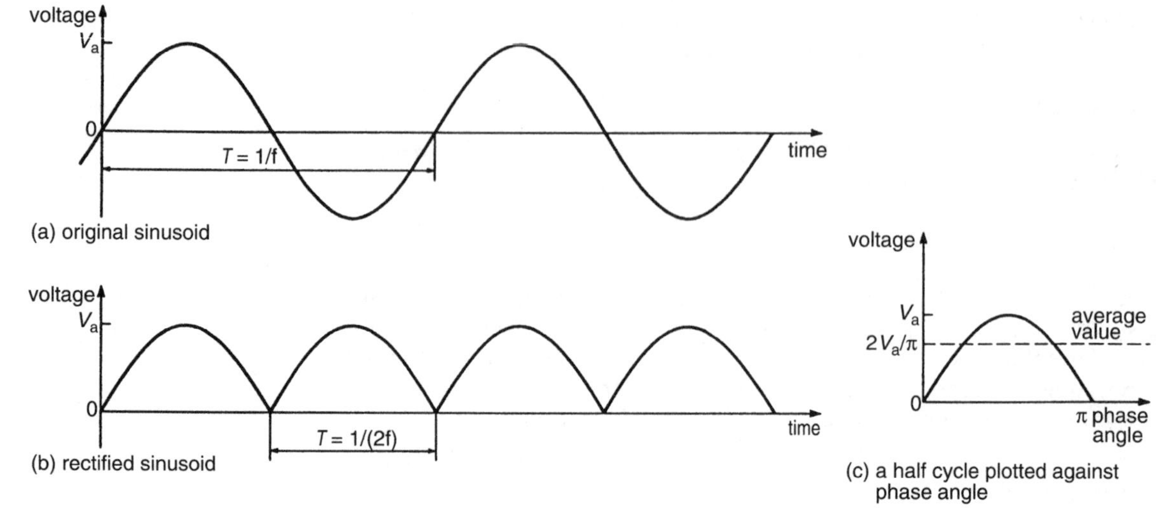

Fig. 2.10 The rectified sinusoid.

Self-assessment question

6 Suppose that only the positive half-cycles of the sinusoid $V_a \sin \omega t$ appear in the rectified waveform, and that during the negative half-cycles the waveform is zero. (This is called 'half-wave' rectification.) What will be the period of this rectified waveform? What will be its average value?

Mean-square value of a periodic waveform

As explained in Chapter 1, a d.c. voltage source of magnitude V driving a current I in a circuit of resistance R, delivers to the circuit a steady power of value given by any one of the three inter-related formulae

$$P = VI, \quad P = I^2R, \quad P = V^2/R$$

A voltage may be changing with time but we can express the *instantaneous power* which it delivers as v^2/R where v is the instantaneous voltage (or as vi or i^2R). The *average power* P_{av} which it delivers, over a period of time, will be the average value of (v^2/R) over that time; so

$$P_{av} = \left(\frac{v^2}{R}\right)_{av} = \frac{(\text{average value of } v^2)}{R} = \frac{\overline{v^2}}{R}$$

The average value of v^2 is usually written $\overline{v^2}$ and is referred to as the 'mean-square voltage'. Since the terms 'mean' and 'average' are interchangeable I can say that:

The mean-square voltage of a waveform is given by the mean (or average) of the squared values of the waveform.

Consider the case of a sinusoidal waveform. Figure 2.11a shows a sinusoid voltage of amplitude V_a, and Fig. 2.11b is a graph of the voltage squared. You will notice that when the sinusoid goes negative the sine-squared waveform is still positive and, in fact, the sine-squared waveform repeats itself every half-cycle of the original sinusoid. This can be shown algebraically:

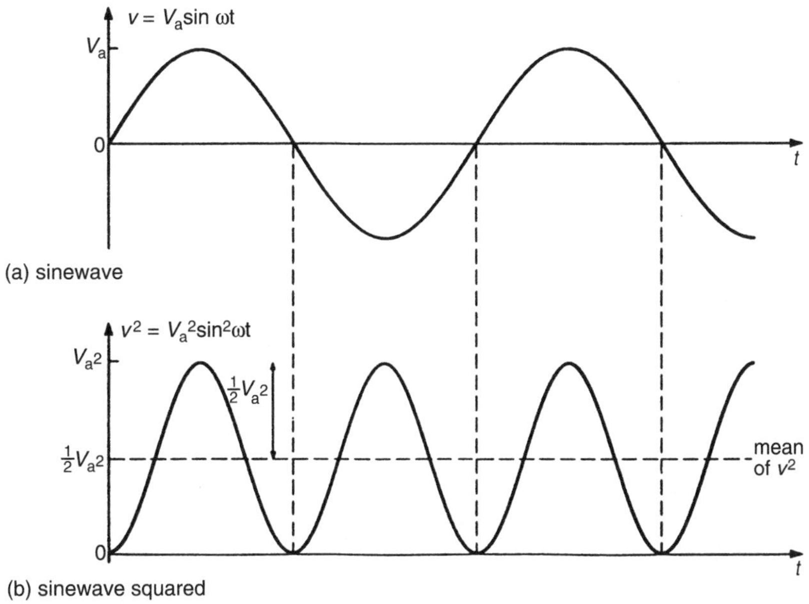

Fig. 2.11 The mean-square value of a sinusoid.

$$v^2 = (V_a \sin \omega t)^2 = V_a^2 \sin^2 \omega t$$

But $\sin^2\theta = \frac{1}{2}(1 - \cos 2\theta)^*$ so $\sin^2 \omega t = \frac{1}{2}(1 - \cos 2\omega t)$. You can see this in Fig. 2.1lb. The waveform of v^2 can be considered as an inverted cosine wave of twice the frequency, with amplitude $\frac{1}{2}V_a^2$ (and zero mean), added to a d.c. voltage of $\frac{1}{2}V_a^2$. This d.c. voltage is the mean value of v^2. So, in this case, we do not need to integrate to find the mean. Provided we average over an integral number of cycles:

$$\overline{v^2} = \frac{1}{2}V_a^2 \qquad \text{for a sinusoid of amplitude } V_a$$

Following from the above, the average power delivered to a resistor is

$$P_{av} = \frac{V_a^2}{2R} \qquad \text{for a sinusoid of amplitude } V_a$$

Strictly, this is the average power over a time interval which is an exact number of half periods of the sinusoidal wave, but provided the time interval is

*This trigonometric expression and its dual $\cos^2 \theta = \frac{1}{2}(1 + \cos 2\theta)$, which are both very useful in signal analysis, can be derived from

$$\cos(A + B) = \cos A \cos B - \sin A \sin B$$

Substituting $A = B = \theta$, then $\cos 2\theta = \cos^2 \theta - \sin^2 \theta$. Now, $\cos^2 \theta + \sin^2 \theta = 1$. Hence, $\cos 2\theta = (1 - \sin^2 \theta) - \sin^2 \theta$ which gives

$$\sin^2 \theta = \frac{1}{2}(1 - \cos 2\theta)$$

many periods long, it makes little difference if it is not an exact number of half periods.

Similar analyses lead to three expressions for the average power dissipated in a resistor, analogous to those for the d.c. case. Thus, for a *sinusoidal* voltage of amplitude V_a, and a corresponding *sinusoidal* current of amplitude I_a:

$$P_{av} = V_a^2/2R \quad P_{av} = I_a^2R/2 \quad P_{av} = V_aI_a/2$$

R.m.s. values

The square root of the mean-square value of any waveform (not just sinusoidal) is called its 'root mean square' or r.m.s. value. Since the mean square voltage is $\overline{v^2}$, the r.m.s. voltage is $V_{rms} = \sqrt{(\overline{v^2})}$.

The average power in a resistor is $\overline{v^2}/R$, so

$$P_{av} = \frac{\overline{v^2}}{R} = \frac{(V_{rms})^2}{R}$$

It is important to use the correct sequence of operations when calculating r.m.s. values, otherwise incorrect results will be obtained. Thus:

> The r.m.s. value of a waveform is found by taking the square *r*oot of the *m*ean of the *s*quare of the waveform.

Mean-square current $\overline{i^2}$ is defined in a similar way to the mean-square voltage, so the r.m.s. current $I_{rms} = \sqrt{(\overline{i^2})}$. The average power is

$$P_{av} = (I_{rms})^2R$$

It follows that the average power dissipated in a resistor can be expressed as

$$P_{av} = V_{rms}I_{rms} \quad P_{av} = (I_{rms})^2R \quad P_{av} = (V_{rms})^2/R$$

In other words:

> The equations for average power (in a resistor) using r.m.s. voltages and currents have the same form as the equations for constant power using d.c. voltages and currents.

R.m.s. values are consistent with Ohm's relationship, that is, in a resistor of value R, $V_{rms} = RI_{rms}$ whatever the waveform. *For sinusoids*, $\overline{v^2} = V_a^2/2$, so

> For sinusoids, $V_{rms} = \sqrt{(\overline{v^2})}V_a/\sqrt{2} \approx 0.707\ V_a$

When a numerical value is quoted for a sine wave voltage without qualification, it is usually the r.m.s. value; thus a quoted a.c. mains voltage of 240 V is its r.m.s. value.

Self-assessment question

7 What is the voltage amplitude of 240 V mains? What is the resistance (assumed constant) of a mains operated 100 W light bulb at its working temperature? (Its 'cold' resistance is much smaller than its resistance when lit.)

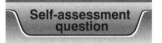

The r.m.s. value of a sinusoid can be measured using a moving-coil meter. The waveform is first rectified to produce the waveform of Fig. 2.10b. The rectified waveform is then connected to the meter via a resistor appropriate for that voltage range, and the meter measures the average value $2/\pi$, or 0.637, times the amplitude V_a. Multimeters using moving-coil meters are arranged so that the meter scale *indicates* the r.m.s. value, which is $V_a/\sqrt{2}$, or 0.707 times V_a, on its a.c. ranges. This is done simply by using somewhat different scales for a.c. measurements, calibrated to show the r.m.s. value, or by using different values of range resistors on the a.c. ranges.

Self-assessment question

8 Suppose a sinusoidal waveform of 100 V r.m.s. is rectified, and the rectified version is measured with a moving-coil multimeter switched to its 200 V d.c. range.

(a) What is the amplitude of the sinusoidal waveform?

(b) What is the average of the rectified waveform? What voltage will the meter indicate?

(c) By what factor must the range resistor be changed to make the meter read the r.m.s. value of the sinusoid correctly?

The next Self-assessment question will remind you that mean-square and r.m.s. values are not restricted to sinusoids.

Self-assessment question

9 From the definitions given, calculate the mean-square and r.m.s. values of the periodic waveforms in Fig. 2.12. (*Hint*: Notice that the waveform in Fig. 2.12c can be represented by $10t/T$ over the interval $t = 0$ to $t = T$.)

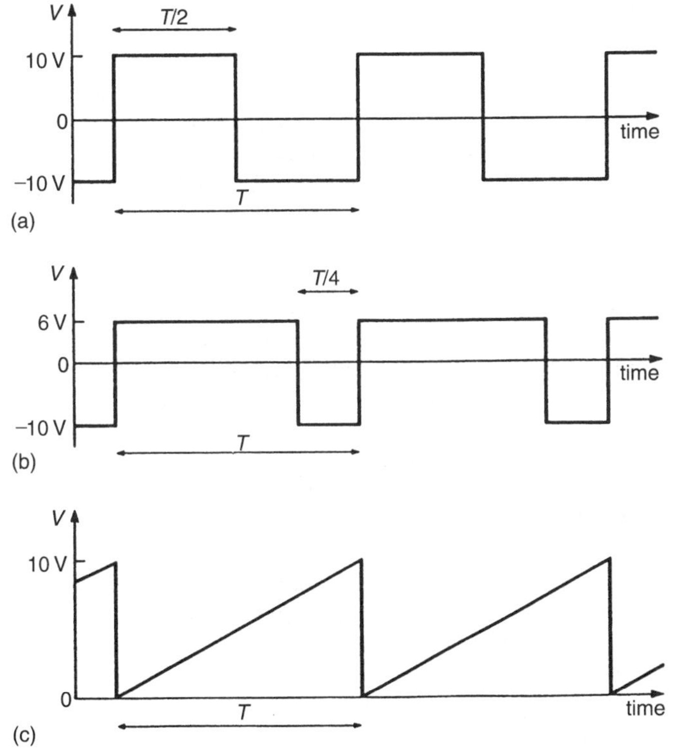

Fig. 2.12 Periodic waveforms.

2.3.2 Non-periodic waveforms: signals and noise

Signals and noise

The term 'waveform' is used in a fairly general way to refer to time-varying voltages and currents. The term 'signal' is reserved for waveforms that convey a message or information from one place to another. In order to carry information, a waveform must keep changing its shape and must therefore be non-periodic. There is an element of unpredictability in this: after all, if we knew what was coming in a news broadcast we would not need to listen to it or watch it! Waveforms with this unpredictable behaviour are said to be **random**, and all signals in the real world vary in a non-periodic random fashion.

When we come to design an amplifier for audio-frequency signals, such as speech or music, we need to know the voltage and frequency range of those signals, to make sure the amplifier will amplify them without distortion. The problem to be tackled in this section is, how do we measure or quantify the voltage and frequency range of a randomly-varying signal?

There is another type of random waveform, called **noise**. The word 'noise' is used in electronics, rather loosely, to describe a waveform which interferes with, or corrupts, a signal. Thus the difference between a signal and noise is simply whether or not we are interested in it, and can interpret it. Noise can be compared to weeds in the garden: just as weeds are unwanted plants, so noise is an unwanted waveform. An example of interference is the reception by a radio receiver of another broadcast signal in addition to the wanted one. Another example is a car ignition system radiating electrical pulses which affect radio or television reception.

Sources of interference have one thing in common: in each case the interference originates *outside* the electronic circuits that we are concerned with. Another type of noise is generated *internally* by all circuits. You can sometimes hear it in an audio amplifier if you remove the signal source. Then, if there is no audible external interference, you may hear a hissing noise from the loudspeaker when you turn up the volume control. This is sometimes called **internal noise** to distinguish it from external noise and sometimes just 'noise'. Sources of internal noise are described in Chapter 4.

In the specification and design of an audio amplifier for example, internally-generated noise has to be considered. The amount of noise which is acceptable at the amplifier output is first specified. The designer must then calculate the noise likely to be generated by the chosen circuit, to see if the design will meet the specification.

If the effects of noise are to be calculated, the same problem arises that was mentioned earlier in connection with signals. That is, how do we measure, or quantify, the voltage and frequency range of a randomly-varying waveform such as noise? Figure 2.13 shows the waveforms of a signal and of some noise. Clearly, with such waveforms, the terms amplitude, frequency and phase have no clear, unambiguous meaning. In their place, the terms r.m.s. voltage (or mean square voltage) and power density spectrum are used. These concepts are explained in the following sections.

The basic ideas in the preceding sections on voltage and power of periodic waveforms apply also to signals and noise. But the ideas have to be modified, because of the non-periodic nature of signals and noise.

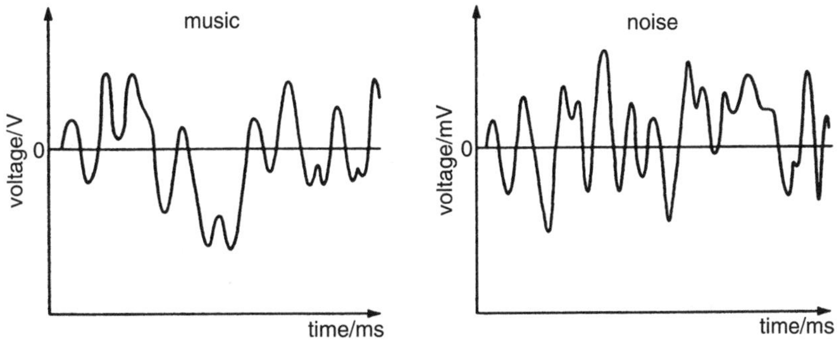

Fig. 2.13 Waveforms of music and electrical noise.

Average value

As you saw at the start of this chapter, it is possible that a signal, such as that from a microphone, can exist as voltage variations with zero mean superimposed on a d.c. voltage level. Clearly, in such a case, the waveform's average, or mean, value is the value of the d.c. voltage.

Mean-square value

Just as with periodic waveforms, the instantaneous power delivered to a resistor by a non-periodic waveform is v^2/R, where v is the instantaneous voltage of the waveform. This is true for both signals and noise. In the same way as before, the average power input by a non-periodic waveform is the average value of v^2/R:

$$P_{\text{av}} = \left(\frac{v^2}{R}\right)_{\text{av}} = \frac{(\text{average value of } v^2)}{R} = \frac{\overline{v^2}}{R}$$

As before, the average value of v^2 is usually written as $\overline{v^2}$, and is called the mean-square voltage.

Now, however, we are faced with non-periodic signal and noise waveforms of the type shown in Fig. 2.14a. There are two problems in finding the mean-square value of such a waveform: the shape is not simple, and with no repetition period it is difficult to know what interval to average over.

In one method of measuring the mean-square voltage of a waveform, an analogue circuit is used which has an output proportional to the square of its input voltage. So its output waveform is like that in Fig. 2.14b. This waveform is then connected to an averaging meter, such as a moving-coil voltmeter, which responds only to slowly changing waveforms. Thus, if all the fluctuations of the waveform are faster than the meter's response, the meter responds only to the mean, or d.c. component, and the mean-square voltage is displayed.

An alternative method uses a digital technique as shown in Fig. 2.15. This is the method used by digital 'true-r.m.s.' meters. The input waveform is sampled, and each sample is fed in turn to an analogue-to-digital converter. The digital output values are then squared by a digital circuit, and these squared digital values are averaged. The digital output from the averaging process represents

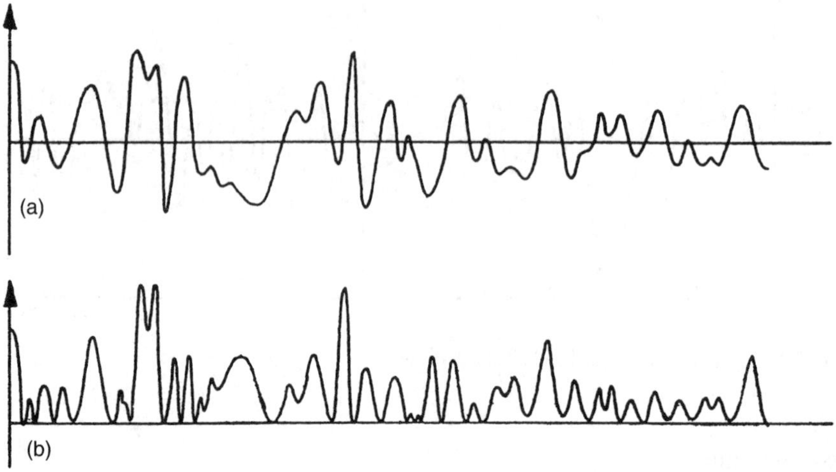

Fig. 2.14 A signal or noise (non-periodic) waveform, and its squared version.

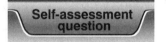

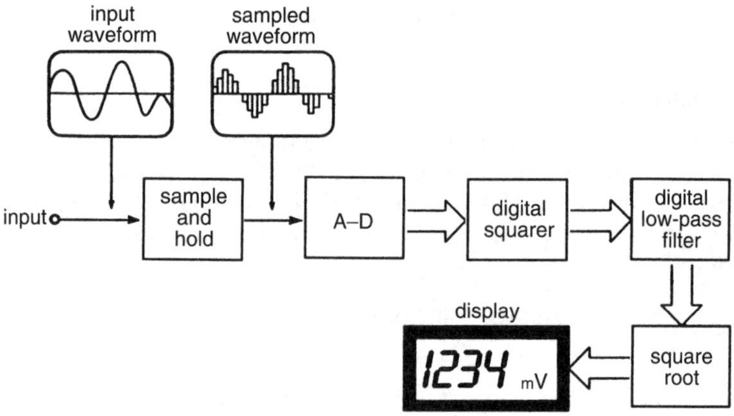

Fig. 2.15 Block diagram of a digital true-r.m.s. voltmeter.

the mean-square value of the input waveform. The final step is to take its square root, and display this as the root mean square, or r.m.s., value of the input. This process should become clear if you answer the following question.

Self-assessment question

10 The waveform of Fig. 2.16 has sample voltages of 0, 2, 1, 0, −2, −3, −1, 0, 1, 3, 4, 1, 0, −1, −2, −4 (measured to the nearest volt), over the interval shown. Find:

(a) the number of samples;
(b) the sum of the sample values;
(c) the average voltage;
(d) the sum of the squares of the sample values;
(e) the mean-square voltage;
(f) the root mean square (r.m.s.) voltage.

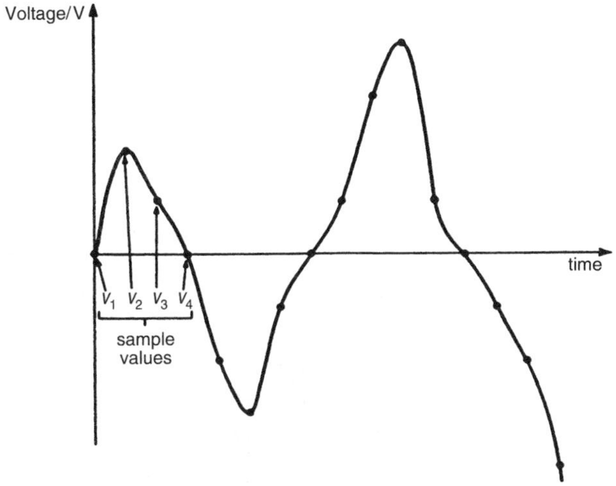

Fig. 2.16 A portion of non-periodic waveform, with sample values.

For the samples chosen for this Self-assessment question, the average value is not quite zero. In fact, with such a small number of samples, both the average value and the r.m.s. value depend on the number of samples and on the particular section of the waveform chosen.

In practice, it is usually assumed that the estimates of the average and r.m.s. values of a random waveform can be progressively improved by taking more samples over longer intervals. For most random noise sources the average value then tends towards zero as the averaging time is increased, while the r.m.s. value tends to a well-defined fixed value.

The Gaussian distribution

The sample values of the waveform of Fig. 2.16 are shown plotted in Fig. 2.17a. The horizontal axis represents voltage ranges. For instance, 1 V represents the centre of the range from 0.5 V to 1.5 V. In the original waveform there are three samples whose values are each 1 V to the nearest volt, that is each lies in the range 0.5 V to 1.5 V. These three samples are represented by a column of height 3. There are four samples in the range of –0.5 V to +0.5 V, centred on 0 V, and so on. In the study of statistics, this type of plot is called a **frequency distribution**, because it shows the frequency with which samples of similar value occur. (Not to be confused with the frequency of a waveform!)

If many more samples of the noise waveform were taken, over a longer time interval, and the sample voltages were measured to a finer resolution (say 1 mV in this case), the corresponding frequency distribution would tend more and more to the shape shown in Fig. 2.17b. This classic bell-shaped curve is known as the **normal**, or **Gaussian distribution**. This is a good approximation to the type of voltage distribution of many internal noise source waveforms. Notice that the r.m.s. value is shown on the curve. A property of the Gaussian distribution is that about 68% of all the samples lie within the area under the curve bounded by $-V_{rms}$ and $+V_{rms}$. In other words, 68% of the voltage samples of the waveform have voltage magnitudes less than or equal to the r.m.s. value V_{rms}.

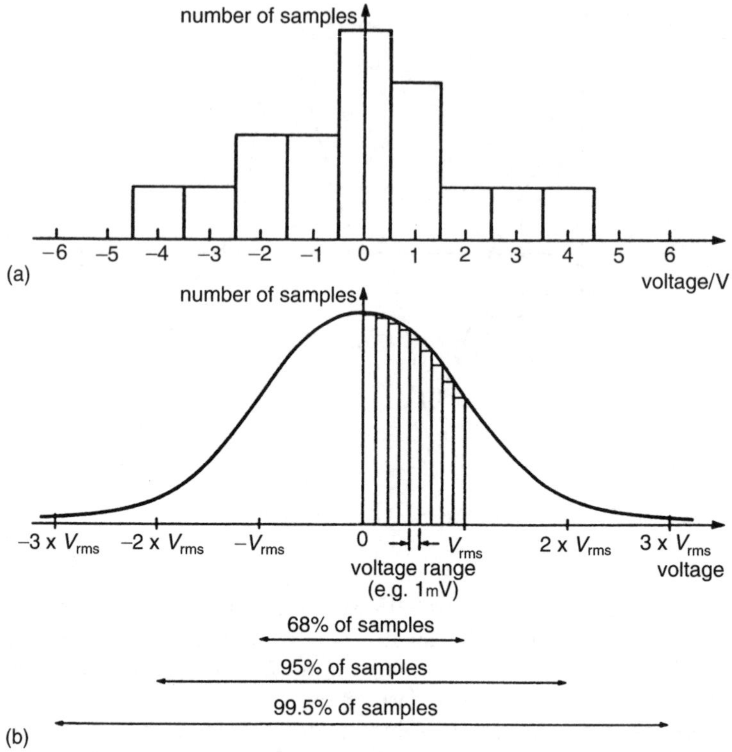

Fig. 2.17 Statistical frequency distributions. (a) Histogram of samples from noise waveform. (b) The Gaussian, or normal distribution.

This means that, at any instant during the waveform's existence, there is a 0.68 probability that the noise voltage will lie between these limits. Putting it another way, the probability of the noise voltage *exceeding* V_{rms} in magnitude (either greater than $+V_{rms}$ or less than $-V_{rms}$) is only 0.32, since the two probabilities must add to 1. This is why expressing a noise waveform as an r.m.s. value is useful. If the noise distribution is normal (Gaussian), an amplifier designer knows that, on average, the noise voltage magnitude will exceed the r.m.s. value for only 32% of the time. In Chapter 4, you will see that the electrical noise generated by an op-amp is specified in this way.

The region under the Gaussian curve between the sample values $-2 \times V_{rms}$ and $+2 \times V_{rms}$ contains about 95% of all the sampled values. More importantly, 99.5% of the samples lie between $-3 \times V_{rms}$ and $+3 \times V_{rms}$. In principle, a very small percentage of samples could have very large values, but in practice their maximum values are usually limited by the electronic circuit.

Worked example

What is the probability of the noise exceeding $+2\times V_{rms}$?

Sample values outside the range $-2\times V_{rms}$ to $+2\times V_{rms}$ constitute 5% of the total. So, because the curve is symmetrical, 2.5% of these occur above $+2\times V_{rms}$. The required probability is thus 2.5% or 0.025.

Worked example

The r.m.s. value of a voltage can be specified for certain waveforms for which the amplitude cannot be specified. Can you state which these are?

The amplitude of a waveform applies only to sinusoidal waveforms. It does not have a meaning for other kinds of waveforms. However, ***all*** *waveforms of voltage or current dissipate power in resistors, so an r.m.s. value of current or voltage can be stated for them.*

2.3.3 Symbols for voltages and currents

Several terms are used to describe the magnitudes of waveforms and d.c. levels. Chapter 1 deals exclusively with d.c. voltages and currents. This chapter deals with instantaneous values of waveforms, amplitudes or peak values of sinusoids, voltages and currents which are the sum of both d.c. and a.c. components, and r.m.s. values. Now is the point to clarify the distinction between the various measures of electrical quantities that are normally used.

Figure 2.18a shows a diagram of a voltage which begins as a d.c. voltage, to which a sinusoid is soon added, so that the waveform can be thought of as the sum of a d.c. and an a.c. voltage. This is similar to Fig. 2.2b except that a sinusoid is the waveform being added. The arrows on the diagram pick out the principal voltages which can be referred to in connection with such a waveform.

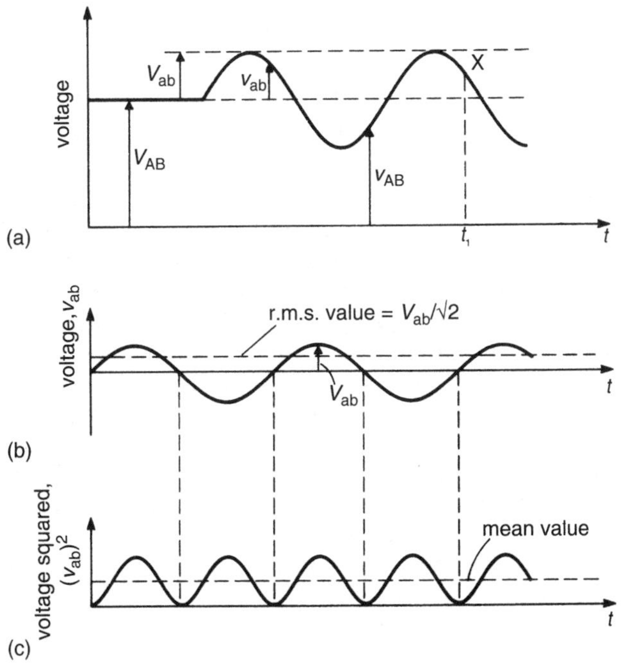

Fig. 2.18 Measures of voltage waveforms. (a) A d.c. voltage plus a sinusoid with the appropriate letter symbols indicated. (b) A sinusoid voltage, $v_{ab} = V_{ab} \sin \omega t$, showing its r.m.s. value of $V_{ab}/\sqrt{2}$. (c) A graph of $(V_{ab})^2$ and its average value.

Beside each arrow is the letter symbol used to refer to each quantity. The letter(s) used in the subscript identify the terminals between which the voltage is measured. In this illustration the terminals referred to are called A and B, so that V_{AB} is the voltage of terminal A with respect to terminal B. When we come to consider transistors, for example, V_{BE} would refer to the voltage of the base terminal relative to the emitter terminal. But which kind of voltage does each symbol refer to?

The code used for **letter symbols** is as follows:

- A capital letter with a capital subscript refers to a *d.c. quantity* (e.g. V_{AB}).
- A capital letter with a lower-case subscript usually refers to the *amplitude of the sinusoid* (e.g. V_{ab}). This symbol is also usually used to refer to the r.m.s. value of any waveform, as explained in a moment.
- A lower-case letter with a lower-case subscript refers to the *instantaneous value* of the *variable* quantity (e.g. v_{ab}).
- A lower-case letter with a capital subscript refers to the *instantaneous value* of the *total* voltage (e.g. v_{AB}).

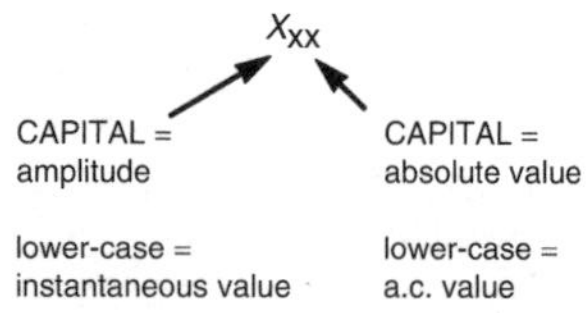

So, for example, at a point such as X in the figure, when the time is t_1, the total voltage can be written as $v_{AB} = V_{AB} + v_{ab} = V_{AB} + V_{ab} \sin \omega t_1$. Obviously a similar coding of subscripts applies to currents, except that normally only one subscript letter is needed to identify a current.

Worked example

What do the symbols I_g, i_g and I_G mean? The subscript refers to a 'generator'.

The symbols refer respectively to the amplitude of the sinusoidal current flowing through the generator, the instantaneous value of this current, and finally to the d.c. current flowing in the generator.

In general it is important when labelling a circuit with voltages or currents, or when writing equations, to use the right symbol, especially when phase changes occur in the circuit. In resistive circuits however, in which there are no phase differences, it is often possible to use either the symbol for amplitude or the symbol for instantaneous value, because there is a one-to-one relationship between them. So, where phase differences do not occur you can use either the symbols for amplitude or the symbols for instantaneous values without being ambiguous. The *ratio of amplitudes* of a waveform in two parts of the circuit is the same as the *ratio of instantaneous values* of the waveforms at those points at any instant. But if there is a phase difference between current and voltage, or between two voltages or between two currents, the ratio of amplitudes is not in general the same as the ratio of instantaneous values, so it is essential to use the correct symbols. For example, if you look back at Fig. 2.7, you will see that the amplitudes of the two waveforms there are the same, but that the ratio of v/i can be anything from minus infinity to plus infinity!

Worked example

In a particular circuit you calculate that at nodes 2 and 3, $v_2 = 2v_3$. If the waveforms at nodes 2 and 3 are sinusoidal, what does this tell you about their phase difference. Do you get the same information from the statement that $V_2 = 2V_3$?

If $v_2 = 2v_3$ you know that the instantaneous values of the voltages at the two nodes vary in synchronism, even though one is always twice the other. So there is no phase difference between them. If you know that $V_2 = 2V_3$ you only know the amplitude ratio, so there could be any phase difference between them.

The fact that V_{ab} is used to stand for both the amplitude of a sinusoid as well as for its r.m.s. value can cause confusion, so it is essential to be clear about which is meant. The symbol V_{abm} is sometimes used for the amplitude if there is danger of ambiguity. Equally $V_{ab(rms)}$ is sometimes used for the r.m.s. voltage between A and B.

Although this all looks very complicated, it is usually obvious from the context what is being referred to.

2.4 FREQUENCY SPECTRA

2.4.1 Periodic waveforms

This section shows how it is that **periodic waveforms**, such as those in Fig. 2.6, can be regarded as being composed of the addition of sinusoids of different frequencies. Sinusoids themselves are a special case of periodic waveforms in that they are composed of only one sinusoid. All other continuous periodic waveforms are in effect composed of more than one sinusoid.

As already remarked, the duration of the repeating wave shape of a periodic waveform is called the period of the waveform. The reciprocal of the period is called the **fundamental frequency** of the waveform because it is the lowest frequency sinusoid that the waveform contains.

Consider, for example, the same note played by different musical instruments. They all have the same *pitch* but they do not sound the same. That is, they all have the same **fundamental frequency** (or period) but they have different waveforms. The different waveforms can be thought of as resulting from different mixtures of sinusoids being added together. We recognize the characteristic sounds of different musical instruments by being able to detect the different frequencies making up the sounds they produce. Notes of a different pitch on the same instrument have different fundamental frequencies but tend to produce the same characteristic mix of sinusoids.

Worked example

What is the fundamental frequency of each of the waveforms shown in Fig. 2.6?

In Fig. 2.6a the period is 2 ms, so the fundamental frequency is 500 Hz. In Fig. 2.6b the period is 1.0 s, so the fundamental frequency is 1 Hz or 60 pulses per minute. The waveform of Fig. 2.6c—which is like the voltage waveform that might be applied to the deflection plates of an oscilloscope—has a period of 2 μs, so the fundamental frequency is 500 kHz.

The frequencies of the sinusoids that make up a periodic waveform are all **harmonics** of the fundamental frequency. That is, their frequencies are

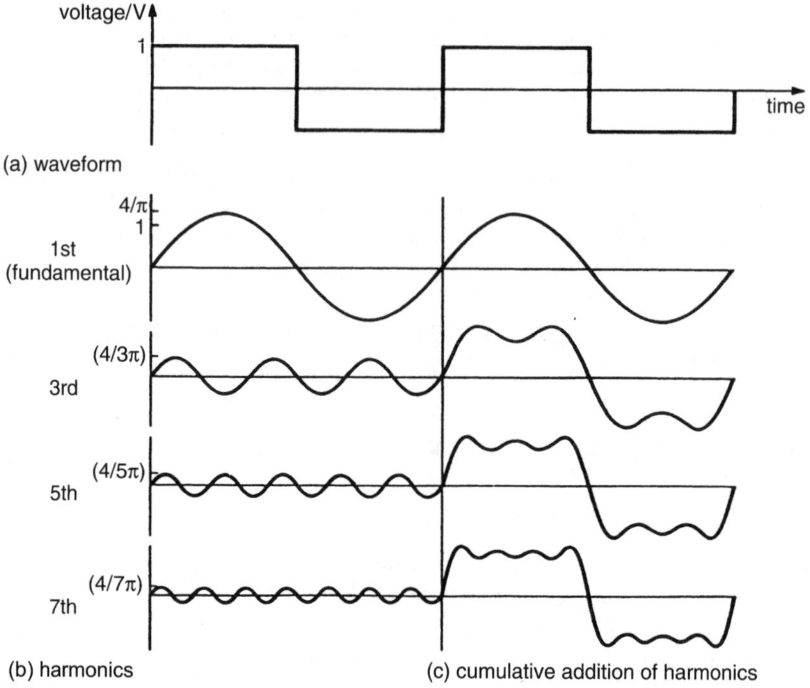

Fig. 2.19 A squarewave and its first few harmonics.

integer (whole number) multiples of the fundamental frequency. In order to understand how it is that periodic waveforms (other than sinusoidal ones) contain more than one frequency, consider what happens when harmonics of the fundamental are added together. Figure 2.19 shows an example, a squarewave.

Figure 2.19a shows the waveform, and Fig. 2.19b shows the first few sinusoidal frequency components, or *harmonics*. Each of these has a frequency which is an exact integer times the waveform's repetition frequency f_0 ($=1/T$). Figure 2.19c shows how the waveform could be built up by successively adding harmonics. You can see that a good approximation to the waveform can be made by representing it with a limited number of harmonics, providing these have the correct amplitudes and phases.

The **Fourier series** for the squarewave is a mathematical expression which shows it as the sum of an infinite series of harmonically-related sinusoids:

$$v = \frac{4V_1}{\pi}(\sin\omega_0 t + \frac{1}{3}\sin 3\omega_0 t + \frac{1}{5}\sin 5\omega_0 t + \cdots)$$

These sinusoids are called the **Fourier components** of the waveform because this technique of representing a waveform by the sum of different sinusoids was discovered by the French mathematician Joseph Fourier (1768–1830).

Here, v is the instantaneous voltage, V_1 is the peak value of the waveform (1 V in Fig. 2.19a) and ω_0 ($=2\pi f_0$) is the angular frequency of the first harmonic. The first harmonic, or *fundamental*, has the same frequency as the waveform's repetition frequency f_0 ($= 1/T$). Its amplitude is $4V_1/\pi$. The other harmonics have frequencies of 3, 5, 7, etc. times the fundamental, that is all harmonics are 'odd'. There are no even harmonics in this case. The amplitudes fall in inverse proportion to the harmonic number, in the ratios 1/3, 1/5, 1/7, etc. of the fundamental's amplitude. In this particular case, all the harmonics have zero phase at $t = 0$.

In general:

- The sinusoidal components of a periodic waveform are all at multiples of the fundamental frequency (the repetition frequency). That is, they are all harmonics.
- Adding successive components improves the approximation but does not give an exact representation. An infinite number of harmonics may be required if an exact representation of an 'ideal' version of a waveform such as the squarewave is needed. In practice, real waveforms do not have infinite bandwidth, but many harmonics may still be needed to represent them.

Any periodic waveform can be represented as a sum of sinusoids by the general **Fourier series**:

$$v = k_0 + k_1 \sin(\omega_0 t + \phi_1) + k_2 \sin(2\omega_0 t + \phi_2) + k_3 \sin(3\omega_0 t + \phi_3) + \cdots + k_n \sin(n\omega_0 t + \phi_n) + \cdots$$

The first term, k_0, is a constant which represents the average value or d.c. level of the waveform.

The squarewave of Fig. 2.19 has zero d.c. level. This is obvious from its waveform, which is symmetrical about the time axis. Its Fourier series has a constant term, k_0, of zero.

Self-assessment question

11 Suppose the squarewave of Fig. 2.19 is modified to switch between 0 V and 4 V, like a simple clock waveform in a digital system. How does this affect its Fourier series?

The constants k_0, k_1, k_2, etc. are called **Fourier coefficients**. The constants ϕ_1, ϕ_2, ϕ_3, etc. are the respective phase angles at $t = 0$. Finding the frequency components of a waveform means finding these coefficients and phase angles. This can be done by integration for waveforms whose shape can be defined mathematically, but this technique is beyond the scope of this book. To measure the coefficients of the harmonics of a real signal, a spectrum analyser can be used. The analogue spectrum analyser picks out each harmonic of the waveform and displays its amplitude. The digital spectrum analyser performs the so-called discrete Fourier transform (DFT), usually using a version of this called the fast Fourier transform (FFT). It can provide a display of harmonic amplitudes like that in Fig. 2.20, together with a display of the phase angles. Figure 2.20 is an example of a **line spectrum**, which is the type of spectrum always obtained for a periodic waveform.

By adding harmonics of different frequencies, amplitudes and phases to the fundamental (not only odd harmonics of course) it is possible to generate any periodic waveform. This is essentially the way in which music synthesizers

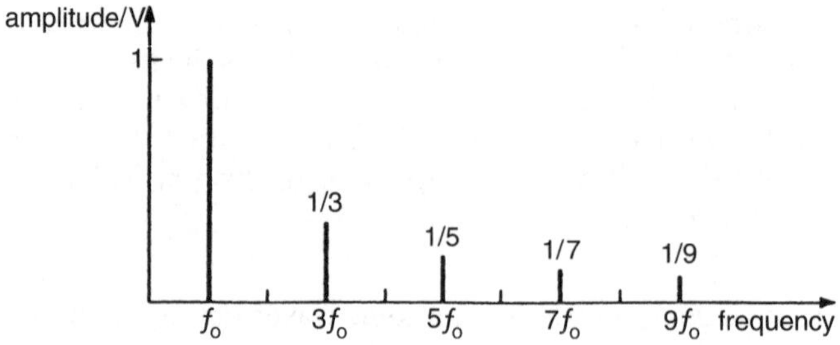

Fig. 2.20 The spectrum of a squarewave.

work; all the harmonics of each pitch are generated, and then added in different proportions to imitate different musical instruments.

Figure 2.21 illustrates the effect of just changing the phase of the fundamental. The waveshape resulting from adding the first few harmonics has changed completely as compared with Fig. 2.19, although the frequencies and amplitudes are the same. Rather surprisingly, the human ear is not very sensitive to the phases of sounds, so the very different waveforms of Fig. 2.19c and Fig. 2.21c would actually sound almost the same if the waveforms they represent were used to drive a loudspeaker. So as far as hearing is concerned, it is the frequencies and amplitudes of the Fourier components of constant-pitch notes that matter, not their phases. For other applications

Use your analogue circuit analysis package to draw the waveform of exactly one cycle of a sine wave of 1 kHz. Then perform spectral analysis (Fourier analysis) on this. Note that, for an accurate spectrum, the number of 'steps' or 'samples' taken by the computer in drawing the waveform should be a power of 2, e.g. 64 or 256, so you must choose the step and sweep times appropriately.

Now repeat this for a squarewave of 1 kHz, and check that the spectral lines have the predicted amplitudes.

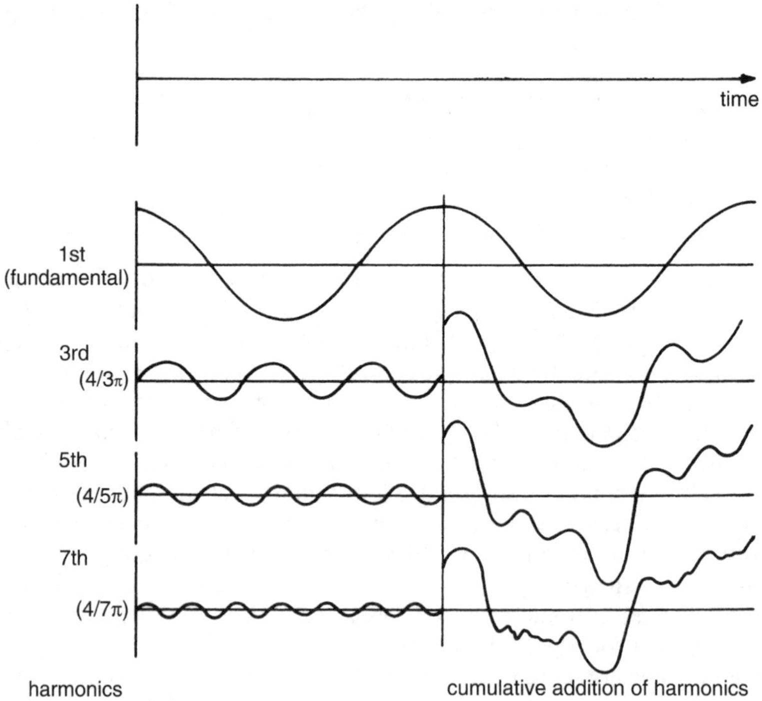

Fig. 2.21 The harmonics of Fig. 2.20 shifted in phase.

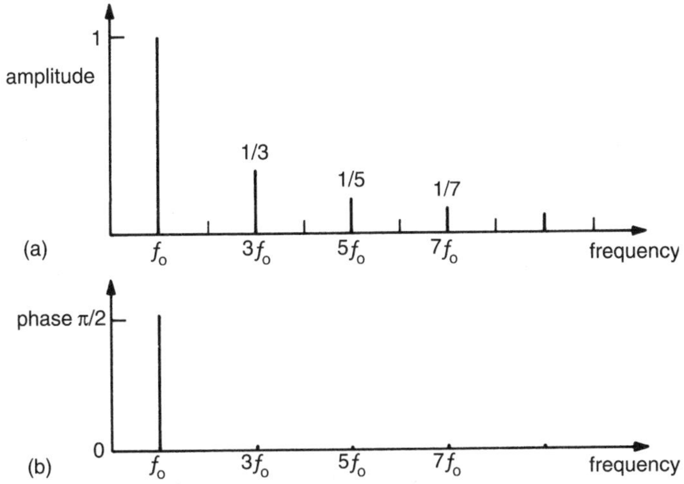

Fig. 2.22 The amplitude and phase spectra of the harmonics of Fig. 2.21.

of signals (e.g. for television transmission), however, all three characteristics are important.

The range of frequencies containing all the Fourier components of a periodic waveform is called its **bandwidth**. So bandwidth is the frequency of the highest-frequency component minus the frequency of the lowest-frequency component, but it includes both extreme frequencies. The **frequency spectrum** is all the Fourier components which make up the signal. So, in Fig. 2.19, the *bandwidth* of the waveform represented by curve (c) is 6 kHz (7 kHz minus 1 kHz) and the *frequency spectrum* of curve (c) is the set of sinusoids of frequencies 1 kHz, 3 kHz, 5 kHz and 7 kHz, which all have zero phase at $t = 0$.

The spectrum can be sub-divided into parts: the **amplitude spectrum** and the **phase spectrum**. The *amplitude spectrum* for both Figs. 2.19 and 2.21 is shown in the bar chart of Fig. 2.22a. The length of the bar at the frequency indicated on the x-axis represents the harmonic amplitude at that frequency. The *phase spectrum* at the particular instant chosen as $t = 0$ can also be represented by a bar chart, as shown in Fig. 2.22b for the waveform of Fig. 2.21.

As you will be aware, the human hearing system automatically carries out a very sophisticated form of spectral analysis on the sounds it hears. When listening to music we can distinguish within the single waveform recorded on a disc or tape the different musical instruments that are being played. The electronic instruments called *frequency analysers*, or *spectrum analysers* perform the much simpler function of picking out the *sinusoids* that are present in a periodic waveform and will display the spectrum as a line spectrum on an oscilloscope screen. Human beings are able to go further than that and combine the sinusoids they hear into the recognizable combinations produced by different musical instruments. Furthermore they can do this job better if the waveform is not just a periodic one but is actual music or speech (i.e. if it is a signal). How the brain does this is very far from being understood.

Self-assessment question

12 The note of middle C (about 250 Hz) is sounded on a musical instrument and produces a waveform which has the following Fourier components:

Frequency	*Amplitude/units*
fundamental	10
2nd harmonic	2
3rd harmonic	4
4th harmonic	1
5th harmonic	2
6th harmonic	negligible
7th harmonic	2
all the rest	negligible

(a) Draw a bar chart representing the amplitude spectrum of this waveform on the axis provided on Fig. 2.23.

(b) What is the bandwidth of the waveform?

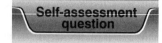

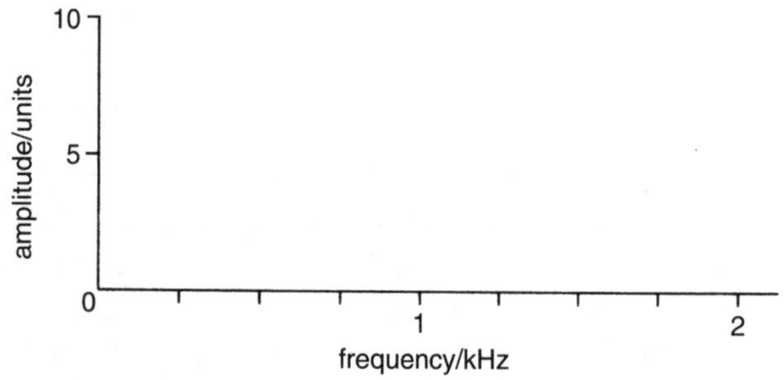

Fig. 2.23 A blank chart for plotting the spectrum of Self-assessment question 12.

2.4.2 Spectra of signals and noise

Continuous spectra

Consider a signal which starts at time zero, whose instantaneous voltage varies in an irregular way, and which ends after a time T (which might be, say, 10 minutes) without ever having repeated itself. Imagine that Fig. 2.24a is a graph of its waveform. This signal could well be the voltage output of a temperature transducer measuring fluctuations of temperature in an industrial process, the whole process lasting 10 minutes. If we wanted to design a circuit to amplify this signal satisfactorily, it would be convenient to represent the signal by a spectrum of sinusoidal components. We could then design the circuit to amplify satisfactorily over the whole of this frequency spectrum.

Figure 2.24b shows part of an infinite succession of repeats of the waveform of Fig. 2.24a. We know from Fourier's theorem that it is possible to find a series of sinusoids (plus a d.c. term) which will add up to give the periodic waveform of Fig. 2.24b. This same Fourier series, therefore, will represent the actual signal over the interval $t = 0$ to $t = T$, and may be thought of as the spectrum of the signal.

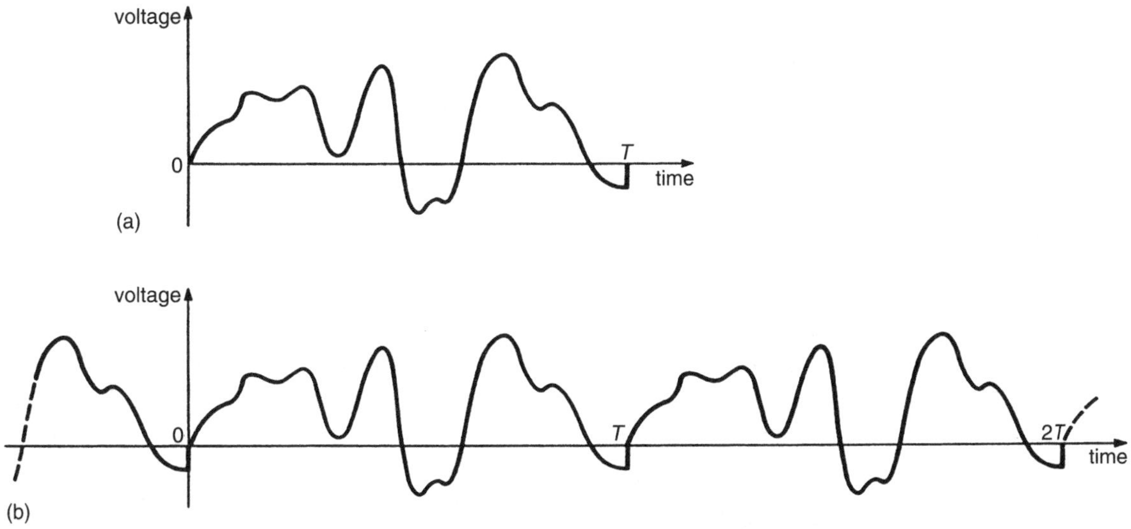

Fig. 2.24 A signal waveform, and the waveform repeated.

The fundamental component of this spectrum has frequency $1/T$ which, if T is 10 minutes, equals 1/600 Hz. All the other components are harmonics of this fundamental, so they are separated in frequency by this amount. The spectrum may extend to very high frequencies, but, as is always the case for practical signals, there will be a frequency above which ignoring all higher harmonics has an acceptably small effect on the representation of the waveform (this defines the effective bandwidth of the signal). Suppose this frequency limit were 1 kHz; a graph or visual display of the line spectrum would then contain 600 000 individual vertical lines and this would be impossible to resolve on a display of normal size. A way that such a spectrum can be drawn is as the outline, or envelope, of the tops of the component amplitudes, like that shown in Fig. 2.25. The shading under the curve represents the components of the spectrum, which cannot be shown individually. A spectrum in which it is a practical impossibility to resolve the individual components is called a **continuous spectrum**.

Suppose a non-repetitive message signal goes on for much longer than 10 minutes; for instance, if the temperature of a continuous industrial process were being monitored, the apparatus might be switched on for a year or more. Is it possible to find the spectrum without taking measurements over the whole year? In practice, you take a section of the message waveform, find its

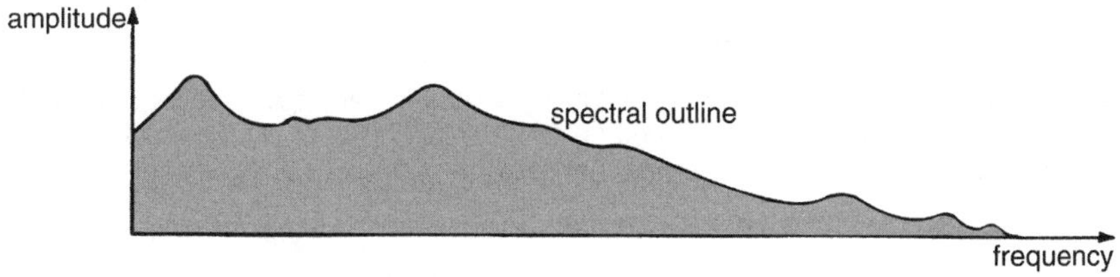

Fig. 2.25 A continuous spectrum of a signal.

spectrum, and assume that this is representative of the spectrum of the whole message signal. Thus, if you measure the spectrum of the signal represented by Fig. 2.24a, you do so on the assumption that the next time the equipment is used the next signal will have a spectrum which is of the same general shape as that of the first signal; otherwise it would not be possible to design electronic circuits to process the signal.

Power density spectra

Think now of the ten-minute sample-section from the signal discussed previously. Its sinusoidal components are 1/600 Hz apart in frequency, so any section of the spectrum occupying 1 Hz of frequency contains 600 components. If you could calculate the mean-square voltage of each and add them together you should obtain the mean-square voltage for the whole 1 Hz block. If you did this for each 1 Hz section of the spectrum in turn, you could draw a graph of mean-square volts per hertz against frequency by plotting, in the middle of each 1 Hz interval, the value for the interval, and joining the points together. Such a graph is called a **power density spectrum**. Some spectrum analysers measure this directly by measuring the average power delivered to a resistor by (and hence the mean-square voltages of) blocks of the spectrum selected in turn by some sort of filter.

Figure 2.26 shows a spectral power density spectrum. The area under the curve over any 1 Hz interval is equal to 1 Hz times the voltage-squared per hertz at that frequency, so it represents the mean-square voltage over that interval. Similarly, the area under the curve between any two frequencies in Fig. 2.26 is equal to the total mean-square voltage of all the components in that part of the spectrum. It is thus *proportional* to the *power* in that part of the spectrum. For example, the mean-square voltage between 7.5 kHz and 10 kHz is found from the area under the curve over this range. This triangular area is $\frac{1}{2}\times \text{height} \times \text{base} = \frac{1}{2}\times 0.5\times 10^{-3}\ \text{V}^2/\text{Hz}\times 2.5\ \text{kHz} = 0.6\ \text{V}^2$.

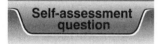

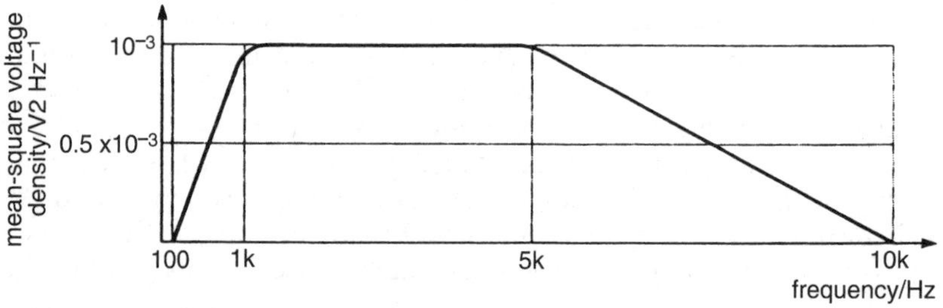

Fig. 2.26 The power density spectrum of a music signal.

Self-assessment question

13 For the spectrum of Fig. 2.26, estimate the total mean-square voltage of the signal components lying between (a) 100 Hz and 5 kHz, (b) 5 kHz and 10 kHz. Hence calculate the approximate r.m.s. signal voltage.

Figure 2.26 is a typical power density spectrum of music. This is the sort of spectrum which is obtained by a spectrum analyser averaging over several minutes, so that all the musical instruments' frequency components are heard, from the lowest-frequency drum beat to the highest-frequency harmonic of the triangle. This particular spectrum extends from about 100 Hz to 10 kHz. Notice that the signal power, or mean-square voltage, is much reduced at the highest frequencies. The major part of the signal is confined to the band up to, say, 7 kHz. In fact, acceptable results are obtained in medium-wave radios with an audiofrequency bandwidth of only 4 or 5 kHz.

For high-fidelity ('hi-fi') reproduction of music, all the audible frequency components should be reproduced. Our hearing falls in sensitivity at higher frequencies. When we are young, about 16 kHz is the highest frequency we can hear. As we get older, our 'passband' reduces at the top end. Even though the higher-frequency components of music are of lower power, and in spite of our sensitivity to these components being lower, it is still important to reproduce them. Their absence, even at a low perceived audible level, robs the music of much of its quality. So, for hi-fi, frequency components from less than 100 Hz up to about 15 kHz should be reproduced. VHF FM radio has an audiofrequency passband extending to 15 kHz for this reason.

The frequency spectra of noise can extend to much higher frequencies. Indeed, the noise generated by a resistor extends to frequencies much higher than those used in electronics. Another type of noise, called 'flicker noise', or $1/f$ noise, has a power density spectrum concentrated at low frequencies. Noise sources and their spectra are described together, in Chapter 4.

2.5 A.C. COMPONENTS

2.5.1 Capacitors

Capacitors consist of two conducting films or surfaces separated by a thin layer of insulation often called a **dielectric** as shown in Fig. 2.27a. The dielectric may be flexible, like polyethylene, in which case the conductor/insulator sandwich can be rolled up so that the capacitor appears cylindrical in shape, as in Fig. 2.27b. Otherwise the dielectric is a solid like mica or ceramic in which case the capacitor might be a multi-layer pack, as in Fig. 2.27c. In integrated circuits, capacitors are usually formed by depositing a layer of metallization on top of the film of silicon dioxide (or silicon nitride) on the surface of silicon, as in Fig. 2.27d, so that the silicon substrate is one of the conductors and the dioxide, which is a very good insulator, is the dielectric.

Capacitors work as follows. When a voltage is applied to the plates of a capacitor, for example by closing the switch in the circuit of Fig. 2.28, the battery voltage is transferred to the plates of the capacitor. You might think that this would take place immediately, but it does not. To create this potential difference between the capacitor plates electrons have to be supplied to the more negative plate and removed from the more positive one, and this takes a little time depending on the current carrying the electrons. The electrons supplied to one plate repel electrons from the other one, leaving a net positive charge of ionized atoms on it. Thus a current is required in both halves of the circuit. Suppose that at a particular instant, v is the voltage that has been built up between the plates and q is the magnitude of the quantity of charge that has been supplied—either of electrons supplied to one plate of the capacitor or

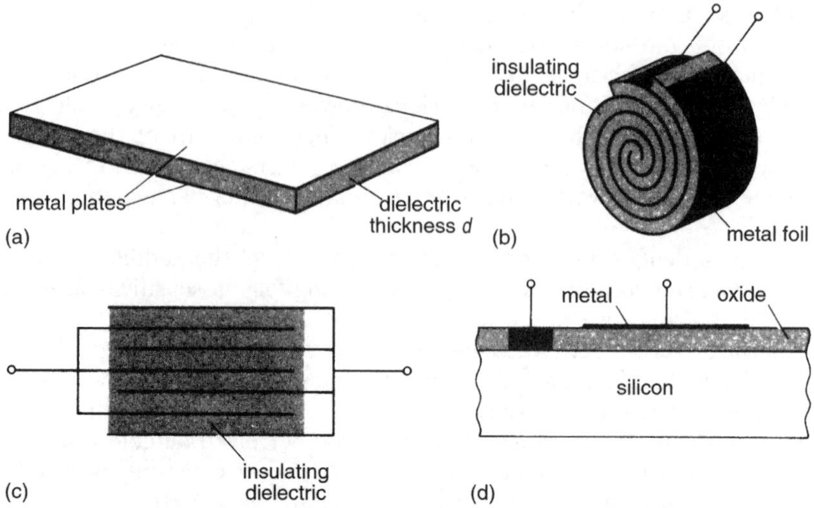

Fig. 2.27 Capacitors—comprising two conductors separated by an insulator. (a) A parallel-plate capacitor of area A and dielectric thickness d has a capacitance of $C = \varepsilon A/d$, where ε is the permittivity of the dielectric. (b) A tubular capacitor made from metal foil and a flexible dielectric such as polythene. (c) A flat-pack capacitor with interleaved electrodes. (d) An integrated circuit capacitor with a metal film as one electrode and silicon as the other.

of positive charge left on the other plate. Then, the capacitance C of the capacitor is defined as

$$C = q/v \tag{2.1}$$

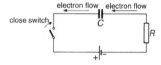

Fig. 2.28 The flow of current as a capacitor is charged up on closing the switch.

The unit of capacitance is the **farad** (symbol F). That is, if q is measured in coulombs and potential difference is measured in volts, the capacitance will be in farads. Typical capacitors range in value from a few picofarads (pF) to a farad or so. Examples are illustrated above. Capacitances associated with integrated circuits are likely to be measured in picofarads, or in fractions of a picofarad.

Now, as already indicated, supplying electrons to one plate and repelling or displacing them from the other, as in Fig. 2.28, implies that a current is flowing towards one plate and away from the other. Indeed, these currents continue to flow—at a decreasing rate—until the voltage across the capacitor has become equal to the battery voltage and the current has fallen to zero. But since the currents in the two halves of the circuit are the same, the flow of electrons is just as if a current was *actually* flowing *round* the circuit, despite the presence of the insulating layer in the capacitor. The flow of electrons towards one plate and away from the other, as in Fig. 2.28, is indistinguishable in the rest of the circuit from a current flowing round the circuit, except that current only flows when the voltage across the capacitor is *changing*.

If the voltage source in the circuit is *an a.c. voltage* instead of a d.c. one, so that the voltage across the capacitor is changing continuously, a corresponding, continuously changing a.c. current will *apparently* flow through the capacitor. This apparent current is called the *displacement current*, and even though there may be a perfect insulator separating the two plates of the capacitor it is helpful to think of this alternating current actually flowing around the a.c. circuit as a whole.

Suppose a small change of voltage Δv is applied to a capacitor so that a charge of Δq is supplied to one plate of the capacitor and removed from the other. (If the change of charge on one plate is Δq, the change on the other plate is $-\Delta q$.) Then

$$\Delta q = C\,\Delta v$$

Dividing these changes of charge and voltage by the small interval of time over which the changes occur gives an expression for the rate of change of charge as a function of the rate of change of voltage, and provides an alternative mathematical model of a capacitor. Thus, dividing by Δt gives

$$\Delta q/\Delta t = C\,\Delta v/\Delta t \qquad (2.2)$$

but current is equal to the rate of change of charge, so $\Delta q/\Delta t = i$, and Equation 2.2 can be written as

$$i = C\,\Delta v/\Delta t$$

In the limit of very small changes in very short intervals of time, $\Delta v/\Delta t$ tends to $\mathrm{d}v/\mathrm{d}t$, the first derivative of v with respect to time, so

$$i = C\frac{\mathrm{d}v}{\mathrm{d}t} \qquad (2.3)$$

An alternative definition of the capacitance of a capacitor is therefore *the ratio of the instantaneous current through the capacitor to the rate of change of voltage across it*. This is sometimes called the small-signal capacitance to distinguish it from the ratio q/v.

The magnitude of the capacitance of a capacitor is proportional to the area A of the two conducting 'plates' and is inversely proportional to the distance of separation, d, between the plates. That is, C is proportional to A/d. The constant of proportionality is the **permittivity** ε of the dielectric. See Fig. 2.27a again. Thus:

> Absolute permittivity $\varepsilon_0 =$ 8.854 pF m^{-1}. Relative permittivity $\varepsilon_r = 1$ for vacuum and air, 2 to 10 for many insulators, and $\geqslant 100$ in ceramic capacitors.

$$C = \frac{\varepsilon A}{d} = \frac{\varepsilon_0 \varepsilon_r A}{d} \qquad (2.4)$$

The permittivity ε is usually written as the *relative permittivity* ε_r multiplied by the *absolute permittivity* ε_0 so that $\varepsilon = \varepsilon_r \varepsilon_0$. The absolute permittivity ε_0 is sometimes also called the *permittivity of free space*. It is a physical constant of magnitude 8.854 picofarads per metre (pF m^{-1}). The relative permittivity ε_r ranges from 1 for free space and air, to between 3 and 10 for most solid insulators. Many commercially available capacitors are made from special ceramics that have much higher relative permittivities, of the order of 100 or more.

Worked example

What is the capacitance of a capacitor whose plates have the dimensions 10 mm × 5 mm, and which are separated by a dielectric of relative permittivity $\varepsilon_r = 200$ whose thickness is 10 μm?

$$\varepsilon = \varepsilon_0 \varepsilon_r = 8.854 \text{ pF m}^{-1} \times 200 = 1770.8 \text{ pF m}^{-1}$$

$$C = \frac{A}{d}\varepsilon = \frac{10^{-2} \text{ m} \times 5 \times 10^{-3} \text{ m}}{10 \times 10^{-6} \text{ m}} \times 1770.8 \text{ pF m}^{-1}$$

$$= 8.854 \text{ nF}$$

Capacitors in which the capacitance is independent of the magnitudes of q or v are called linear capacitors. With some capacitors their capacitance varies with the magnitude of the charge or voltage on them. This is true of ceramic capacitors and of the capacitances associated with junction diodes. These are therefore non-linear capacitors.

Self-assessment question

14 How could you test whether a capacitor was a linear one? (Assume that you know how to measure the small-signal capacitance of any capacitor under any circumstances.)

Circuits always contain stray capacitances between wires and components, because a circuit is nothing if not a lot of conductors separated by insulation! These stray capacitances contribute a good deal to the slowing down of the speed of response of digital circuits or to limiting the frequency range of analogue circuits. This can be understood by looking again at the equation $i = C\,\mathrm{d}v/\mathrm{d}t$. This equation indicates that the rate of change of voltage across a capacitance is limited by the current, i, available to charge it up and discharge it. So the smaller the signal current the slower the possible rate of change of voltage across stray capacitances. To achieve fast circuits—circuits in which voltage changes can occur rapidly—it is therefore necessary to have either large signal currents or small stray capacitances or both.

Capacitive reactance

Suppose a sinusoidal voltage is applied to a capacitor as shown in Fig. 2.29a, so that

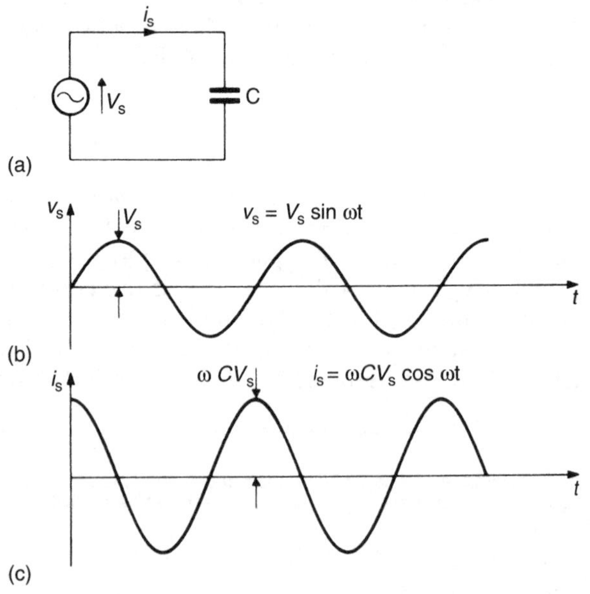

Fig. 2.29 The relationship between the current through a capacitor and the voltage applied across it. In an ideal capacitor (with infinite d.c. resistance) the voltage always lags in phase behind the current by $\pi/2$ rad.

$$v_s = V_s \sin \omega t \qquad (2.5)$$

The current can be calculated from Equation 2.3 with v_s substituted for v. Thus,

$$i_s = C\frac{dv_s}{dt}$$

Differentiating Equation 2.5 gives $dv_s/dt = \omega V_s \cos \omega t$, so

$$i_s = \omega C V_s \cos \omega t \qquad (2.6)$$

From this result we can draw two conclusions.

Firstly, *the phase of the sinusoidal voltage waveform across a capacitor lags by $\pi/2$ radians behind the sinusoidal current flowing through it*, as shown in Figs. 2.29b and 2.29c. You can visualize why the voltage phase lags behind the current phase by noting that current must flow 'through' the capacitor *in order to bring about* the change in voltage across it; so voltage changes only begin to occur when the current flow causing them has started to flow.

The **reactance of a capacitor**, X_C, is the ratio of the amplitudes of voltage and current, so the second conclusion is that *the reactance of a capacitor is* $1/\omega C$. That is,

Capacitive reactance, $1/\omega C$.

$$X_C = \frac{V_S}{\omega C V_S} = \frac{1}{\omega C}$$

This result is obtained by taking the ratio of the amplitudes in Equations 2.5 and 2.6. Thus the reactance of a capacitor *decreases* with increasing frequency.

Reactance is measured in ohms, just like resistance, but you should remember that with reactances the voltage and current are not in phase.

Self-assessment question

15 Calculate the reactance of a 1 μF capacitor at 60 Hz. Hence calculate the current which it would draw when connected across an a.c. supply of 120 V at 60 Hz.

A.C. response of *CR* networks

Figures 2.30 and 2.31 show two single *CR* (capacitor–resistor) networks driven by sinusoidal input signals. In each case, the capacitor and resistor form a potential divider, so that the output signal is an attenuated version of the input signal.

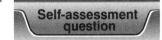

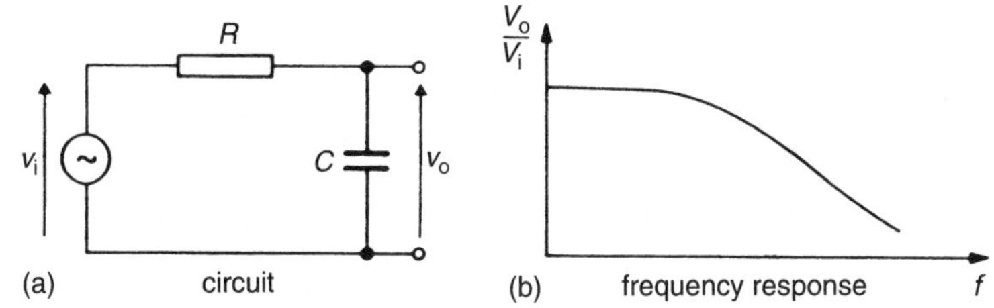

Fig. 2.30 The low-pass *CR* circuit.

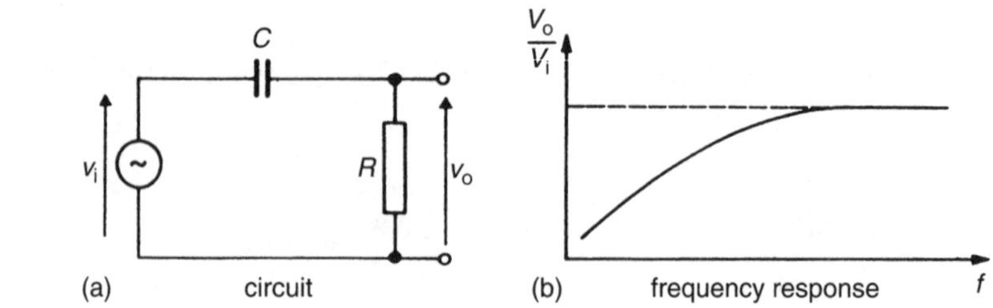

Fig. 2.31 The high-pass *CR* circuit.

In Fig. 2.30a, the capacitor's reactance is much greater than the resistor's resistance at low frequencies, so there is little attenuation at low frequencies. At high frequencies, the capacitive reactance falls with frequency, and attenuates the output signal. So this is a **low-pass** filter, which 'passes' low-frequency sinusoids but attenuates high frequencies. Its frequency response is shown in Fig. 2.30b.

In Fig. 2.31a, the positions of the capacitor and resistor are swapped. Now the high reactance of the capacitor at low frequencies attenuates the output, but at high frequencies its reactance is negligible and there is little attenuation. So this is a **high-pass filter**. Its frequency response is shown in Fig. 2.31b. The full a.c. analysis of these two circuits is left to Chapter 3, using phasors.

Step response of *CR* networks

The effect of a *CR* network on a digital waveform, or a squarewave test signal, is best analysed by considering the input as a series of voltage 'steps'. The low-pass filter is considered first.

Low-pass RC network

See Fig. 2.32. The voltage step is represented by closing the switch to connect the battery to the circuit. When the voltage step is applied at $t = 0$, the capacitor is uncharged, so at that instant the voltage across it ($v_o = v_c$) is zero. Current V_i/R flows through the resistor *R* and, as charges of opposite signs accumulate on the plates of the capacitor, the voltage across it gradually rises. This leaves less voltage across the resistor, so the current gradually falls, and this results in the capacitor charging at a lower rate. Eventually, the capacitor voltage effectively reaches the value V_i, and no further current flows.

A formula for the output voltage curve can be obtained as follows. At any instant of time *t* after the input 'step' is applied, the instantaneous current *i* and the instantaneous capacitor voltage are related by

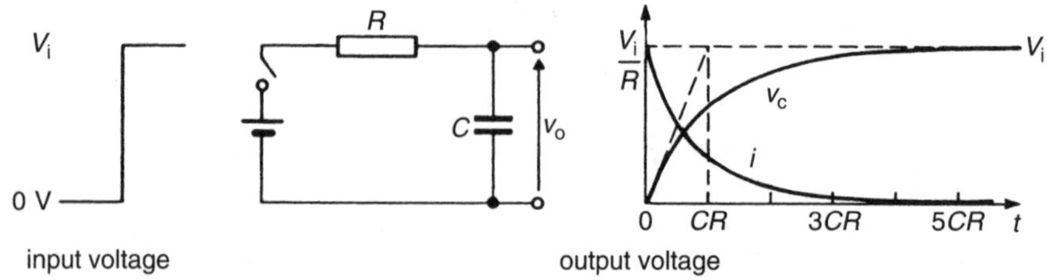

Fig. 2.32 Step response of low-pass *CR* network.

$$V_i = iR + v_c \qquad (2.7)$$

where V_i is the constant value of the step input.

The displacement current in a capacitor is given by

$$i = C\,dv_c/dt$$

Thus, substituting the value of i in Equation 2.7 gives

$$V_i = CR\,dv_c/dt + v_c \qquad (2.8)$$

It can be shown that the solution for v_c as a function of time is

$$v_c = V_i[1 - \exp(-t/CR)] \qquad (2.9)$$

You can check this by differentiating v_c with respect to time:

$$dv_c/dt = v_i(1/CR)\exp(-t/CR)$$

and then substituting for dv_c/dt and v_c in the right-hand side of Equation 2.8 to give

$$CR\,V_i(1/CR)\exp(-t/CR) + V_i[1 - \exp\,(-t/CR)]$$

which is equal to V_i. Hence Equation 2.9 is a solution of Equation 2.8.

Figure 2.32 shows the waveforms of the current and the capacitor voltage. Notice how the current starts with a value of V_i/R, because the capacitor voltage is zero. Thereafter, the current gradually falls to zero whilst the capacitor voltage rises to the input voltage V_i.

Time constant, *CR*.

The product of capacitance and resistance has dimensions of time. Thus the factor ($-t/CR$), which is the exponent in the exponential expression, is dimensionless. The product *CR* is known as the *time constant* of the circuit. The time constant affects the rate of change of the output waveform; the greater *CR*, the slower the output rises in response to the input step.

In Fig. 2.32 you can see that the initial slope of the capacitor voltage curve is such that, if this rate of charge were maintained, the capacitor would charge up in one time constant. Because the rate of change gradually falls, the capacitor actually charges to about 63% of V_i in this time. After three time constants, the capacitor voltage reaches about 95% of V_i. After five time constants, charging is nearly complete at 99% of V_i.

The 'rise time' of the circuit is commonly taken as the time for the output to rise from 0.1 V_i to 0.9 V_i (from 10% to 90% of V_i). The times at which these output voltages are reached can be calculated by substituting in the formula for v_c (Equation 2.9):

The expressions for the output voltages of the *CR* networks are: $v_c = V_i \times [1 - \exp(-t/CR)]$ and $v_r = V_i \exp(-t/CR)$. Both contain the term $\exp(-t/CR)$. To help remember which is which, remember that v_c must start at zero, and that v_r must be equal to V_i immediately after the input step. Putting $t = 0$ gives $\exp(-t/CR) = 1$ so, at $t = 0$, $v_c = V_i[1-1] = 0$ and $v_r = V_i \times 1 = V_i$.

For t_1:

$$0.1\,V_i = V_i[1 - \exp\,(-t_1/CR)]$$
$$0.9 = \exp\,(-t_1/CR)$$

Taking natural logs, we obtain

$$ln\,0.9 = -t_1/CR$$

Therefore,

$$t_1 = -CR\,ln\,0.9 = 0.105\,CR$$

Similarly, for t_2:

$$0.9\,V_i = V_i[1 - \exp(-t_2/CR)]$$

Proceeding as before, we obtain

$$t_2 = 2.303\,CR$$

Thus, the risetime $t_2 - t_1$ is

$$t_r = 2.2\,CR$$

Self-assessment question

16 A 4 V step is applied to a low-pass *CR* circuit, with the capacitor initially discharged. (i) Calculate the output voltage after two time constants.

If $C = 100$ pF and $R = 1$ kΩ, calculate (ii) the 10% to 90% rise time, (iii) the time for the output to reach 2 V.

High-pass CR network

The step response of the high-pass network is shown in Fig. 2.33. The response to the rising edge of the input step is a rising edge of the output. The capacitor voltage cannot change instantaneously, so the output follows the input initially. Thereafter, the capacitor charges up and the output voltage 'droops' as the current through the resistor begins to falls.

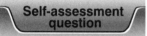

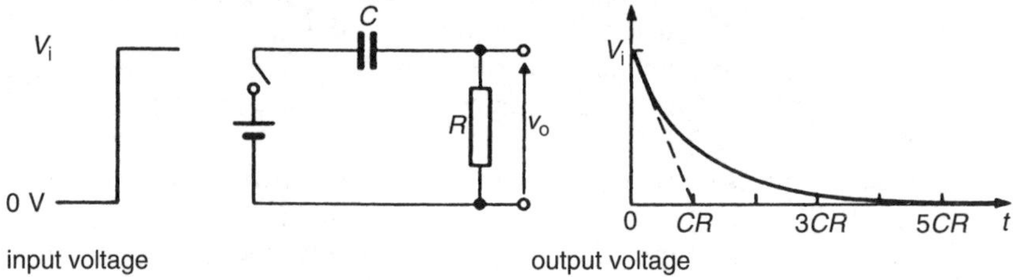

Fig. 2.33 Step response of high-pass *CR* network.

In this case, the output voltage is given by substituting the value of v_c from Equation 2.9:

$$v_r = V_i - v_c = V_i \exp(-t/CR)$$

2.5.2 Inductors

An inductor is a coil of wire such as that illustrated in Fig. 2.34a. A varying current flowing through the wire of the coil induces a voltage in this same wire by the process of electromagnetic induction, so that a voltage–current relationship is set up which has nothing to do with the resistance of the wire. The voltage between the ends of the coil is dependent on the *rate of change* of the current through it and on the number of turns of wire in the coil; it does not depend on the actual value of the current. These properties depend on

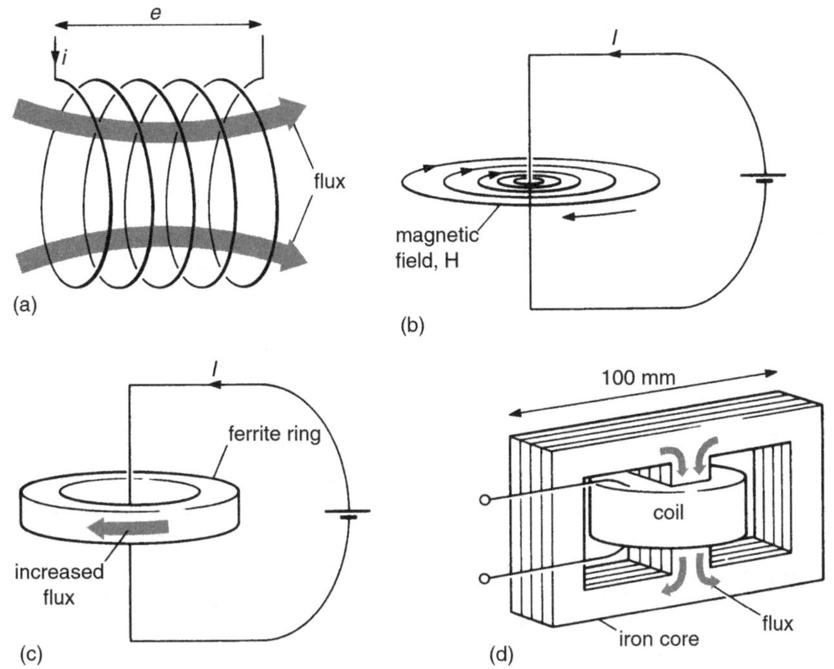

Fig. 2.34 The magnetic effect of an electric current. (a) The magnetic field in a coil created by the flow of current through the coil. (b) The magnetic field created by the current in a single straight wire. (c) A magnetic field applied to a ring of ferromagnetic material produces much more magnetic flux than the same field would produce in air (e.g. as in (b)). (d) The structure of an inductor of several henries inductance. The coil and ferromagnetic core surround each other.

the principles of electromagnetism. The ones that concern us here can be summarized as follows.

(1) An electric current creates a **magnetic field** similar to that produced by magnets. This magnetic field encircles the wire carrying the current, as shown in Fig. 2.34b. The rings drawn round the wire in this figure are intended to indicate the existence of the field. The arrows show its direction for the current shown—reverse the current and you reverse the direction of the field. The closeness of the lines indicates the strength of the field. Evidently then, from this diagram, the field decreases with distance from the wire. The strength of **magnetic field**, H, can be calculated from Ampère's law, which states, for the case shown in Fig. 2.34b, that

$$I = H \times (\text{circumference of the circular path})$$

Since the circumferences is 2π times the distance from the wire, the field strength for a given current I is inversely proportional to the distance from the wire.

When a current-carrying wire is coiled up as in Fig. 2.34a, the field is concentrated within the coil, each turn adding to the strength of the field.

(2) This magnetic field produces **magnetic flux** which follows the *direction* of the field but the *magnitude* of the flux depends on the material surrounding the wire. The flux per unit cross-section—called **flux density**,

B—depends on the **permeability** μ of the material through which the field passes; that is

$\mu = \mu_0\mu_r$. Free-space permeability $\mu_0 = 1.257\ \mu H\ m^{-1}$. Relative permeability $\mu_r = 1$ for air and most materials, $\geqslant 1000$ for ferromagnetics.

Flux density = permeability × magnetic field strength

or

$$B = \mu H$$

The permeability μ is usually expressed as the product of the permeability of free space, μ_0, and the relative permeability μ_r which is specific to different materials. $\mu_0 = 1.257$ microhenries per metre ($\mu H\ m^{-1}$). The relative permeabilities of most materials are close to 1; but there is one class of materials, the **ferromagnetic materials**, which have permeabilities which are very much greater. For ferromagnetic materials μ_r may be 1000 or more. Ferromagnetic materials include iron, steel and various oxides of iron called *ferrites*. Because of its high permeability, the flux produced in a ferromagnetic material by a given magnetic field is much larger than the flux produced in air. For example the flux in the ring of ferrite in Fig. 2.34c is 1000 or more times the flux in the same cross-section of air in Fig. 2.34b even though the magnetic field H is the same in both cases.

(3) A *changing* magnetic flux induces an electrical voltage or e.m.f. in a wire placed in this field. The magnitude of the e.m.f. can be calculated from Faraday's law, which states that *the induced e.m.f. in a wire is equal to the rate of change of flux linked with the wire*. So, returning to Fig. 2.34a again, a current flowing in the coil will create a magnetic flux through the coil as indicated by the broad arrows. If the current through the coil is now varied in magnitude, the flux will vary correspondingly, with the result that an e.m.f. will be induced in the coil. This process is called **self-induction**. Applied to a coil of wire, Faraday's law can be expressed as

Faraday's law.

$$e = L\frac{di}{dt}$$

where e is the induced e.m.f. and L is the **inductance** of the coil.

The induced e.m.f. is sometimes called the *back e.m.f.* because it 'opposes' the voltage applied to the coil.

The unit of inductance is the **henry** (symbol H). Thus the inductance of a coil is $L = 1$ H if an e.m.f. of 1 V is induced in it when the change of current through it is $di/dt = 1\ A\ s^{-1}$. The inductance of a coil depends upon the number of turns in the coil and the nature of the magnetic path. For a tightly wound coil, whose diameter is much greater than its length,

L is proportional to the square of the number of turns.

L is greatly increased by giving the coil a ferromagnetic core as illustrated in Fig. 2.34d. The greatest effect is achieved if the coil and the core encircle each other as shown.

Worked example

Why does a ferromagnetic core increase the inductance of the coil?

Because, for a given current in the coil, the flux within it is increased by the high permeability of the ferromagnetic core. Thus, for a given rate of change of current, the rate of change of flux is increased too.

All conductors possess some inductance; even straight wires possess some. A 10 mm length of wire in air has an inductance of the order of 1 nH: a small value, but significant in amplifiers for television signals. Inductors for use in electronic circuits are likely to be in the range of millihenries or microhenries. The inductor in Fig. 2.34d is likely to be a few henries.

The permeability of a ferromagnetic core varies somewhat with the strength of the applied magnetic field—the larger the field the smaller the permeability. This means that the inductance of a coil with a ferromagnetic core will depend somewhat upon the magnitude of current in the coil. In other words, coils with ferromagnetic cores are usually non-linear. Air, however, has a constant permeability, so air-cored coils are linear inductors.

Note that the ferromagnetic materials used in coils are not the same as those used to make permanent magnets. Permanent magnets hold their flux after the magnetic field causing it has been removed. But the flux in the materials used in coils should ideally follow the variations of the field as precisely as possible.

Self-assessment question

17 On which of the above electromagnetic principles does the operation of (a) electric generators or dynamos and (b) electromagnets depend?

Inductive reactance

Consider what happens when a sinusoidal current flows in an ideal inductor in which the wire has zero resistance as in Fig. 2.35a. (Note that it is possible

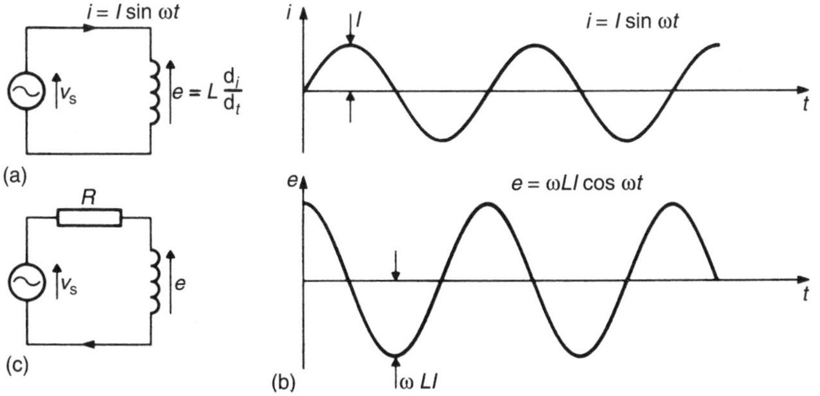

Fig. 2.35 A sinusoidal voltage applied to an inductor. (a) The relationship between current and voltage: $e = L(di/dt)$. With no resistance in the circuit it is also true that $e = v_S$. (b) Waveforms illustrating the phase relationship between current and voltage across an inductor of zero resistance; the current always lags $\pi/2$ rad behind the voltage. (c) When there is resistance in the circuit $v_S = iR + L\,(di/dt)$.

nowadays to make such an ideal inductor at low temperatures using superconducting wire.)

By Faraday's law an e.m.f. will be induced in the inductor which will be proportional to the rate of change of the current.

Worked example

(Revision) What can you say about the waveform of this induced e.m.f. bearing in mind that it is proportional to the rate of change of the current?

The rate of change of a sinusoid is again a sinusoid. It is of the same frequency as the original but of different phase. In fact its phase is always $\pi/2$ radians in advance of the original waveform.

If the current is given by $i = I \sin \omega t$, then since $e = L\, di/dt$,

$$e = \omega L I \cos \omega t$$

So the amplitude of the sinusoidal e.m.f. induced in the coil is ωLI, as shown in Fig. 2.35b, and it has a phase lead of $\pi/2$ rad.

Inductive reactance, ωL.

The ratio of the amplitude of the e.m.f. to the amplitude of the current is called the **reactance** of the coil, or sometimes the **inductive reactance** to distinguish it from capacitive reactance. The usual symbol for an inductive reactance is X_L. So here

$$X_L = \frac{\omega L I}{I} = \omega L$$

Self-assessment question

18 A tightly wound inductor is constructed with an additional connection made to the mid-point of the coil, so that either half the coil can be used as an inductor or the whole coil can be used. If the inductance of the whole coil is 0.3 H and its resistance is negligible, what is its reactance when $f = 500$ Hz? What is the reactance of half the coil at 1000 Hz?

A.C. response of *LR* networks

The inductor–resistor circuit of Fig. 2.36a acts as a high-pass filter to sinusoidal inputs. At low frequencies, where the inductive reactance ωL is much lower than R, the output voltage is much smaller than the input. But at high frequencies, where $\omega L \gg R$, the output voltage is nearly equal to the input. The frequency response is shown in Fig. 2.36b.

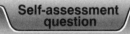

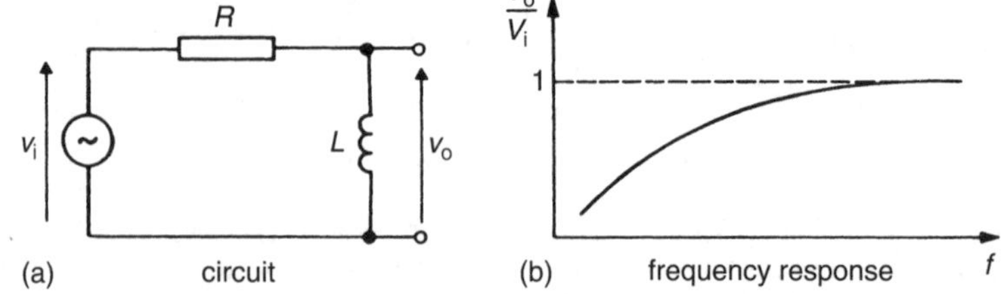

Fig. 2.36 The high-pass *LR* circuit.

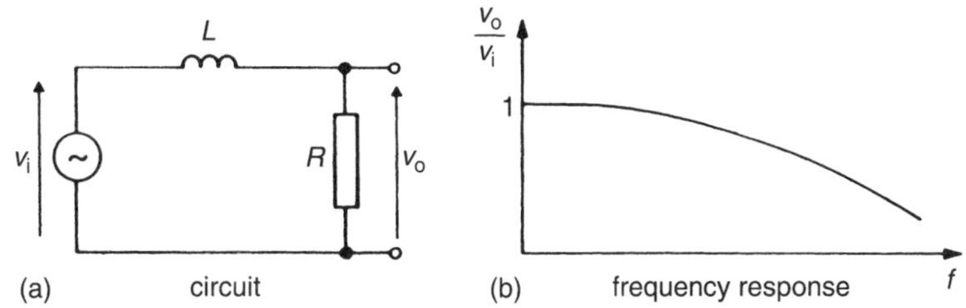

Fig. 2.37 The low-pass *LR* circuit.

The *LR* circuit of Fig. 2.37a acts as a low-pass filter, with the inductive reactance attenuating high frequencies. Its frequency response is shown in Fig. 2.37b.

The full analysis of both these filter circuits is left until Chapter 3, using phasors.

Step response of *LR* circuits

High-pass LR circuit

Figure 2.38a represents the application of a voltage step to the circuit. After the switch is closed, the input voltage is equal to the sum of the voltage drops in the circuit:

$$V_S = iR + L\,di/dt \qquad (2.10)$$

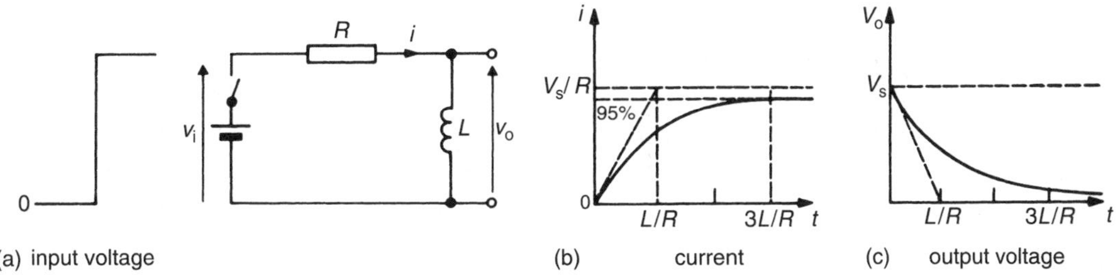

Fig. 2.38 Step response of the high-pass *LR* circuit.

At switch-on, the current cannot change abruptly, because that means an infinite value for di/dt, implying an infinite value for V_S. So i starts at zero, producing no voltage drop across R, and with an initial rate of increase determined by $V_S = L(di/dt)$. As the current increases, iR increases, leaving progressively less voltage across the inductor. So $L(di/dt)$ decreases and the rate of increase of current decreases. This is shown in Fig. 2.38b. Eventually, when the current has stopped increasing, $L(di/dt)$ is zero, there is no voltage across the inductor, and the current is then $i = V_S/R$.

The solution of Equation 2.10,

$$V_S = iR + L\,di/dt$$

is

$$i = (V_S/R)\,[1 - \exp(-tR/L)] \qquad (2.11)$$

As in the *CR* circuit, you can check this solution by substituting for *i* and d*i*/d*t* in Equation 2.10. First,

$$\mathrm{d}i/\mathrm{d}t = (V_S/R)(R/L)\,[\exp(-tR/L)] = (V_S/L)\,[\exp(-tR/L)] \qquad (2.12)$$

Substituting for *i* and d*i*/d*t* in Equation 2.10:

Time constant, *L*/*R*.

$$V_S = V_S[1 - \exp(-tR/L)] + V_S \exp(-tR/L)$$
$$= V_S, \quad \text{as required}$$

Thus the current rises exponentially, as shown in Fig. 2.38b. The exponent is $-tR/L$. The ratio L/R has dimensions of time and is called the time constant of the *LR* circuit. The greater the value of the time constant, the longer the current takes to build up to its final value. The greater value of *R*, the less time it takes.

The output voltage of the circuit is

$$v_o = v_L = L\,\mathrm{d}i/\mathrm{d}t$$

Substituting for d*i*/d*t* from Equation 2.12.

$$v_L = V_S e^{-tR/L}$$

At $t = 0$, $e^{-tR/L} = 1$, so $v_L = V_S$. Thereafter, the output voltage falls exponentially, as shown in Fig. 2.38b.

Low-pass LR circuit

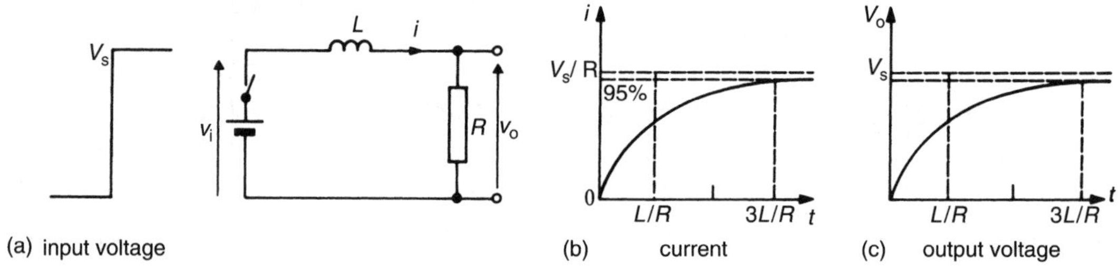

Fig. 2.39 Step response of the low-pass *LR* circuit.

This is shown in Fig. 2.39a. As in the high-pass case, after the switch is closed the input voltage is equal to the sum of the voltage drops in the circuit. (Equation 2.10 again):

$$V_S = iR + L\,\mathrm{d}i/\mathrm{d}t \qquad [2.10]$$

The solution for the current is, as before (Equation 2.11 again):

$$i = (V_S/R)(1 - e^{-tR/L}) \qquad [2.11]$$

Thus the output voltage is

$$v_R = iR = V_S(1 - e^{-tR/L})$$

At $t = 0$, $e^{-tR/L} = 1$, so $V_R = 0$. Thereafter, the output voltage rises exponentially, as shown in Fig. 2.39c.

Self-assessment question

19 A 12 V car ignition coil has an inductance of 10 mH and resistance of 2 Ω (so its equivalent circuit is a 10 mH inductor in series with a 2 Ω resistor). Calculate how long it takes the current to build up to 95% of its maximum value after a 12 V battery is connected to the coil.

Use your circuit analysis package to build *CR* and *LR* networks, both low-pass and high-pass. Check their frequency responses. Do they have the predicted shapes? Check their step responses (transient responses) using a series of positive and negative steps — that is a squarewave. Explain the output waveforms. Do the networks have the predicted time constants?

2.5.3 Transformers

Transformers consist of two inductors placed close together, as indicated diagrammatically in Fig. 2.40a. They are placed sufficiently close that the rate of change of current in one coil not only induces a voltage in itself by electromagnetic induction, as already described, but also induces a voltage in the other coil. The magnetic flux produced by the current in the 'primary coil'—the one connected to the source—affects both coils, and so a sinusoidal current in the primary coil induces sinusoidal voltages in both the primary and secondary coils. The more closely the two coils are linked the more flux from the primary coil reaches the secondary coil. In Fig. 2.40a the broad arrows are intended to indicate that the magnetic flux created by the current in the primary is linked with the secondary coil. Figure 2.40b shows a diagram of a

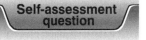

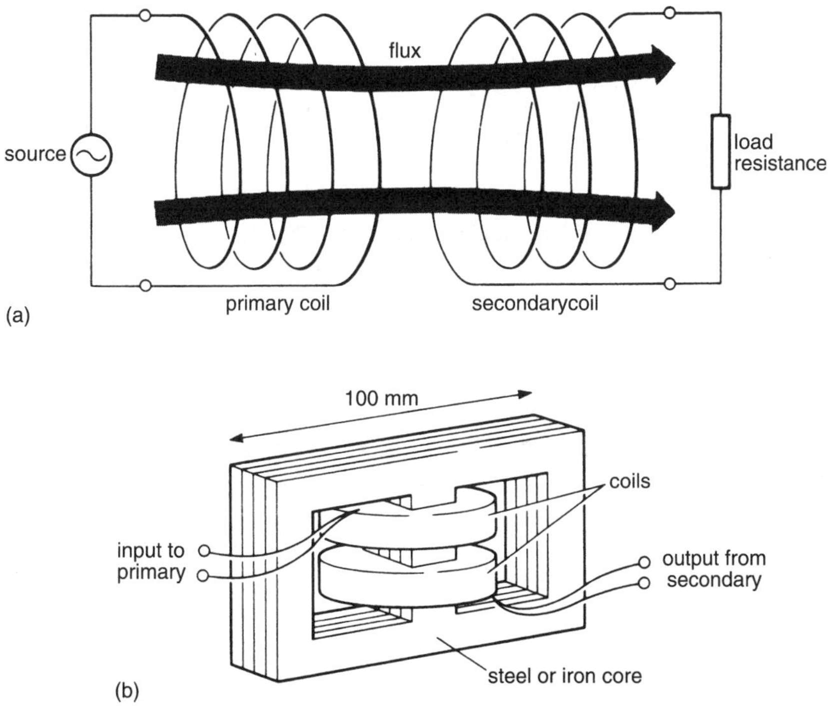

Fig. 2.40 (a) A diagram of a transformer showing the flux produced by the primary winding passing through the secondary winding too. Any *change* in the primary current will cause a *change* in the flux, which then induces a voltage in both the primary and secondary coils. (b) A typical construction of a mains transformer. The iron core encircling the coils increases the inductance of the primary and improves the flux linkage between the two coils.

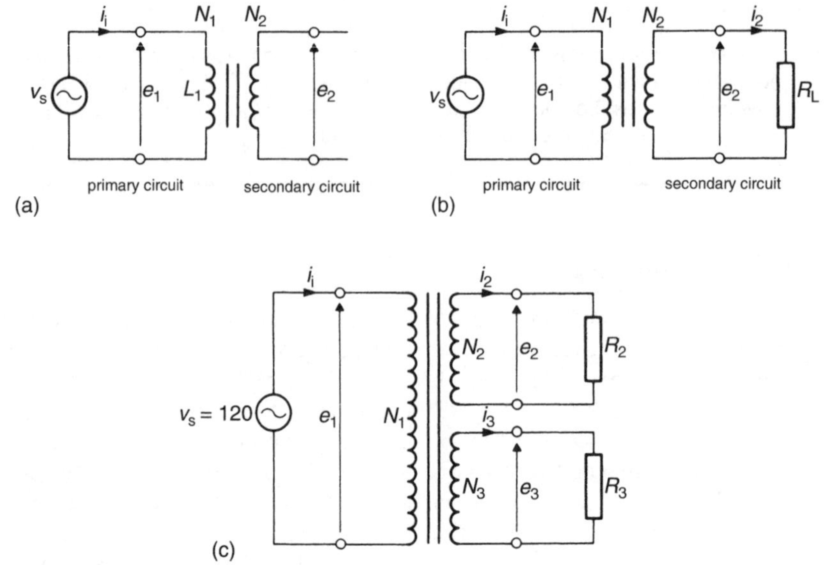

Fig. 2.41 Transformer circuit showing the numbers of turns in the coils as N_1, N_2 and N_3. (a) A transformer with open-circuit secondary, where $e_2/N_2 = e_1/N_1$ and $e_1 = v_S = L_1\ (di_1/dt)$. This primary current is called the magnetizing current i_m. By making L_1 large, i_m can be kept small. (b) A transformer with a load resistance, where $e_2/N_2 = e_1/N_1$, $i_2 = e_2/R_L$ and $i_1N_1 = i_2N_2$ if i_m is small. (c) A transformer with two secondary windings.

transformer of the kind that might be used in a mains power supply. As with the inductor, the ferromagnetic core has the property of increasing the amount of flux, and by encircling both coils it ensures that any variations in flux affect both coils equally.

The result is that the same e.m.f. is induced in each turn of wire in each coil, so the total e.m.f. induced in each coil is proportional to its number of turns. If N_1 and N_2 are the numbers of turns in the primary and secondary coils, as shown in Fig. 2.41a, and e_1 and e_2 are the instantaneous values of the back e.m.f. induced in them, then

$$\frac{e_1}{e_2} = \frac{N_1}{N_2} \tag{2.13}$$

If we suppose that the coil resistances are negligibly small, the back e.m.f. in the primary must exactly equal the applied voltage, as explained in Section 2.5.2, so $e_1 = v_S$. Thus,

$$\frac{v_S}{e_2} = \frac{N_1}{N_2} \quad \text{or} \quad \frac{v_S}{N_1} = \frac{e_2}{N_2} \tag{2.14}$$

If $N_2 > N_1$ the transformer is called a *step-up transformer*, giving a larger voltage output than the input voltage, or a *voltage gain* of N_2/N_1. If $N_2 < N_1$ the device is a *step-down transformer* with a voltage gain of less than 1.

One use of a step-down transformer is in a low-voltage power supply (see Chapter 8), in which the supply mains voltage is reduced to a lower voltage (and then converted to d.c.) for use in electronic circuits. Such transformers also improve the safety of mains-driven equipment because the two windings are electrically isolated from one another.

Worked example

How can a step-up transformer be converted into a step-down one?

By turning it round, by making the winding with the large number of turns the primary winding.

Note the graphical symbol for a transformer with a ferromagnetic core shown in Fig. 2.41a. If the transformer does not have such a core, then the vertical lines in the symbol are omitted.

If the secondary coil of a transformer is not connected to a load, as in Fig. 2.41a, and is left open-circuit, the secondary current must be zero. Under these circumstances the secondary coil can be ignored and the primary coil simply behaves like an inductor in which $v_S = L_1(di_m/dt)$ where L_1 is the primary inductance. The primary current i_m, when the secondary current is zero, is called the **magnetizing current**. Transformers are usually designed to have relatively large inductances so that i_m is small compared with the current that flows when the secondary is connected to its intended load. So we can ignore it for the moment.

Figure 2.41b shows the transformer connected to a load resistor whose resistance is R_L. Evidently, since the voltage appearing across the secondary winding is e_2, the current in R_L, and in the secondary circuit as a whole, is e_2 divided by R_L. That is $i_2 = e_2/R_L$.

Now, contrary to what you might expect, in view of the fact that there is no direct connection between the two coils, this does not leave the primary current unaffected. The simplest way to see why this must be so, and to arrive at a value for the current in the primary circuit, is to note that since there is no resistance in the coils, there will be very little power dissipated in the transformer. The law of conservation of energy, therefore, demands that, to a first approximation, the power output of the transformer must equal the input power. But the instantaneous output power is $e_2 i_2$ and the instantaneous input power is $v_S i_1$. So, if there is no power loss,

$$v_S i_1 = e_2 i_2$$

or, substituting for v_S from Equation 2.14,

$$i_1 N_1 = i_2 N_2 \qquad (2.15)$$

The above equations apply to any waveform. If the input waveform is a sinusoid then it follows from these equations that all the waveforms are sinusoids. The input and output voltages are either in phase with each other or completely out of phase, depending on which way round the secondary is connected. Let us suppose that e_2 and e_1 are in phase.

If the load is a resistor, then the output current must be in phase with the output voltage. That is e_2 and i_2 are in phase. Then by Equations 2.14 and 2.15, all the currents and voltages are in phase. The only out-of-phase current is the magnetizing current which lags $\pi/2$ radians behind the input voltage, as is the case with any inductor. The total input current is then the sum of two currents, i_1 and i_m, which are $\pi/2$ radians out of phase with each other. Chapter 3 explains how to calculate the sum of two sinusoidal currents that are not in phase.

Self-assessment question

20 In Fig. 2.41b suppose that $N_2 = 200$ turns, $N_1 = 50$ turns and $R_L = 4$ kΩ. By calculating the values of i_1 and v_1 in terms of i_2 and v_2, deduce the effective input resistance of the transformer and its load as 'seen' by the source. Can you deduce any general relationship between the resistance as seen by the source and the load resistance of a transformer?

The resistance 'seen' by the source is called the **input resistance**. If i_m is negligible, the input resistance can be calculated as follows, using Equations 2.13 and 2.15:

$$\text{Input resistance } R_{in} = \frac{e_1}{i_1} = \frac{e_2 N_1}{N_2} \times \frac{N_1}{i_2 N_2} = \frac{e_2}{i_2}\left(\frac{N_1}{N_2}\right)^2$$

So if the load is a resistor of resistance R_L, then $e_2/i_2 = R_L$ and the input resistance is

$$R_{in} = (N_2/N_1)^2 \times R_L \quad (2.16)$$

A step-up transformer therefore apparently reduces the resistance of a load resistor by the square of the turns ratio.

By the same reasoning it follows that if a transformer has a capacitor as a load, the source 'sees' a capacitance rather than a resistance. That is, the phase of the current i_1 leads the input voltage v_S by $\pi/2$ rad. But the capacitance it sees is $(N_2/N_1)^2$ times the capacitance of the actual load. So a transformer can be used to increase or decrease the effective capacitance of a capacitor. In general, therefore, the phase difference between the input current and voltage (assuming i_m is negligible) is determined by the load, not by the transformer.

Self-assessment question

21 (a) Suppose, as is commonly the case, that two secondary coils are wound on the same core as the primary of a transformer so that the same flux links all three coils, as indicated in the diagram of Fig. 2.41c. If the numbers of turns on the three coils are N_1, N_2 and N_3, what is the relationship between the voltages induced in the three coils?

(b) The two secondary coils are connected to two load resistors so that currents i_2 and i_3 flow in them. What is the relationship between i_1, i_2 and i_3 and the numbers of turns of the coils? Assume that the magnetizing current is negligible, and use the same energy argument as was used in the text in your answer.

Self-assessment question

22 In Fig. 2.41c, $N_1 = 600$ turns, $N_2 = 300$ turns and $N_3 = 200$ turns. If the input is a sinusoidal voltage of 120 V amplitude, and if the load resistors are $R_2 = 1$ kΩ and $R_3 = 500$ Ω, calculate (a) the amplitudes I_2 and I_3 of the output currents and (b) the total power input to the transformer. (Assume the magnetizing current is negligible).

The physical explanation, in terms of magnetic flux, of why the flow of secondary current causes an increase of primary current can be found in most textbooks on electrical engineering, and need not concern us here.

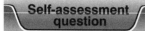

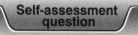

2.6 CONCLUSION AND SUMMARY

Much of this text has been concerned with the meanings of new terms. These are listed at the beginning. The following important points were also explained.

1. Digital waveforms are made up of discrete levels rather than a continuum of levels as in analogue signals. Typically the discrete levels are different voltages (e.g. 5 V and 0 V).
2. Useful information is conveyed only by waveforms whose shape is not wholly predictable. These are called signals and must be, to some extent, *non-periodic*. But because signals can be regarded as being made up of *sinusoids*, most *analogue* circuit analysis and design is in terms of sinusoidal waveforms.
3. A voltage sinusoid, for example, can be wholly defined by the equation $v = V\sin(\omega t + \phi)$ where v is the *instantaneous voltage*, V is the *amplitude*, ω is the *angular frequency*, which equals $2\pi f$, and ϕ is the *phase*. A leading phase is positive and a lagging phase is negative. *Phase difference* is a comparison between the phases of two sinusoids of the same frequency.
4. The mean-square value of a simple periodic waveform can be calculated from one cycle of the waveform. The square root of this is the root mean square (r.m.s.) value. The power dissipated in a resistor is $V_{\mathrm{rms}}^2/R = I_{\mathrm{rms}}^2 R = V_{\mathrm{rms}} I_{\mathrm{rms}}$. These expressions are analogous to those for d.c. voltage and current.
5. The r.m.s. value of a sinusoidal voltage of amplitude V_{a} is $V_{\mathrm{rms}} = V_{\mathrm{a}}/\sqrt{2}$.
6. The r.m.s. value of both periodic and non-periodic waveforms, such as signals and noise, can be measured either by an analogue meter or by a digital 'true r.m.s.' multimeter which takes samples, squares them, takes the average, and displays the square root.
7. Symbols for voltage are:

   V_{AB} D.C. voltage of node A with respect to node B. Where voltages are quoted with respect to ground (earth), V_{A} may be used instead.

   V_{ab} Amplitude of sinusoid (also used for r.m.s. value; extra subscripts are used when the difference is not clear, as in (5) above).

   v_{AB} Instantaneous value of total voltage.

   v_{ab} Instantaneous value of the variable (signal) component of the voltage.

   Current symbols are I and i with similar interpretations (except that subscripts are usually only single letters).

8. The frequency spectrum of a periodic waveform is the Fourier series, or line spectrum. Each line is a harmonic of the fundamental frequency, which is the reciprocal of the repetition period.

   The mean-square value of the waveform can be calculated by adding the squares of the r.m.s. values of its Fourier components.
9. The frequency spectra of signals and noise are continuous spectra, or power density spectra, in which the power density in V^2/Hz is plotted against frequency.
10. A capacitor has a capacitance defined as $C = Q/V$, where Q is the charge stored when the voltage is V. The capacitance can be calculated from $C = \varepsilon_0 \varepsilon_r A/d$ (in farads (F)), where ε_0 is the permittivity of free space, ε_r is

the relative permittivity of the dielectric, A is the area of the plates, and d is the distance between them.

11. The instantaneous current in a capacitor is $i = C(\mathrm{d}v/\mathrm{d}t)$. You cannot change the voltage suddenly without providing a large current flow.
12. The reactance of a capacitor is $1/\omega C$. Sinusoidal current through it leads the voltage across it by $\pi/2$ radians (90°).
13. Low-pass and high-pass filters can be made with simple CR networks. The capacitor's reactance approaches infinity at low frequencies, and approaches zero at high frequencies.
14. The step responses of CR networks are:
    low-pass: $v_O = V_i[1 - \exp(-t/CR)]$
    high-pass: $v_O = V_i\exp(-t/CR)$
    CR is the time constant.
15. The instantaneous voltage across an inductor is $v = L(\mathrm{d}i/\mathrm{d}t)$, where L is its inductance, measured in henrys (H). You cannot change the current suddenly without generating a large back e.m.f.
16. The reactance of an inductor is ωL. Sinusoidal current through it lags the voltage across it by $\pi/2$ radians (90°).
17. Low-pass and high-pass filters can be made with simple LR networks. The inductor's reactance approaches zero at low frequencies, and approaches infinity at high frequencies.
18. The step responses of simple LR networks are:
    low-pass: $v_O = V_i[1 - \exp(-tR/L)]$
    high-pass: $v_O = V_i\exp(-tR/L)$
    L/R is the time constant. Note that these expressions are the same as for the simple CR networks, with the time constant CR replaced by L/R.
19. Transformers consist of two or more coils of wire (inductors) placed close together around a common core so that the flux generated by the magnetizing current i_m in the primary coil (called the primary winding) is shared by all the coils. In an ideal transformer with infinite primary inductance (so that $i_m = 0$), $e_1/N_1 = e_2/N_2 = e_3/N_3$, etc. With finite primary inductance L_1, the magnetizing current i_m, whose value is given by $e_1 = L_1(\mathrm{d}i_m/\mathrm{d}t)$, has to be added to i_1. Where a sinusoidal voltage is applied to the primary, the input resistance of a transformer with a secondary load R_L is $(N_1/N_2)^2R_L$.

Answers to self-assessment questions

1. To determine the frequency of a sinusoidal waveform you must first determine the period of the repeating wave, in seconds, and then calculate the reciprocal. The answers are, therefore:
   (a) Period = 0.1 s. So frequency = 10 Hz. Amplitude = 1 V
   (b) Period = 10 ms = (10)/1000 s = 0.01 s. So frequency = 100 Hz. Amplitude = 2 V
   (c) Period = 20 ms = (20)/1000 s = 0.02 s. So frequency = 50 Hz. Amplitude = 3 V
   (d) Period = 2.5 ms = (2.5)/1000 s = 0.0025 s. So frequency = 400 Hz. Amplitude = 20 V

   All the waveforms are sine waves, because $v = 0$ when $t = 0$ and the voltage initially increases as t increases.

2. In this book the term 'signal' is reserved for those voltage or current waveforms that carry some new information. Simple, repeating waveforms, like sinewaves, are not being called signals. The three waveforms in Fig. 2.6 carry a limited amount of information during one cycle of the waveform, but then, since each subsequent cycle is a repetition of the first, no new information is added. You can argue endlessly about whether, say, the waveform showing a completely regular heart beat conveys new information as the waveform continues, but it is not really worth the trouble. Certainly you display such a waveform only because it may, at some time, change its waveform. Also in *information theory*—a topic not discussed in this book—periodic signals are said to convey very little information after the first cycle because you can recode all the rest of the waveform in one symbol meaning 'no change'. In this book we will continue to use the term 'waveform' to refer to waveforms like those of Fig. 2.6, but we will not be very concerned about it if you do not follow suit!

3. The waveform of Fig. 2.8c is 0.4 ms in advance of that of Fig. 2.8a. This time difference is 0.4 ms/2.0 ms = 0.2 cycle of the waveform. So $\phi = 0.2 \times 2\pi = 0.4\pi$ rad = 72°. The phase difference between waveforms (b) and (c) is the sum of those already calculated, including that in the worked example. So waveform (c) is $(0.4 + 0.25)\pi = 0.65\pi$ rad = 117° in advance of waveform (b).

4. For the 2 MHz waveform the period of one cycle is 0.5 μs, so the delayed sinusoid lags the original one by a phase angle of $\phi = (0.4/0.5)2\pi = 1.6\pi$ rad or 288°. For the 5 MHz waveform the period is 0.2 μs. So $\phi = (0.4/0.2)2\pi = 4\pi$ rad or 720°.

5. Waveform (b) differs in amplitude only (Answer B).
   Waveform (c) differs in amplitude and phase (Answer D).
   Waveform (d) differs in phase only (Answer A).
   Waveform (e) differs in frequency only (Answer C).
   Waveform (f) differs in amplitude and frequency (Answer E). It also differs in phase at time $t = 0$, but phase difference as such is not meaningful for waveforms of different frequency.

6. (a) The half-wave rectified sinusoid is shown in Fig. 2.42b. Its waveform has a repetition interval T equal

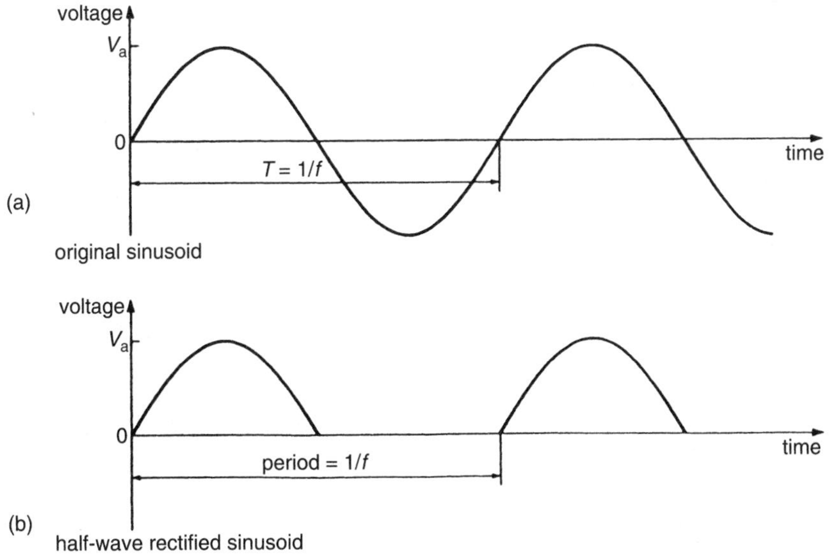

Fig. 2.42 A half-wave rectified sinusoid.

to that of the original sinusoid in the upper half of the figure. We can write

$v = V_a \sin \omega t$ for values of ωt from 0 to π

$v = 0$ for values of ωt from π to 2π

Denoting ωt by θ as before, the period is now equal to 2π and the area contained by the waveform and the horizontal axis over one period is

$$A = \int_0^{\pi} V_a \sin\theta \, d\theta = 2V_a$$

To find the average value over one period we must divide the area by the period, giving

$$v = 2V_a/2\pi = V_a/\pi$$

7. The r.m.s. value $V_{rms} = V_a/\sqrt{2}$, so the amplitude $V_a = \sqrt{2}\ V_{rms} = \sqrt{2} \times 240\ \text{V} \approx 340\ \text{V}$.
Power $P = (V_{rms})^2/R$, so $R = (V_{rms})^2/P = (240^2/100)\Omega = 576\ \Omega$.

8. (a) For a sinusoid, the r.m.s. value is (amplitude)/$\sqrt{2}$. Thus a 100 V r.m.s. sinusoid has an amplitude of $\sqrt{2} \times 100\ \text{V} \approx 141.4\ \text{V}$.
(b) The meter indicates the average value of the rectified sinusoid, which is (amplitude) $\times 2/\pi = 141.4 \times 2/\pi\ \text{V} = 90\ \text{V}$.
(c) The range resistor must be modified by the factor 0.9. For instance, a rectified average of 90 V would result in a reading of 90 V/0.9 = 100 V, which is the correct r.m.s. value.

9. In each case, the mean-square voltage is calculated by finding the area under the waveform of the voltage-squared, over one period, and then dividing by the period:

(a) From $t = 0$ to $t = T$, $v^2 = 100\ \text{V}^2$, so the area is $100\ T$
Mean-square voltage $\overline{v^2} = 100\ T/T = 100\ \text{V}^2$.

$$V_{rms} = \sqrt{100} = 10\ \text{V}$$

(b) From $t = 0$ to $t = 3T/4$, $v^2 = 36$ so the area is $36 \times 3T/4 = 27T$. From $t = 3T/4$ to $t = T$, $v^2 = 100$, so the area is $100T/4 = 25T$. Thus the total area is $52T$. Thus the mean-square voltage is $52T/T = 52\ \text{V}^2$.

$$V_{rms} = \sqrt{52}\ \text{V} = 7.2\ \text{V}.$$

(c) $v = 10t/T$, so $v^2 = [10t/T]^2$. In this case we have to integrate to find the average value:

$$\overline{v^2} = \frac{1}{T}\int_0^T (10t/T)^2 dt$$
$$= \frac{100}{T^3}\int_0^T t^2 dt = \frac{100}{T^3}[t^3/3]_0^T$$
$$= (100/3)\ \text{V}^2 = 33.3\ \text{V}^2$$
$$V_{rms} = \sqrt{(33.3\ \text{V}^2)} \approx 5.77\ \text{V}$$

10. (a) $n = 16$. (b) The sum = -1 V. (c) Average = -1 V/16 = -62.5 mV. (d) Sum of the squares = 67 V^2. (e) Mean-square value = 67 V^2/16 = 4.19 V^2. (f) R.M.S. value = $\sqrt{(4.19\ \text{V}^2)} = 2.05$ V.

11. The average value, or d.c. level, is now 2 V, because the waveform has simply been 'lifted' by 2 V. The new waveform can be considered as a symmetrical square-wave of 4 V peak-to-peak, or 2 V peak, added to a d.c. level of 2 V. So the new k_0 is 2 V, and the amplitudes of all the harmonics are doubled, to $8/\pi$ V, $8/(3\pi)$ V, $8/(5\pi)$ V, etc.

12. Figure 2.43 shows the required bar chart. The bandwidth is the highest frequency minus the lowest frequency, but includes both frequencies, namely $7 \times 250 - 250 = 6 \times 250 = 1500$ Hz.

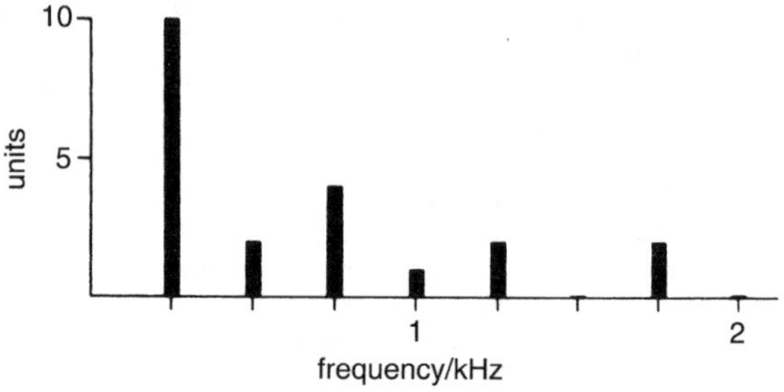

Fig. 2.43

13. (a) From 100 Hz to 1 kHz, spectral voltage density ranges from zero to $10^{-3}\ \text{V}^2/\text{Hz}$, so total $\overline{v^2}$ = area under triangular curve:

$$\overline{v^2} = \frac{1}{2} \times \text{base} \times \text{height}$$
$$= \frac{1}{2} \times 900\ \text{Hz} \times 10^{-3}\ \text{V}^2/\text{Hz} = 0.45\ \text{V}^2$$

From 1 kHz to 5 kHz, spectral voltage density = $10^{-3}\ \text{V}^2/\text{Hz}$, so total $\overline{v^2} = 10^{-3}\ \text{V}^2/\text{Hz} \times 4\ \text{kHz} = 4\ \text{V}^2$.
Overall $\overline{v^2}$ from 100 Hz to 5 kHz = $0.45\ \text{V}^2 + 4\ \text{V}^2 = 4.45\ \text{V}^2$.
(b) From 5 kHz to 10 kHz, the spectral voltage density ranges from $10^{-3}\text{V}^2/\text{Hz}$ to zero. So that $\overline{v^2}$ = area under triangular curve:

$$\overline{v^2} = \frac{1}{2} \times \text{base} \times \text{height}$$
$$= \frac{1}{2} \times 5\ \text{kHz} \times 10^{-3}\ \text{V}^2/\text{Hz}$$
$$= 2.5\ \text{V}^2$$

Overall $\overline{v^2} = 4.45\ \text{V}^2 + 2.5\ \text{V}^2 = 6.95\ \text{V}^2$
R.M.S. voltage = 2.6 V.

14. The simplest way to test for linearity is to measure the small-signal capacitance when a d.c. voltage as well as a small test signal is applied to the capacitor. If the capacitance varies as the d.c. bias is varied the capacitor is a non-linear one.

Alternatively, though less sensitively, and providing your measuring equipment is itself 'linear' (i.e. gives the same results with a linear capacitor over a range of signal levels), you can vary the level of your signal and check whether the capacitance is constant. If it is not constant, the capacitor is non-linear.

15. Capacitive reactance is

$$X_C = 1/\omega C = 1/(2\pi fC) = 1/(2\pi \times 60 \text{ Hz} \times 1 \ \mu\text{F})$$
$$= 2.653 \text{ k}\Omega$$

$$I = V/X_C = 120\,\text{V}/2.652 \text{ k}\Omega$$
$$\approx 45.2 \text{ mA}$$

16. (i) $v_c = V_i[1 - \exp(-t/CR)]$
$= 4 \text{ V } [1 - \exp(-2)] = 4 \text{ V} \times 0.8647 = 3.46 \text{ V}$

(ii) Time constant $CR = 100 \text{ pF} \times 1 \text{ k}\Omega = 100$ ns. So 10% to 90% rise time $t_r = 2.2\ CR = 220$ ns.

(iii) $v_c = V_i[1 - \exp(-t/CR)]$. Rearranging: $\exp(-t/CR) = (V_i - v_c)/V_i$. Taking natural logs:

$$t/CR = -\ln[(V_i - v_c)/V_i]$$
$$t = CR \ln[V_i/(V_i - V_c)]$$
$$= 100 \text{ ns} \times \ln 2$$
$$= 100 \text{ ns} \times 0.693 = 69.3 \text{ ns}$$

17. (a) Electric generators are machines that produce an e.m.f. by rotating coils of wire within a field of magnetic flux. The magnitude of the e.m.f. depends on Farady's law. If the magnetic field giving rise to the flux is created by another current, then the first principle is also invoked. And if the generator has an iron core, the second principle is involved.

(b) Electromagnets, such as those used for lifting scrap iron, nearly always have iron cores, so the first two principles are applied in their design.

18. The reactance of a coil is equal to ωL, so in this case $X_L = 2\pi \times 500 \text{ Hz} \times 0.3 \text{ H} = 942\ \Omega$.

Since the inductance of a tightly wound coil is proportional to the square of the number of turns in it, the inductance of half the coil will be a quarter that of the whole coil, namely 0.075 H. At 1000 Hz, which is double the previous frequency, the reactance will therefore be half that of the whole coil at 500 Hz, namely 471 Ω.

19. The time constant is $L/R = 10 \text{ mH}/2\ \Omega = 5$ ms. As shown in Fig. 2.39, it takes 3 time constants for an exponential rise to reach 95% of its final value. (This applies whether it is a *CR* circuit, an *LR* circuit, or any other exponential.) Thus the current in the coil takes $3 \times 5 \text{ ms} = 15$ ms.

20. By Equation 2.13, $e_1 = (N_1/N_2)\ e_2$; and by Equation 2.15 $i_1 = (N_2/N_1)\ i_2$. Therefore,

$$i_1/e_1 = (N_2/N_1)^2 \times i_2/e_2$$

But $N_1/N_2 = 1/4$ and $e_2/i_2 = 4000\ \Omega$, so

$$e_1/i_1 = 4000/16 = 250\ \Omega$$

This is evidently the resistance as seen by the source. In general therefore the 'input resistance'—as it is called—of a transformer with a load is the load resistance multiplied by $(N_1/N_2)^2$.

21. (a) If the flux is the same through all three coils, then, as before, the induced e.m.f.s are proportional to the numbers of turns. So $e_1/N_1 = e_2/N_2 = e_3/N_3$.

(b) Since the output power is equal to the input power if the transformer dissipates no power itself, it follows that $e_1 i_1 = e_2 i_2 + e_3 i_3$. (Notice the plus sign rather than an equals sign on the secondary side of the equation.) By substituting e_1N_2/N_1 for e_2, and e_1N_3/N_1 for e_3 in this equation, it reduces to $i_1N_1 = i_2N_2 + i_3N_3$.

22. (a) Let the amplitudes of the output voltages be V_2 and V_3. Then from Equation 2.14: $120/600 = V_2/300 = V_3/200$. So $V_2 = 60$ V and $V_3 = 40$ V.
Since $R_2 = 1$ kΩ and $I_2 = V_2/R_2$, then $I_2 = 60$ mA.
Similarly, with $R_3 = 500\ \Omega$, $I_3 = 40/500 = 80$ mA.

(b) From Equation 2.15—replacing instantaneous values by amplitudes—the total current into the transformer is given by $I_1N_1 = I_2N_2 + I_3N_3$. So $600\ I_1 = 300 \times 0.06 + 200 \times 0.08$.
Therefore $I_1 = (18 + 16)/600 = 56.7$ mA.
From Section 2.3 you know that to calculate power you must use r.m.s. values of voltage and current, which are the amplitudes divided by $\sqrt{2}$. The total input power is therefore

$$V_{in(rms)} \times I_{1(rms)} = \frac{120}{\sqrt{2}} \times \frac{0.0567}{\sqrt{2}} = 3.402 \text{ W}$$

3 Phasor analysis of a.c. circuits

AIMS

To explain the mathematical techniques which are used to analyse linear circuits containing resistive and reactive components and sinusoidal sources. This will involve the following subsidiary aims.

1. To introduce phasor diagrams.
2. To introduce the operator j.
3. To explain the use of complex numbers to represent phasors.
4. To explain how a.c. circuits can be analysed using phasors.
5. To introduce decibels.
6. To describe how Bode plots are used to display the frequency-response properties of linear circuits and amplifiers.
7. To show how to sketch approximations to the Bode plots of individual linear circuits and for circuits buffered by an ideal voltage amplifier.
8. To describe the frequency-response properties of electrical circuits containing resistance, capacitance and inductance.
9. To introduce the principle of duality and to give examples of dual circuits.

GENERAL OBJECTIVES

After studying this chapter you should be able to understand the meanings of the following terms:

- a.c. circuit analysis
- admittance
- amplitude-response plot
- Argand diagram
- bel
- Bode plot
- break point
- complex algebra
- complex numbers
- conductance
- corner frequency
- decibel
- double-lag circuit
- dual circuit
- dual quantity
- dynamic impedance
- imaginary part
- impedance
- phase-response plot
- phasor
- phasor addition
- phasor and operator division
- phasor and operator multiplication
- phasor components
- phasor diagram
- phasor magnitude and angle

- phasor operator
- Q-factor
- real part
- resonant frequency
- series and parallel resonant circuits
- single-lag circuit
- single-lead circuit
- susceptance
- the operator j
- voltage transfer function
- 3 dB point

SPECIFIC OBJECTIVES

1. Understand the operation of elementary low-pass and high-pass networks containing a resistor and a capacitor or a resistor and an inductor.
2. Apply the techniques of phasor analysis to circuits containing resistance, reactance and sinusoidal sources, to find:
   (i) the voltage transfer function;
   (ii) the voltages and currents;
   (iii) the input impedance and input admittance.
3. Remember and use the formulae for expressing power ratios, sinusoidal voltage ratios and sinusoidal current ratios in decibels and vice versa.
4. Remember and use the formulae for the general voltage transfer functions of low-pass single-lag and high-pass single-lead circuits.
5. Plot and/or take values from the Bode plots of low-pass and high-pass circuits.
6. Work out the overall transfer function of circuits buffered by an ideal voltage amplifier and sketch linear approximations to their overall Bode plots.
7. Calculate the resonant frequency, Q-factor and bandwidth of a series resonant circuit.
8. Arrange the transfer function of a resonant circuit to enable comparison with the standard form.
9. Describe the main features of the Bode plots for a band-pass series resonant circuit.
10. Deduce the properties of a parallel resonant circuit from a knowledge of the properties of its dual circuit.

3.1 PHASORS

In Chapter 2, simple capacitor–resistor (CR) and inductor–resistor (LR) filters are considered, but their full a.c. analysis is deferred to this chapter.

Consider again the low-pass CR network, shown again in Fig. 3.1. The instantaneous voltages around the circuit are related by

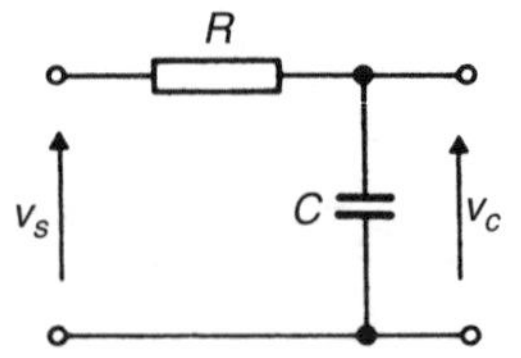

Fig. 3.1

$$v_S = iR + \frac{1}{C}\int i\mathrm{d}t$$

With a sinusoidal input voltage, a sinusoidal current flows, so we can substitute

$$i = I\sin \omega t$$

(taking the phase of the current as the reference) and

$$\int i\mathrm{d}t = -\frac{I}{\omega}\cos \omega t$$

Thus,

$$\begin{aligned} v_S &= IR\sin \omega t - \frac{I}{\omega C}\cos \omega t \\ &= I(R\sin \omega t - X_C \cos \omega t) \\ &\qquad \text{(where } X_C = 1/\omega C \text{ as explained in Chapter 2)} \\ &= I[R\sin \omega t + X_C \sin(\omega t - \pi/2)] \end{aligned}$$

This expression for the sum of sinusoidal voltages across the resistor and the capacitor is not very helpful; it is complicated by the phase lag of the voltage across the capacitor. Certainly, v_S must be sinusoidal, although this is not obvious from the expression. In fact, it turns out that the sum of any two (or more) sinusoids *of the same frequency* is always another sinusoid, at that frequency. The question that arises is: what is amplitude, and what is its phase angle?

The next problem is: even if we know the amplitude and phase of v_S in terms of i (and hence i in terms of v_S), how can we find the amplitude and phase of the output voltage (across C)?

This task is greatly simplified by combining the two parameters, amplitude and phase, in a single parameter known as a **phasor**. One of the advantages of the phasor approach is that we can make use of elementary line diagrams—**phasor diagrams**—which show at a glance the relationship between the various sinusoidal voltages and currents within a circuit.

3.1.1 Phasor diagrams

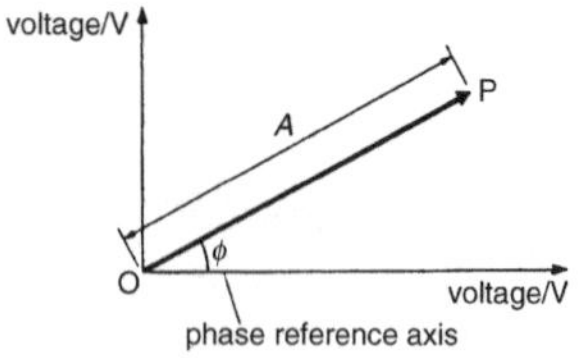

Fig. 3.2 A phasor.

The line OP in Fig. 3.2 is the phasor representing a sinusoidal voltage. We can draw an arrow at the phasor tip to emphasize that a phasor is like a *vector* in that it can be visualized as a line whose length and direction both have a meaning. Notice that because OP is a voltage phasor, the axes in Fig. 3.2 are labelled in units of volts.

In the diagram the length between O and P is proportional to the amplitude of the sinusoid; its value is called the **magnitude** of the phasor. The phase of the sinusoid relative to a chosen reference is represented by the angle ϕ, which is measured from the horizontal or **phase reference** axis. In a phasor analysis, ϕ is known as the **phase angle**, or simply the **angle** of the phasor. A *positive* phase angle is measured in an *anticlockwise* direction on a phasor diagram.

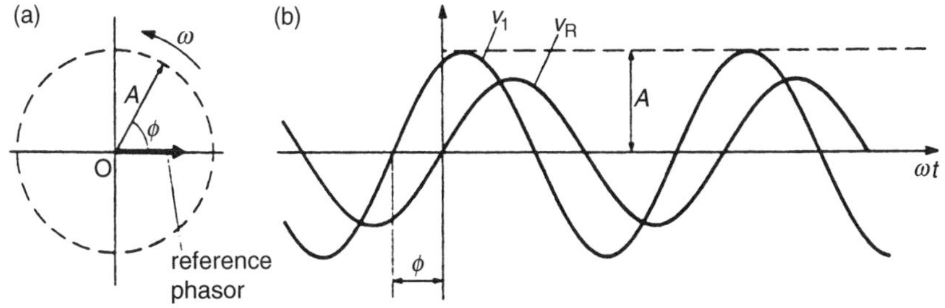

Fig. 3.3 (a) A rotating phasor. (b) The projection of the rotating phasor on the y-axis is the sinusoidal waveform V_1 which leads the reference sinusoid V_R by the phase angle ϕ.

In order to generate the sinusoid represented by the phasor in Fig. 3.2 we can rotate the phasor in an anticlockwise direction about the point O. The projection of the phasor on the vertical axis as it rotates with constant angular frequency will then vary as the sinusoid v_1 shown in Fig. 3.3b. The waveform that is generated has an amplitude that is equal to the phasor magnitude, A, and its phase—as compared with the reference sinusoid v_R—is represented by the angle ϕ. Notice that the reference sinusoid is the waveform that would be generated by rotating a phasor directed along the phase reference axis in Fig. 3.3a. Since both phasors would be rotated at the same angular frequency, the phase angle between them remains fixed. It is evident from the direction of rotation that sinusoid v_1 must lead the reference v_R by the angle ϕ.

In this interpretation, therefore, the phasor diagram is rather like a 'snapshot', illustrating the position of a rotating radius vector at a particular instant of time. The frequency of the sinusoid is not part of the phasor model, although the magnitude and angle of a phasor will usually be functions of the frequency under consideration.

Worked example

At a particular frequency, the sinusoidal voltage input to a linear circuit has an amplitude of 5 V. The steady-state output has an amplitude of 3 V and lags the input by $\pi/3$ radians. Draw a phasor diagram to show the relationship between the input and output voltages.

Because the phase of the output has been measured with respect to the input, we can use the input as a phase reference. The input phasor of magnitude 5 V therefore points along the phase reference axis as shown in Fig. 3.4.

Because the phase of the output lags that of the input, the output phasor must be drawn at an angle of $\pi/3$ or 60°, measured clockwise from the phase of the reference axis.

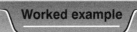

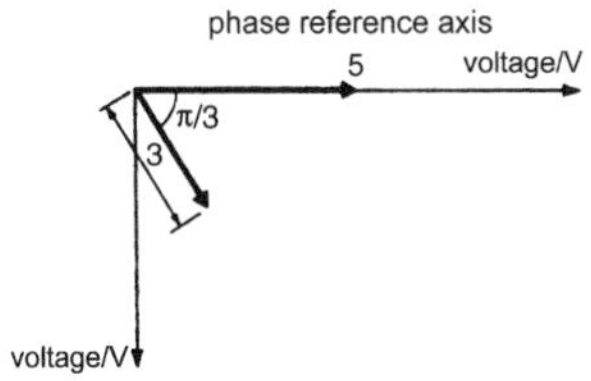

Fig. 3.4

When two or more sinusoids have the same frequency they can all be represented as phasors on the same diagram so as to show the relationship between them.

Self-assessment question

1 (a) The following voltages are measured in a linear circuit. Draw a phasor diagram to show how they are related to the input sin ωt V:

$v_1 = 3 \sin \omega t$ V, $v_2 = 2 \sin(\omega t + \pi/3)$ V

$v_3 = 4 \sin(\omega t - \pi)$ V, $v_4 = \sin(\omega t - 5\pi/6)$ V

$v_5 = 2 \sin(\omega t + 5\pi/6)$ V

(b) Use the phasor diagram to find the phase difference between v_4 and v_5. Does v_4 lead v_5 or vice versa?

(c) Where would the phasors representing $v_6 = \cos \omega t$ V and $v_7 = \cos(\omega t - \pi/4)$ V be located on this diagram?

3.1.2 Phasor notation

The phasors used in circuit analysis always represent sinusoidal voltages and currents. They can be conveniently referred to in writing as a magnitude followed by an angle. If a single symbol is used to represent a phasor, then it is printed in ***bold italic*** type. For example, in Fig. 3.4, the input phasor points along the phase reference axis and can be written as $\boldsymbol{V}_{\text{in}} = 5 \angle 0$. The output phasor, which I shall call $\boldsymbol{V}_{\text{out}}$, is given by $\boldsymbol{V}_{\text{out}} = 3 \angle -\pi/3$ or $3 \angle -60°$.

Worked example

The voltage at the input to a linear circuit is a sinusoid represented by the phasor $V_{\text{in}} = 5\angle 0$ V. The output voltage has an amplitude equal to one-tenth of the input and leads the input by 82°. What is the output phasor $\mathbf{V}_{\text{out}}$?

The amplitude of the output is 5/10 = 0.5 V. The phase of the output is 82°, measured with respect to the input. With the input represented by the phasor 5∠0, the output phasor will be

$V_{\text{out}} = 0.5 \angle 82°$ V

In handwritten notes and calculation you can use the tilde (˜) to distinguish between phasor and non-phasor quantities. Thus, whenever this text uses the bold italic $\boldsymbol{V}$ to denote phasor, you should use $\tilde{\text{V}}$. $\boldsymbol{I}$ becomes $\tilde{\text{I}}$, and so on. Thus you now have three notations which serve to represent different aspects of a sinusoidal voltage: a lower-case italic v represents the instantaneous value, an upper-case italic V denotes the amplitude and $\boldsymbol{V}$ or $\tilde{\text{V}}$ represents a phasor whose amplitude and phase are both specified. A final point: remember that phasors, like instantaneous values and amplitudes, are properly expressed in units of volts or amps.

Self-assessment question

2 The following voltages and currents were all measured with respect to sinewave references. Represent them in magnitude and angle form and sketch the corresponding phasor diagrams.

(a) $v = 6 \sin 10t$ V.
(b) $i = 7.5 \sin(1050t + \pi/4)$ A.
(c) $i = \omega CV \cos \omega t$.
(d) $v = 3.6 \sin(500\pi t - 3\pi/4)$.

Self-assessment question

3 The sinusoidal voltage across a circuit component and the current flowing through it are represented by phasors $\boldsymbol{V} = V\angle\phi$ and $\boldsymbol{I} = I\angle\theta$. Draw phasor diagrams to show the relationship between $\boldsymbol{V}$ and $\boldsymbol{I}$ when the component is (a) a resistor, (b) a capacitor and (c) an inductor.

3.1.3 Phasor addition

Suppose that two sinusoids v_1 and v_2 of the same frequency are added together to give a third sinusoid v_3:

$$v_3 = v_1 + v_2$$

If v_1 and v_2 are represented by phasors $\boldsymbol{V}_1$ and $\boldsymbol{V}_2$, then v_3 will have a phasor $\boldsymbol{V}_3$ given by the **phasor sum**:

$$\boldsymbol{V}_3 = \boldsymbol{V}_1 + \boldsymbol{V}_2$$

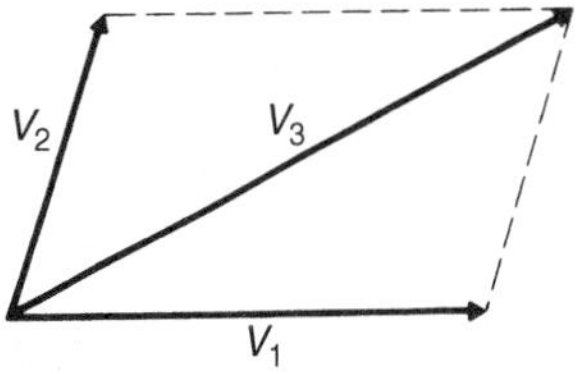

Fig. 3.5 Parallelogram construction to find the phasor sum $\boldsymbol{V}_3$ of two phasors $\boldsymbol{V}_1$ and $\boldsymbol{V}_2$.

The sum of two phasors, known as the **resultant**, can be found graphically using the *parallelogram rule* illustrated in Fig. 3.5. You may recognize that this is the same construction that is used in vector analysis to combine two vectors.

We will not dwell too long on Fig. 3.5 because one of the features of circuit analysis is that we can usually arrange to work with phasors at *right angles*. The low-pass filter discussed earlier is an example of this. The parallelogram then becomes a rectangle and calculations are greatly simplified. A particular example of two phasors differing in phase by 90° is shown in Fig. 3.6.

Worked example

Using your knowledge of geometry and trigonometry, what is the amplitude and phase, expressed in terms of V_1 and V_2, of the sinusoidal voltage represented by phasor V_3, in Fig. 3.6?

The amplitude is given directly by Pythagoras' theorem:

$$V_3 = \surd(V_1^2 + V_2^2)$$

The phase angle has a tangent equal to V_2/V_1. So $\tan\phi = V_2/V_1$ and the angle is given by

$$\phi = \tan^{-1}(V_2/V_1)$$

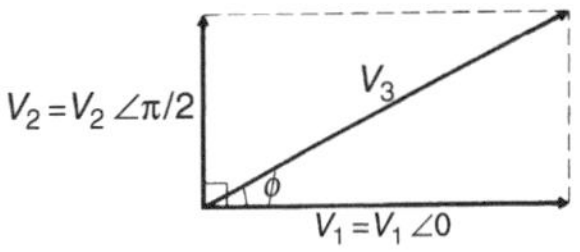

Fig. 3.6

When using phasors $\boldsymbol{V}$ and $\boldsymbol{I}$ it is sometimes helpful to use the notation $|\boldsymbol{V}|$ and $|\boldsymbol{I}|$ to refer to their magnitudes. For example, if $\boldsymbol{I} = 5\angle\pi/4$ A, then $|\boldsymbol{I}| = 5$ A. Also, the symbol $\angle$ can be placed in front of a phasor to denote its angle. Thus, if $\boldsymbol{V} = 4\angle\pi/3$, then $\angle\boldsymbol{V} = \pi/3$. Using this notation for the phasors in Fig. 3.6, we have

The **magnitude** of a sinusoidal voltage or current is the same as its **amplitude**.

$$V_1 = |\boldsymbol{V}_1| \quad \text{and} \quad V_2 = |\boldsymbol{V}_2|$$

$$|\boldsymbol{V}_3| = \surd(V_1^2 + V_2^2)$$

$$\angle\boldsymbol{V}_3 = \phi = \tan^{-1}(V_2/V_1)$$

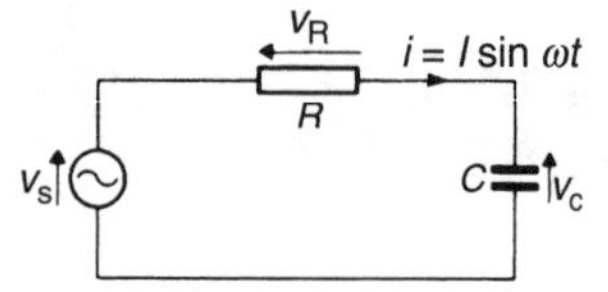

Fig. 3.7

3.1.4 Phasor analysis of the low-pass *CR* circuit

The *CR* circuit is shown again in Fig. 3.7. As shown at the start of this chapter, the input voltage is given by:

$$v_S = IR \sin \omega t + IX_C \sin(\omega t - \pi/2) \tag{3.1}$$

For phasor analysis, the three voltages v_s, v_R and v_C will be represented by the phasors $\boldsymbol{V}_S$, $\boldsymbol{V}_R$ and $\boldsymbol{V}_C$. Because the current, i, is common to both components in the circuit, we can use it to define a phase reference for the calculations and represent it by the phasor $\boldsymbol{I} = I \angle 0$.

Now, the phasor magnitudes $|\boldsymbol{V}_R|$ and $|\boldsymbol{V}_C|$ are equal to IR and IX_C respectively, these being the amplitudes of v_R and v_C.

The resistor voltage v_R is in phase with the current. This means that if we use the current as the phase reference, phasor $\boldsymbol{V}_R$ will point along the phase reference axis, as shown in Fig. 3.8. Since the capacitor voltage v_C lags the current by $\pi/2$ radians, its phasor $\boldsymbol{V}_C$ points downwards in Fig. 3.8.

Referring to Equation 3.1 above, we see that the phasor $\boldsymbol{V}_s$ will be equal to the phasor sum or resultant of the other two phasors:

$$\boldsymbol{V}_S = \boldsymbol{V}_R + \boldsymbol{V}_C$$

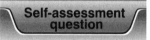

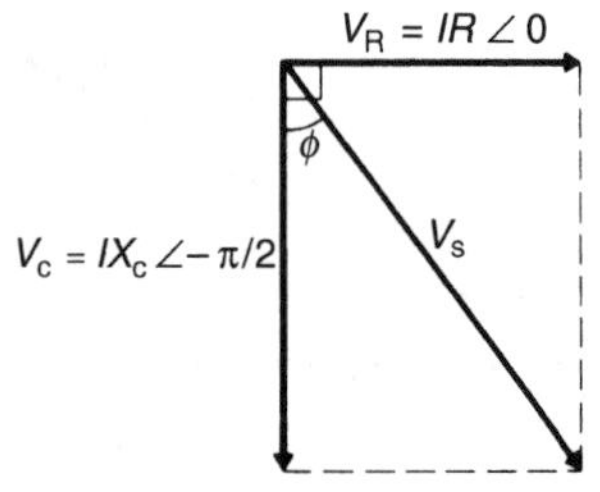

Fig. 3.8 Phasor diagram for the *CR* circuit.

The graphical construction to find $\boldsymbol{V}_S$ is shown in Fig. 3.8. However, you have already seen how this calculation can be carried out directly.

Self-assessment question

4 Using the rules for the addition of two phasors at right angles, work out (a) the magnitude of phasor $\boldsymbol{V}_S$, representing the amplitude of the input voltage v_S; (b) the angle ϕ by which the output voltage v_C lags the input.

Using results from Self-assessment question 4 we can now find the voltage transfer ratio of the low-pass *RC* network, expressed as the ratio V_C/V_S together with the phase angle ϕ.

The amplitude ratio V_C/V_S

The amplitude ratio is

$$\frac{V_C}{V_S} = \frac{IX_C}{I\sqrt{(X_C^2 + R^2)}} = \frac{X_C}{\sqrt{(X_C^2 + R^2)}}$$

An alternative form of this expression can be obtained by substituting $X_C = 1/\omega C$. First, divide top and bottom of the expression by X_C (which means divide by X_C^2 inside the square root), giving

$$\frac{V_C}{V_S} = \frac{1}{\sqrt{(1 + R^2/X_C^2)}} = \frac{1}{\sqrt{(1 + \omega^2 C^2 R^2)}}$$

This expression should represent the graph of amplitude ratio versus frequency which is shown in Chapter 2. This graph is repeated here in more exact form in Fig. 3.9. Note that the frequency is plotted on a logarithmic scale, and so is the amplitude ratio. Note especially the following features.

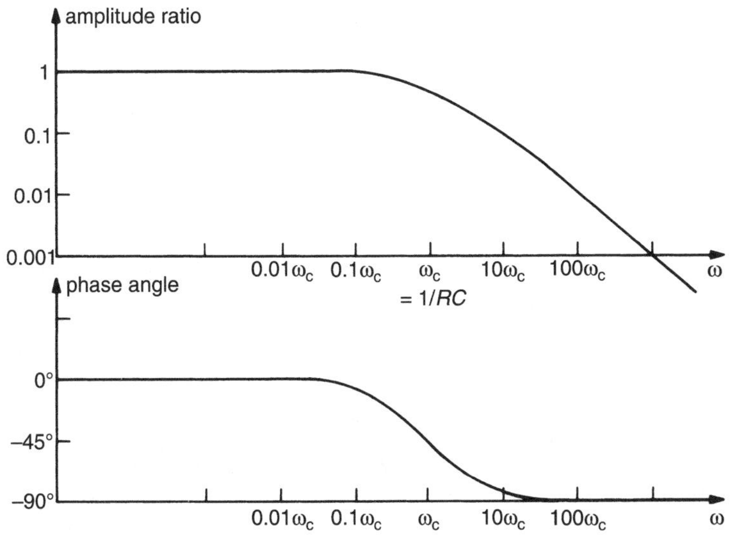

Fig. 3.9

(a) At frequencies which are sufficiently low, such that $\omega^2C^2R^2 >> 1$, the amplitude ratio ≈ 1.
(b) At frequencies which are sufficiently high, such that $\omega^2C^2R^2 >> 1$, the amplitude ratio approaches $1/\omega CR$, which is inversely proportional to frequency. (Taking logs: $\log(V_C/V_S) = \log(1/\omega CR) = -\log(\omega CR)$, so the log–log plot produces a straight line.)
(c) At the frequency at which $\omega^2C^2R^2 = 1$, that is when $\omega = 1/CR$, the amplitude ratio is $1/\sqrt{2}$. This is called the *cut-off* frequency of the network.

The angular cut-off frequency, denoted by the symbol ω_c, is defined by $\omega_c = 1/RC$. It corresponds to the frequency at which the reactance of the capacitor becomes equal to the circuit resistance. Thus, when $\omega = \omega_c$, we have the relationship

$$R = 1/\omega_c C$$

The phase shift of the low-pass *CR* circuit

You should have found in Self-assesment question 4 that the output voltage lags the input voltage by the angle $\phi = \tan^{-1}(\omega CR)$.

Worked example

What is the phase difference between the input and the output, (a) at the cut-off frequency and (b) at frequencies much higher than the cut-off frequency?

(a) At cut-off, $\omega = 1/CR$ and $\omega CR = 1$. The phase difference is therefore

$$\phi = \tan^{-1}1 = \pi/4 \text{ or } 45°$$

Because the output voltage lags the input voltage, the phase of the output will be $-45°$ with respect to the input.

(b) At frequencies much higher than the cut-off frequency, we have

$$\phi = \tan^{-1}(\omega CR), \quad \text{where } \omega CR \gg 1$$

Now, the inverse tangent of a large number is always close to (but less than) $\pi/2$ or 90°. As ω increases, therefore, the phase of the output will approach −90° with respect to the input.

You will see that this behaviour is in agreement with the graph of phase shift versus frequency in Fig. 3.9.

Example

The resistor and capacitor in the low-pass *CR* network are 15 kΩ and 10 nF, respectively. A sinusoidal voltage of amplitude 3 V and frequency 2 kHz is applied to the input. What is the amplitude of the output voltage, and by what phase angle does it lag the input?

When the frequency f is 2 kHz, the angular frequency ω is $2\pi \times 2000 = 12\,566$ rad s^{-1}. The input voltage can therefore be expressed as

$$v_S = 3\sin(12\,566t)\text{V}$$

The output will then have the form

$$v_c = V_C \sin(12\,566t - \phi)$$

Now, $\omega CR = 12\,566 \times 10 \times 10^{-9} \times 15 \times 10^3 = 1.885$.

Notice that ωCR has no dimensions. Its value is a simple number and requires no units.

The phase angle $\phi = \tan^{-1}(\omega CR)$, so

$$\phi = \tan^{-1}1.885 = 1.08 \text{ rad or } 62°$$

The amplitude ratio is

$$\frac{V_C}{V_S} = \frac{1}{\sqrt{(1 + \omega^2 C^2 R^2)}}$$

We have already calculated ωCR to be 1.885, so

$$\frac{V_C}{V_S} = \frac{1}{\sqrt{(1 + 1.885^2)}} = 0.469$$

We thus obtain

$$V_C = 0.469 \times V_S = 0.469 \times 3 \text{ V} = 1.41 \text{ V}$$

The form of the output voltage v_C is therefore

$$v_C = 1.41\sin(12\,566t - 1.08) \text{ V}$$

Notice from this example that although the phasor diagram was used to obtain the general expression for V_C/V_S and for ϕ, the expressions themselves are used when we have specific values of ω, C and R. We do not construct a phasor diagram for each calculation.

Self-assessment question

5 Calculate the amplitude ratio and the phase shift associated with the low-pass *CR* network of Fig. 3.7 at a frequency of 1 kHz if the values of the resistor and capacitor are 4.7 kΩ and 0.47 μF, respectively.

If the input amplitude is 4 V, what is the output voltage amplitude at 1 kHz?

Self-assessment question

6 The low-pass *CR* circuit of Fig. 3.7 has a resistor of 10 kΩ. Calculate the value of the capacitor that must be used if the phase shift through the circuit is to be 30° at a frequency of 1kHz.

At what frequency will the phase shift be 45°?

The following Self-assessment question is rather a long one. It will give you the opportunity to apply your knowledge of phasors to the high-pass *LR* circuit considered earlier in Chapter 2 (Figure 2.36).

Self-assessment question

7 (a) Draw a phasor diagram to show the relationship between the voltages in the high-pass *LR* circuit. Hence, find expressions for the voltage amplitude ratio V_L/V_S and for the phase shift ϕ in the circuit.

(b) The angular cut-off frequency ω_c corresponds to the frequency at which the inductive reactance becomes equal to the resistance in the circuit.

Identify the angular cut-off frequency and show that the expression for the voltage amplitude ratio can be written in the form

$$\frac{V_L}{V_S} = \frac{\omega/\omega_c}{\sqrt{(1+\omega^2/\omega_c{}^2)}}$$

What are the values of the voltage amplitude ratio and the phase shift when $\omega = \omega_c$?

(c) Find the approximate expressions for V_L/V_s in the following frequency ranges, (i) $\omega \gg \omega_c$; (ii) $\omega \gg \omega_c$, and confirm that the circuit acts as a high-pass filter.

3.2 PHASOR MULTIPLICATION AND DIVISION

3.2.1 Multiplication: voltage gain

The ratio of the output voltage to the input voltage of a circuit is sometimes called the voltage gain of the circuit. This applies particularly to amplifier circuits designed to produce an output voltage amplitude greater than their input amplitude. However, even amplifiers have an amplitude ratio less than 1 outside their frequency range, so the term voltage gain can be applied to non-amplifying circuits too. For example, the low-pass *CR* circuit has a voltage gain (from Section 3.1.4):

A gain of less than 1 is often referred to as **attenuation**, and circuits which are designed to reduce signal voltage are called **attenuators**. Attenuation is the reciprocal of gain; for instance, a gain of $\frac{1}{2}$ is an attenuation of 2.

$$\mathbf{A} = \frac{V_{\text{out}}}{V_{\text{in}}} = \frac{1}{\sqrt{(1+\omega^2C^2R^2)}}\angle\phi$$

where $\phi = -\tan^{-1}(\omega CR)$.

Note that bold-faced type is used for all quantities with both magnitude and angle. Phasors such as ***V*** are in bold italic type, and **phasor operators** such as **A** are in bold upright type.

Clearly, **A** is a phasor quantity because it represents both the amplitude ratio and the phase shift between output and input. The voltage gain **A** is a phasor operator. We can write

$$\boldsymbol{V}_{\text{out}} = \mathbf{A} \times \boldsymbol{V}_{\text{in}}$$

This can be read as 'The gain A *operates* on the input voltage phasor $\boldsymbol{V}_{\text{in}}$ to produce the output voltage phasor $\boldsymbol{V}_{\text{out}}$'.

In the case of the *CR* circuit, the phasor multiplication represented by this operator thus involves *multiplication* of the input amplitude by the amplitude ratio, $1/\sqrt{(1 + \omega^2 C^2 R^2)}$, and a *shift of phase* by the angle $\phi = -\tan^{-1}(\omega CR)$.

If the input phasor is represented by $\boldsymbol{V}_{\text{in}} = V_{\text{in}} \angle \theta$, with a relative phase angle of θ with respect to some reference, then the circuit shifts the phase by an additional angle ϕ, producing a total relative phase of $\theta + \phi$. So, in general:

Phasor multiplication of two quantities produces the *product of their magnitudes* and the *sum of their phase angles*:

$$V_1 \angle \phi_1 \times V_2 \angle \phi_2 = V_1 V_2 \angle (\phi_1 + \phi_2)$$

Self-assessment question

8 The voltage gain of an amplifier at 1 kHz is $\mathbf{A} = 100 \angle -60°$. The input voltage phasor at this frequency is 10 mV $\angle 30°$. What is the output voltage?

Suppose that a circuit is being tested to measure its voltage gain **A**. The input and output voltage phasors $\boldsymbol{V}_{\text{in}}$ and $\boldsymbol{V}_{\text{out}}$ are measured, and the voltage gain is given by

$$\mathbf{A} = \frac{\boldsymbol{V}_{\text{out}}}{\boldsymbol{V}_{\text{in}}}$$

Clearly, the magnitude of **A** is given by the amplitude ratio $V_{\text{out}}/V_{\text{in}}$.

The phase of the output voltage, ϕ_{out}, is due partly to the input phase ϕ_{in} and partly to the circuit, so the circuit contributes a phase shift of $(\phi_{\text{out}} - \phi_{\text{in}})$. Thus,

$$\mathbf{A} = \frac{\boldsymbol{V}_{\text{out}}}{\boldsymbol{V}_{\text{in}}} = \frac{V_{\text{out}} \angle \phi_{\text{out}}}{V_{\text{in}} \angle \phi_{\text{in}}} = \frac{V_{\text{out}}}{V_{\text{in}}} \angle (\phi_{\text{out}} - \phi_{\text{in}})$$

So, in general:

Phasor division is done by division of the magnitude of the numerator (top) phasor by the magnitude of the denominator (bottom), and the subtraction of the angle of the denominator from the angle of the numerator:

$$V_1 \angle \phi_1 \div V_2 \angle \phi_2 = V_1 / V_2 \angle (\phi_1 - \phi_2)$$

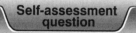

Self-assessment question

9 An amplifier has an input

$v_{in} = 0.1 \cos 3142t$ V

and an output

$v_{out} = 11.5 \sin(3142t + \pi/3)$ V

Express the gain of the amplifier in the form $\mathbf{A} = A\angle\phi$.

3.2.2 Division: impedance

The concept of impedance is derived from the phasor equivalent of Ohm's relationship. In an a.c. circuit both voltage and current are sinusoids, so their instantaneous values are continuously and periodically varying. Unless the voltage and current are in phase, the ratio of their instantaneous values v/i will also be continuously and periodically changing. The amplitude and relative phase of a sinusoid at any particular frequency, however, does not change with time. We can thus adapt Ohm's relationship for a.c. circuits by considering how the amplitudes and phases of currents and voltages are related. We use the phasor operator **impedance**, represented by the symbol **Z**, where

$$\mathbf{Z} = \frac{\boldsymbol{V}}{\boldsymbol{I}}$$

This includes the special case of *resistance*, where $\boldsymbol{V}$ and $\boldsymbol{I}$ are in phase, and **reactance**, where $\boldsymbol{V}$ and $\boldsymbol{I}$ have a phase difference of $\pi/2$. Like resistance and reactance, impedance is measured (and quoted) in ohms. In a general impedance the phase difference between voltage and current can be any angle.

Remember that phasor division is carried out by:

1. Dividing the magnitude of the top phasor by the magnitude of the bottom phasor—this gives the magnitude $|\mathbf{Z}|$ of the impedance (the magnitude of the voltage divided by the magnitude of the current).

2. Subtracting the phase angle of the bottom phasor from the phase angle of the top phasor—this gives the phase ϕ of the impedance (the angle by which the voltage leads the current).

Example

If a network has an input sinusoidal voltage and current represented by the phasors $\boldsymbol{V} = 3 \angle 0°$ V and $\boldsymbol{I} = 6.4 \angle 62°$ μA, then

$$\begin{aligned}\mathbf{Z} &= Z\angle\phi \\ &= \frac{3}{6.4 \times 10^{-6}}\angle(0° - 62°)\,\Omega \\ &= 4.7 \times 10^{5}\angle -62°\,\Omega = 470\,\text{k}\Omega\angle -62°\end{aligned}$$

The voltage amplitude in volts is 4.7×10^5 times larger than the current amplitude in amps, and the voltage lags the current by 62°.

The phasor form of Ohm's relationship for a.c. circuits is analogous to that for d.c. circuits and, as for d.c., can be written in three different ways:

$$I = \frac{V}{\mathbf{Z}}, \quad V = \mathbf{Z}I, \quad \mathbf{Z} = \frac{V}{I}$$

Self-assessment question

10 Calculate the impedance **Z** of:

(a) A 47 kΩ resistor.
(b) A 1.8 nF (1.8×10^{-9} F) capacitor at a frequency of 5 kHz.
(c) A 1 mH inductor at a frequency of 100 kHz.

Self-assessment question

11 At 30 kHz the input impedance of an *CR* low-pass network is 5 kΩ $\angle -30°$. What is the input current if the input voltage at 30 kHz is 2.5 V $\angle 40°$?

3.3 PHASOR MANIPULATIONS USING THE OPERATOR j

The combining of phasors and phasor operators in previous sections has required either some complicated mathematical manipulations or the use of diagrams and geometrical construction. The algebraic manipulations required to add, subtract, multiply and divide phasors or phasor operators can be made much less cumbersome by the introduction of an operator j.

3.3.1 The operator j

Suppose a network had the characteristic shown in Fig. 3.10, that is, the output-voltage amplitude was equal to the input-voltage amplitude, and the phase of the output, relative to the input, was $\pi/2$ radians (90°). Then the voltage gain V_{out}/V_{in} of the circuit could be represented by the phasor operator $1 \angle \pi/2$.

Some special circuits have this property, but only over a limited frequency range.

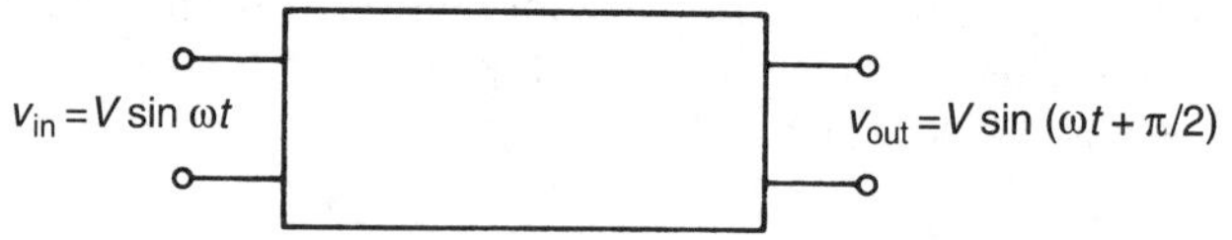

Fig. 3.10

Although there is no practical circuit which has this characteristic for all frequencies of input voltage, the concept of a unity-gain, $\pi/2$ radian phase-shift operator is extremely useful. It is so useful that we give it a special symbol. We say it is the phasor operator j. That is, if

$$v_{in} = V \sin \omega t \quad \text{and} \quad v_{out} = V \sin(\omega t + \pi/2)$$

then

$$V_{out} = jV_{in}$$

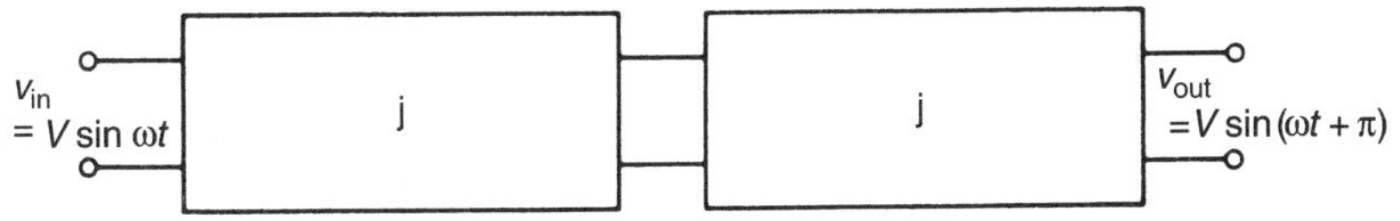

Fig. 3.11

In this one special case, by convention, we do not use the bold capital typeface which has previously been used for phasors.

If we cascade two such networks (and assume that there is no loading effect by the second on the first) as in Fig. 3.11, then the output-voltage amplitude will be equal to the input-voltage amplitude, and the total phase shift of the output relative to the input will be $+\pi$ radians. That is, if $v_{in} = V \sin \omega t$ and $v_{out} = V \sin(\omega t + \pi)$, then using phasor notation, $\mathbf{V}_{out} = jj\mathbf{V}_{in} = j^2\mathbf{V}_{in}$.

On a phasor diagram, the effect of the operator j is to move a phasor through $\pi/2$ radians (90°) anticlockwise, leaving its length constant. Figure 3.12a is the phasor diagram of the single network of Fig. 3.10 and Fig. 3.12b is the phasor diagram for the double network of Fig. 3.11. In Fig. 3.12b, $\mathbf{V}_{out}$ is simply $\mathbf{V}_{in}$ reversed in direction, and is equally validly represented as $-\mathbf{V}_{in}$. (At any instant of time, the value of v_{out} is equal and opposite to that of v_{in}.) The operator j^2 is therefore the same as the operation of multiplying by -1, or *negating*:

$$j^2 \equiv -1$$

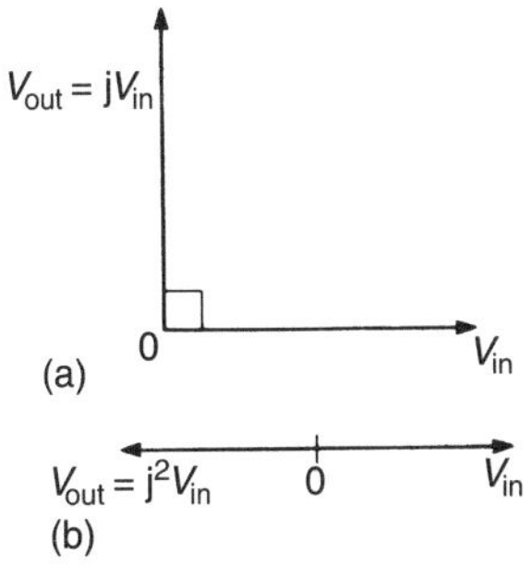

Fig. 3.12

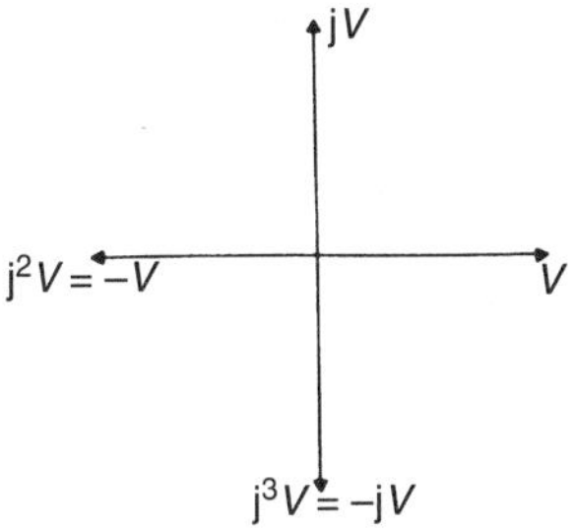

Fig. 3.13

Worked example

What is an alternative representation of the operator j^3?

The operator j^3 *can be written* $j^3 = j^2j = -1j = -j$

We therefore have the property:

$$j^3 \equiv -j$$

Figure 3.13 summarizes the use of the operator j:

- The operator j moves a phasor through $\pi/2$ radians (90°) anticlockwise.
- The operator $j^2(=-1)$ moves a phasor through π radians (180°) anticlockwise.
- The operator $j^3(=-j)$ moves a phasor through $3\pi/2$ radians (270°) anticlockwise, or $\pi/2$ radians (90°) clockwise.

$$j = 1\angle 90°.$$
$$j^2 = 1 \times 1\angle 90° + 90° = 1\angle 180°.$$
$$j^3 = 1 \times 1 \times 1\angle 90° + 90° + 90° = 1\angle 270° = 1\angle -90°.$$

Worked example

What effect does the operator j^4 have on a phasor?

Nothing, because it moves the phasor clockwise round a full circle back to where it started.

$$j^4 \equiv 1$$

This is what you ought to expect, because

$$j^4 = j^2 \times j^2 = (-1) \times (-1) = 1$$

3.3.2 Using the operator j

The operator j is just a shorthand way of indicating a phase shift of $+90°$. Similarly, $-$j indicates a phase shift of $-90°$. We can use this new notation immediately to write down the impedance of a capacitor. We know that a voltage $v = V \sin \omega t$ across a capacitor C will give rise to a current

$$i = \omega C V \sin(\omega t + \pi/2)$$

(see Chapter 2, Section 2.5.1).

In phasor notation the voltage and current are

$$\boldsymbol{V} = V\angle 0°, \quad \boldsymbol{I} = \omega C V \angle 90° = (V/X_C)\angle 90°$$

where $X_C = 1/\omega C$ is the capacitive reactance. So the impedance of a capacitor is

$$\mathbf{Z} = \frac{\boldsymbol{V}}{\boldsymbol{I}} = \frac{V\angle 0°}{\omega C V \angle 90°} = \frac{V}{\omega C V}\angle(0° - 90°)$$
$$= \frac{1}{\omega C}\angle - 90°$$

Using $-$j to represent the $-90°$ phase shift we can write $\mathbf{Z} = -\mathrm{j}/\omega C$. Now, because of the property $\mathrm{j} \times \mathrm{j} = -1$, we can write $\mathrm{j} = -1/\mathrm{j}$, which gives

$$-\mathrm{j} = \frac{1}{\mathrm{j}}$$

$\mathbf{Z} = \frac{1}{\mathrm{j}\omega C}$

Using this result the impedance of a capacitor becomes

$$\mathbf{Z} = \frac{-\mathrm{j}}{\omega C} = \frac{1}{\mathrm{j}\omega C}$$

Self-assessment question

12 Show that the impedance **Z** of inductor L using j notation is jωL.

$\mathbf{Z} = \mathrm{j}\omega L$

Merely using j to write the impedance of a capacitor or an inductor in a compact form may not seem particularly useful. This apparently trivial step, however, leads us to a method that allows us to work out voltages, currents, impedances and gains in an a.c. circuit without using phasor diagrams at all. We can do all the necessary calculations algebraically. This is especially useful for analysing complicated circuits, where the phasor diagram can be very difficult to construct, and to interpret. The trick, of course, is to be able to interpret what our equations mean when they contain j, and to do this a simple sketch on a phasor diagram is sometimes useful.

The operator j allows us to represent phasors of any magnitude and angle. Suppose we describe a current phasor using j notation as $\boldsymbol{I} = (2 + \mathrm{j}3)$ mA. If we take the terms 2 and j3 individually, we can write

$$2 \equiv 2\angle 0° \quad \text{and} \quad \mathrm{j}3 \equiv 3\angle 90°$$

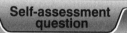

So $\boldsymbol{I}$ is simply a sum of phasors

$$\boldsymbol{I} = (2\ \angle 0^\circ + 3\ \angle 90^\circ)\ \text{mA}$$

Figure 3.14a shows the phasors on a phasor diagram. The phasor $2 \angle 0^\circ$ is of length 2 and lies along the positive horizontal axis of the diagram. The phasor $3 \angle 90^\circ$ is of length 3 and lies along the positive vertical axis. The equivalent representation $I \angle \phi$ is found by constructing the sum, or resultant, as shown in Fig 3.14b. By calculation, the length of the resultant is

$$I = \sqrt{(2^2 + 3^2)}\ \text{mA} = 3.6\ \text{mA}$$

The phasor is at an angle of $\tan^{-1}(3/2) = 56.3^\circ$ anticlockwise from the horizontal. The angle is thus positive and we can write

$$I = 3.6\ \angle 56.3^\circ\ \text{mA}$$

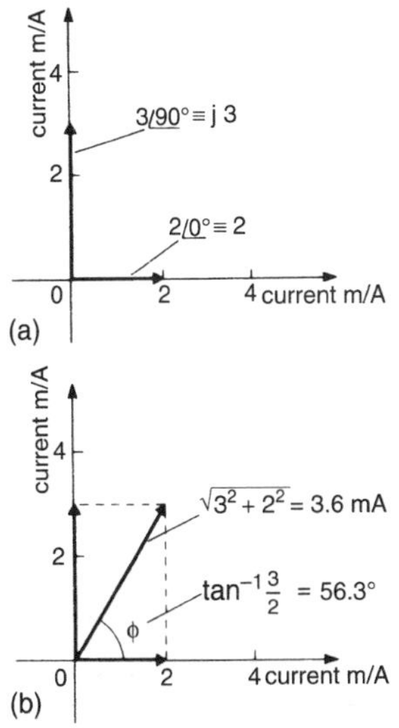

Fig. 3.14

In other words using j notation leads to the definition of a phasor in terms of its horizontal and vertical components, whereas the $I \angle \phi$ notation leads to the definition of a phasor directly in terms of its magnitude and phase. In general, a phasor can be represented in j notation as $a + jb$. From Fig. 3.15 the corresponding $m \angle \phi$ notation is

$$\sqrt{(a^2 + b^2)}\ \angle \tan^{-1}(b/a)$$

Self-assessment question

13 Express in $m \angle \phi$ form the phasors $1 + j$, $3 + j4$, $2.6 + j8.7$.

When both a and b are positive, the evaluation of the phase angle is straightforward. It is an angle between 0 and $\pi/2$ radians (0° and 90°) and can be obtained from tables or by using the inverse tangent function on a calculator. However, when either a or b (or both) is negative, the interpretation of the phasor angle is not as simple. In this case you should *always* draw a phasor diagram as an aid to calculation. To see the sort of pitfalls that can arise, take the simple example $\boldsymbol{I} = -1 - j$. A careless calculation gives

> The $a + jb$ notation has rectangular coordinates. The $m \angle \phi$ notation has polar coordinates. Most scientific calculators can convert from rectangular to polar (R → P) and from polar to rectangular (P → R).

$$\phi = \tan^{-1}(-1/-1) = \tan^{-1}1 = \pi/4$$

whereas the correct answer is $\phi = \pi + \pi/4 = 5\pi/4$.

This case is covered by part (c) of Fig. 3.16; the diagram shows the four possibilities:

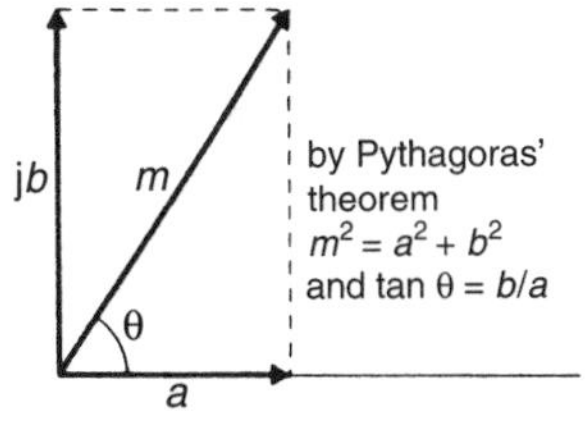

Fig. 3.15

(a) a and b both positive, $\phi = \tan^{-1}(b/a) = \tan^{-1}|b/a|$.
(b) a negative and b positive, $\phi = \pi - \tan^{-1}|b/a|$.
(c) a negative and b negative, $\phi = \pi + \tan^{-1}|b/a|$.
(d) a positive and b negative, $\phi = 2\pi - \tan^{-1}|b/a|$.

(For this explanation $\tan^{-1}|b/a|$ is assumed to be always an angle between 0° and 90° as would be obtained from tables or a calculator having the inverse tangent function.)

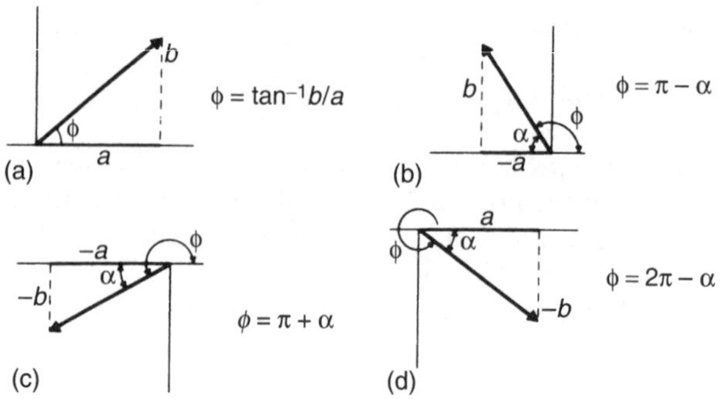

Fig. 3.16 Phasor angle ϕ.

Self-assessment question

14 Express in $m \angle \phi$ form the phasors $3 - j4$, $-4 + j3$, $-4 - j3$.

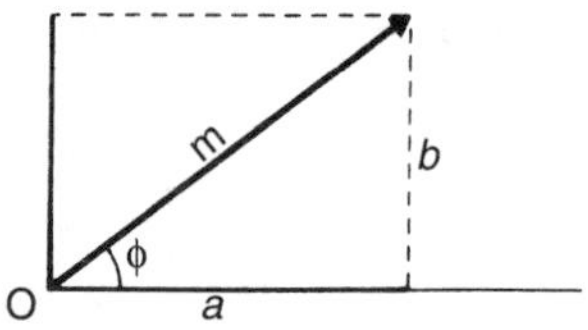

Fig. 3.17 $a = m \cos \phi$ and $b = m \sin \phi$.

Sometimes it is necessary to convert a phasor in $m \angle \phi$ form into its $a + jb$ form. This is known as *resolving* the phasor into two components. For angles ϕ between 0 and $\pi/2$ radians, the conversion is straightforward. Looking at Fig. 3.17 and using the simple trigonometric relations

$$\sin \phi = b/m \quad \text{and} \quad \cos \phi = a/m$$

we can write $b = m \sin \phi$ and $a = m \cos \phi$, so

$$m \angle \phi = m\cos \phi + jm\sin \phi$$

Self-assessment question

15 Express in $a + jb$ form the phasors: $10 \angle 60°$, $0.4 \angle 12°$, $15 \times 10^3 \angle \pi/6$ rad.

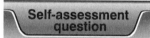

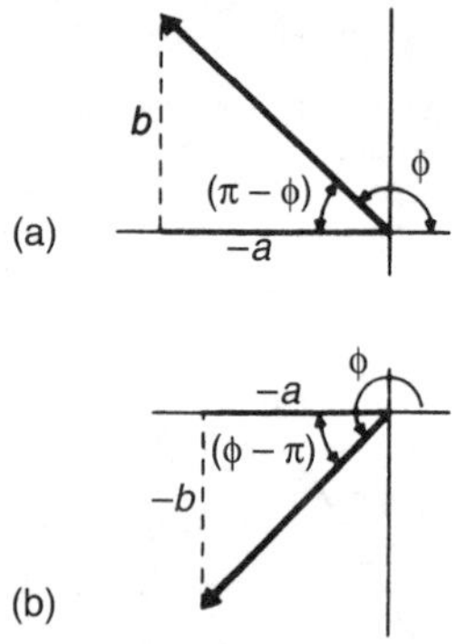

Fig. 3.18

If ϕ is an angle greater than $\pi/2$ radians then care is required with resolution of the phasor into its component form $a + jb$. A sketch of the phasor is always helpful, but the general rule can be seen from Fig. 3.18.

(a) If ϕ is between $\pi/2$ and π radians, the phasor has the form $(-a + jb)$, where a and b are positive and the angle used to evaluate a and b is the angle $(\pi - \phi)$.
(b) If ϕ is between π and $3\pi/2$ radians, the phasor has the form $(-a - jb)$ and the angle used to evaluate a and b is the angle $(\phi - \pi)$.
(c) If ϕ is between $3\pi/2$ and 2π radians, the phasor has the form $(a - jb)$ and the angle used to evaluate a and b is the angle $(2\pi - \phi)$.

Self-assessment question

16 Express in $a + jb$ form the phasors $10 \angle -45°$, $20 \angle 120°$, $30 \angle 210°$, $30 \angle -150°$.

3.3.3 Complex number terminology

If you have ever studied complex numbers you will have realized by now that phasor algebra has a similar structure to the algebra of complex numbers. For this reason some of the terminology of complex number theory has found its way into phasor algebra.

In this terminology a component measured along the datum line of zero angle is referred to as the **real part**, and the component which has been operated on by j the **imaginary part** of a phasor operator. So when a phasor is represented by $a + jb$, a is the 'real part' and b the 'imaginary part'.

Similarly, a phasor diagram or an impedance diagram with real and imaginary axes shown, like Figs. 3.12, 3.13 and 3.14, can be referred to as an **Argand diagram**.

If you have not studied complex numbers, do not worry; everything you require for sinusoidal circuit analysis is included here.

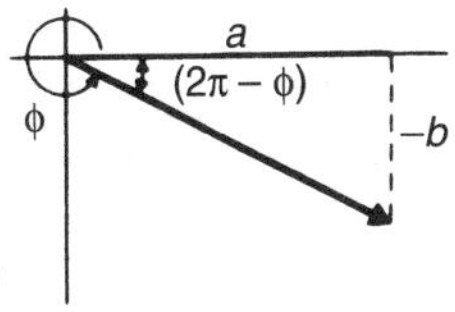

(c)

Fig. 3.18 (Continued)

3.3.4 Addition and subtraction of phasors in component form

Figure 3.19 shows two impedances in series. At a particular frequency, $\mathbf{Z}_1 = 6 + j7\ \Omega$ and $\mathbf{Z}_2 = 3 - j2\ \Omega$. If we need to find the combined impedance, $\mathbf{Z}_T$, we must add the individual impedances because they are in series:

$$\mathbf{Z}_T = \mathbf{Z}_1 + \mathbf{Z}_2 = (6 + j7) + (3 - j2)\ \Omega$$

To add together these two impedance phasors, we group the terms without j ('real parts') and the terms containing j ('imaginary parts') separately, so

$$\mathbf{Z}_T = (6 + 3) + (j7 - j2)\Omega = 9 + j5\ \Omega$$

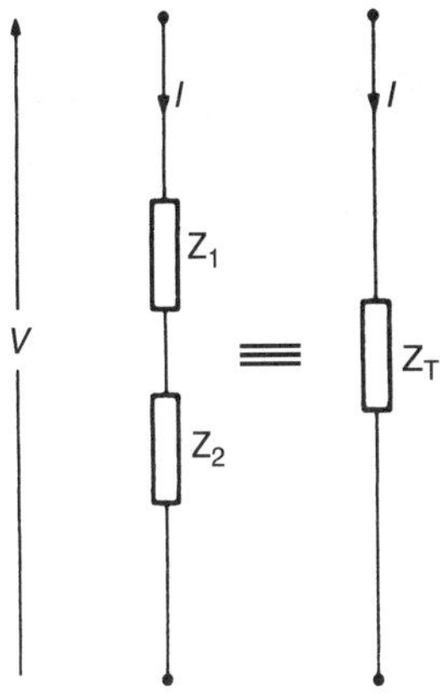

Fig. 3.19

Worked example

Figure 3.20 shows two impedances $\mathbf{Z}_1$ and $\mathbf{Z}_2$ in parallel. If the total input current $\mathbf{I}_T$ is 2 − j5 mA and the current $\mathbf{I}_1$ through $\mathbf{Z}_1$ is 4 + j4 mA, what current flows in $\mathbf{Z}_2$?

Using Kirchhoffs current law,

$$I_T = I_1 + I_2, \quad \text{so} \quad I_2 = I_T - I_1$$

Therefore,

$$I_2 = (2 - j5) - (4 + j4)\ \text{mA}$$

Combining real and imaginary parts separately as before,

$$I_2 = (2 - 4) + j(-5 - 4)\ \text{mA} = -2 - j9\ \text{mA}$$

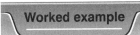

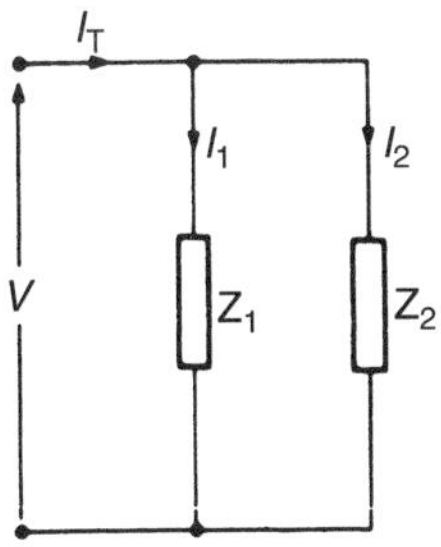

Fig. 3.20

The general result is that when two phasors $a_1 + jb_1$ and $a_2 + jb_2$ are added the resulting phasor $a_r + jb_r$ has the real and imaginary parts

$$a_r = (a_1 + a_2), \quad b_r = (b_1 + b_2)$$

For the subtraction of two phasors $(a_1 + jb_1) - (a_2 + jb_2)$, the general result is

$$a_r = (a_1 - a_2), \qquad b_r = (b_1 - b_2)$$

Self-assessment question

17 Perform the following additions and subtractions of complex numbers: (a) (3 + j2) + (6 + j9); (b) (9 − j10) − (9 + j10); (c) (−2 −j) − (−6 + j9); (d) (1 + j0) − (0 + j); (e) (7 + j) − (−7 + j).

Before leaving the topic of addition and subtraction, note that numbers given in the form $m \angle \phi$ must *always* be converted to component form before the rules can be applied. The following Self-assesment question will give you extra practice in conversion, besides addition and subtraction.

Self-assessment question

18 Perform the following additions and subtractions of complex numbers. Express your results in both $a + jb$ and $m \angle \phi$ form.
(a) (5 + j6) + 5 ∠ 53°.
(b) (3 − j4) + 10 ∠ 127°.
(c) (−2 + j3) − 8 ∠ −37°.
(d) (−4 −j8) − 14 ∠ 30°.

3.3.5 Multiplication and division of complex numbers

The addition and subtraction of complex numbers is always carried out with the numbers in component form. When we multiply and divide complex numbers we often have a choice; we can work with the component form or with the magnitude–phase representation. You will get some guidance in the following, to enable you to make a choice that is appropriate to the problem in hand.

Multiplication of complex numbers

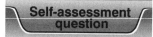

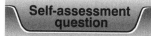

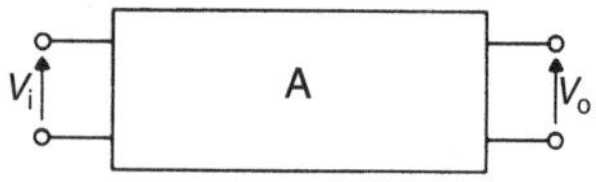

Fig. 3.21

Figure 3.21 shows a network having a voltage gain specified as a phasor operator **A**. To find the output voltage, the input-voltage phasor must be multiplied by the voltage gain

$$V_o = \mathbf{A}V_i$$

If both **A** and V_i are given in $m\angle\phi$ form then V_o can be found immediately: we multiply their magnitudes and add their phases as described in Section 3.2.1.

Now, suppose that both **A** and V_i have been specified in component form, for example, $V_i = (1 + j2)$ V and $\mathbf{A} = (8 + j3)$. We then have

$$V_o = (8 + j3)(1 + j2)\ \text{V}$$

We can evaluate the product without first converting to $m \angle \phi$ form. To do this, multiply each term in the second bracket by each term in the first, then add the terms, as follows:

$$V_o = (8 \times 1 + j2 \times 8 + j3 \times 1 + j3 \times j2)\text{V}$$
$$= (8 + j16 + j3 + j^2 6)\text{V}$$

Remembering that $j^2 = -1$, we can collect real and imaginary parts of the expression as follows:

$$\begin{aligned} V_o &= (8-6) + j(16+3)\text{V} \\ &= 2 + j19\text{V} \end{aligned}$$

In general, the product rule for numbers in component form is

$$(a_1 + jb_1) \times (a_2 + jb_2) = (a_1a_2 - b_1b_2) + j(a_1b_2 + a_2b_1)$$

The result is always in component form.

Self-assessment question

19 Perform the following multiplications of pairs of complex numbers: (a) (6 + j) × (3 + j4); (b) (2 + j5) × (2 – j5); (c) (–2 –j) × (–6 –j9); (d) j × (3 + j4); (e) (8 + j) × (3 + j0).

In later work you will find calculations involving the product of more than two complex numbers (when calculating the impedance of several components in parallel, for example.). In this case you may find it advantageous to make use of the following extended product rule for **complex multiplication**:

> First express each number in $m \angle \phi$ form. The magnitude of the product is given by multiplying the magnitudes of the individual numbers. The phase angle of the product is the sum of the phase angles of the individual numbers.

Thus, if $\mathbf{A}_1 = 4 \angle 70°$, $\mathbf{A}_2 = 3 \angle -40°$ and $V = 2 \angle 5°$ V, the product is

$$\begin{aligned} \mathbf{A}_1 \times \mathbf{A}_2 \times V &= (4 \times 3 \times 2)\angle(70° - 40° + 5°) \\ &= 24\angle 35°\text{V} \end{aligned}$$

Obviously this rule proves more effective when

(a) most of the numbers given are in $m \angle \phi$ form to begin with;
(b) the desired form of the product is the $m \angle \phi$ form;
(c) either the magnitude or the angle of the product is required, rather than both.

Division of complex numbers

Suppose now that we need to find the total impedance, $\mathbf{Z}_T$ of two impedances $\mathbf{Z}_1$ and $\mathbf{Z}_2$ in parallel, where $\mathbf{Z}_1 = (6 + j2)\ \Omega$ and $\mathbf{Z}_2 = (10 - j4)\ \Omega$.

In a.c. circuits we handle impedance in parallel in the same way as we handle resistances in parallel in d.c. circuits, so the formula for $\mathbf{Z}_T$ can be expressed either as

$$\frac{1}{\mathbf{Z}_T} = \frac{1}{\mathbf{Z}_1} + \frac{1}{\mathbf{Z}_2}$$

or in the alternative but equivalent form:

$$\mathbf{Z}_T = \frac{\mathbf{Z}_1\mathbf{Z}_2}{\mathbf{Z}_1 + \mathbf{Z}_2}$$

So, in substituting for Z_1 and Z_2 in this expression for $\mathbf{Z}_T$,

$$\mathbf{Z}_T = \frac{(6 + j2) \times (10 - j4)}{(6 + j2) + (10 - j4)}\,\Omega$$

Using the rules for multiplication and addition of complex numbers we can evaluate the numerator and denominator of this expression:

$$\mathbf{Z}_T = \frac{(60 + 8) + j(20 - 24)}{16 - j2}\,\Omega$$

$$= \frac{68 - j4}{16 - j2}\,\Omega$$

To complete the calculation of $\mathbf{Z}_T$, we need to be able to divide one complex number by another. In the following the aim is a result in the form $a + jb$. Here one needs a trick borrowed from complex number algebra called *rationalizing the denominator* (i.e. getting rid of the imaginary term in the denominator).

If we multiply the expression for $\mathbf{Z}_T$ by $(16 + j2)/(16 + j2)$, we are merely multiplying by 1, and therefore not changing the value of the expression. However, this significantly changes the *form* of the expression:

$$\mathbf{Z}_T = \frac{(68 - j4)}{(16 - j2)} \times \frac{(16 + j2)}{(16 + j2)}\,\Omega$$

Taking the numerator first gives

$$\begin{aligned}(68 - j4) \times (16 + j2) &= (68 \times 16 + 4 \times 2) + j(68 \times 2 - 4 \times 16)\\ &= 1096 + j72\end{aligned}$$

For the denominator we have

$$\begin{aligned}(16 - j2) \times (16 + j2) &= (16^2 + 2^2) + j(32 - 32)\\ &= 260 + j0\end{aligned}$$

The result of multiplying the top and bottom of the expression by $(16 + j2)$ *has been to make the imaginary term in the denominator zero.*

Therefore, we can write

$$\begin{aligned}\mathbf{Z}_T &= \frac{1096}{260} + j\frac{72}{260}\\ &= 4.2 + j0.28\ \Omega\end{aligned}$$

(Remember this is the total impedance of $\mathbf{Z}_1 = (6 + j2)\ \Omega$ and $\mathbf{Z}_2 = (10 - j4)\ \Omega$ in parallel.)

The general method for the division of two complex numbers is as follows. If you have

Rationalize the denominator

$$\frac{a + jb}{c + jd}$$

then multiply top and bottom by $c - \mathrm{j}d$. The resultant is

$$\frac{(a + \mathrm{j}b)(c - \mathrm{j}d)}{(c + \mathrm{j}d)(c - \mathrm{j}d)}$$

and the denominator becomes

$$(c + \mathrm{j}d)(c - \mathrm{j}d) = c^2 - \mathrm{j}cd + \mathrm{j}cd - \mathrm{jj}d^2 = c^2 + d^2$$

The denominator has lost its j term and has become, in complex number parlance, a 'real' number. (If the denominator of the division had been $c - \mathrm{j}d$, we should have multiplied top and bottom by $c + \mathrm{j}d$, which would again make the denominator $c^2 + d^2$.)

The numerator $(a + \mathrm{j}b)$ $(c - \mathrm{j}d)$ is equal to $(ac + bd)$ + $\mathrm{j}(bc - ad)$, so the rationalized complex number is

$$\frac{ac + bd}{c^2 + d^2} + \mathrm{j}\left(\frac{bc - ad}{c^2 + d^2}\right)$$

Self-assessment question

20 Perform the following complex number divisions and obtain the result in $a + \mathrm{j}b$ form:

(a) $\dfrac{1 + \mathrm{j}}{2 - \mathrm{j}}$ (c) $\dfrac{3 + \mathrm{j}4}{1 + \mathrm{j}3}$

(b) $\dfrac{1 - \mathrm{j}}{2 + \mathrm{j}}$ (d) $\dfrac{2 - \mathrm{j}5}{2 - \mathrm{j}3}$

Self-assessment question

21 A voltage $\mathbf{V} = 3 + \mathrm{j}2$ V is applied across an impedance $\mathbf{Z} = 2 + \mathrm{j}5\ \Omega$. Using the relationship $\mathbf{I} = \mathbf{V}/\mathbf{Z}$ work out, in $a + \mathrm{j}b$ form, the current flowing in the impedance. By sketching the voltage and current on a phasor diagram, state whether the voltage leads the current, or vice versa, at this particular frequency.

Note that the sole objective in rationalizing the denominator in a complex division is to ensure that the quotient is expressed in the form $a + \mathrm{j}b$.

Returning to the earlier example, if all we required was the magnitude of Z_T or its phase, then it would be very wasteful of time and effort to work through the rationalization procedure, given the following rule for **complex division**:

> The magnitude of the quotient is given by dividing the magnitude of the numerator by the magnitude of the denominator. The phase angle of the quotient is found by subtracting the phase angle of the denominator from the phase angle of the numerator.

This rule is identical to that used earlier for the division of phasor quantities.

Self-assessment question

22 For each of the examples in Self-assessment question 20, work out directly (a) the magnitude, and (b) the phase angle of the quotient.

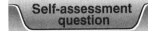

3.4 EQUIVALENT IMPEDANCE AND ADMITTANCE

3.4.1 Impedance

You have seen in previous sections that the impedance $\mathbf{Z}_{eq}$, which is equivalent to any number of *impedances in series*, is the phasor sum of those impedances; that is

$$\mathbf{Z}_{eq} = \mathbf{Z}_1 + \mathbf{Z}_2 + \mathbf{Z}_3 + \ldots$$

For example, if we take a resistor R, and inductor L and a capacitor C in series, the equivalent impedance is

$$\begin{aligned}\mathbf{Z}_{eq} &= R + j\omega L - j/\omega C \\ &= R + j(\omega L - 1/\omega C)\end{aligned}$$

For *impedances in parallel* an argument similar to that used with resistors in Chapter 1 shows that

$$\frac{1}{\mathbf{Z}_{eq}} = \frac{1}{\mathbf{Z}_1} + \frac{1}{\mathbf{Z}_2} + \frac{1}{\mathbf{Z}_3} + \cdots$$

For the particular case of two impedances in parallel, we can use the 'product over sum' rule:

$$\mathbf{Z}_{eq} = \frac{\mathbf{Z}_1\mathbf{Z}_2}{\mathbf{Z}_1 + \mathbf{Z}_2}$$

The equivalent impedance of *any linear* two-terminal network, however complicated the network, can always be expressed in the form

$$\mathbf{Z}_{eq} = R + jX$$

R is called the resistive part (also the 'real' part) of $\mathbf{Z}_{eq}$ and jX the reactive part (or the 'imaginary' part).

If, at a given frequency, the reactive part has a positive sign then the circuit is said to be *inductive* at that frequency, even though there may be both inductors and capacitors in the circuit. This is because, *at that frequency* the circuit could be modelled simply by a resistor and inductor in series.

Similarly, if the reactive part has a negative sign, for example (10 – j5) Ω, the circuit is said to be *capacitive* at that frequency.

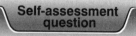

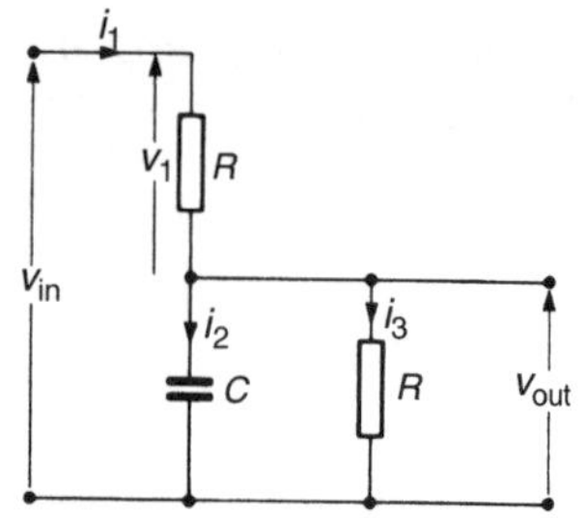

Fig. 3.22

Self-assessment question

23 Consider the input impedance of the circuit of Fig. 3.22. State by inspection of the circuit what the input impedance will tend towards

(i) at very low frequencies:
(ii) at very high frequencies.

What do you think the impedance does in between?

3.4.2 Admittance

Admittance is the reciprocal of impedance; it is a phasor operator which relates the phasor current in a two-terminal network to the phasor voltage across it. It is defined by the equation

$$\mathbf{Y} = \frac{I}{V}$$

where **Y** is the admittance in siemens.

You will see from its definition that, for a particular network,

$$\mathbf{Y} = 1/\mathbf{Z}$$

The real part of **Y** is called **conductance** and given the symbol G; the imaginary part is called **susceptance** and is represented by B:

$$\mathbf{Y} = G + \mathrm{j}B$$

You should be able to show that:

The admittance of a resistor of value R is $1/R$.

The admittance of a capacitor of value C is $\mathrm{j}\omega C$.

The admittance of an inductor of value L is $1/\mathrm{j}\omega L$, or $-\mathrm{j}/\omega L$

When working with numerical values, the conversion from **Z** to **Y** and vice versa is best done using the form $m \angle \phi$.

Worked example

At a particular frequency, the impedance of a circuit is $(3+\mathrm{j}4)\ \Omega$. What is the equivalent admittance?

$\mathbf{Z} = (3+\mathrm{j}4) = 5 \angle 53.1°\ \Omega$
$\mathbf{Y} = 1/\mathbf{Z} = 0.2 \angle -53.1° = (0.12 - \mathrm{j}0.16)\ S$

In each case the sign of the reactive and the susceptive part shows that the circuit is inductive at the frequency considered.

At times, however, it will be necessary to work with Z and Y in component form.

Worked example

What is the admittance of a resistor of value R in series with an inductor of value L?

We have

$$\mathbf{Z} = R + \mathrm{j}\omega \mathbf{L} \quad \text{so} \quad \mathbf{Y} = \frac{1}{R + \mathrm{j}\omega L}$$

Rationalizing the denominator gives

$$\mathbf{Y} = \frac{R - \mathrm{j}\omega L}{R^2 + \omega^2 L^2}$$

So, in $G + jB$ form:

$$\mathbf{Y} = \frac{R}{R^2 + \omega^2 L^2} - \mathrm{j}\frac{\omega L}{R^2 + \omega^2 L^2}$$

Notice that if $\mathbf{Z} = R + \mathrm{j}X$ and $\mathbf{Y} = 1/\mathbf{Z} = G + \mathrm{j}B$, then if both R and X are present, G is not equal to $1/R$ and B is not equal to $1/X$.

The rules for finding the equivalent admittance of a number of admittances connected in series and in parallel are analogous to those given in Chapter 1 for combining conductances.

For *admittances in parallel* we have

$$\mathbf{Y}_{\text{eq}} = \mathbf{Y}_1 + \mathbf{Y}_2 + \mathbf{Y}_3 + \ldots$$

For *admittances in series* the result is

$$\frac{1}{\mathbf{Y}_{\text{eq}}} = \frac{1}{\mathbf{Y}_1} + \frac{1}{\mathbf{Y}_2} + \frac{1}{\mathbf{Y}_3} + \ldots$$

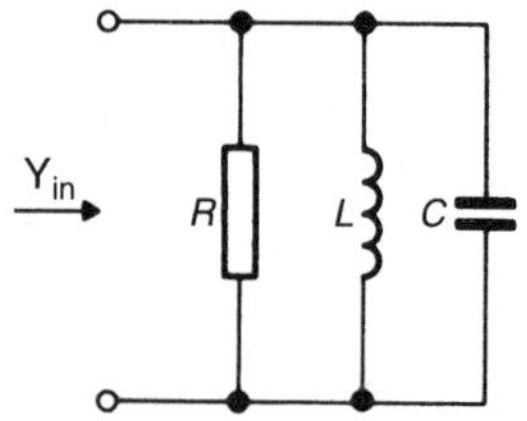

Fig. 3.23

The 'product over sum' rule can also be applied to the particular case of two admittances in series:

$$\mathbf{Y}_{\text{eq}} = \frac{\mathbf{Y}_1\mathbf{Y}_2}{\mathbf{Y}_1 + \mathbf{Y}_2}$$

Comparing the expressions for combining impedances and combining admittances, you will see that it is usually easiest to represent a parallel circuit by its admittance rather than by its impedance.

Self-assessment question

24 Derive an expression for the input admittance of the parallel *RLC* circuit in Fig. 3.23. Hence:

(a) Find the equivalent conductance and the equivalent susceptance of the circuit at frequency ω.

(b) Find the frequency ω_0 at which the susceptance becomes zero.

(c) Derive approximate expressions for the input admittance in each of the following frequency ranges and state, in each case, whether the circuit is capacitive or inductive.

(i) $\omega \gg \omega_0$; (ii) $\omega \gg \omega_0$.

3.5 CIRCUIT ANALYSIS USING THE OPERATOR j

Many of the ideas of d.c. circuit theory can be applied to the steady-state solution of linear circuits containing sinusoidal sources. If the voltages and currents are represented by their phasors and the components by their impedances or admittances, then Kirchhoff's law, the Thévenin and Norton theorems, nodal analysis and the voltage and current divider rules can all be used to set up equations relating the voltage and current phasors. The rules of complex algebra can then be used to obtain solutions containing both amplitude and phase information.

This is the advantage of the operator j: it enables us to perform complicated a.c. analysis using the simple methods used for d.c.

3.5.1 A.C. analysis of the low-pass *CR* circuit

The low-pass *CR* circuit is shown again in Fig. 3.24a with the instantaneous values of the current, input voltage and output voltage identified using lowercase letters, i, v_{in} and v_{out}.

The circuit is redrawn in Fig. 3.24b. Each of the sinusoids in the circuit is now represented by its phasor and the 'black boxes' represent the impedances of the resistor and capacitor, denoted by $\mathbf{Z}_R$ and $\mathbf{Z}_C$ respectively. Since we are now dealing with phasor quantities it is now helpful to regard the reference arrows as *phase reference* arrows. In choosing their direction, the circuit is treated as if it were purely resistive, with no phase differences anywhere. The results of the calculations will then show by how much the phases of the voltages and current differ from the phase of the input.

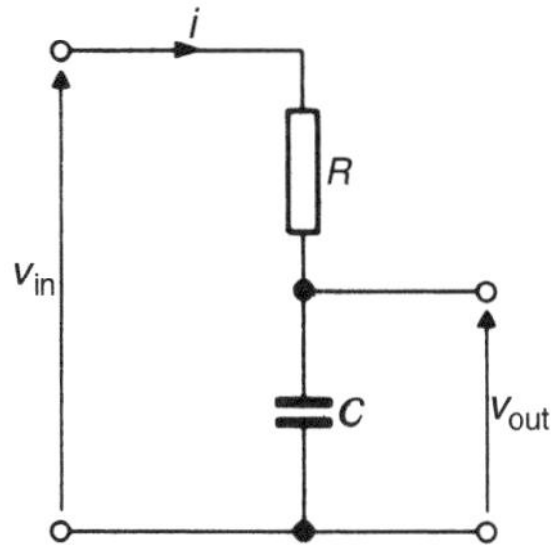

(a)

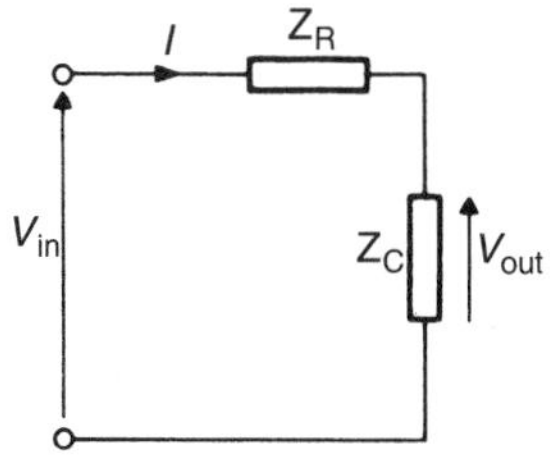

(b)

Fig. 3.24 Phasor analysis of the low-pass *CR* circuit.

The voltage transfer function

The voltage transfer function of a circuit is defined as the input–output phasor ratio $\mathbf{A} = \mathbf{V}_{out}/\mathbf{V}_{in}$. It is another name for the voltage gain, defined in Section 3.2.1. This complex ratio has magnitude

$$|\mathbf{A}| = \left|\frac{\mathbf{V}_{out}}{\mathbf{V}_{in}}\right| = \frac{|\mathbf{V}_{out}|}{|\mathbf{V}_{in}|}$$

which is equal to the magnitude of the output voltage divided by the magnitude of the input voltage. The magnitude of **A** is therefore the ratio of the *amplitudes* of the sinusoidal output and input voltages v_{out} and v_{in} at a frequency ω.

The angle of **A**, which we can call ϕ, is the *difference* between the angles of $\mathbf{V}_{out}$ and $\mathbf{V}_{in}$:

$$\phi = \angle\mathbf{V}_{out} - \angle\mathbf{V}_{in}$$

and therefore represents the phase shift introduced by the circuit at a particular frequency ω.

Now, we can relate $\mathbf{V}_{in}$ and $\mathbf{V}_{out}$ in Fig. 3.24b by using the phasor form of the voltage divider rule:

$$\mathbf{V}_{out} = \mathbf{V}_{in} \times \frac{\mathbf{Z}_C}{\mathbf{Z}_C + \mathbf{Z}_R}$$

The impedance $\mathbf{Z}_R$ is simply R since the current through and the voltage across the resistor are in phase.

Section 3.3.2 shows that $\mathbf{Z}_C = -\mathrm{j}/\omega C$, or the alternative form $\mathbf{Z}_C = 1/\mathrm{j}\omega C$ which is often more convenient to use. The voltage transfer function is therefore

$$\mathbf{A} = \frac{\mathbf{V}_{out}}{\mathbf{V}_{in}} = \frac{\mathbf{Z}_C}{\mathbf{Z}_C + \mathbf{Z}_R} = \frac{1/\mathrm{j}\omega C}{1/\mathrm{j}\omega C + R}$$

This expression can be simplified by multiplying top and bottom by $\mathrm{j}\omega C$, so that

$$\mathbf{A} = \frac{1}{1 + \mathrm{j}\omega CR}$$

The form of the denominator $(1 + \mathrm{j}\omega CR)$ is very common in analysis carried out in many branches of electronics, and you are likely to find it and slight variations of it cropping up over and over again.

The rationalization procedure can be used to express **A** as a complex number in real and imaginary form, but this is not necessary if all we require is its magnitude and angle. Taking the magnitude first we have

$$|\mathbf{A}| = \left|\frac{1}{1 + \mathrm{j}\omega CR}\right|$$

To evaluate an expression of this form, we divide the magnitude of the numerator by the magnitude of the denominator, giving

$$|\mathbf{A}| = \frac{|1|}{|1 + \mathrm{j}\omega CR|}$$
$$= \frac{1}{\surd(1 + \omega^2 C^2 R^2)}$$

Notice that to work out the denominator the relationship $|a + \mathrm{j}b| = \sqrt{(a^2 + b^2)}$ is used.

To find ϕ, the angle of **A**, we subtract the phase of the denominator from the phase of the numerator

$$\phi = 0 - \angle(1 + \mathrm{j}\omega CR)$$
$$= -\tan^{-1}(\omega CR)$$

which follows from the rule $\angle(a + \mathrm{j}b) = \tan^{-1}(b/a)$.

These results should be familiar to you from the analysis in Section 3.3.2.

Self-assessment question

25 Figure 3.25 shows a voltage divider circuit. Work out an expression for the voltage transfer function $\mathbf{A} = \mathbf{V}_0/\mathbf{V}_1$ in terms of the impedances $\mathbf{Z}_1$ and $\mathbf{Z}_2$, assuming that no current is drawn from the circuit.

(a) If $\mathbf{Z}_1$ and $\mathbf{Z}_2$ are resistors of value 1 kΩ and 10 kΩ, respectively, what is the output voltage when the input V_1 is 6 + j4 V?

(b) If $\mathbf{Z}_1$ is a 10 kΩ resistor and $\mathbf{Z}_2$ is a 15 nF capacitor, what is the output voltage in a + jb form when the input $\mathbf{V}_1$ is 6 + j4 V at 5 kHz?

(c) If $\mathbf{Z}_1$ and $\mathbf{Z}_2$ are capacitors of value 10 nF and 15 nF, respectively, work out the voltage transfer function **A** of the circuit. How will the amplitude ratio and the phase shift associated with the circuit vary with frequency?

(d) Suppose that $\mathbf{Z}_1$ is a capacitor C and $\mathbf{Z}_2$ is a resistor R. Work out an expression for the voltage transfer function. What are the amplitude ratio and the phase shift associated with the circuit:

(i) at very high frequencies ($\omega \gg 1/CR$);
(ii) at very low frequencies ($\omega \gg 1/CR$);
(iii) when $\omega = 1/CR$?

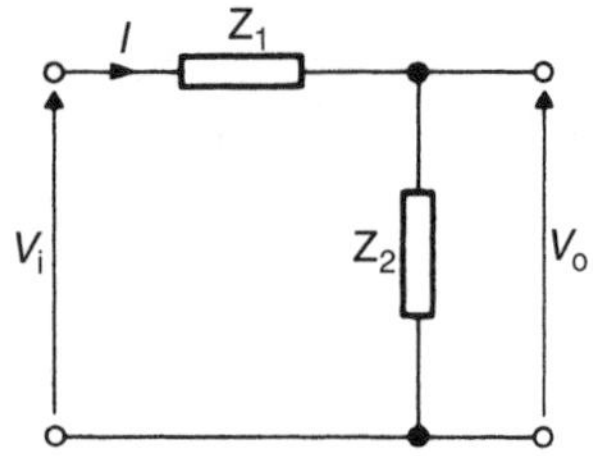

Fig. 3.25

Self-assessment question

26 Calculate the voltage transfer function of the *RL* high-pass circuit shown in Chapter 2.

Your results should agree with those found in Self-assessment question 7 when you analysed the circuit using a phasor diagram.

The current in the low-pass *CR* circuit

Since we are dealing with a series circuit we can find the current phasor $\boldsymbol{I}$ from the relationship $\boldsymbol{I} = V_{\text{in}}/\mathbf{Z}$, where $\mathbf{Z}$ is the series impedance of the resistor and capacitor, given by

$$\mathbf{Z} = \mathbf{Z}_\mathrm{R} + \mathbf{Z}_\mathrm{C} = R + 1/\mathrm{j}\omega C$$

The current phasor is then

$$I = V_{in}/Z = \frac{V_{in}}{R + 1/j\omega C}$$

Multiplying top and bottom by $j\omega C$ as before, we obtain

$$I = \frac{j\omega C V_{in}}{1 + j\omega CR} \qquad (3.1)$$

To find the sinusoid which I represents, we can use the input voltage as the phase reference and express its phasor in the form

$$V_{in} = V_{in} + j0 = V_{in}$$

The current amplitude is given by $|I|$, which can be found by dividing the magnitude of the numerator in Equation 3.1 by the magnitude of the denominator. Taking these separately, we have

$$|j\omega C V_{in}| = \omega C V_{in}$$
$$|1 + j\omega CR| = \surd(1 + \omega^2 C^2 R^2)$$

The magnitude of I is therefore

$$|I| = \frac{\omega C V_{in}}{(1+\omega^2 C^2 R^2)} \qquad (3.2)$$

Because we are using the input voltage as the phase reference, the angle of I will represent the phase of the current relative to that of the input voltage. This can be found by subtracting the angle of the denominator from the angle of the numerator in Equation 3.1, giving

$$\begin{aligned}\phi &= \angle(j\omega C V_{in}) - \angle(1 + j\omega CR) \\ &= \pi/2 - \tan^{-1}(\omega CR)\end{aligned} \qquad (3.3)$$

NOTE
To make sense of this result, recall that the voltage across the capacitor lags the input voltage by an angle $= \tan^{-1}(\omega CR)$. Since the current leads the capacitor voltage by $\pi/2$ radians, its phase with respect to the input must be as given by Equation 3.3. For an input voltage $v_{in} = V_{in} \sin \omega t$, the current in the CR circuit will therefore be of the form

$$i = |I|\sin(\omega t + \phi)$$

where $|I|$ and ϕ are given by Equations 3.2 and 3.3.

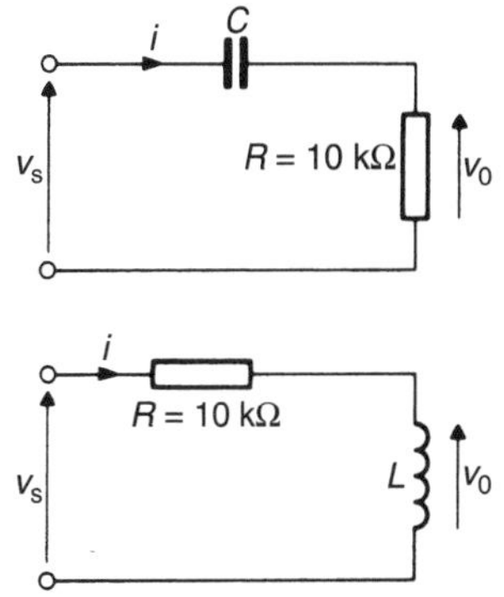

Fig. 3.26

Self-assessment question

27 The voltage transfer functions of the two circuits in Fig. 3.26 are considered in Self-assessment questions 25 and 26.

(a) Choose values of C and L so that both circuits have a high-pass cut-off frequency of 100 kHz.

(b) Will the currents in these circuits now be the same, assuming that the same voltage is applied to each?

To check your conclusions, derive expressions for the current in both circuits for an input voltage of the form $v_{in} = V_{in} \sin \omega t$ and work out the amplitude and phase of each current at the cut-off frequency ω_C.

3.6 THE DECIBEL

The bel is a logarithmic unit of power ratio, originally devised to express power ratios of sounds, but subsequently extended in use into the field of electronics. It is defined as the logarithm to base 10 of the power ratio, so we can use it for an electrical network to express the power gain of the output over the input:

$$\text{Power gain in bels} = \log_{10}\left(\frac{P_o}{P_i}\right)$$

where P_i is the electrical input power and P_o electrical output power.

The bel proves to be an inconveniently large unit (1 bel represents a power ratio of 10), so it is divided into decibels. The power gain in decibels (dB) is given by the formula

$$\text{Power gain in decibels} = 10\log_{10}\left(\frac{P_o}{P_i}\right) \qquad (3.4)$$

and there are 10 decibels in 1 bel.

Worked example

What power ratio does 1 dB represent?

$1 = 10\log_{10}(P_o/P_i)$, so $P_o/P_i = \text{antilog}(1/10) = 1.26$.

Decibels can also be used to express ratios of sinusoidal voltages (r.m.s. values or amplitudes). Suppose that a network has an input resistance of value R, and that the output feeds a load resistor of the same value, then

$$\text{Input power} = \frac{V_i^2}{R}; \quad \text{Output power} = \frac{V_o^2}{R}$$

where V_i and V_o are r.m.s. values. Substituting these expressions in the formula for power gain in decibels, we get

$$\text{Power gain in decibels} = 10\log_{10}\left(\frac{V_o^2}{V_i^2}\right) \qquad (3.5)$$

and, because $\log_{10}x^2 = 2\log_{10}x$, Equation 3.5 can be expressed as

$$\text{Voltage gain in decibels} = 20\log_{10}\left(\frac{V_o}{V_i}\right) \qquad (3.6)$$

Because sinusoidal amplitudes are proportional to r.m.s. values, V_o/V_i in the above formula can also be the ratio of amplitudes.

So long as the resistance values at input and output are the same, a gain quoted as X dB will give either the power ratio, by inverting Equation 3.5,

$$P_o/P_i = \text{antilog}\ (X/10)$$

or the voltage amplitude ratio from Equation 3.6:

$$V_o/V_i = \text{antilog}\ (X/20)$$

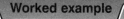

It has, however, become the practice to quote *voltage* gain in decibels, defined by Equation 3.6, even when the resistance fed by the output is not the same as the input resistance. Where this is the case, care must be taken not to use the quoted value of decibels as a direct indication of power ratio.

Current gain can be stated similarly in decibels. The appropriate formula is

$$X = 20\log_{10}\left(\frac{I_o}{I_i}\right)$$

where I_o and I_i can be either both r.m.s. values or both amplitude values. Again, this will only correspond with power gain if the resistances are the same at the input and output.

The following approximate values are worth remembering. You should check them from the formula using a calculator. In terms of voltage amplitude ratios:

$+3$ dB is equivalent to $\sqrt{2}$ times $\simeq 1.4$ times
$+6$ dB is equivalent to 2 times
$+20$ dB is equivalent to 10 times exactly
-3 dB is equivalent to $1/\sqrt{2}$ times $\simeq 0.7$ times
-6 dB is equivalent to 1/2 times

References here to gain do not mention gain *magnitudes*, but this must be understood whenever a gain is expressed in decibels. A true gain is a phasor operator which has angle as well as magnitude, but the decibel value cannot tell us anything about the phase angle.

Self-assessment question

28 Convert voltage ratios of 3, 6, 40, 100 and 2000 to decibels.

Self-assessment question

29 Convert 12 dB, 18 dB, 26 dB, -12 dB, -18 dB, -26 dB to voltage ratios.

Worked example

Two amplifiers with voltage gains of 13 dB and 7 dB, at a particular frequency, are connected in cascade, with the output of one feeding the input of the other. What is the overall voltage gain, assuming that connecting the second amplifier does not affect the gain of the first, or vice versa?

When two (or more) networks are cascaded, provided they do not interact to affect each other's gains, the overall gain, expressed as a voltage ratio, will be the product of the individual gains. This can be seen by reference to Fig. 3.27. The voltage gains G_1 *and* G_2 *are given by*

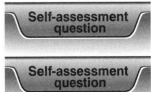

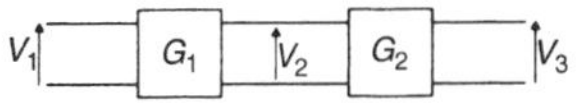

Fig. 3.27 V_1, V_2 and V_3 are r.m.s. values.

$$G_1 = \frac{V_2}{V_1};\quad G_2 = \frac{V_3}{V_2}$$

The overall gain is $V_3/V_1 = G_1 \times G_2$.

Now, the overall gain in decibels will be

$$20\log_{10}(V_3/V_1) = 20\log_{10}(G_1G_2)$$

which can be expressed as

$$20\log_{10}(V_3/V_1) = 20\log_{10}G_1 + 20\log_{10}G_2$$

Thus:

The overall gain in decibels for non-interacting circuits in cascade is the sum of the individual gains in decibels.

For the values quoted in the worked example above, we have:

$$\text{Overall gain} = 20\,\text{dB}$$

which is equivalent to an input–output voltage amplitude ratio of 10.

You can check this result as follows. First of all, the gain of 13 dB is equivalent to a voltage ratio

$$\text{antilog}(13/20) = \text{antilog}\,0.65 = 4.47\,\text{times}$$

For the gain of 7 dB we have the voltage ratio

$$\text{antilog}(7/20) = \text{antilog}\,0.35 = 2.24\,\text{times}$$

The overall voltage ratio is therefore $4.47 \times 2.24 = 10$, as before.

Worked example

What will the overall voltage ratio be if the two amplifiers have a 6 dB attenuator (a network with a gain of −6 dB) connected between them?

The overall gain is now 14 dB which is equivalent to a voltage ratio of 5. We could find the answer using a calculator but it is easier to argue that subtracting 6 dB divides the gain by 2, so the answer must be 5.

To work in decibels the practising engineer hardly ever needs to use a calculator. For example, suppose you need to know what voltage ratio is represented by 17 dB.

Now 20 dB would be 10 times. Subtracting 3 dB would be equivalent to dividing by $\sqrt{2}$. Hence 17 dB represents $10/\sqrt{2}$ times or about 7 times. Look at Self-assessment question 29 again; it could have been answered by this means without bothering to use the formula! If you have to work in decibels a great deal you will soon become adept at this technique.

Self-assessment question

30 Two amplifiers having voltage gains of 100 and 50 at 1 MHz are connected by 2 metres of attenuating cable. If the cable attenuation is 1.5 dB per metre at 1 MHz, and assuming that the amplifier gains are not affected by the connections, find the overall gain in decibels at 1 MHz.

3.7 BODE PLOTS

Earlier in this chapter, the amplitude ratio and phase shift of simple filter circuits are plotted against frequency. These plots use a logarithmic scale for the frequency axis, a linear scale for the phase-shift axis and a logarithmic scale for the amplitude ratio axis. An alternative approach is commonly used, in which the amplitude ratios are expressed in *decibels*, while the linear phase-shift scale and the logarithmic frequency scale are retained. When these conventions are followed, the graphs showing the amplitude/frequency response and the phase/frequency response are known as **Bode plots**.

You will now see how Bode plots can be plotted directly from the voltage transfer function of a linear circuit, and how to make quick sketches of Bode plots using linear (or 'straight-line') approximations.

3.7.1 Bode plots for simple low-pass networks

You will recall that the low-pass circuit shown in Fig. 3.28 has a voltage transfer function

$$\frac{V_o}{V_i} = \frac{1}{1 + j\omega CR}$$

The circuit has a low-pass cut-off frequency

$$\omega_c = 2\pi f_c = 1/CR$$

Substituting $CR = 1/\omega_c$, we obtain

$$\frac{V_o}{V_i} = \frac{1}{1 + j\omega/\omega_c}$$

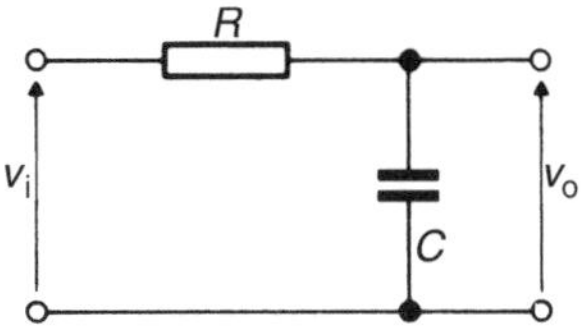

Fig. 3.28 *CR* circuit, low-pass.

The mathematical derivation of a transfer function almost always leads to an expression in terms of the angular frequency ω. However, when working 'at the bench', we invariably work with the frequency, *f*, given in hertz. To make for an easier comparison between theoretical and experimental results, therefore, we can substitute $\omega = 2\pi f$ and $\omega_c = 2\pi f_c$ in the transfer function and use the equivalent expression

$$\frac{V_o}{V_i} = \frac{1}{1 + jf/f_c}$$

where f_c is the cut-off frequency in hertz.

Self-assessment question

31 Identify the cut-off frequency of the *RL* network shown in Fig. 3.29 and hence show that its voltage transfer function can be written in the form

$$\frac{V_o}{V_i} = \frac{1}{1 + jf/f_c}$$

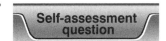

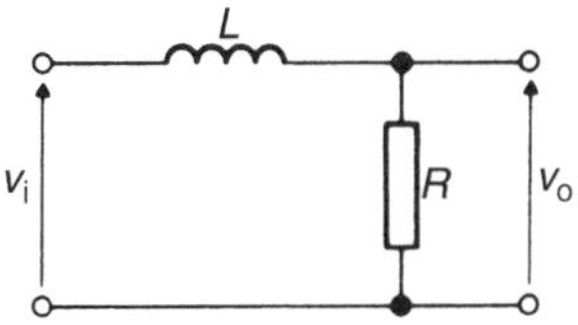

Fig. 3.29 *LR* circuit, low-pass.

The results of Self-assessment question 31 show that the transfer function derived for the low-pass *CR* network is a standard form which can be related to any low-pass network containing a resistor and a single reactive component.

The general amplitude/frequency response curve of simple low-pass networks

The magnitude of the voltage transfer function for the low-pass circuits considered above is the amplitude ratio:

$$\frac{V_o}{V_i} = \frac{1}{\sqrt{[1 + j(f/f_c)^2]}}$$

This can be expressed as a voltage gain in decibels:

$$\begin{aligned} G(\text{dB}) &= 20\log_{10}|V_o/V_i| \\ &= 20\log_{10}[1 + (f/f_c)^2]^{-1/2} \\ &= -10\log_{10}[1 + (f/f_c)^2] \end{aligned}$$

Self-assessment question

32 Calculate the values of the voltage amplitude ratio, in decibels, at the following frequencies;
(i) $f = f_c/100$ (ii) $f = f_c/10$ (iii) $f = f_c$
(iv) $f = 10f_c$ (v) $f = 20f_c$ (vi) $f = 100\,f_c$

Your 'spot-frequency' values from Self-assessment question 32 should lie on the curve shown in Fig. 3.30. This graph, in which voltage gain in decibels is plotted against log frequency is the amplitude/frequency response of a simple low-pass circuit. It is shown as a **Bode amplitude** or **Bode gain plot**, where the 'gain' is strictly the gain *magnitude* of the circuit at different frequencies. Notice that the frequency axis is labelled in multiples of f_c. At the frequency f_c the gain magnitude is $1/\sqrt{2}$, which is equivalent to -3 dB very nearly; for this reason the cut-off frequency is also referred to as the **3 dB point**.

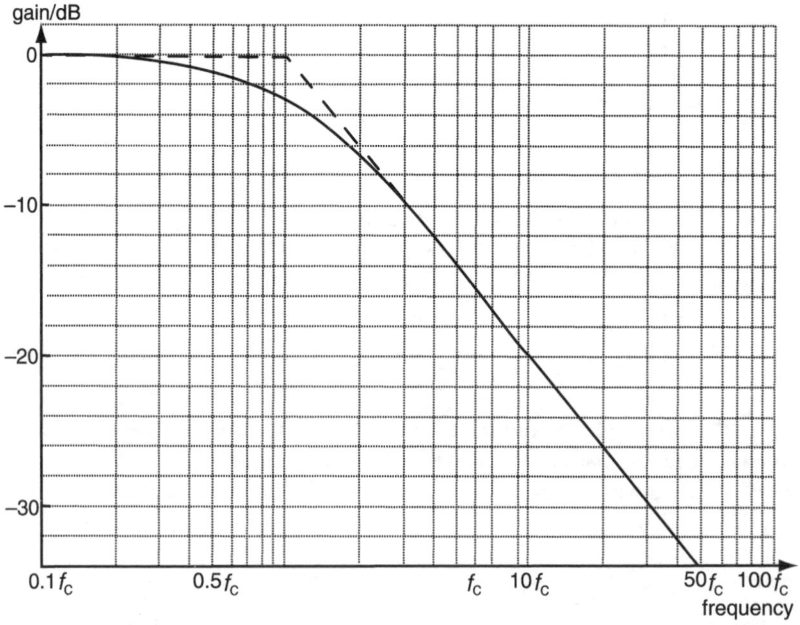

Fig. 3.30 Amplitude–frequency response of the *CR* network, low-pass.

Above $f \simeq 5f_c$, the Bode plot is a straight line, corresponding to the fact that $(f/f_c)^2 >> 1$ in the denominator of the voltage transfer function. Similarly, below $f \simeq 0.2f_c$, $(f/f_c)^2 << 1$ so that the gain is effectively unity, i.e. 0 dB. These straight parts are extended on the diagram and you will see that they meet at the 3 dB frequency. It is often convenient in design to approximate the low-pass response curve by the two straight lines shown. This sort of linearizing approximation can very often be applied to amplitude/frequency response curves; for this reason a 3 dB point is often also referred to as a **break point or corner frequency**. You will notice that the greatest error in this linear approximation occurs at the corner frequency, and is an error of +3 dB.

If you look at the falling part of the linear approximation to the response curve (at frequencies above the break point) you will see that the gain is inversely proportional to the frequency, so each time the frequency is doubled the gain falls by 6 dB. This is called a roll-off of *6 dB per octave* (an octave rise in pitch in music represents a doubling of frequency). When the frequency increases ten-fold the fall in gain is 20 dB, so the roll-off can also be expressed as *20 dB per decade*.

Finally, notice that we now have four different terms describing the frequency f_c: (i) cut-off frequency, (ii) 3 dB point, (iii) break point, or (iv) corner frequency. You should regard these terms as being equivalent. They can be interchanged without altering the meaning of a statement or sentence.

The general phase/frequency response curve of simple low-pass networks

To find how the phase shift of a low-pass network varies with frequency, we must calculate the angle of its voltage transfer function, given by

$$\phi = \angle\frac{V_o}{V_i}$$

Using the standard form

$$\frac{V_o}{V_i} = \frac{1}{1 + jf/f_c}$$

we obtain

$$\begin{aligned}\phi &= \angle(1) - \angle(1 + jf/f_c) \\ &= 0 - \tan^{-1}(f/f_c) \\ &= -\tan^{-1}(f/f_c)\end{aligned}$$

A Bode plot of this function is shown in Fig. 3.31. It is also possible to make a linearizing approximation to this curve. In the approximation, all the change of phase lag is assumed to occur between 1/10 times the cut-off frequency and 10 times the cut-off frequency. The linearized curve is shown as a broken line in Fig. 3.31.

You will see from Fig. 3.31 that, as the frequency increases beyond cut-off, the total phase shift tends to $\pi/2$ radians or 90°.

A low-pass network which produces a phase lag increasing with frequency and tending towards a value of 90° at very high frequencies is sometimes called a **single-lag circuit**. The lag is usually associated with one capacitor or one inductor, and the single-lag circuit produces a roll-off of 6 dB per octave.

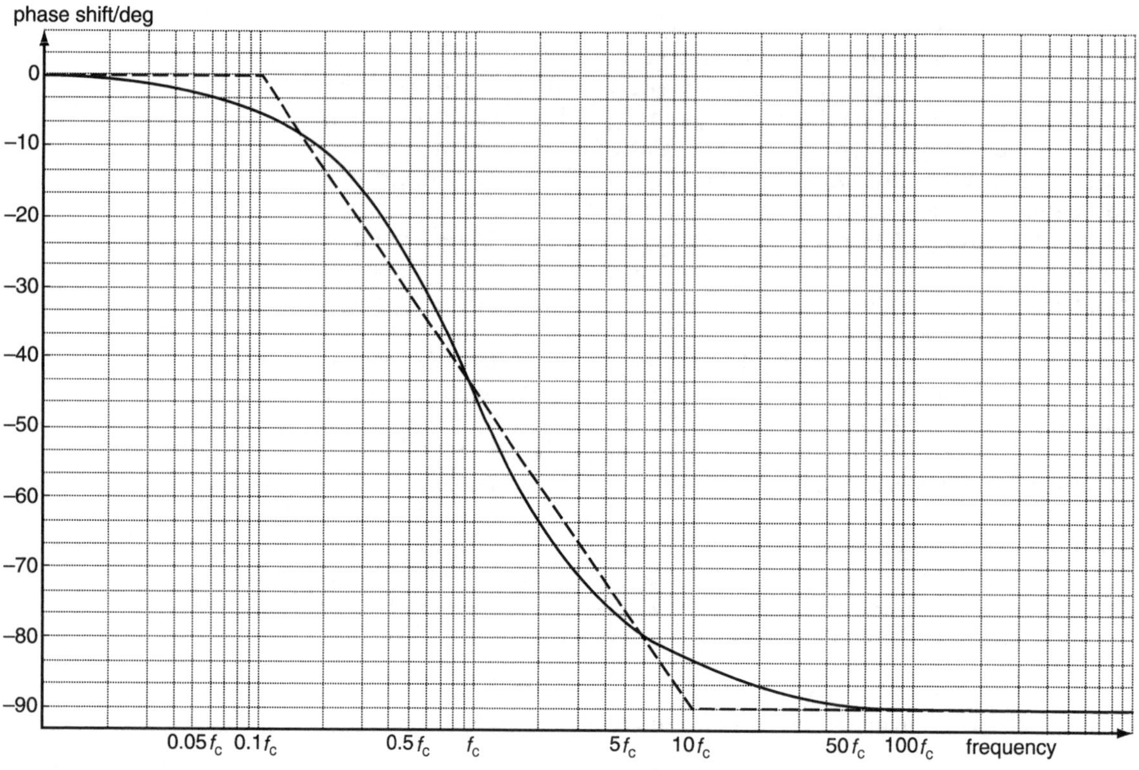

Fig. 3.31 Phase response of the *CR* network, low-pass.

Self-assessment question

33 In a particular low-pass *CR* network *R* is 1 kΩ and *C* is 2 μF. At what frequency is the break point? At what frequency is the gain ratio 1/10? What is the phase shift in degrees at each of these frequencies?

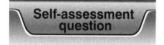

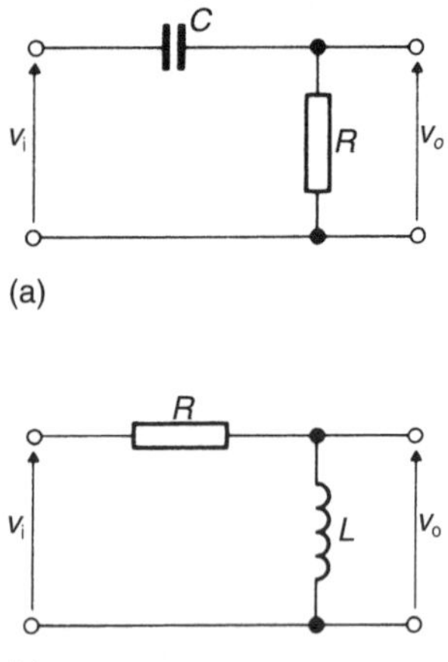

Fig. 3.32 (a) *CR* circuit, high-pass. (b) *LR* circuit, high-pass.

3.7.2 Simple high-pass networks

You should recognize the circuits in Fig. 3.32 as high-pass circuits. Their transfer functions can be written as

$$\frac{V_o}{V_i} = \frac{R}{R + 1/j\omega C} = \frac{1}{1 + 1/j\omega CR}$$

and

$$\frac{V_o}{V_i} = \frac{j\omega L}{R + j\omega L} = \frac{1}{1 + R/j\omega L}$$

You have already worked with these high-pass transfer functions when answering Self-assessment questions 26 and 27 in this chapter. If we substitute the high-pass angular cut-off frequencies $\omega_c = 1/RC$ and $\omega_c = R/L$ for the *CR* and *LR* circuits, respectively, we obtain

$$\frac{V_o}{V_i} = \frac{1}{1 - j\omega_c/\omega}$$

The result is a standard form of the transfer function which applies to both types of high-pass circuit. For the reasons given earlier it is preferable to substitute $\omega = 2\pi f$ and $\omega_c = 2\pi f_c$ giving

$$\frac{V_o}{V_i} = \frac{1}{1 - jf_c/f}$$

At f_c, $V_o/V_i = 1(1 - j1) = 1/(\sqrt{2}\angle{-45°}) = 0.707\angle{+45°}$, so f_c is the 3 dB point. At higher frequencies, where $f > f_c$,

$$V_o/V_i = 1(1 - j0) \approx 1$$

so the circuit passes higher frequencies with little attenuation. At lower frequencies, where $f < f_c$,

$$V_o/V_i = 1/(-jf_c/f) = +jf/f_c$$

so the output rises with frequency, at 6 dB/octave, and has a *leading* phase shift of 90° at low frequencies.

The overall result in Fig. 3.33 is a Bode plot which rises from low frequencies at 6 dB/octave and then levels out at 0 dB above the cut-off frequency. You should recognize this as the amplitude frequency response of a high-pass network.

Use your circuit analysis package to check again the frequency responses of the low-pass and high-pass *CR* and *LR* networks which you constructed in Chapter 2. (Build them now if you did not before.) This time, choose logarithmic frequency axes and a decibel scale for amplitude, so you have Bode plots. Check that the slopes are 6 dB/octave (or 20 dB/decade) and that the predicted cut-off frequencies are obtained.

Self-assessment question

34 Starting from the standard form of the transfer function, sketch an approximation to the phase frequency response of the high-pass network.

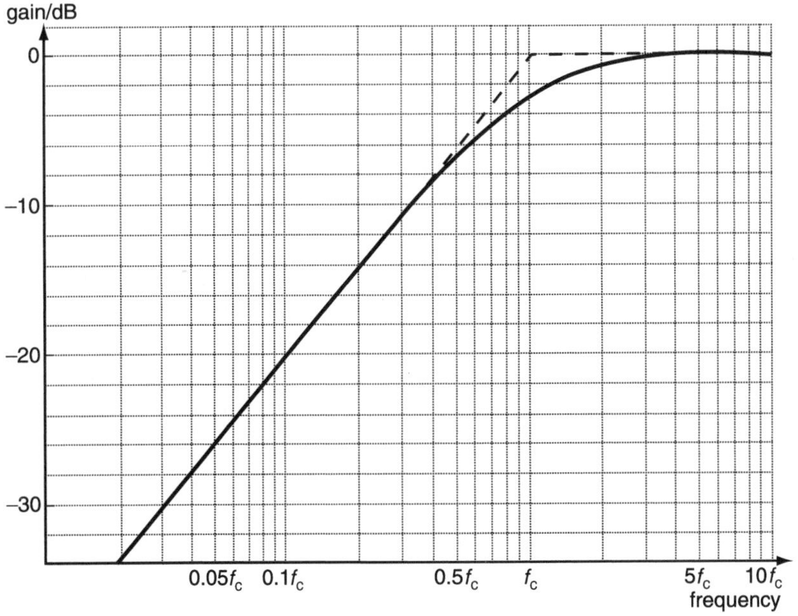

Fig. 3.33 Amplitude–frequency response of the *CR* network, high-pass.

You should have found in Self-assessment question 34 that the high-pass network introduces a phase lead which tends towards 90° at low frequencies and towards 0° at high frequencies, as in Fig. 3.34. We say that the high-pass *CR* and *LR* circuits are examples of **single-lead networks**.

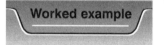

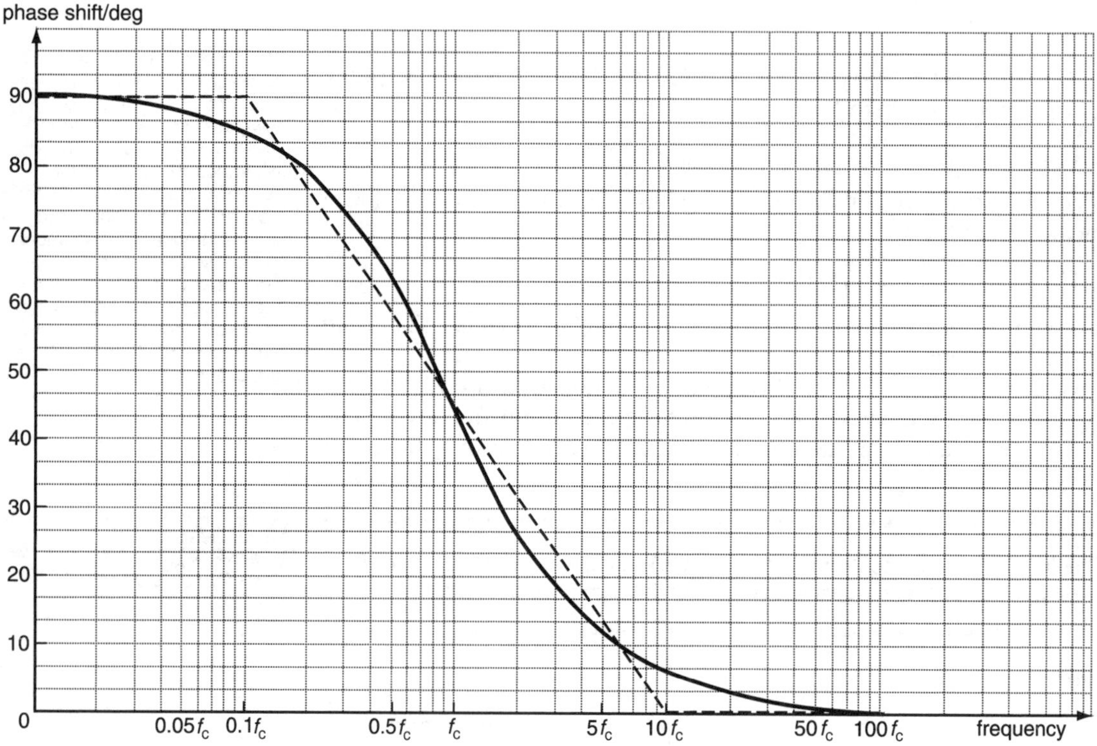

Fig. 3.34 Phase response of the *CR* network, high-pass.

3.7.3 Networks with two break points

Suppose that two different low-pass *CR* networks with cut-off frequencies f_1 and f_2 are cascaded with a buffer amplifier between them, so that the second network does not affect the operation of the first. The circuit is shown in Fig. 3.35. In order to simplify the analysis we can assume that the buffer acts as an ideal voltage amplifier with unity gain and zero phase shift *over all frequencies of interest*. This means that its transfer function will be approximated by $\mathbf{G} = 1 \angle 0$.

Worked example

Under what conditions is this assumption likely to be valid?

When the actual cut-off frequency of the buffer, f_c, is very much greater than the cut-off frequencies f_1 and f_2 of the two CR circuits.

Now, referring to Fig. 3.35, we have

$$\mathbf{A}_1 = \frac{V_B}{V_i} = \frac{1}{1 + jf/f_1}$$

and

$$\mathbf{A}_2 = \frac{V_o}{V_B} = \frac{1}{1 + jf/f_2}$$

and so

$$\frac{V_o}{V_i} = \mathbf{A}_1 \times \mathbf{A}_2 = \frac{1}{(1 + jf/f_1)(1 + jf/f_2)}$$

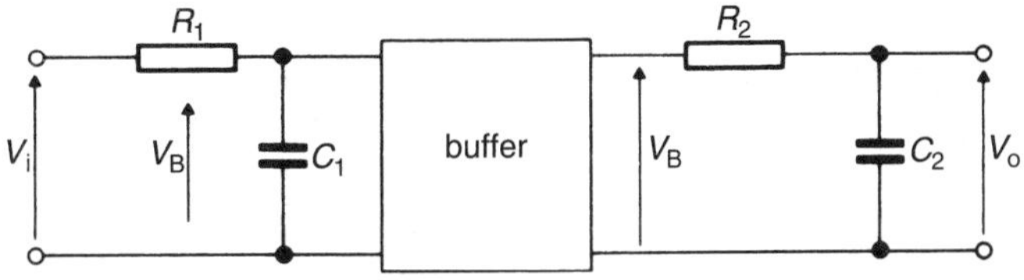

Fig. 3.35 Two *CR* networks in cascade (buffered).

Amplitude response of the two *CR* networks

The overall frequency response of the buffered *CR* circuits can be found by adding together the Bode plots for the individual transfer functions $\mathbf{A}_1$ and $\mathbf{A}_2$.

Self-assessment question

35 Figure 3.36 shows linear approximations for the amplitude/frequency responses of two low-pass *CR* circuits Sketch an approximation for the response of the two circuits in cascade, assuming that they are buffered by an ideal voltage amplifier with unity gain.

You should have found in Self-assessment question 35 that the composite plot, representing the sum of the linear approximations rolls-off at 12 dB/octave above f_2. The two single-lag circuits in cascade act as a **double-lag circuit**.

Figure 3.37 shows the effect for the specific values $f_1 = 1$ kHz and $f_2 = 20$ kHz. You can see the accurate graph, and the linear approximation obtained by drawing straight lines between the break points. The error involved in taking the linear approximation is greatest at the break points. Because the break points are well separated, the error is still about 3 dB. For break points closer together it is greater, rising to 6 dB when the two break points coincide.

Phase response of the two *CR* networks

To find the overall phase shift we must add together the phase shifts of the individual networks, giving the following expression:

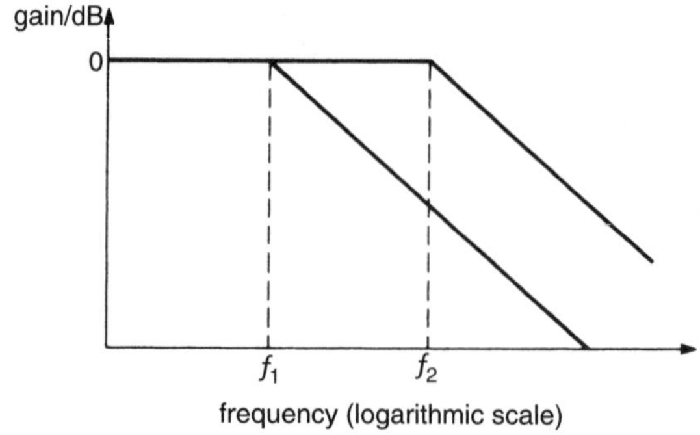

Fig. 3.36 Bode plots for the two *CR* networks.

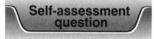

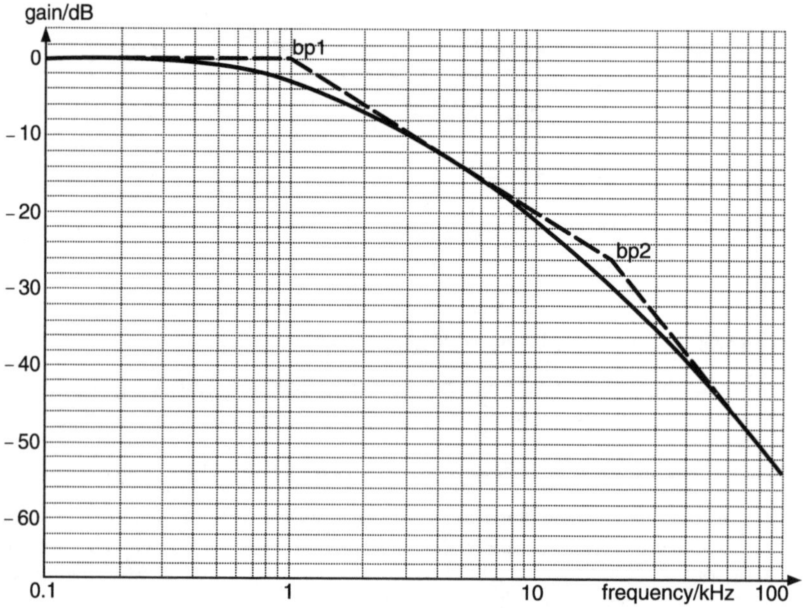

Fig. 3.37 Composite amplitude–frequency response for two low-pass networks in cascade.

$$\angle(V_o/V_i) = -\tan^{-1}(f/f_1) - \tan^{-1}(f/f_2)$$

The curve, for $f_1 = 1$ kHz and $f_2 = 20$ kHz, is plotted out in Fig. 3.38. The maximum phase variation over all frequencies is now π radians (180°). A linearized approximation to the overall phase response is shown in Fig. 3.39, where the linearized responses of the two individual *CR* networks have been added together.

Self-assessment question

36 Figure 3.40 shows a circuit in which the signal passes through a high-pass and a low-pass network. Write down the overall voltage transfer function and sketch the amplitude response curve and phase response curve using linear approximations. The component values are: $C_1 = 1$ μF; $C_2 = 10$ nF; $R_1 = 1500$ Ω; $R_2 = 1500$ Ω.

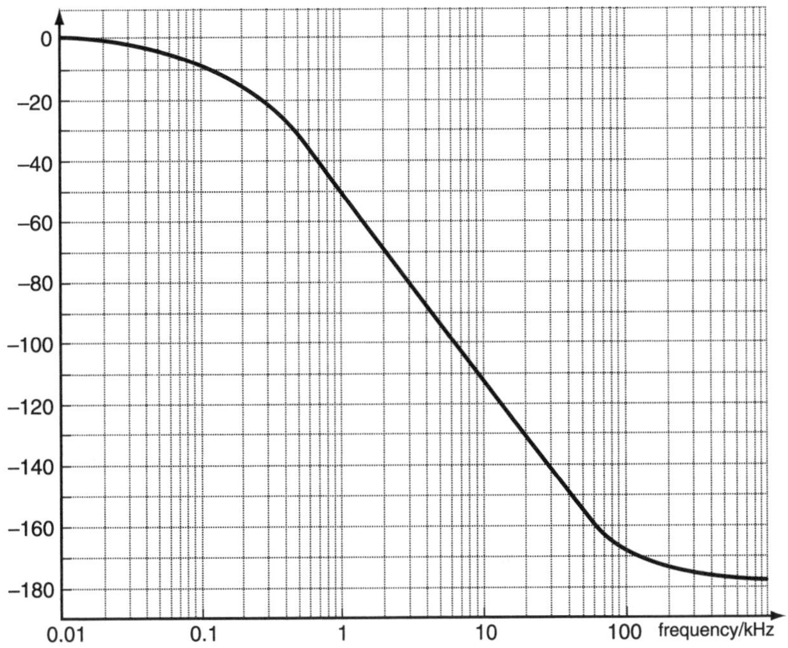

Fig. 3.38 Composite phase response for two low-pass networks in cascade.

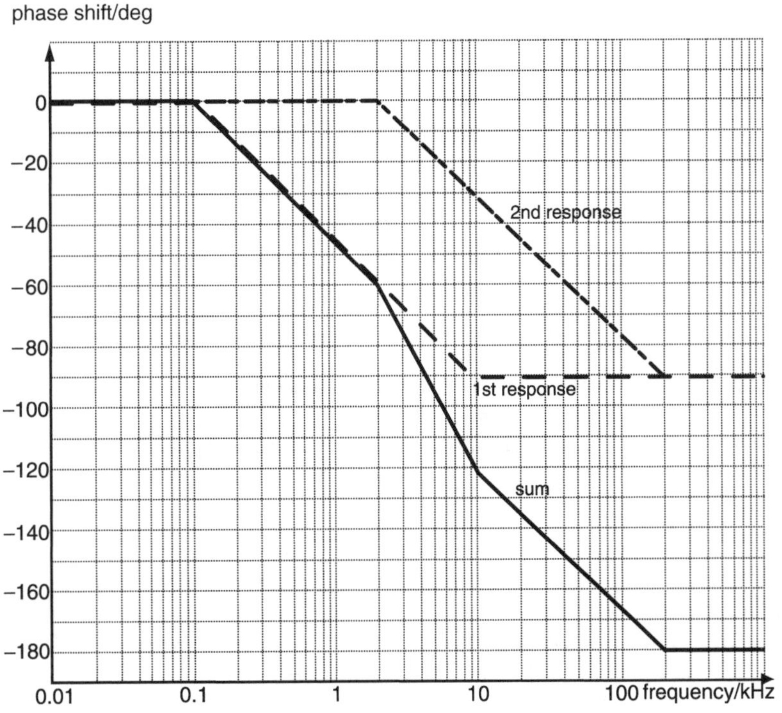

Fig. 3.39 Linear approximation to the phase response of two low-pass networks in cascade, obtained by adding the linear approximations to their individual responses.

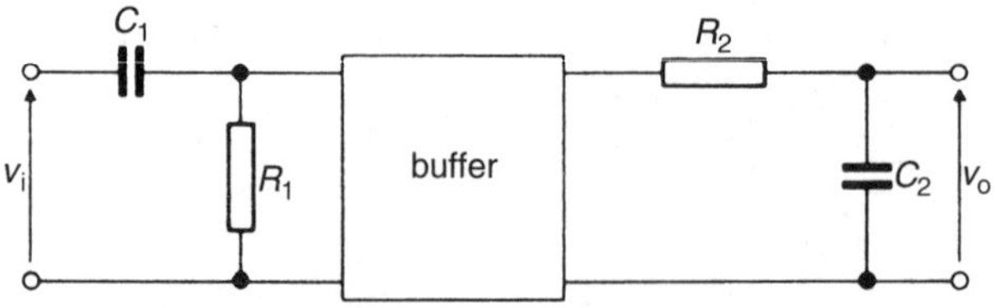

Fig. 3.40 A high-pass and a low-pass network in cascade (buffered).

3.8 THE FREQUENCY RESPONSE OF *RLC* CIRCUITS

So far we have dealt mainly with circuits containing resistors and just one type of reactive component. Now we shall look at some examples of *RLC* circuits, that is circuits which contain resistors, capacitors and inductors.

Such circuits generally exhibit what is known as *resonant* behaviour, which will be explained first.

3.8.1 The series *RLC* circuit

Figure 3.41 shows a series *RLC* circuit. The resistive losses in the inductor and the output resistance of the voltage source are all included in the series resistance *R*.

The impedance of the *RLC* circuit can be found by adding the impedances of the individual components in series, giving

$$\begin{aligned}\mathbf{Z} &= R + \mathrm{j}\omega L - \mathrm{j}/\omega C \\ &= R + \mathrm{j}(\omega L - 1/\omega C)\end{aligned}$$

Since the inductive reactance increases with frequency while the capacitive reactance decreases, there will be some frequency ω_0 at which the two are the same. We then have $\omega_0 L = 1/\omega_0 C$, giving

$$\omega_0 = 1/\sqrt{LC}$$

ω_0 is known as the **resonant frequency** of the series *RLC* circuit.

Worked example

What is the impedance of the *RLC* circuit at resonance?

The phasor impedances $\mathrm{j}\omega L$ *and* $1/\mathrm{j}\omega C$ *become equal in magnitude at resonance. The total impedance of the reactive components is then zero and the impedance is purely resistive, with* $\mathbf{Z} = R$.

The impedance of the *RLC* circuit at resonance is sometimes known as the dynamic impedance.

At a frequency ω, the phasor voltages V_L and V_C shown in Fig. 3.41 are related to the input phasor V_s by the voltage divider rule:

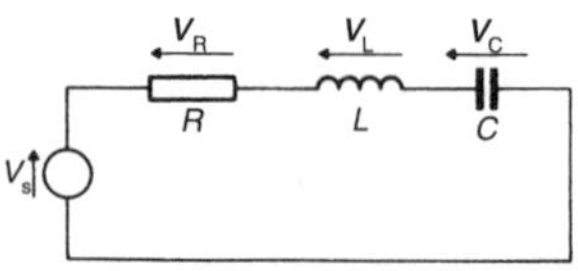

Fig. 3.41 Series *RLC* circuit.

$$V_L = \frac{\mathrm{j}\omega L}{\mathbf{Z}} V_s \quad \text{and} \quad V_C = \frac{1/\mathrm{j}\omega C}{\mathbf{Z}} V_s$$

where **Z** is the series impedance. At resonance, we can substitute $\omega = \omega_0$ and $\mathbf{Z} = R$ and obtain the relationships

$$V_L = \frac{j\omega_0 L}{R} V_S \quad \text{and} \quad V_C = \frac{1}{j\omega_0 CR} V_S$$

In terms of voltage amplitudes we have

$$V_L = \frac{\omega_0 L}{R} V_S \quad \text{and} \quad V_C = \frac{1}{\omega_0 CR} V_S$$

Now, at the resonant frequency ω_0, the reactance of the inductor becomes equal to that of the capacitor, that is

$$\omega_0 L = 1/\omega_0 C$$

Dividing both sides of this expression by R we see that $\omega_0 L/R = 1/\omega_0 CR$, which means that the amplitudes of V_L and V_C will be identical at resonance. We can then write

$$V_L = V_C = QV_S$$

where

$$Q = \frac{\omega_0 L}{R} = \frac{1}{\omega_0 CR}$$

is known as the ***Q*-factor** or **quality factor** of the circuit.

Self-assessment question

37 A radio-frequency *RLC* circuit has the component values $L = 60\ \mu H$, $C = 30$ pF, $R = 10\ \Omega$. Refer to Fig. 3.41 and find:

(a) The resonant frequency and *Q*-factor.

(b) The phase of the circuit current at resonance, relative to that of the input voltage.

(c) The values of the voltage phasors $\mathbf{V_L}$ and $\mathbf{V_C}$ at resonance when the input phasor is $1 \angle 0$ V.

Self-assessment question 37 shows that if the resonant frequency is sufficiently high and the series resistance R is sufficiently low then the Q-factor can take very large values. In a circuit where R accounts for the series winding resistance of the coil, a low value of R denotes a coil of high 'quality', hence the use of the term 'quality-factor' in this context.

Now, if Q is greater than 1, the inductor and capacitor voltage amplitudes will be bigger than the input voltage amplitude at resonance. To see how this can happen, look at the relationship

$$V_S = V_R + V_L + V_C$$

given by Kirchhoff's voltage law.

At resonance $|V_L| = |V_C|$. However, V_L leads V_S by 90° while V_C lags V_S by 90°. In other words V_L and V_C have identical amplitudes but they are opposed in phase by 180°. The total voltage across the inductor and capacitor in series is therefore zero, leaving the entire source voltage developed across the resistor.

Thus, at resonance, the circuit current is in phase with the applied voltage and we have

$$V_R = V_S, \quad I = V_S/R$$
$$V_L = V_C = QV_S$$

The Q-factor represents the ratios of the magnitudes of the capacitor impedance (or the inductor impedance) to the total impedance of the circuit at resonance. In Self-assessment question 37, $Q = 141$, which accounts for the large voltage 'magnification' which is obtained.

Self-assessment question

38 (a) Using the equations for the resonant frequency and Q-factor of the series RLC circuit, show that the Q-factor can be expressed in the form

$$Q = \frac{1}{R}\sqrt{L/C}$$

(b) A series RLC circuit has $L = 10\ \mu\text{H}$, $C = 1$ nF. What is the resonant frequency of the circuit, in radians/second and in hertz? What is the maximum value of R for which the capacitor voltage magnitude will be greater than the input voltage magnitude at the resonant frequency?

Self-assessment question

39 A coil having an inductance of 100 μH and a series resistance of 5 Ω is connected in series with a capacitor. The resonant frequency of the circuit is measured to be 30 kHz. What is

(a) The value of the capacitor?
(b) The Q-factor of the circuit?
(c) How will the resonant frequency and the Q-factor change if the value of the capacitor is reduced by a factor of 10?

In summary: for a series RLC circuit, at resonance the dynamic impedance of the circuit is purely resistive and given by the series resistance, R. The current in the circuit is then in phase with the applied voltage and the voltage across the inductor and capacitor are equal in amplitude and opposite in phase.

At resonance the amplitudes of the capacitor and inductor voltages are equal to the amplitude of the input voltage multiplied by the Q-factor. The Q-factor increases as the series resistance is reduced, while the resonant frequency is unchanged.

3.8.2 The series *RLC* bandpass circuit

RLC circuits are used extensively in radio tuners to select a narrow band of frequencies in preference to other signals picked up by the radio aerial.

In the RLC circuit of Fig. 3.42 the output is the voltage developed across the resistor. The important features of the frequency response can be inferred by looking at the impedance of the circuit. Remember that the frequency response magnitude shows how the amplitude ratio V_R/V_S varies with frequency.

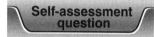

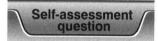

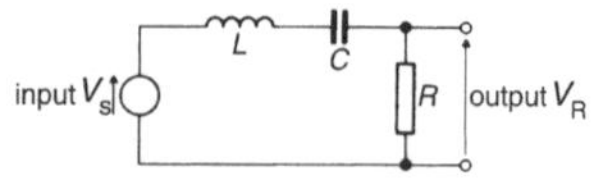

Fig. 3.42 Series *RLC* circuit, output taken across the resistor.

The circuit impedance is

$$\mathbf{Z} = R + j\omega L - j/\omega C$$

Worked example

What is the impedance of the *RLC* circuit at the low and high extremes of frequency?

'Low' and 'high' frequencies in the present context mean with respect to the resonant frequency ω_0. Now, for $\omega \ll \omega_0$, the capacitor impedance is very high compared to the inductor impedance. Thus, at low frequencies, the circuit behaves like the CR *high-pass circuit drawn in Fig. 3.43a. When $\omega \gg \omega_0$, the situation is reversed and the circuit behaves like the* LR *circuit in Fig. 3.43b. You should be able to see that this works like a low-pass circuit; both of these effects together make the* RLC *circuit a bandpass circuit, cutting off both low and high frequencies.*

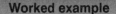

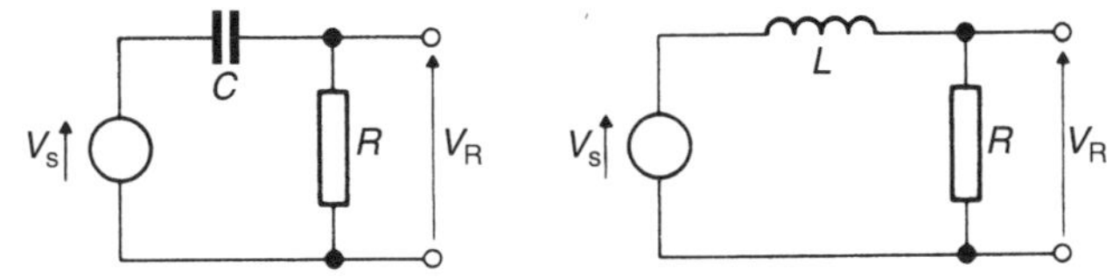

Fig. 3.43 Behaviour of the series *RLC* circuit for (a) $\omega \ll \omega_0$ and (b) $\omega \gg \omega_0$.

A rough approximation to the amplitude frequency response is shown in Fig. 3.44. The response 'peaks' in the region of the resonant frequency, ω_0. Since $|V_R| = |V_S|$ at resonance, the amplitude ratio must have a maximum value, $|V_R/V_S| = 1$ for $\omega \simeq \omega_0$ Notice that the diagram also shows the two frequencies ω_1 and ω_2 for which the amplitude ratio falls to -3 dB ($\times$ 0.707) of its maximum value. The difference between these values defines the **3 dB bandwidth** of the bandpass *RLC* circuit.

Self-assessment question

40 The voltage transfer function of the series bandpass circuit is given by

$$\frac{\mathbf{V}_R}{\mathbf{V}_S} = \frac{R}{R + j\omega L - j/\omega C}$$

(a) Derive expressions for the amplitude ratio and for the phase shift in the circuit.
(b) Confirm that your results give $V_R/V_S = 1$ and phase angle $\phi = 0$ when $\omega = \omega_0$.
(c) Show that the amplitude ratio is equal to $1/\sqrt{2}$ at the frequencies for which $(\omega L - 1/\omega C)^2 = R^2$

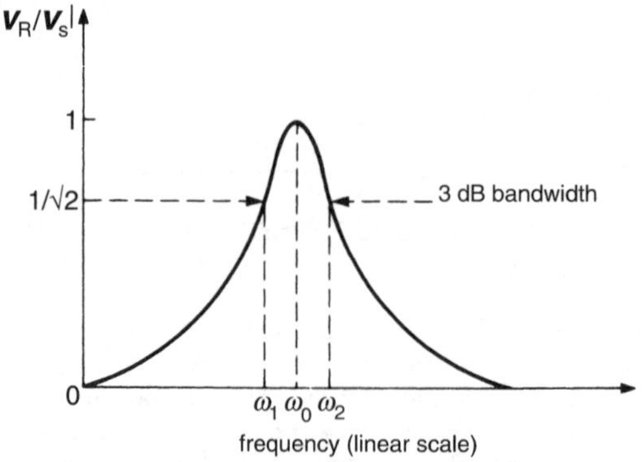

Fig. 3.44 Approximate frequency response magnitude for the bandpass *RLC* circuit.

3.8.3 Bandwidth of the series bandpass circuit

Following on from Self-assessment question 40, there are two positive values of ω which satisfy the equation

$$(\omega L - 1/\omega C)^2 = R^2$$

These values correspond to the upper and lower 3 dB frequencies ω_2 and ω_1 identified in Fig. 3.44. Taking the difference between the 3 dB frequencies gives the 3 dB bandwidth in rad s^{-1}:

$$W = \omega_2 - \omega_1$$

In order to save tedious calculations you can simply note the following relationship which you should try to remember:

$$W = \frac{\omega_0}{Q}$$

In other words, the bandwidth of the series *RLC* circuit is equal to the resonant frequency divided by the *Q*-factor.

We can also state this important result in the form

$$B = \frac{f_0}{Q}$$

where $B = W/2\pi$ is the bandwidth and $f_0 = \omega_0/2\pi$ is the resonant frequency, both given in hertz.

These results show that the bandwidth of the *RLC* circuit becomes smaller as the *Q*-factor is increased. We therefore associate a 'high-quality' circuit with a narrow bandwidth and a correspondingly 'sharp' resonant peak in its frequency response. Notice that the relationship

$$Q = \frac{\omega_0}{W} = \frac{f_0}{B}$$

may also be used to define the *Q*-factor of an *RLC* circuit. In practice, the actual value of *Q* might be found by dividing the measured value of the resonant frequency by the measured value of the 3 dB bandwidth.

Self-assessment question

41 (a) Using the equations defining W and Q, show that the bandwidth of the series *RLC* circuit can be written as

$W = R/L$

(b) What will be the smallest possible bandwidth when the coil in Self-assessment question 39 is connected into a series resonant circuit?

(c) If the coil is connected in series with a capacitor of 10 nF, what will be: (i) the resonant frequency; (ii) the Q-factor?

3.8.4 Bode plots for the bandpass *RLC* circuit

We now have virtually all the information we need concerning the frequency response of the bandpass *RLC* circuit. It can all be summarized by drawing Bode plots from the transfer function

$$\frac{V_R}{V_S} = \frac{R}{R + j\omega L - j/\omega C}$$

When you drew Bode plots for *LR* and *CR* circuits, the frequency scale was calibrated in multiples of the upper or lower cut-off frequency f_c. In this way they became universal curves which applied to any circuit of the appropriate type, no matter what the particular values of R and C (or L). It would also be useful to plot universal curves for the *RLC* circuit. To do this we need to express the frequency response in terms of the resonant frequency f_0 and the Q-factor. This can be done by making appropriate substitutions in the results derived previously.

Self-assessment question

42 By using the results

$Q = \omega_0 L/R = 1/\omega_0 CR$

show that the transfer function

$$\frac{V_R}{V_S} = \frac{R}{R + j\omega L - j/\omega C}$$

can be expressed in the form

$$\frac{V_R}{V_S} = \frac{1}{1 + jQ(f/f_0 - f_0/f)}$$

The expression $1 + jQ(f/f_0 - f_0/f)$ is known as a **standard form** of the denominator for the transfer function of a circuit containing resistance, capacitance and inductance. Each of the terms in this denominator becomes more or less important, depending on the value of the applied frequency f.

For example, *at resonance* $f = f_0$ the transfer function then becomes

$$\frac{V_R}{V_S} = 1\angle 0 \qquad \text{for } f = f_0$$

The resonant gain of the circuit has a magnitude of unity as expected, and we see that the phase angle of the circuit is zero.

At *very low frequencies,* when $f \ll f_0$ the magnitudes of the first two terms in the denominator become negligible compared with Qf_0/f. We can then write, to a good approximation:

$$\frac{V_R}{V_S} \approx \frac{1}{jQf_0/f}$$

$$= \frac{1}{Q}\left(\frac{f}{f_0}\right) \angle \pi/2 \qquad \text{for } f \ll f_0$$

For a given value of Q, the gain magnitude falls as f is reduced with respect to f_0 while the phase angle approaches $+\pi/2$ radians. This is consistent with the idea that, at low frequencies, the circuit behaves like a high-pass *RC* filter.

At *very high frequencies*, when $f \gg f_0$, the second term in the denominator is the dominant one and we have the approximation

$$\frac{V_R}{V_S} \approx \frac{1}{jQf/f_0}$$

$$= \frac{1}{Q}\left(\frac{f_0}{f}\right) \angle -\pi/2 \qquad \text{for } f \gg f_0$$

For a given value of Q, the gain magnitude is reduced as the applied frequency is increased with respect to f_0. The circuit thus operates as a low-pass filter to high frequencies with a high-frequency phase shift of $-\pi/2$ radians.

Worked example

What is the roll-off of the bandpass filter to low and high frequencies, expressed in dB/octave?

The magnitude of the voltage ratio decreases as 6 dB/octave, or 20 dB/decade, for frequencies well above and well below f_0. *The slope of –6 dB/octave is the same as for the* CR *and* LR *circuits. This is to be expected, since it is the capacitor which causes the low-frequency stop band and the inductor which causes the high-frequency stop band.*

The points covered in this informal discussion should help you to interpret the Bode plots shown in Fig. 3.45. Curves are shown for three different values of Q. You should be able to summarize the effects of the Q-factor on the frequency response in the following manner. The higher the Q-factor:

(a) the sharper is the frequency response in the vicinity of f_0;
(b) the narrower is the bandwidth for a given value of f_0;
(c) the lower is the response magnitude for a given value of f well outside the passband.

We also have the following additional information:

(a) The Bode plots of the bandpass *RLC* circuit roll off at a rate of 6 dB/octave at frequencies well away from the resonance frequency. It is clear from Fig. 3.45 that the overall attenuation at higher and lower frequencies increases as the Q-factor is increased, while the roll-off rate remains constant.

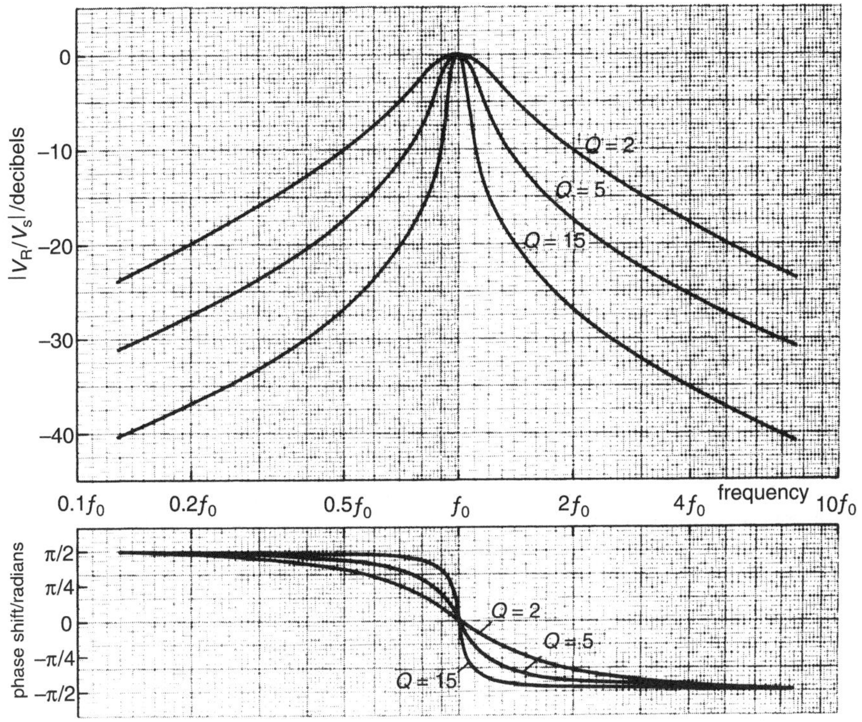

Fig. 3.45 Bode plots for the bandpass *RLC* circuit.

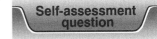

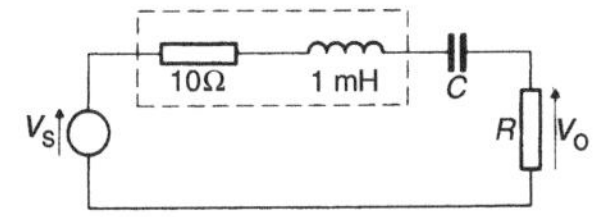

Fig. 3.46 A bandpass circuit using an inductor with losses represented by a series resistance.

(b) The phase angle varies from $+\pi/2$ at low frequencies to $-\pi/2$ at high frequencies, becoming zero at $f = f_0$. Notice from Fig. 3.45 that the extreme values of phase shift and the phase at $f = f_0$ do not change with the Q-factor. However, the transition from $+\pi/2$ to $-\pi/2$ radians becomes more abrupt as the Q-factor is increased.

Self-assessment question

43 Suppose you have a 1 mH inductor which has losses that can be represented by a series resistance of 10 Ω. Use this inductor to devise a bandpass circuit, as in Fig. 3.46, with a centre frequency of 10 kHz and choose R to give a bandwidth of 3 kHz. (Notice that these frequencies are in hertz, not radians/second.)

How will the coil resistance affect the maximum voltage ratio?

Is it possible to achieve a bandwidth of 1 kHz with this inductor?

Self-assessment question

44 A series bandpass resonant circuit is required to transmit a signal component at a frequency of 5 MHz with negligible attenuation and provide 40 dB of attenuation to an unwanted signal at 35 MHz.

(a) Calculate the Q-factor required, using the approximations derived above.

(b) By considering the circuit bandwidth, find how much the wanted signal is allowed to 'drift' from its nominal frequency of 5 MHz before the response falls to −3 dB of its maximum value.

3.8.5 Parallel resonance

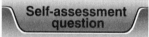

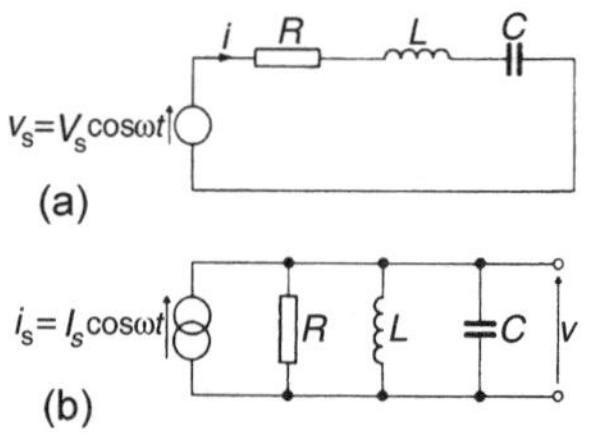

Fig. 3.47 Series and parallel *RLC* circuits.

Consider the two circuits shown in Fig. 3.47a and 3.47b. Circuit (a) contains a *voltage* source in *series* with the circuit elements. Circuit (b) contains a *current* source in *parallel* with the circuit elements. For every *voltage* in circuit (a) there is a corresponding *current* in circuit (b). Also, notice that the current i flows through each of the components in the series circuit while in circuit (b) it is the voltage v that is common to each of the components.

Admittance of the parallel resonant circuit

The current source in Fig. 3.47b supplies a current i_s with an amplitude I_s that remains constant over all frequencies of interest. In considering the response of this circuit to the current input it is convenient to work with the circuit *admittance* rather than with impedance.

Self-assessment question

45 Write down an expression for the admittance **Y** of the circuit in Fig. 3.47b. Hence derive expressions for the conductance G and the susceptance B of the circuit.

At the resonant frequency ω_0 the admittance of the circuit will be a pure conductance. Using results from Self-assessment question 45 we have, at the resonant frequency ω_0:

$$\omega_0 C = 1/\omega_0 L$$

giving

$$\omega_0 = 1/\sqrt{LC}$$

Current magnification in the parallel circuit

At resonance the voltage v developed across the circuit is in phase with the current i_S and is given by

$$V = \mathbf{I}_S/G = \mathbf{I}_S R$$

Worked example

Will the amplitude of the voltage be a maximum or a minimum at resonance?

A maximum, because the magnitude of the admittance is smallest at resonance, where $|\mathbf{Y}| = G = 1/R$.

In the series tuned circuit we looked at the voltages developed across each of the components. Here, in the parallel circuit we are interested in the distribution of the currents in the three branches of the circuit.

Self-assessment question

46 Derive expressions for the capacitor and inductor phasor currents $\boldsymbol{I}_C$ and $\boldsymbol{I}_L$ at resonance, in terms of the input current $\boldsymbol{I}_s$.

In Self-assessment question 46 you should have found that the capacitor current leads the applied current by 90° at resonance, while a similar current having opposite sign flows in the inductance. At resonance, the amplitudes of i_C and i_L are greater than the amplitude of i_s by an amount

$$Q = \omega_0 CR = R/\omega_0 L$$

In a manner similar to the series resonant circuit, we thus find that *current* magnification can occur in the parallel tuned circuit. Notice, however, that the Q-factor is the inverse of that obtained for the series tuned circuit. To obtain a high Q-factor we now require the value of R to be as large as possible. Alternatively, we can say that the conductance $G = 1/R$ should be as small as possible. It is important to remember that R is now the equivalent *parallel* resistance of the coil and quite different from the relatively small series resistance considered earlier.

3.9 THE PRINCIPLE OF DUALITY

Section 3.8.5 begins by pointing out the close correspondence between the series and parallel RLC circuits. In fact, this analogy can be taken much further; it is an example of the **principle of duality** and comes from the basic laws of the circuit elements and their interconnection. This important principle is stated as follows:

Given any circuit containing sources, resistors, capacitors and inductors, you can construct another circuit model with identical behaviour if you replace (a) series connections with parallel connections and vice versa; (b) voltage sources with current sources and vice versa; (c) resistances with conductances and vice versa; (d) capacitors with inductors and vice versa; (e) voltage waveforms with current waveforms and vice versa.

The importance of this principle is that the properties of a circuit can be found simply by analogy if the properties of its dual circuit are known. In the present example this means that the current ratio $\boldsymbol{I}_R/\boldsymbol{I}_S$ in the parallel circuit has the same form as the voltage ratio $\boldsymbol{V}_R/\boldsymbol{V}_S$ in the series circuit. Thus, if we replace $|\boldsymbol{V}_R/\boldsymbol{V}_S|$ in Fig. 3.43 with $|\boldsymbol{I}_R/\boldsymbol{I}_S|$ we find that the parallel circuit is a bandpass circuit for currents. Similarly, the Q-factor relates the resonant frequency to the bandwidth of the parallel circuit response in just the same way as for the series tuned circuit.

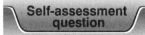

Construct a series RLC circuit, driven by a sinusoidal voltage source. Construct a parallel RLC circuit, driven by a current source.

Check the frequency response of the series circuit by monitoring the voltage across R, so in effect the series *current* is measured.

Check the frequency response of the parallel circuit by measuring the voltage across it.

In both cases, is the resonant frequency what you would expect? And the bandwidth?

Try the effect of varying R in both cases.

Self-assessment question

47 For each of the following statements concerning the series RLC circuit, provide an equivalent statement for the parallel RLC circuit.

(a) The Q-factor is the ratio of the magnitude of the capacitive or inductive impedance to the magnitude of the circuit impedance at resonance.

(b) The voltage generator is in series with the LC combination and must accordingly have a low resistance if a high voltage magnification factor is to be obtained.

(c) The current in the circuit is a maximum at the resonant frequency and in phase with the input voltage.
(d) The resonant frequency $\omega_0 = 1/\sqrt{LC}$ and the 3 dB bandwidth is $W = R/L$.

3.10 SUMMARY

Phasors

In this chapter, phasors are used to solve the problem of finding the voltage transfer properties of simple low-pass and high-pass circuits with sinusoidal inputs. The following are the highlights of the phasor analysis development:

1. Circuit analysis using trigonometrical formulae to add together sinusoidal voltages or currents can be mathematically taxing even for the simplest linear circuit.
2. It is easier to add together sinusoidal voltages or currents if they are represented by phasors. A phasor is a line from a fixed point on a diagram, or a mathematical expression representing that line. A phasor represents the sinusoid $A \sin(\omega t + \phi)$ by having a length or magnitude proportional to A and an angle (relative to a fixed line on the diagram) equal to ϕ.
3. The next step is to represent properties of the circuit and of its components (principally gain and impedance) by phasor operators, which, like phasors, have magnitude and angle. In the process of analysis it is necessary to add, subtract, multiply and divide phasors and operators.
4. The combination and manipulation of phasors and phasor operators is made less cumbersome by the introduction of the operator j. When a phasor or phasor operator is multiplied by j its angle is increased by 90°, leaving its magnitude unaltered. Multiplying by j twice gives a 180° phase shift, leading to the important result

$$\mathrm{j} \times \mathrm{j} = -1$$

A phasor or a phasor operator can be represented by either (a) its magnitude and angle: $m \angle \phi$; or (b) its components: $a + \mathrm{j}b$.

Borrowing terminology from the algebra of complex numbers, in the representation $a + \mathrm{j}b$, a is often called the 'real part' and b the 'imaginary part'. The formulae for converting from $m \angle \phi$ form to $a + \mathrm{j}b$ form are

$$a = m \cos\phi, \quad b = m \sin\phi$$

and the converse

$$m = \sqrt{(a^2 + b^2)}, \phi = \tan^{-1}(b/a) \text{ (selecting one of the two distinct solutions)}$$

Rather than remembering these formulae, you were encouraged to think them out by making a sketch on a diagram with real and imaginary axes called an Argand diagram.

5. Phasor analysis consists largely of algebraic operations carried out with phasors or operators represented in one or other of the forms $m \angle \phi$ or

$a + jb$. In an equation involving real and imaginary terms, the real terms and the imaginary terms must be equated separately.

Phasors and operators of the same sort can be added by either

(a) geometry, using the parallelogram construction; or
(b) expressing in $a + jb$ form and adding components.

Phasors and operators can be multiplied together in either

(a) $m \angle \phi$ form, by multiplying magnitudes and adding angles; or
(b) $a + jb$ form by complex algebra.

Phasors and operators can be divided in either

(a) $m \angle \phi$ form, by dividing magnitudes and subtracting the denominator angle from the numerator angle; or
(b) $a + jb$ form by rationalizing the denominator.

Impedance and admittance

Impedance is the ratio of the phasors representing the sinusoidal voltage between a pair of terminals and the sinusoidal current that flows.

In a resistance the voltage and current are in phase and the impedance is R, or $R \angle 0$.

In a capacitor with sinusoidal voltage and current, the current leads the voltage across the capacitor by 90°. The ratio of their amplitudes is $V/I = 1/\omega C$ and the impedance is $\mathbf{Z} = 1/\mathrm{j}\omega C$ or $1/\omega C \angle -90°$.

In a pure inductance with sinusoidal voltage and current, the voltage across the inductance leads the current by 90°. The ratio of their amplitudes is $V/I = \omega \mathrm{L}$ and the impedance is $\mathbf{Z} = \mathrm{j}\omega \boldsymbol{L}$ or $\omega \boldsymbol{L} \angle 90°$.

For impedances in series,

$$\mathbf{Z}_{\mathrm{eq}} = \mathbf{Z}_1 + \mathbf{Z}_2 + \mathbf{Z}_3 + \ldots$$

For impedances in parallel

$$\frac{1}{\mathbf{Z}_{\mathrm{eq}}} = \frac{1}{\mathbf{Z}_1} + \frac{1}{\mathbf{Z}_2} + \frac{1}{\mathbf{Z}_3} + \ldots$$

Admittance is the reciprocal of impedance. Admittance, like impedance, is a phasor operator. The admittance of a resistor is $1/R$, of a capacitor $\mathrm{j}\omega C$, and of an inductor $1/\mathrm{j}\omega L$.

For admittances in series

$$\frac{1}{\mathbf{Y}_{\mathrm{eq}}} = \frac{1}{\mathbf{Y}_1} + \frac{1}{\mathbf{Y}_2} + \frac{1}{\mathbf{Y}_3} + ..$$

For admittances in parallel

$$\mathbf{Y}_{eq} = \mathbf{Y}_1 + \mathbf{Y}_2 + \mathbf{Y}_3 + \ldots$$

It is usually easier to represent a parallel network by its admittance rather than by its impedance.

Voltage transfer functions

An important application of phasor analysis is to deduce the phasor ratio V_{out}/V_{in} for a linear network. The impedances of a capacitor and an inductor are each functions of frequency, so that if a network contains reactive components the ratio will also be a function of frequency. The ratio—known as the voltage transfer function—is usually left in the form of a quotient containing real and imaginary terms. The magnitude of the function gives the input–output voltage amplitude ratio and its angle the phase shift of the network. The voltage transfer function of the low-pass *CR* circuit is

$$1/(1 + \mathrm{j}\omega CR)$$

Terms like $(1 + \mathrm{j}\omega CR)$ frequently occur in the analysis of circuits containing resistance and reactance. It is common practice to express complex transfer functions as combinations of terms of this form.

Bode plots

1. The information contained in a voltage transfer function can be displayed in graphical form, showing how the amplitude ratio and phase vary with frequency. Amplitude ratios or gains are first converted to decibels, phase is expressed in degrees or radians, and a logarithmic scale is used for the frequency axis. When these conventions are followed, the resulting graphs are known as Bode plots.
2. The transfer function of a simple low- or high-pass circuit (containing a resistor and a single reactive component) can be written in a universal form. Their Bode plots can be sketched using linear (or 'straight-line') approximations, involving the minimum of calculation. The low-pass circuit introduces a maximum phase lag of 90°, while the high-pass circuit introduces a maximum phase lead of 90°. They are known as single-lag and single-lead circuits and always roll off at 6 dB per octave to high or low frequencies.
3. When two circuits are buffered by an ideal voltage amplifier, the overall Bode plot can be found by adding together the Bode plots for the individual circuits. When two single-lag circuits are connected in this way, the result is a double-lag circuit; the amplitude response rolls off at 12 dB per octave and the phase lag increases to 180° at high frequencies.

RLC circuits

4. In a series circuit with resistance, capacitance and inductance, the impedance can become a pure resistance at a particular frequency known as the resonant frequency. By combining the behaviour of a low-pass *LR* circuit and a high-pass *CR* circuit, the series *RLC* circuit acts as a band-pass circuit centred on the resonant frequency

$$\omega_0 = 1/\sqrt{LC}$$

The sharpness of the resonance (and the 'narrowness' of the response) depends on the ratio of the impedances of the inductor and capacitor to that of the resistance near the resonant frequency. It is characterized by

$$Q = \frac{\omega_0 L}{R} = \frac{1}{\omega_0 CR} = \frac{1}{R}\sqrt{(L/C)}$$

Bode plots can be constructed for the bandpass *RLC* circuit by first expressing the voltage transfer function in terms of Q and f_0.

Duality

5. The parallel *RLC* circuit with a current source input is known as the circuit dual of the series *RLC* circuit. According to the principle of duality, the behaviour of any circuit can be deduced from a knowledge of the properties of its dual counterpart.

Answers to self-assessment questions

1. The phasors representing the three voltages v_1, v_2 and v_3 are shown together in Fig. 3.48a. Notice that because v_3 and v_1 differ in phase by π radians or 180°, their phasors appear in direct opposition on the phasor diagram.

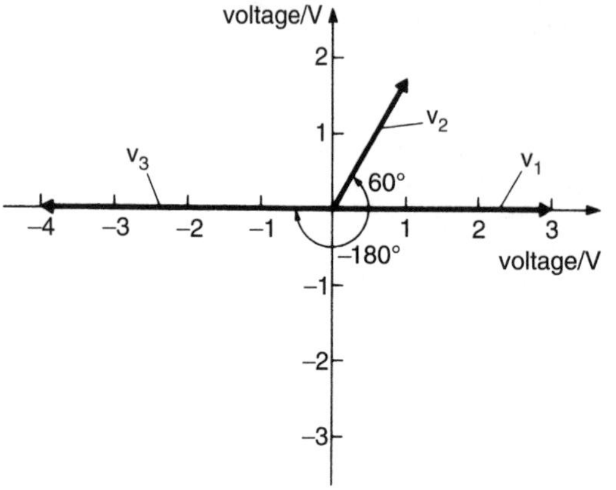

Fig. 3.48a

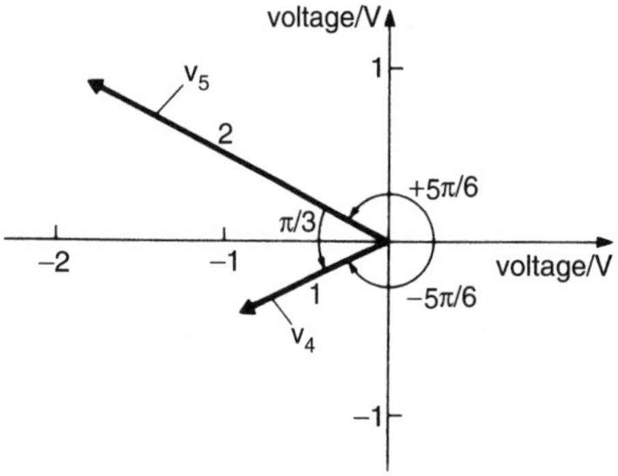

Fig. 3.48b

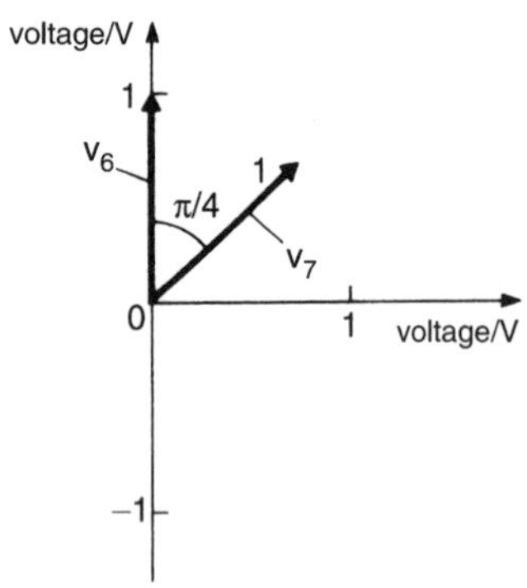

Fig. 3.48c

For convenience the phasors representing v_4 and v_5 are shown separately in Fig. 3.48b. From the diagram, the phase difference between these voltages is $\pi/3$ radians or 60°. Since an *increase* in phase angle corresponds to *anticlockwise* rotation on a phasor diagram, voltage v_4 must *lead* v_5 by $\pi/3$ radians.

Figure 3.48c shows v_6 leading by $\pi/2$ radians, or 90°. v_7 lags this by $\pi/4$ radians, or 45°.

2. (a) $v = 6 \sin 10t$ V, so $\boldsymbol{V} = 6 \angle 0$ V.

(b) $i = 7.5 \sin (1050t + \pi/4)$ A, so $\boldsymbol{I} = 7.5 \angle \pi/4$ A, or $7.5 \angle 45°$ A.

(c) $i = \omega CV \sin (\omega t + \pi/2)$, so $\boldsymbol{I} = \omega CV \angle \pi/2$, or $\omega CV \angle 90°$ A.

(d) $v = 3.6 \sin (500\pi t - 3\pi/4)$ V, so $\boldsymbol{V} = 3.6 \angle -3\pi/4$ V, or $3.6 \angle -135°$V.

Figures 3.49a–d show the corresponding phasor diagram representations of the above voltages and currents.

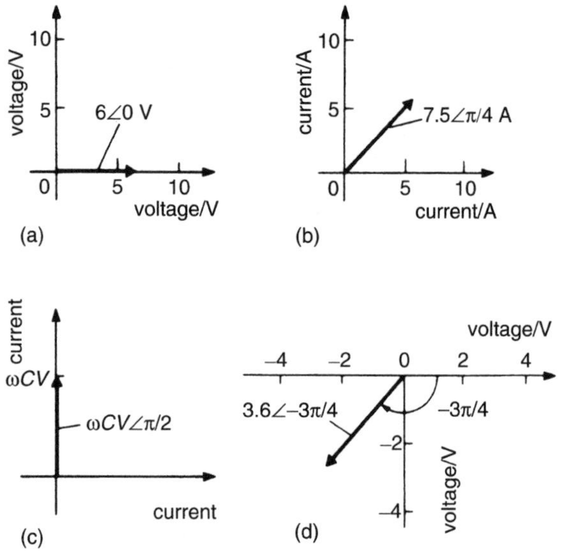

Fig. 3.49

3. (a) The voltage and the current are in phase in a resistor R. If the voltage is represented by the phasor $\boldsymbol{V} \angle \phi$, the current will have amplitude $I = V/R$ and the same phase, $\theta = \phi$, giving

$$\boldsymbol{I} = (V/R)\angle\phi$$

as shown in Fig. 3.50a.

(b) If a voltage of amplitude V is applied to a capacitance C, the current amplitude will be $I = \omega CV$. The current *leads* the voltage by $\pi/2$ radians or 90°. Hence, if the voltage is represented by the phasor $\mathbf{V} = V \angle \phi$, the current phasor will be of the form

$$\mathbf{I} = \omega C V \angle(\phi + \pi/2)$$

as shown in Fig. 3.50b.

(c) If a voltage of amplitude V is applied to an inductance L, the current amplitude will be $I = V/\omega L$. The current *lags* the voltage by $\pi/2$ radians or 90°. Hence, if the voltage is represented by the phasor $\mathbf{V} = V \angle \phi$, the current phasor will be of the form

$$\mathbf{I} = (V/\omega L)\angle(\phi - \pi/2)$$

as shown in Fig. 3.50c.

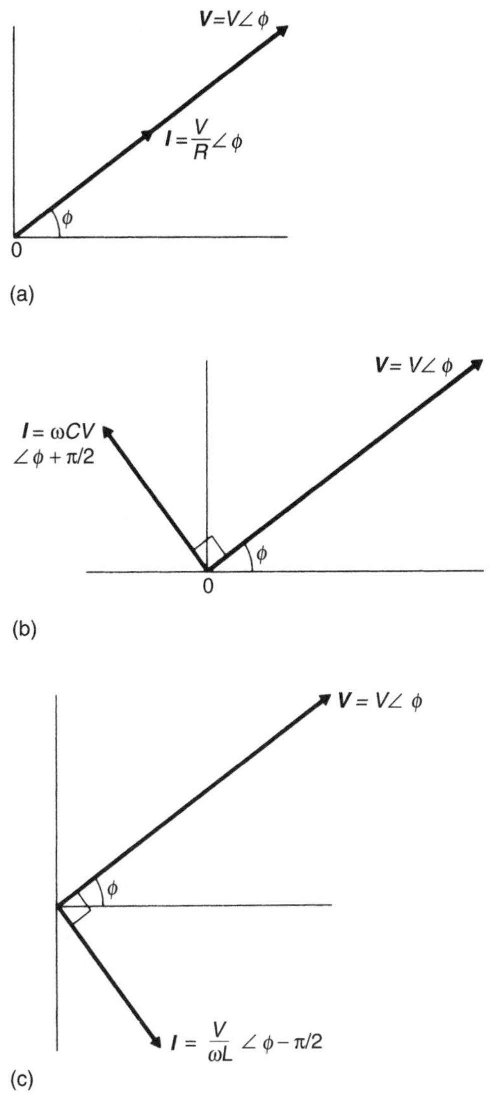

Fig. 3.50

4. Phasor $\mathbf{V}_S$ is the phasor sum or resultant of $\mathbf{V}_C$ and $\mathbf{V}_R$:

$$\mathbf{V}_S = \mathbf{V}_C + \mathbf{V}_R = IX_C\angle -\pi/2 + IR\angle 0$$

$\mathbf{V}_S$ has magnitude

$$|\mathbf{V}_S| = \sqrt{(I^2 X_C{}^2 + I^2 R^2)} = I\sqrt{(X_C{}^2 + R^2)}$$

The output phasor $\mathbf{V}_C$ lags the input phasor $\mathbf{V}_S$ by the angle ϕ. Referring to the phasor diagram in Fig. 3.8, we see that

$$\tan\phi = \frac{IR}{IX_C} = R/X_C$$

If we put $X_C = 1/\omega C$, then $\tan\phi = \omega CR$ and the phase lag ϕ is given by

$$\phi = \tan^{-1}(\omega CR).$$

5. The amplitude ratio of the CR network is

$$\frac{V_C}{V_S} = \frac{1}{\sqrt{(1 + (\omega CR)^2)}}$$

In this case $C = 0.47 \times 10^{-6}$ F, $R = 4.7 \times 10^3\ \Omega$ and $\omega = 2\pi \times 1000$ rad s^{-1}, or 6283 rad s^{-1}. So

$$\omega CR = 6283 \times 0.47 \times 10^{-6} \times 4.7 \times 10^3 = 13.879$$

Hence

$$\frac{V_C}{V_S} = \frac{1}{\sqrt{(1 + (13.879)^2)}} = 0.0719$$

The phase shift is

$$\phi = \tan^{-1}(\omega CR) = \tan^{-1} 13.879 \approx 1.5 \text{ rad, or about } 86°$$

At 1 kHz the amplitude ratio of the CR network is $V_C/V_S = 0.0719$. If $V_i = 4$ V then the amplitude of the output voltage is $V_0 = 0.0719 \times 4\ \text{V} \approx 0.29$ V.

6. The phase shift introduced by the CR network is $\phi = \tan^{-1}(\omega CR)$ or $\tan\phi = \omega CR$. So

$$C = \frac{1}{\omega R} \times \tan\phi$$

$$= \frac{1}{2\pi \times 1000 \times 10 \times 10^3} \times \tan 30°\ \text{F}$$

$$= 9.189 \times 10^{-9}\ \text{F}$$

Capacitance required ≈ 9.2 nF.

The phase shift of the network will be 45° when $\omega CR = \tan\phi = 1$. In this case the frequency is

$$\omega = \frac{1}{CR} = \frac{1}{9.189 \times 10^{-9} \times 10 \times 10^{-3}} = 10\,882 \text{ rad s}^{-1}$$

$$f \approx 1732 \text{ Hz}$$

7. The high-pass LR circuit is shown in Fig. 2.36 and again in Fig. 3.32b. We can define the current in the circuit to be of the form

$$i = I \sin \omega t$$

The resistor voltage will be in phase with the current and is therefore

$v_R = IR \sin \omega t$

The voltage across the inductor will lead the current by $\pi/2$ radians and its amplitude will be IX_L where $X_L = \omega L$ is the reactance of the inductor:

$v_L = IX_L \sin(\omega t + \pi/2)$

Using Kirchhoff's voltage law, at any instant

$v_S = v_R + v_L = IR \sin \omega t + IX_L \sin(\omega t + \pi/2)$

(a) Representing the three voltages by their phasors, we have

$\mathbf{V}_S = \mathbf{V}_R + \mathbf{V}_L$

Using current as the phase reference, then $\mathbf{I} = I\angle 0$ and

$\mathbf{V}_S = IR\angle 0 + IX_L \angle \pi/2$

In the phasor diagram of Fig. 3.51, $\mathbf{V}_R$ points along the phase reference axis and $\mathbf{V}_L$ has an angle of $\pi/2$ radians, measured clockwise from the phase reference axis. We can now use the phasor diagram to find the voltage transfer ratio of the high-pass *LR* network. The first step is to find the magnitude of the input phasor $\mathbf{V}_S$. This is

$$|\mathbf{V}_S| = \sqrt{(I^2 X_L{}^2 + I^2 R^2)}$$
$$= I\sqrt{(X_L{}^2 + R^2)}$$

The input–output voltage amplitude ratio is

$$\frac{V_L}{V_S} = \frac{IX_L}{I\sqrt{(X_L{}^2 + R^2)}} = \frac{X_L}{\sqrt{(X_L{}^2 + R^2)}}$$

If we put $X_L = \omega L$, we obtain

$$\frac{V_L}{V_S} = \frac{\omega L}{\sqrt{(\omega^2 L^2 + R^2)}}$$

This result can be expressed in the form

$$\frac{V_L}{V_S} = \frac{1}{\sqrt{(1 + R^2/\omega^2 L^2)}} \qquad (1)$$

The phasor diagram in Fig. 3.51 shows that the output phasor $\mathbf{V}_L$ leads $\mathbf{V}_S$ by the angle ϕ, where

$$\tan \phi = \frac{IR}{IX_L} = R/X_L = R/\omega L$$

The phase lead introduced by the circuit is therefore

$$\phi = \tan^{-1}(R/\omega L) \qquad (2)$$

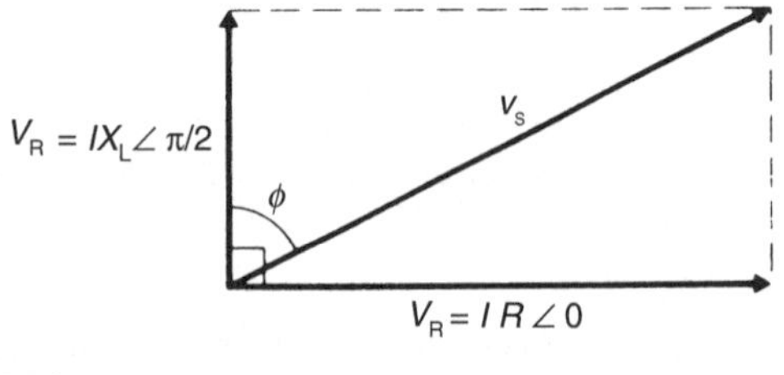

Fig. 3.51

(b) At the cut-off frequency, ω_c, the resistance in the circuit is equal to the reactance of the inductor,

$R = \omega_c L$

so that the angular cut-off frequency is defined by

$\omega_c = R/L$

If we substitute $R/L = \omega_c$ in Equations 1 and 2, the voltage amplitude ratio and the phase shift then become

$$\frac{V_L}{V_S} = \frac{1}{\sqrt{(1 + \omega_c^2/\omega^2)}} \qquad (3)$$
$$\phi = \tan^{-1}(\omega_c/\omega) \qquad (4)$$

When the angular input frequency ω is equal to the angular cut-off frequency ω_c, Equation 3 shows that

$$\frac{V_L}{V_S} = \frac{1}{\sqrt{(1+1)}} = 1/\sqrt{2}$$

Similarly, in Equation 4, we have

$\phi = \tan^{-1} 1 = \pi/4$ radians or 45°

showing that, at the cut-off frequency, the output leads the input by 45°.

(c) At low frequencies, when $\omega \ll \omega_c$, the second term in the denominator of Equation 3 is very much more than 1 and we have the approximate expression

$V_L/V_S \approx \omega/\omega_c$

When $\omega \gg \omega_c$, the second term in the denominator of Equation 3 is very much less than 1, giving

$$\frac{V_L}{V_S} \approx 1$$

The first of these results shows that for frequencies well below cut-off the voltage amplitude ratio is proportional to frequency. The output amplitude thus falls to zero as the input frequency is reduced.

If the input frequency is increased beyond the cut-off frequency, the output amplitude approaches the same value as the input amplitude. The circuit thus behaves as a high-pass filter.

8.
$$\mathbf{V}_{out} = \mathbf{A} \times \mathbf{V}_{in} = 100\angle -60° \times 10 \text{ mV} \angle 30°$$
$$= 100 \times 10 \text{ mV} \angle -60° + 30°$$
$$= 1 \text{ V}\angle -30°$$

9. First, express the input and output voltages in magnitude-angle form: $v_{in} = 0.1 \cos 3142t$ V, so its phasor is $\mathbf{V}_{in} = 0.1 \angle \pi/2$ V if $\sin 3142t$ is taken as the reference. $\mathbf{V}_{out} = 11.5 \sin(3142t + \pi/3)$ V, which is a sinusoid at the same frequency, but leading the reference by $\pi/3$ radians. Thus $\mathbf{V}_{out} = 11.5 \angle \pi/3$ V.

The gain is

$$\mathbf{A} = \mathbf{V}_{out}/\mathbf{V}_{in} = (11.5\angle\pi/3)/(0.1\angle\pi/2)$$
$$= 115\angle -\pi/6$$

10. (a) In a resistor the current and the voltage sinusoids are in phase. So if the phasor voltage across the resistor is $V \angle \phi$ the phasor current will be $I \angle \phi$. The impedance of the resistor is

$$\mathbf{Z}_R = \frac{V\angle\phi}{I\angle\phi} = \frac{V}{I}(\phi - \phi) = \frac{V}{I}\angle 0$$

But $V/I = R$, so

$$\mathbf{Z}_R = R\angle 0 = 47 \times 10^3 \angle 0\,\Omega$$

(b) If the phasor voltage across a capacitor is $V \angle \phi$ then the phasor current will be $I \angle (\phi + 90°)$. The impedance is

$$\mathbf{Z}_c = \frac{V\angle\phi}{I\angle(\phi + 90°)} = \frac{V}{I}\angle(\phi - (\phi + 90°))$$

$$= V/I \angle -90°$$

But in a capacitor $I = \omega CV$, so

$$\mathbf{Z}_c = \frac{V}{\omega CV}\angle -90° = \frac{1}{\omega C}\angle -90°$$

The reactance of the 1.8 nF capacitor at 5 kHz is

$$\frac{1}{\omega C} = \frac{1}{2\pi \times 5 \times 10^3 \times 1.8 \times 10^{-9}}\,\Omega$$

$$= 18\,000\,\Omega$$

The impedance of the capacitor at 5 kHz is thus

$$\mathbf{Z}_c = 18\,000 \angle -90°\,\Omega$$

(c) If the phasor voltage across an inductor is $V \angle \phi$, then the phasor current will be $I \angle (\phi - 90°)$. The impedance is

$$\mathbf{Z}_L = \frac{V\angle\phi}{I\angle(\phi - 90°)} = \frac{V}{I}\angle(\phi - (\phi - 90°))$$

$$= V/I \angle 90°$$

Now, in an inductor, $V = I\omega L$, so

$$\mathbf{Z}_L = (I\omega L/I)\angle 90° = \omega L \angle 90°$$

The reactance of the 1 mH inductor at 100 kHz is

$$\omega L = 2\pi \times 100 \times 10^3 \times 10^{-3}\,\Omega \approx 628\,\Omega$$

The impedance of the inductor at 100 kHz is thus

$$\mathbf{Z}_L = 628\angle 90°\,\Omega$$

11. Impedance is the ratio of the phasor voltage to the phasor current. At the input terminals of the *CR* network

$$\mathbf{Z}_{in} = \frac{\mathbf{V}_{in}}{\mathbf{I}_{in}}$$

so the phasor input current is

$$\mathbf{I}_{in} = \frac{\mathbf{V}_{in}}{\mathbf{Z}_{in}} = \frac{2.5 \angle 40°}{5 \times 10^3 \angle -30°}\text{A}$$

$$= \frac{2.5}{5 \times 10^3}\angle(40° - (-30°))\text{A}$$

$$= 0.5 \angle 70°\,\text{mA}$$

The input current leads the input voltage by 30°.

12. The impedance of an inductor is $\mathbf{Z} = \mathbf{V}/\mathbf{I}$. From Self-assessment question 10, $\mathbf{Z} = \omega L \angle 90°$. Using j to represent the 90° phase shift, we obtain $\mathbf{Z} = j\omega L$.

13. The phasor 1 + j is sketched in Fig. 3.52a. (Notice that j is really 1 × j).

$$m = \sqrt{(a^2 + b^2)} = \sqrt{(1^2 + 1^2)} = \sqrt{2} = 1.41$$

and

$$\phi = \tan^{-1}(b/a) = \tan^{-1}(1/1) = 45°$$

The phasor is 1.41 ∠ 45°.
The phasor 3 + j4 is sketched in Fig. 3.52b.

$$m = \sqrt{(3^2 + 4^2)} = 5$$

and

$$\phi = \tan^{-1}(4/3) = 53°$$

The phasor is 5 ∠ 53°.
For the phasor 2.6 + j8.7,

$$m = \sqrt{(2.6^2 + 8.7^2)} = \sqrt{(6.76 + 75.69)} = \sqrt{(82.45)} \approx 9.1$$

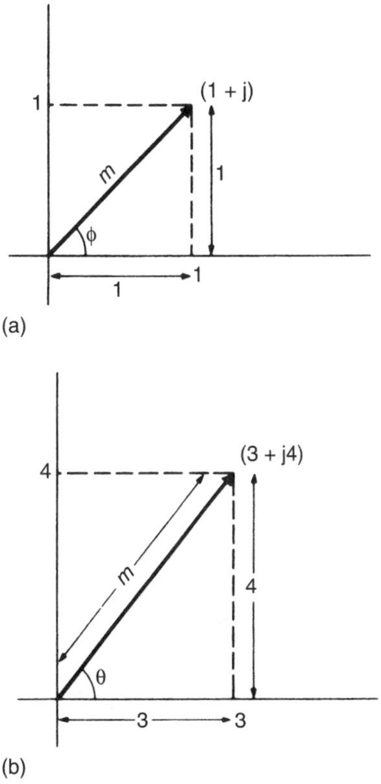

Fig. 3.52

and

$\phi = \tan^{-1}(8.7/2.6) = \tan^{-1}3.35 \approx 73°$

The phasor is 9.1 $\angle$ 73°.

14. The phasor 3 – j4 is sketched in Fig. 3.53a.

$m = \sqrt{(3^2 + 4^2)} = 5$

$\phi = 360° - \theta,\quad$ where $\theta = \tan^{-1}(4/3) = 53°$

Therefore, $\phi = 360° - 53° = 307°$. Angle ϕ can alternatively be expressed as −53°, that is, an angle of 53° measured *clockwise*. The phasor is 5 $\angle$ 307° or 5 $\angle$ −53°.

The phasor −4 + j3 is sketched in Fig. 3.53b.

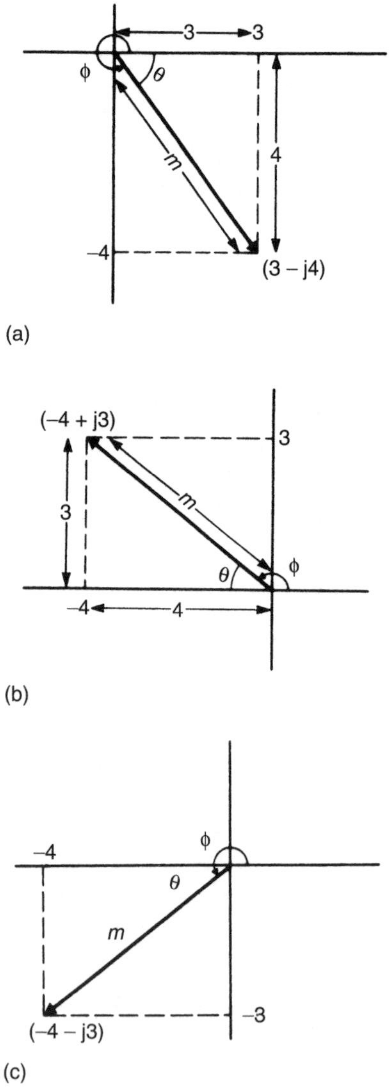

Fig. 3.53

$m = \sqrt{(4^2 + 3^2)} = 5$

$\phi = 180° - \theta,\quad$ where $\theta = \tan^{-1}(3/4) = 36.9°$

Therefore, $\phi = 180° - 36.9° = 143.1°$. The phasor is 5 $\angle$ 143.1°.

The phasor −4 – j3 is sketched in Fig. 3.53c.

$m = \sqrt{(4^2 + 3^2)} = 5$

$\phi = 180° + \theta,\quad$ where $\theta = \tan^{-1}(3/4) = 36.9°$

Therefore, $\phi = 180° + 36.9° = 216.9°$. The phasor is 5 $\angle$ 216.9°.

15. For the phasor 10 $\angle$ 60° = $a + jb$,

$a = m \cos\phi = 10 \cos 60° = 10 \times 0.5 = 5$

$b = m \sin\phi = 10 \sin 60° = 10 \times 0.866 = 8.66$

The phasor is 5 + j8.66.

For the phasor 0.4 $\angle$ 12°,

$a = 0.4 \cos 12° = 0.4 \times 0.98 = 0.392$

$b = 0.4 \sin 12° = 0.4 \times 0.21 = 0.084$

The phasor is 0.392 + j0.084.

For the phasor $15 \times 10^3 \angle \pi/6$,

$a = 15 \times 10^3 \cos \pi/6 = 15 \times 10^3 \times 0.866 = 13 \times 10^3$

$b = 15 \times 10^3 \sin \pi/6 = 15 \times 10^3 \times 0.5 = 7.5 \times 10^3$

The phasor is $13 \times 10^3 + j7.5 \times 10^3$.

16. The four phasors given in modulus–angle form are shown in Fig. 3.54; 30 $\angle$ 210° and 30 $\angle$ −150° are, in fact, the same.

$10 \angle -45° \equiv 10 \cos 45° - j10 \sin 45° = 7.07 - j7.07$

$20 \angle 120° \equiv -20 \cos 60° + j20 \sin 60° = -10 + j17.3$

$30 \angle 210° \equiv 30 \angle -150° = -30 \cos 30° - j30 \sin 30°$

$= -26 - j15$

17. (a) (3 + j2) + (6 + j9) = (3 + 6) + j(2 + 9) = 9 + j11

(b) (9 – j10) – (9 + j10) = (9 – 9) + j(−10 −10) = −0 −j20 = −j20

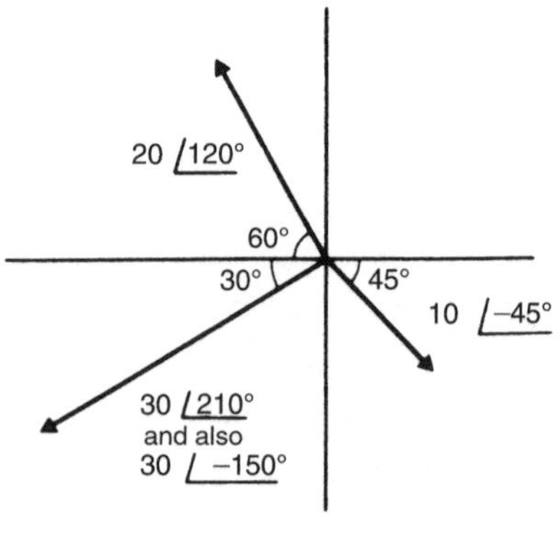

Fig. 3.54

(c) $(-2-j)-(-6+j9) = (-2+6)+j(-1-9) = 4-j10$
(d) $(1+j0)-(0+j) = 1-j$
(e) $(7+j)-(-7+j) = (7+7)+j(1-1) = 14+j0$
$= 14$

18. (a) $5\angle 53° = 5\cos 53° + j5\sin 53° = 3+j4$ $(5+j6)$
$+ 5\angle 53° = (5+j6)+(3+j4) = 8+j10$
$= 12.8\angle 51.3°$
(b) $10\angle 127° = 10\cos 127° + j10\sin 127° = -6+j8$
$(3-j4)+(-6+j8) = -3+j4 = 5\angle 127°$
(c) $8\angle -37° = 8\cos(-37°) + j8\sin(-37°)$
$= 6.4-j4.8$ $(-2+j3)-(6.4-j4.8) = -8.4+j7.8$
$= 11.5\angle 137°$
(d) $14\angle 30° = 14\cos 30° + j14\sin 30° = 12.1+j7$
$(-4-j8)-(12.1+j7) = -16.1-j15 = 22\angle 223°$

19. (a) $(6+j)\times(3+j4) = 6\times 3 + j3 + j24 + j^2 4$
$= (18-4)+j(3+24) = 14+j27$
(b) $(2+j5)\times(2-j5) = 4+j10-j10-j^2 25$
$= (4+25)+j(10-10) = 29+j0 = 29$
(c) $(-2-j)\times(-6-j9) = 12+j6+j18+j^2 9$
$= (12-9)+j24 = 3+j24$
(d) $j\times(3+j4) = j3+j^2 4 = -4+j3$
(e) $(8+j)\times(3+j0) = 24+j3$

20. (a) $\dfrac{1+j}{2-j} = \dfrac{(1+j)(2+j)}{(2-j)(2+j)} = \dfrac{1+j3}{(4+1)} = 0.2+j0.6$

(b) $\dfrac{1-j}{2+j} = \dfrac{(1-j)(2-j)}{(2+j)(2-j)} = \dfrac{1-j3}{5} = 0.2-j0.6$

(c) $\dfrac{3+j4}{1+j3} = \dfrac{(3+j4)(1-j3)}{(1+j3)(1-j3)} = \dfrac{15-j5}{(1+9)} = 1.5-j0.5$

(d) $\dfrac{2-j5}{2-j3} = \dfrac{(2-j5)(2+j3)}{(2-j3)(2+j3)} = \dfrac{19-j4}{13} = 1.46-j0.31$

21. The current phasor is

$$\boldsymbol{I} = \frac{\boldsymbol{V}}{\boldsymbol{Z}} = \frac{3+j2}{2+j5}\ \text{A}$$

Multiplying top and bottom by $2-j5$ gives

$$\boldsymbol{I} = \frac{(3+j2)(2-j5)}{(2+j5)(2-j5)}\ \text{A} = \frac{16-j11}{(4+25)}\ \text{A}$$
$$\approx 0.55-j0.38\ \text{A}$$

Figure 3.55 shows the voltage and current phasors. The voltage leads the current.

22. (a) $\dfrac{|1+j|}{|2-j|} = \dfrac{\sqrt{(1^2+1^2)}}{\sqrt{(2^2+1^2)}} = \dfrac{\sqrt{2}}{\sqrt{5}} \approx 0.63$

$\angle(1+j) - \angle(2-j) = 45° - (-26.6°) \approx 71.6°$

(b) $\dfrac{|1-j|}{|2+j|} = \dfrac{\sqrt{(1^2+1^2)}}{\sqrt{(2^2+1^2)}} = \dfrac{\sqrt{2}}{\sqrt{5}} \approx 0.63$

$\angle(1-j) - \angle(2+j) = -45° - 26.6° \approx -71.6°$

(c) $\dfrac{|3+j4|}{|1+j3|} = \dfrac{\sqrt{(3^2+4^2)}}{\sqrt{(1^2+3^2)}} = \dfrac{\sqrt{25}}{\sqrt{10}} \approx 1.58$

$\angle(3+j4) - \angle(1+j3) = 53.1° - 71.6° \approx -18.5°$

(d) $\dfrac{|2-j5|}{|2-j3|} = \dfrac{\sqrt{(2^2+4^2)}}{\sqrt{(2^2+3^2)}} = \dfrac{\sqrt{29}}{\sqrt{13}} \approx 1.49$

$\angle(2-j5) - \angle(2-j3) = -68.2° - (-56.3°)$
$\approx -11.9°$

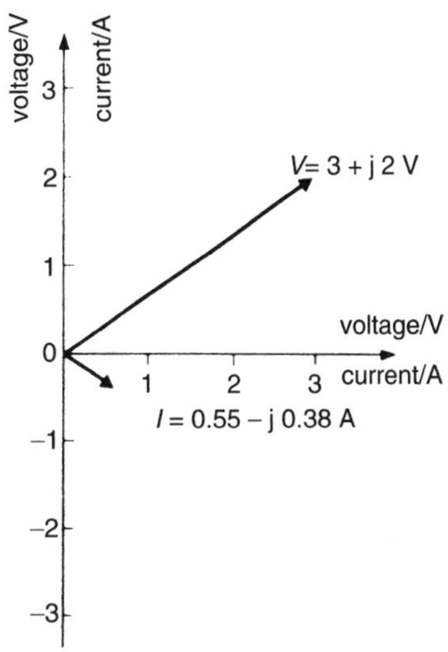

Fig. 3.55

23. At very low frequencies the capacitor is effectively an open circuit to signal currents, so the impedance seen at the input is virtually R in series with R, that is $\mathbf{Z} \approx 2R$.

At very high frequencies the capacitor is effectively a short circuit, so the input impedance tends to $\mathbf{Z} \approx R$.

In both cases the input impedance is very nearly resistive (zero angle). At intermediate frequencies the capacitor will make the input impedance capacitive (negative angle).

This sort of 'by inspection' analysis is very useful in gaining a first idea of how a circuit functions.

24. (a) The input admittance, $\mathbf{Y}_{in}$, is equal to the admittance of the resistor, inductor and capacitor in parallel:

$$\mathbf{Y}_{in} = \mathbf{Y}_R + \mathbf{Y}_L + \mathbf{Y}_C = \frac{1}{R} + \frac{1}{j\omega L} + j\omega C$$
$$= \frac{1}{R} + j(\omega C - 1/\omega L)$$

The input admittance has the form $(G + jB)$. The equivalent conductance is

$G = 1/R$

The equivalent susceptance is

$B = (\omega C - 1/\omega L)$

(b) When the input frequency $\omega = \omega_0$, the susceptance of the parallel circuit becomes zero. We then have

$$\omega_0 C - 1/\omega_0 L = 0$$

For this condition to be satisfied, $\omega_0 C = 1/\omega_0 L$, which gives

$$\omega_0 = \sqrt{(1/LC)}$$

(c) At very low frequencies, when $\omega \ll \omega_0$, the susceptance of the capacitor, ωC, will be very much less than the susceptance of the inductor $1/\omega L$. The input admittance then takes the form

$$Y_{in} = \frac{1}{R} - j/\omega L, \quad \text{for } \omega \ll \omega_0$$

Since the angle of the admittance is *negative*, the circuit is *inductive* in this frequency region.

At higher frequencies, when $\omega \gg \omega_0$, the situation is reversed and the input admittance is approximated by

$$Y_{in} = \frac{1}{R} + j\omega C, \quad \text{for } \omega \gg \omega_0$$

The angle of the admittance is now *positive*, showing that the circuit becomes *capacitive* at high frequencies.

If we had considered the input impedance of the circuit, $\mathbf{Z}_{in}$, we would have found the angle of the impedance to be positive at low frequencies and negative at high frequencies, corresponding to an inductive circuit and a capacitive circuit as before. However, the expressions for the impedance would have been far more complicated than those obtained for the admittance.

25. Suppose a current **I** flows into the network. Since we assume that negligible current is drawn from the network, all the current flowing through $\mathbf{Z}_1$ towards the junction of $\mathbf{Z}_1$ and $\mathbf{Z}_2$ must flow away from the junction through $\mathbf{Z}_2$.

The current flowing through $\mathbf{Z}_1$ is $(\mathbf{V}_1 - \mathbf{V}_0)/\mathbf{Z}_1$, which is equal to the current $\mathbf{V}_0/\mathbf{Z}_2$ flowing through $\mathbf{Z}_2$. So

$$\frac{\mathbf{V}_1 - \mathbf{V}_0}{\mathbf{Z}_1} = \frac{\mathbf{V}_0}{\mathbf{Z}_2} \quad \text{or} \quad \mathbf{V}_i - \mathbf{V}_o = \frac{\mathbf{Z}_1}{\mathbf{Z}_2}\mathbf{V}_o$$

So $$\mathbf{V}_i = \left(1 + \frac{\mathbf{Z}_1}{\mathbf{Z}_2}\right)\mathbf{V}_o$$

Therefore, $$\frac{\mathbf{V}_o}{\mathbf{V}_i} = \frac{1}{1 + \mathbf{Z}_1/\mathbf{Z}_2} = \frac{\mathbf{Z}_2}{\mathbf{Z}_1 + \mathbf{Z}_2}$$

(a) $\mathbf{Z}_1 = 1\ \text{k}\Omega$, $\mathbf{Z}_2 = 10\ \text{k}\Omega$ and $\mathbf{V}_i = 6 + j4$ V. The output voltage is

$$\mathbf{V}_o = \frac{\mathbf{Z}_2}{\mathbf{Z}_1 + \mathbf{Z}_2}\mathbf{V}_i = \frac{10 \times 10^3}{10^3 + 10 \times 10^3} \times (6 + j4)\ \text{V}$$
$$= \frac{10}{11} \times (6 + j4)\text{V} = 5.45 + j3.64\ \text{V}$$

(b) $\mathbf{Z}_1 = 10\ \text{k}\Omega$, $\mathbf{Z}_2$ is the impedance of a 15 nF capacitor at 5 kHz. If we write $\mathbf{Z}_1 = R$ and $\mathbf{Z}_2 = 1/j\omega C$ the voltage transfer function of the network is

$$\frac{\mathbf{V}_o}{\mathbf{V}_1} = \frac{\mathbf{Z}_2}{\mathbf{Z}_1 + \mathbf{Z}_2} = \frac{1/j\omega C}{R + 1/j\omega C} = \frac{1}{1 + j\omega CR}$$

At 5 kHz the voltage transfer ratio is

$$\frac{\mathbf{V}_o}{\mathbf{V}_1} = \frac{1}{1 + j(2\pi \times 5 \times 10^3 \times 15 \times 10^{-9} \times 10^4)}$$
$$= \frac{1}{1 + j4.712}$$

If the input-voltage phasor $\mathbf{V}_i$ is 6 + j4 V, the output-voltage phasor is

$$\mathbf{V}_o = \frac{1}{1 + j4.712} \times \mathbf{V}_i = \frac{6 + j4}{1 + j4.712}\ \text{V}$$
$$= \frac{(6 + j4)(1 - j4.712)}{(1 + j4.712)(1 - j4.712)}\ \text{V}$$
$$= \frac{(6 + 18.85) + j(4 - 28.27)}{1 + 22.2}\ \text{V}$$
$$= 1.07 - j1.05\ \text{V}$$

(c) $\mathbf{Z}_1 = 1/j\omega C_1$, $\mathbf{Z}_2 = 1/j\omega C_2$. The voltage transfer function of the network is

$$\frac{\mathbf{V}_o}{\mathbf{V}_1} = \frac{\mathbf{Z}_2}{\mathbf{Z}_1 + \mathbf{Z}_2} = \frac{1/j\omega C_2}{1/j\omega C_1 + 1/j\omega C_2}$$

Multiplying top and bottom by $j\omega C_1$,

$$\frac{\mathbf{V}_o}{\mathbf{V}_1} = \frac{C_1/C_2}{1 + C_1/C_2} = \frac{C_1}{C_1 + C_2}$$

In this case the voltage transfer ratio does not depend upon frequency. The ratio **A** is a constant of value:

$$\mathbf{A} = \frac{10 \times 10^{-9}}{10 \times 10^{-9} + 15 \times 10^{-9}} = \frac{10}{25} = 0.4$$

The network will reduce the amplitude of an input sinewave voltage by 0.4 times, irrespective of the frequency of the sinewave. The phase shift associated with the network is zero at all frequencies.

(d) $\mathbf{Z}_1 = 1/j\omega C$, $\mathbf{Z}_2 = R$. The voltage transfer function is

$$\frac{\mathbf{V}_o}{\mathbf{V}_1} = \frac{\mathbf{Z}_2}{\mathbf{Z}_1 + \mathbf{Z}_2} = \frac{R}{1/j\omega C + R} = \frac{1}{1 + 1/j\omega CR}$$

When we use impedance, admittance or transfer functions, rather than specific numbers, it is usual to leave the expression in non-rationalized form.

(i) At very high frequencies, when $\omega \gg 1/CR$ or $\omega CR \gg 1$, the denominator approximates to 1, and

$$\frac{\mathbf{V}_o}{\mathbf{V}_1} \approx 1 \quad \text{or} \quad 1\angle 0$$

At very high frequencies, then, the amplitude ratio of the network is 1 and the phase shift is negligibly small.

(ii) At very low frequencies, when $\omega \ll 1/CR$ or $\omega CR \ll 1$, the denominator approximates to $1/j\omega CR$, so the transfer function becomes

$$\frac{\mathbf{V}_o}{\mathbf{V}_1} \approx j\omega CR \quad \text{or} \quad \omega CR\angle 90°$$

At very low frequencies the amplitude ratio associated with the circuit is ωCR and the phase shift is $+90°$. The amplitude ratio at low frequencies is proportional to frequency. This means that if the frequency of an input sinewave voltage is reduced the amplitude of the output sinewave voltage will decrease in proportion. The network thus passes the high-frequency components of signals with little reduction in amplitude and almost zero phase shift. The low-frequency components are reduced in amplitude proportionally to frequency and are shifted in phase by almost 90°. The circuit is a *high-pass CR* network. Notice that the phase shift is $+90°$ indicating that the output voltage *leads* the input voltage by 90°.

(iii) $\omega = 1/CR$ or $\omega CR = 1$

$$\frac{\mathbf{V}_o}{\mathbf{V}_1} = \frac{1}{1+1/\mathrm{j}\omega CR} = \frac{1}{1+1/\mathrm{j}1} = \frac{1}{1-\mathrm{j}1} = \frac{1}{\sqrt{2}\angle-45°} = \frac{1}{\sqrt{2}}\angle 45°$$

The angular frequency $\omega = 1/CR$ is the angular cut-off frequency of the high-pass network. At this frequency the amplitude ratio has fallen to $1/\sqrt{2}$ or 0.707 times its high frequency value of 1, and the phase lead is 45°.

26. The input and output phasors in Fig. 3.56 are related by the voltage divider rule

$$\mathbf{V}_L = \mathbf{V}_S \times \frac{\mathbf{Z}_L}{\mathbf{Z}_R + \mathbf{Z}_L}$$

where $\mathbf{Z}_R = R$ and $\mathbf{Z}_L = \mathrm{j}\omega L$. The voltage transfer function is therefore

$$\mathbf{A} = \frac{\mathbf{V}_L}{\mathbf{V}_S} = \frac{\mathbf{Z}_L}{\mathbf{Z}_R + \mathbf{Z}_L} = \frac{\mathrm{j}\omega L}{R+\mathrm{j}\omega L} = \frac{1}{1-\mathrm{j}R/\omega L} \qquad (1)$$

The voltage transfer function has magnitude

$$|\mathbf{A}| = \frac{1}{|1-\mathrm{j}R/\omega L|} = \frac{1}{\sqrt{(1+R^2/\omega^2 L^2)}}$$

which is the result you should have obtained in Self-assessment question 7 when working from the phasor diagram.

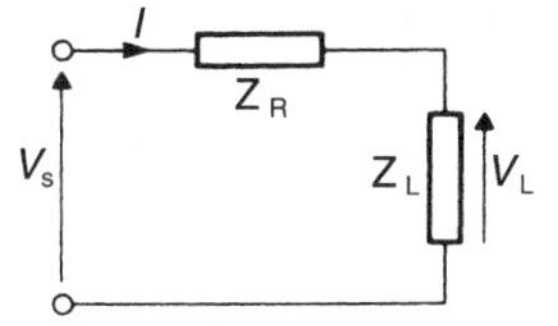

Fig. 3.56

The phase shift of the *LR* circuit is given by

$$\begin{aligned}\phi &= \angle\mathbf{A} \\ &= \angle(1) - \angle(1-\mathrm{j}R/\omega L) \\ &= 0 - \tan^{-1}(-R/\omega L) = \tan^{-1}(R/\omega L)\end{aligned}$$

Using the phasor diagram in Self-assessment question 7 gave the same result.

27. (a) The angular cut-off frequencies of the two-high pass circuits in Fig. 3.26 are $\omega_c = 1/CR$ and $\omega_c = R/L$. The resistance R is 10 kΩ in both circuits. If the high-pass cut-off frequency of each circuit is to be 100 kHz, then

$$C = \frac{1}{2\pi \times 100 \times 10^3 \times 10 \times 10^3} \approx 1.6 \times 10^{-10}\,\mathrm{F} = 0.16\,\mathrm{nF}$$

and

$$L = \frac{10 \times 10^3}{2\pi \times 100 \times 10^3} \approx 0.016\,\mathrm{H} = 16\,\mathrm{mH}$$

(b) For an input $V_S \sin \omega t$, the current in the two circuits will be of the form $i = I \sin(\omega t + \theta)$. Now, the circuit in Fig. 3.26a has a *capacitive* impedance, which means that the phase of the current will *lead* that of the applied voltage. The circuit of Fig. 3.26b is *inductive* and the current will *lag* the applied voltage. This is sufficient to show that the currents in the two circuits will be different when the same voltage is applied to each.

We can now find expressions for the amplitude and phase of the current in the two circuits. Taking the high-pass *CR* circuit first, this has an impedance $\mathbf{Z} = R - \mathrm{j}/\omega C$, which is the same as for the low-pass *CR* circuit considered in Section 3.5.1. The current amplitude I is therefore

$$I = \frac{\omega C\, V_S}{\sqrt{(R^2 + \omega^2 C^2 R^2)}}$$

and the phase is

$$\theta = \pi/2 - \tan^{-1}(\omega CR)$$

At cut-off, when $\omega = 1/CR$, the phase of the current is 45° relative to the input voltage. The high-pass *LR* circuit has an impedance $\mathbf{Z} = R + \mathrm{j}\omega L$. For a phasor voltage input $\mathbf{V}_S$, the phasor current I will be

$$I = \frac{\mathbf{V}_S}{R + \mathrm{j}\omega L}$$

Taking the input voltage as the phase reference, we have $\mathbf{V}_S = V_S + \mathrm{j}0 = V_S$. The current amplitude is now

$$I = |\mathbf{I}| = \frac{|\mathbf{V}_S|}{R+\mathrm{j}\omega L} = \frac{V_S}{\sqrt{(R^2+\omega^2 L^2)}}$$

The phase, θ, is given by

$$\theta = \angle\mathbf{V}_S - \angle(R+\mathrm{j}\omega L) = -\tan^{-1}(\omega L/R)$$

At the cut-off frequency, when $\omega = R/L$, the current lags the input voltage by 45°.

28. $3 = 2 \times 1.5 \simeq 6$ dB + 3.5 dB $\simeq$ 9.5 dB
$6 = 2 \times 3 \simeq 6$ dB + 9.5 dB $\simeq$ 15.5 dB
$40 = 2 \times 2 \times 10 \simeq 6$ dB + 6 dB + 20 dB $\simeq$ 32 dB
$100 = 10^2 = 2 \times 20$ dB = 40 dB
$2000 = 2 \times 10^3 = 6$ dB + 3×20 dB $\simeq$ 66 dB

29. If X is given in dB, then the equivalent voltage ratio is antilog ($X/20$), which is the same as $10^{X/20}$.
Using a calculator or tables, we find that
12 dB = ×3.98 (usually taken as ×4)
18 dB = ×7.94 (taken as ×8)
26 dB = ×19.95 (taken as ×20)
−12 dB = ×0.251 (taken as ×1/4)
−18 dB = ×0.126 (taken as ×1/8)
−26 dB = ×0.05 or ×1/20

30. Since the voltage gains have no units we assume that they are given as voltage ratios. Writing 100 as (10 × 10), the gain of the first amplifier is 20 dB + 20 dB = 40 dB. Similarly, 50 = 100/2, so the gain of the second amplifier is 40 dB − 6 dB = 34 dB. You should check these values on your calculator.

The 2-metre-long connecting cable introduces attenuation of 2 × 1.5 dB, which is an overall gain reduction of 3 dB. Taking account of the effect of the connecting cable, therefore, the gain of the cascaded amplifiers is

40 dB − 3 dB + 34 dB = 71 dB

31. The circuit of Fig. 3.29 has a voltage transfer function

$$\frac{\mathbf{V}_o}{\mathbf{V}_i} = \frac{R}{R + j\omega L} = \frac{1}{1 + j\omega L/R}$$

The transfer function magnitude falls to a value $1/\sqrt{2}$ at the cut-off frequency $\omega_c = R/L$. Putting $\omega = 2\pi f$ and $L/R = 1/2\pi f_c$ in the transfer function, we obtain

$$\frac{\mathbf{V}_o}{\mathbf{V}_i} = \frac{1}{1 + jf/f_c}$$

32. The voltage gain in dB is given by

$$G = -10 \log_{10}[1 + (f/f_c)^2]$$

The minus sign indicates that the output will be less than the input for all frequencies, f, greater than zero. For the first two frequencies given, we have

(i) $f = f_c/100$: $G = -10 \log_{10}[1 + (0.01)^2] \approx 0$ dB
(ii) $f = f_c/10$: $G = -10 \log_{10}[1 + (0.1)^2] = -0.043$ dB
You will see that if $f \ll f_c$, then $(f/f_c)^2$ is very small compared with 1 and the gain is closely approximated by $G \approx -10 \log_{10}(1) = 0$ dB.
(iii) $f = f_c$: $G = -10 \log_{10}[1 + 1] = -3$ dB
The remaining frequencies are much greater than the cut-off frequency. In this region, $(f/f_c)^2$ becomes much greater than 1 and we can make the approximation $G \approx -10 \log_{10}(f/f_c)^2$. The approximation is very close indeed for $f \geqslant 10 f_c$, as you will find by comparing the following results with those obtained from an exact calculation
(iv) $f = 10 f_c$: $G = -10 \log_{10}[10^2] = -20$ dB
(v) $f = 20 f_c$: $G = -10 \log_{10}[20^2] = -26$ dB
(vi) $f = 100 f_c$: $G = -10 \log_{10}[100^2] = -40$ dB
Notice that, for frequencies beyond cut-off, a doubling of the frequency reduces the gain by a factor of 2 or by 6 dB. Similarly, a ten-fold increase in frequency reduces the gain by a factor of 10 or by 20 dB.

33. The break point comes at the frequency $\omega_c = 2\pi f_c = 1/CR$. With the values given we obtain

$$2\pi f_c = \frac{1}{10^3 \times 2 \times 10^{-6}}$$
$$f_c = 79.6 \text{ Hz}$$

At this frequency, the phase of the output relative to that of the input is −45°.

From Fig. 3.30, the amplitude response has fallen to 1/10 or −20 dB at the frequency $10f_c$, which corresponds to the frequency 796 Hz.

From Fig. 3.31, the phase shift at $10f_c$ is a lag of about 84°.

34. The universal form of the transfer function for a high-pass circuit containing a resistor and a single reactive component is

$$\frac{\mathbf{V}_o}{\mathbf{V}_i} = \frac{1}{1 - jf_c/f}$$

The phase shift ϕ of the circuit is minus the phase of the denominator:

$$\phi = \tan^{-1} f_c/f$$

So the phase of the high-pass circuit varies from 90° at low frequencies to zero at high frequencies and has a value of 45° at the cut-off frequency.

35. The linear approximations for the two Bode plots are shown in Fig. 3.36. The figure has three different regions, with boundaries marked by the corner frequencies f_1 and f_2.

Now, both plots have values of 0 dB below f_1. Their sum is thus equal to 0 dB for all frequencies less than f_1.

Between f_1 and f_2, plot 2 (with corner frequency f_2) remains at a constant value of 0 dB. The sum of the two plots in this region is then equal to plot 1 and rolls off at a rate of 6 dB/octave between f_1 and f_2.

The sum of the two straight-line plots above f_2 is itself a straight line. Its slope may be found by adding together the slopes of plot 1 and plot 2, both of which are equal to −6 dB/octave. The sum of these two plots, therefore, rolls off at a rate of 12 dB/octave. The approximation to the overall amplitude/frequency response is shown in Fig. 3.37 for the case $f_1 = 1$ kHz and $f_2 = 20$ kHz.

Remember that this is for the case of two low-pass circuits buffered by an ideal amplifier.

36. The first circuit is high-pass, single lead. Its voltage transfer function is

$$\frac{1}{1 - \mathrm{j}f_1/f}$$

The corner frequency, f_1, is given by $2\pi f_1 = 1/C_1R_1$. Therefore,

$$f_1 = \frac{1}{2\pi \times 0.0015} = 106\,\text{Hz}$$

The second circuit is low-pass, single lag, described by the transfer function

$$\frac{1}{1 + \mathrm{j}f/f_2}$$

Its corner frequency is given by $2\pi f_2 = 1/C_2R_2$. Therefore,

$$f_2 = \frac{1}{2\pi \times 0.000\,015} = 10\,600\,\text{Hz}$$

The overall voltage transfer function is

$$\frac{1}{(1 - \mathrm{j}f_1/f)(1 + \mathrm{j}f/f_2)}$$

Linear approximations for the amplitude/frequency responses of the high-pass and low-pass circuits are shown in Figs. 3.33 and 3.30, with $f_c = 106$ Hz and 10.6 Hz respectively. An approximation to the overall amplitude response is obtained by adding the two together.

Linear approximations for the phase shift of each circuit are given in Figs. 3.34 and 3.31, with f_c taking the previous values again. The sum of the two gives an approximation for the overall phase response.

37. (a) The resonant frequency in rad s^{-1} is $\omega_0 = 1/\sqrt{LC}$. With the component values given:

$$\omega_0 = \frac{1}{\sqrt{(60 \times 10^{-6} \times 30 \times 10^{-12})}} = 23.57 \times 10^6\,\text{rad s}^{-1}$$

The equivalent frequency in hertz is

$$f_o = \omega_0/2\pi = 3.75\,\text{MHz}$$

The Q-factor is given by

$$Q = \frac{\omega_0 L}{R} = \frac{1}{\omega_0 CR}$$

Using the first of these expressions with $\omega_0 = 23.57 \times 10^6$ rad s^{-1} we obtain

$$Q = \frac{23.57 \times 10^6 \times 60 \times 10^{-6}}{10} = 141.4$$

(b) At resonance, the impedance of the series RLC circuit is resistive and equal to the circuit resistance. The circuit current is then in phase with the input voltage and represented by the phasor $\boldsymbol{I} = \boldsymbol{V}_S/R$.

(c) At resonance we have the relationships

$$\boldsymbol{V}_L = \frac{\mathrm{j}\omega_0 L}{R}\boldsymbol{V}_S \quad \text{and} \quad \boldsymbol{V}_c = \frac{\boldsymbol{V}_S}{\mathrm{j}\omega_0 CR}$$

Hence, if $\boldsymbol{V}_S = 1\angle 0$, we have

$$V_L = \frac{\omega_0 L}{R}\angle\pi/2 \quad \text{and} \quad V_c = \frac{1}{\omega_0 CR}\angle -\pi/2$$

At resonance, the inductor and capacitor voltages differ in phase by $\pi/2 - (-\pi/2) = \pi$ radians or 180°. Their amplitudes are equal and are given by the input voltage amplitudes multiplied by the Q-factor. Hence, for a 1 V input voltage, we have

$$\boldsymbol{V}_L = 141.4\angle\pi/2\,\text{V}$$
$$\boldsymbol{V}_C = 141.4\angle -\pi/2\,\text{V}$$

38. (a) The Q-factor of the series RLC circuit is defined by

$$Q = \frac{\omega_0 L}{R} \quad \text{or} \quad Q = \frac{1}{\omega_0 CR}$$

Taking the first of these expressions and substituting for the resonant frequency, $\omega_0 = 1/\sqrt{LC}$, we obtain

$$Q = \frac{L}{\sqrt{(LC)}R} = \frac{1}{R}\sqrt{(L/C)}$$

Similarly, working from the second expression gives the identical result

$$Q = \frac{\sqrt{(LC)}}{CR} = \frac{1}{R}\sqrt{(L/C)}$$

Notice that the Q-factor *increases* as the square root of the inductance and *decreases* as the square root of the capacitance. For given values of L and C, the Q-factor is inversely proportional to the series resistance, R.

(b) $L = 10\,\mu\text{H} = 10^{-5}$ H and $C = 10^{-9}$ F. The resonant frequency is therefore

$$\omega_0 = \frac{1}{\sqrt{(LC)}} = \frac{1}{\sqrt{(10^{-5} \times 10^{-9})}} = 10 \times 10^6\,\text{rad s}^{-1}$$

This is equivalent to the frequency

$$f_0 = \omega_0/2\pi = 1.59\,\text{MHz}$$

If the capacitor (or inductor) voltage magnitude is to be greater than the input voltage magnitude at resonance, then the Q-factor must be greater than 1. Using the values given, we have

$$Q = \frac{1}{R}\sqrt{(L/C)} = \frac{1}{R}\sqrt{(10^{-5}/10^{-9})} = \frac{1}{R} \times 100$$

For Q to be greater than 1, therefore, the resistance R must be *less* than 100 Ω.

39. (a) The relationship $\omega_0 = 1/\sqrt{(LC)}$ can be rearranged to give

$$C = 1/\omega_0^2 L$$

The resonant frequency, $\omega_0 = 2\pi \times 30 \times 10^3$ rad s^{-1} and the inductance, $L = 100 \times 10^{-6}$ H. Therefore

$$C = \frac{1}{(2\pi \times 30 \times 10^3)^2 \times 100 \times 10^{-6}} = 2.81 \times 10^{-7} \text{ F}$$
$$= 281 \text{ nF}$$

(b) We now have three expressions for the Q-factor:

$$Q = \frac{\omega_0 L}{R} = \frac{1}{\omega_0 CR} = \frac{1}{R}\sqrt{(L/C)}$$

Using the first of these expressions with $L = 100 \times 10^{-6}$ and $R = 5\ \Omega$, we obtain

$$Q = \frac{2\pi \times 30 \times 10^3 \times 100 \times 10^{-6}}{5} = 3.77$$

You should verify that the same answer is obtained using the alternative definitions with $C = 281$ nF.

(c) The resonant frequency is inversely proportional to the *square root* of the capacitance (and of the inductance). Thus, if either the capacitor (or the inductor) is *reduced* in value by a factor of 10, the resonant frequency will *increase* by a factor of $\sqrt{10}$. Notice that the resonant frequency does not depend on the value of the series resistance, R.

Using the results from Self-assessment question 38, if C is reduced by a factor of 10, the Q-factor will *increase* by a factor of $\sqrt{10}$. If the inductance is reduced by a factor of 10, leaving the capacitor unchanged, then Q will *decrease* by a factor of $\sqrt{10}$.

40. The voltage transfer function of the series RLC circuit is given by the voltage divider rule:

$$\frac{\mathbf{V}_R}{\mathbf{V}_S} = \frac{R}{\mathbf{Z}} = \frac{R}{R + j\omega L - j/\omega C}$$
$$= \frac{R}{R + j(\omega L - 1/\omega C)}$$

(a) The transfer function magnitude is given by dividing the magnitude of the numerator by the magnitude of the denominator. We obtain

$$\left|\frac{\mathbf{V}_R}{\mathbf{V}_S}\right| = \frac{R}{\sqrt{[R^2 + (\omega L - 1/\omega C)^2]}}$$

The phase shift ϕ is given by subtracting the phase angle of the denominator from that of the numerator:

$$\phi = \angle R - \angle[R + j(\omega L - 1/\omega C)]$$
$$= -\tan^{-1}\frac{\omega L - 1/\omega C}{R}$$

(b) When $\omega = \omega_0$, the transfer function magnitude becomes

$$\left|\frac{\mathbf{V}_R}{\mathbf{V}_S}\right| = \frac{R}{\sqrt{[R^2 + (\omega_0 L - 1/\omega_0 C)^2]}}$$

Now, at resonance, we have the condition

$$\omega_0 L = 1/\omega_0 C$$

and the transfer function magnitude is simply

$$\left|\frac{\mathbf{V}_R}{\mathbf{V}_S}\right| = \frac{R}{\sqrt{[R^2]}} = 1$$

as expected.

Similarly, the phase shift at resonance is given by

$$\phi = -\tan^{-1}\frac{\omega_0 L - 1/\omega_0 C}{R}$$

which gives zero phase shift, since $\tan^{-1} 0 = 0$.

(c) Dividing the numerator and denominator of the transfer function magnitude by R (and remembering to divide by R^2 inside the square-root sign) we obtain

$$\left|\frac{V_R}{V_S}\right| = \frac{1}{\sqrt{[1 + (\omega L - 1/\omega C)^2/R^2]}}$$

$|V_R/V_S|$ will have a value of $1/\sqrt{2}$ (i.e. 3 dB down on its value at $\omega = \omega_0$) when

$$(\omega L - 1/\omega C)^2/R^2 = 1$$

that is, at those frequencies for which $(\omega L - 1/\omega C)^2 = R^2$.

41. (a) We can rearrange the expression $Q = \omega_0 L/R$ to give the bandwidth.

$$W = \frac{\omega_0}{Q} = \frac{R}{L}$$

This is the most simple approach. Alternatively, we can start with the bandwidth $W = \omega_0/Q$ and substitute the values

$$\omega_0 = \frac{1}{\sqrt{LC}} \quad \text{and} \quad Q = \frac{1}{R}\sqrt{(L/C)}$$

to show that $W = R/L$ as before.

(b) Self-assessment question 39 specified a coil of inductance 100 μH and series resistance 5 Ω. If the coil is connected into a series circuit with total series resistance R, the bandwidth, in rad s^{-1}, will be $W = R/L$. The value of R cannot be less than the 5 Ω resistance of the coil winding. The smallest possible bandwidth that can be achieved is therefore

$$W = 5/L = 5/(100 \times 10^{-6}) = 5 \times 10^4 \text{ rad s}^{-1}$$

The minimum bandwidth in hertz is then

$$B = W/2\pi = 7.96 \text{ kHz}$$

(c) If the coil is connected with a capacitor of 10 nF, we have:

(i) $\omega_0 = 1/\sqrt{(LC)}$

$$= \frac{1}{\sqrt{(100 \times 10^{-6} \times 10 \times 10^{-9})}} = 10^6 \text{ rad s}^{-1}$$

$$f_0 = 10^6/2\pi = 159 \text{ kHz}$$

(ii) Assuming that the total circuit resistance is that of the coil winding, then

$$Q = \frac{\omega_0 L}{R} = \frac{10^6 \times 10^{-4}}{5} = 20$$

42. We begin with

$$\frac{\mathbf{V}_R}{\mathbf{V}_S} = \frac{R}{R + j\omega L - j/\omega C}$$

Dividing through by R gives:

$$\frac{\mathbf{V}_R}{\mathbf{V}_S} = \frac{1}{1 + j(\omega L/R - 1/\omega CR)}$$

But $\omega LR = (\omega_0 LR)\omega/\omega_0 = Q\omega/\omega_0$ and $1/\omega CR = (1/\omega_0 CR)\ \omega_0/\omega = Q\omega_0/\omega$. So

$$\frac{\mathbf{V}_R}{\mathbf{V}_S} = \frac{1}{1 + jQ(\omega/\omega_0 - \omega_0/\omega)} = \frac{1}{1 + jQ(f/f_0 - f_0/f)}$$

43. The centre frequency determines the value of the capacitor needed:

$$C = 1/\omega_0^2 L = \frac{1}{(2\pi \times 10^4)^2 \times 10^{-3}} = 0.25\,\mu\text{F}$$

The bandwidth determines the series resistance of the circuit:

$$W = \omega_0/Q = R/L$$

Therefore,

$$R = WL = 2\pi \times 3 \times 10^3 \times 10^{-3} = 18.8\,\Omega$$

The series resistance of the inductor accounts for 10 Ω. Thus R must have a value of 8.8 Ω.

At resonance the impedances of the capacitor and inductor cancel each other, but the series resistance of the inductor remains. Using the voltage divider rule, the maximum voltage ratio is found to be

$$\left|\frac{\mathbf{V}_R}{\mathbf{V}_S}\right|_{(max)} = \frac{8.8}{8.8 + 10} = 0.47$$

If the inductor were ideal, with zero series resistance, the voltage ratio would be unity, as we saw previously.

For a bandwidth of 1 kHz, the total series resistance would have to be 6.3 Ω. Since the series resistance of the inductor is greater than this, a filter with a 1 kHz bandwidth cannot be built using the inductor described.

44. (a) In order to transmit the signal component with negligible attenuation, the bandpass circuit should have a resonant frequency, f_0, of 5 MHz. The unwanted signal is at 35 MHz which is $7f_0$.

Using the standard curves given in Fig. 3.45 we see that a circuit with a Q-factor of 15 would provide slightly more than the required 40 dB of attenuation at $7f_0$.

Since we are working on the linear portion of the Bode plot, where the log–log plot has a constant slope of −6 dB/octave, we can also make use of the approximation derived earlier. Thus, for frequencies well above resonance, we have

$$\left|\frac{\mathbf{V}_R}{\mathbf{V}_S}\right| \approx \frac{1}{Q}(f_0/f) \quad \text{for} \quad f \gg f_0$$

We require the voltage ratio to be −40 dB or 1/100 when $f = 7f_0$. Therefore,

$$\frac{1}{Q} \times \frac{1}{7} = \frac{1}{100}$$

for which $Q = 100/7 = 14.3$.

(b) If the response to the wanted signal is to be within 3 dB of its maximum value, then the signal frequency must lie between the lower and upper 3 dB frequencies f_1 and f_2.

If the signal is initially at a frequency of f_0, therefore, a change in frequency equivalent to about one-half the circuit bandwidth will cause the response to fall by 3 dB. The bandwidth is given by $B = f_0/Q$. Using the exact value, $Q = 14.3$, with $f_0 = 5$ MHz, we obtain

$$B = 5\,\text{MHz}/14.3 = 350\,\text{kHz}$$

The maximum change in frequency that can be tolerated is then $\pm B/2$, or ± 175 kHz.

45. The parallel circuit has an admittance

$$Y = 1/R + j\omega C - j/\omega L$$
$$= G + jB$$

where the conductance $G = 1/R$ and the susceptance B is given by $B = \omega C - 1/\omega L$

46. Following on from Self-assessment question 45, the circuit admittance is

$$\mathbf{R} = G + \mathbf{Y}_L + \mathbf{Y}_C$$

and the voltage V across the circuit is

$$\mathbf{V} = \mathbf{I}_S/\mathbf{Y}$$

The current in each branch of the circuit can be found by applying the current-divider rule for admittances in parallel:

$$\mathbf{I}_R = \frac{G}{\mathbf{Y}}\mathbf{I}_S,\ \mathbf{I}_L = \frac{\mathbf{Y}_L}{\mathbf{Y}}\mathbf{I}_S,\ \mathbf{I}_C = \frac{\mathbf{Y}_C}{\mathbf{Y}}\mathbf{I}_S$$

$$\frac{\mathbf{I}_C}{\mathbf{I}_S} = \frac{\mathbf{Y}_C}{G} = \frac{j\omega_0 C}{G},\ \frac{\mathbf{I}_L}{\mathbf{I}_S} = \frac{1}{j\omega_0 LG}$$

If we write G as $1/R$, these results can be expressed in the form

$$\frac{\mathbf{I}_L}{\mathbf{I}_S} = \frac{R}{\omega_0 L}\angle -\pi/2 \quad \text{and} \quad \frac{\mathbf{I}_C}{\mathbf{I}_S} = \omega_0 CR\angle \pi/2$$

47. The following statements apply to the parallel RLC circuit with a current-source input. Notice the interchange of the dual quantities, impedance/admittance, voltage/current, etc. Other terms such as 'low', 'high', 'maximum', 'minimum', etc., remain unchanged since these apply equally to a circuit and to its dual.

(a) The Q-factor is the ratio of the magnitude of the capacitive or inductive *admittance* to the magnitude of the circuit *admittance* at resonance.

(b) The *current* generator is in *parallel* with the LC combination and must therefore have a low *conductance* if a high *current* magnification is to be obtained.

(c) The *voltage* in the circuit is a maximum at the resonant frequency and in phase with the input *current*.

(d) The resonant frequency $\omega_0 = 1/\sqrt{(LC)}$ and the 3 dB bandwidth is $W = G/C$.

We obtain the bandwidth of the parallel circuit in part (d) by starting with the expression $W = R/L$ for the series circuit. We then replace R with G and L with C to obtain $W = G/C$. This procedure can be used to 'translate' expressions relating to a circuit and to its dual. Notice that, in the case of the resonant frequency, interchanging the symbols L and C has no effect; the same expression applies to both circuits.

Amplifiers and feedback

4

AIMS

1. To introduce amplifier equivalent circuits suitable for a.c. circuit analysis.
2. To provide details of the specification of an integrated-circuit op amp.
3. To explain and analyse the effects of applying different types of feedback to op amps.

GENERAL OBJECTIVES

On completing this part of the course you should understand the meaning of the following terms.

- a.c. amplifier
- closed-loop input and output impedance
- common-mode input impedance
- CR-active filter
- current-derived feedback
- d.c. amplifier
- differential input impedance
- dominant lag
- feedback factor
- feedback fraction
- Frequency compensation
- gain–bandwidth product with frequency independent feedback
- loop gain
- negative feedback
- positive feedback
- series-applied feedback
- shunt-applied feedback
- slew rate
- small-signal input and output impedance
- stability
- summing amplifier
- summing junction
- unity-gain frequency
- virtual-earth approximation for a.c. amplifiers
- voltage-derived feedback

SPECIFIC OBJECTIVES

1. Draw an a.c. equivalent for an op amp, using appropriate values for the equivalent circuit parameters taken from a manufacturer's data sheet.
2. Explain the principle of negative feedback.
3. Describe the characteristics of a frequency-compensated amplifier.
4. Derive the standard form of the feedback equation for a voltage-derived series feedback amplifier.

5. Describe the conditions which can give rise to instability in feedback amplifiers.
6. Analyse the effects of feedback on the amplifier frequency response and equivalent circuit parameters.
7. Using appropriate simplifying assumptions, explain and analyse the operation of some practical feedback circuits.
8. Calculate the output offset voltage of a differential amplifier, given its bias currents, input offset current and input offset voltage.
9. Calculate the output noise voltage of an amplifier, given the power density spectra of its equivalent input noise sources.

4.1 WHAT ARE AMPLIFIERS?

The purpose of an amplifier is to increase the power in a signal without distorting it. This process is called **amplification**. For example, amplifiers are used to increase the power in the signal from a microphone so that it can drive a loudspeaker, or to amplify the control signals produced by the pilot of an aircraft so that they will move the flaps on the wings and other control devices. The extra power must come from a **power supply**. In the case of an electronic amplifier, the power supply is usually a d.c. voltage power supply (which gets its power from the a.c. supply mains) or a battery.

The input signal causes the amplifier to control the flow of current from this voltage supply to the load. Thus more power may be delivered to the load than is taken from the input signal source. In practice, amplification usually means increasing the voltage amplitude of the signal into a given load. The opposite of amplification is called **attenuation**, and usually refers to a decrease in signal voltage.

Voltage amplifier

An electronic system which is designed primarily to give an output voltage proportional to the input signal voltage, without taking a significant amount of signal current, is called a **voltage amplifier**. Its *voltage* gain is *specified*, but its current gain is not, so it *may* give an increase in signal current if the load impedance is low enough.

An example of this is the unity-gain buffer. It is designed so that its output voltage is almost equal to its input voltage, but the output current may be much larger than the input current. So the voltage gain can be specified as nearly 1, but the current gain cannot be specified.

Current amplifier

An amplifier designed primarily to give an output *current* proportional to the input signal current, without requiring a significant input signal voltage, is called a **current amplifier**.

Power amplification

All these examples involve **power amplification** of course.

Passive components

Active components or devices

In electronics a distinction is made between two types of component: those which can only absorb or transfer signal power, such as resistors and transformers, which are called **passive components**, and those, such as transistors, which can accept power from an extra power source and amplify signal power. These are called **active components**, or **active devices**.

Throughout this chapter, there are occasional references to bipolar transistors and field-effect transistors (FET). You are not expected to know anything about transistors at this stage except that they are active devices. They are explained in Chapter 9.

Circuits composed entirely of passive components, are known as **passive networks**, and circuits containing active components are known as **active networks**. An active network is not necessarily an amplifier. It may not be intended as an amplifier. The converse, however, is certainly true. An amplifier must be an active network.

Passive networks

Active networks

Self-assessment question

1 Classify each of the following components as either passive or active on the basis of the above definitions:

resistor	capacitor	inductor
transformer	battery	bipolar transistor
FET (field-effect transistor)		

4.1.1 Equivalent circuits

Amplifiers can be obtained as complete integrated circuits that contain many transistors and other components, but whose overall performance can be quite simply specified. So you can think of them as circuit components with special properties, without having to know much about how they work. This is how they will be considered in this chapter.

Figure 4.1a shows an amplifier inserted between a signal source and load. Let us suppose, initially, that phase shifts are negligible at the frequency of

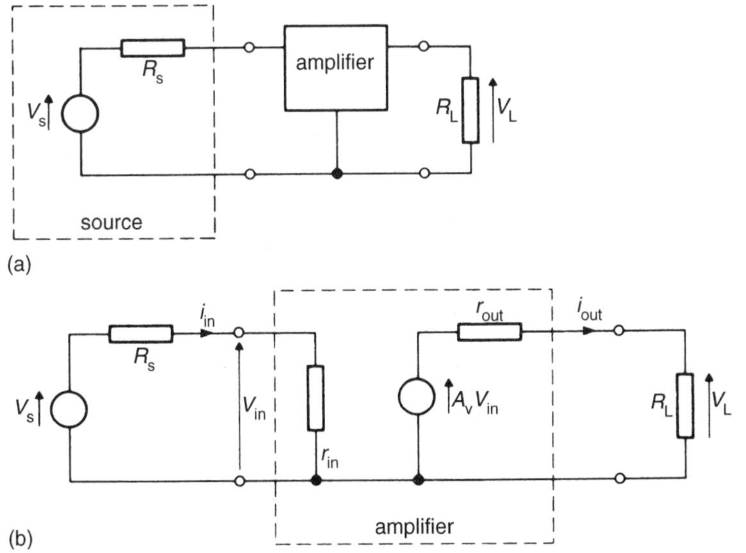

Fig. 4.1 (a) An amplifier inserted between a source and a load. (b) The same but with the equivalent circuit of the amplifier shown, including the input resistance r_{in}, the output resistance r_{out} and the voltage generator A_vV_{in}.

operation. So all quantities in the circuit will be 'real', and phasors will not be necessary. To be able to calculate the signal voltage delivered to the load, we need to know the following amplifier parameters:

(i) *The small-signal input resistance* r_{in} *of the amplifier*. This is the small-signal resistance of the amplifier as seen by the source.

Worked example

Why do you need to know the input resistance?

So that you can calculate the input voltage to the amplifier. If r_{in} *is this input resistance, and if* R_S *is the internal resistance of the source, then* $V_{in} = V_S \times r_{in}/(R_S + r_{in})$. *The two resistors just form a potential divider.*

The input resistance, as well as the other two parameters described below, can be represented by the small-signal equivalent circuit shown in Fig. 4.1b. With r_{in} across the input terminals it is clear that $V_{in} = V_S \times r_{in}/(R_S + r_{in})$. The larger the value of r_{in} the closer the value of V_{in} is to V_S.

(ii) *The open-circuit voltage gain* A_v *of the amplifier*. Just as with the networks considered in Chapter 3, the amplifier has a voltage transfer function. It is commonly called the *open-circuit voltage gain* A_v. It is represented in the equivalent circuit of Fig. 4.1b by a *voltage-dependent* voltage source, which means that the *voltage it produces is a constant,* A_v, *times another voltage*. In this case the voltage it produces is $A_v \times V_{in}$. This voltage source forms part of a small-signal Thévenin equivalent circuit at the output of the amplifier; that is, the voltage source is in series with the effective 'output resistance' r_{out} of the amplifier. $A_v \times V_{in}$ is called the *open-circuit* output voltage because, in order to measure it, the output current must be zero—to avoid any voltage drop in the output resistance. In practice this means that the resistance of the voltmeter used to measure $A_v \times V_{in}$ must be much larger than r_{out}.

(iii) *The small-signal output resistance* r_{out}. This is the resistive part of the Thévenin equivalent circuit representing the output of the amplifier. You can in principle determine its value by measuring the a.c. open-circuit output voltage and the a.c. short-circuit output current. (Achieving a short circuit output is not always easy, however.) Their ratio equals r_{out} as explained for d.c. circuits in Chapter 1. The same principle applies to small signals in a.c. circuits. In many amplifiers the output resistance is rather non-linear, so the value used in the Thévenin equivalent is valid *only* for small signals. For larger signals, the value can only be a poor approximation.

Worked example

Why do you need to know the output resistance r_{out} of the amplifier?

Because the voltage appearing across the load is the open-circuit voltage produced by the voltage source multiplied by $R_L/(R_L + r_{out})$.

It should now be clear how to calculate the signal voltage across the load. Knowing V_S you can calculate V_{in}, then knowing A_v you can calculate V_L.

Self-assessment question

2 In Fig. 4.1b suppose $R_S = 9\ \text{k}\Omega$, $r_{in} = 91\ \text{k}\Omega$, $A_v = 100$, $r_{out} = 100\ \Omega$ and $R_L = 1\ \text{k}\Omega$. If the amplitude of the e.m.f. of the sinusoidal source, V_S, is 2 mV, what is the amplitude of the voltage V_L across the load R_L?

Note that lower-case letters are being used to represent the small-signal resistances of the equivalent circuit. The upper case voltages V_{in}, V_{out}, V_S, etc. refer to the amplitudes of the sinusoidal voltages associated with the circuit. Since the circuit is purely resistive, there are no phase changes in the circuit.

It is important to be quite clear about the meaning of the voltage gain A_v of an amplifier. The *voltage gain A_v of an amplifier is defined for the case of an open-circuit load*. So, *with open-circuit load*, the magnitude of A_v is

$$A_v = \frac{\text{signal voltage amplitude at the amplifier's output terminals}}{\text{signal voltage amplitude at the amplifier's input terminals}}$$

The ratio V_L/V_S in the circuits of Fig. 4.1b is also sometimes referred to as voltage gain, so it is important to be clear which voltage ratio is being referred to.

Self-assessment question

3 In Fig. 4.1b suppose that the amplifier has an input resistance r_{in} that is equal to the source resistance R_S, has a voltage gain A_v of 10, and an output resistance r_{out} that is equal to the load resistance R_L. What signal voltage, expressed in terms of V_S, will be applied to the load?

Self-assessment question

4 Imagine that you have a supply of meters capable of measuring small-signal currents and voltages in any circumstances. How would you measure the equivalent circuit parameters r_{in}, r_{out} and A_v of an amplifier? (Note that real measuring methods may not be quite so amenable as these imaginary ones!)

4.1.2 Inverting and non-inverting amplifiers

Some amplifiers are inverters and some are not. Figure 4.2 shows the difference in output between two amplifiers that have the same voltage gain—namely 3 in this instance—but one circuit is an inverting amplifier and the other is a non-inverting one. Evidently waveform (b) in the figure is the non-inverted output, whilst curve (c) is the inverted output. Another way of expressing this difference is to say that the non-inverting amplifier shows zero phase difference between input and output ($\mathbf{A}_v = 3 \angle 0°$), whilst the inverting amplifier produces a phase shift of 180° or π rad ($\mathbf{A}_v = 3 \angle 180°$). Yet another way of expressing the difference is to say that the non-inverting amplifier has a voltage gain of +3, whilst the inverting amplifier has a gain of −3. In other words the + and − signs are used to distinguish between non-inversion and inversion. Gains of +3 and −3 must not, of course, be confused with gains of +3 dB and −3 dB.

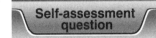

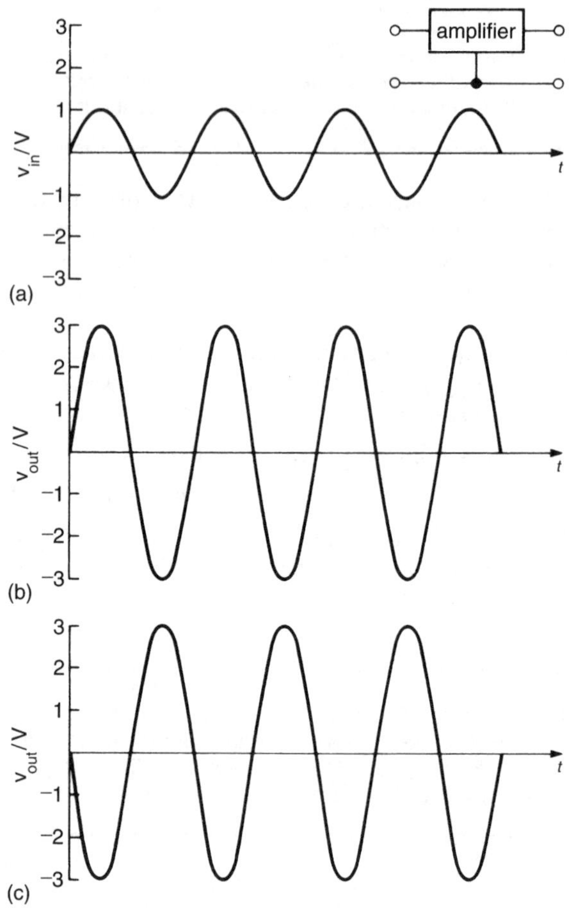

Fig. 4.2 Inverting and non-inverting amplifiers. (a) The input voltage waveform. (b) The output of a non-inverting amplifier of gain +3. (c) The output of an inverting amplifier of gain −3.

Worked example

What is the difference between a voltage gain of −10 and a voltage gain of −10 dB?

A voltage gain of −10 means that the ratio of output to input voltage amplitude is 10, but that the amplifier is an inverting one. The minus sign in the voltage gain statement of −10 dB means that the 'amplifier' attenuates rather than amplifies. The gain is given by the equation

$$20 \times \log(\text{gain}) = -10$$

so $\log(\text{gain}) = -0.5$ *or* $\log(\text{attenuation}) = 0.5$ *so attenuation factor* $= 3.16$.

Note that gain expressed in terms of dB makes no reference to phase, or to inversion. Whether or not the amplifier inverts has to be stated separately if it is an important parameter. Often however it is not an important feature (e.g. in audio amplifiers). The advantage of the dB system is that you can add together the gains in dB of amplifiers and attenuators in cascade to obtain the overall gain in dB. You have to multiply or divide by the actual numerical gains to get the overall numerical gain of such a series combination.

Note that the waveforms in Fig. 4.2 show variations in input and output voltages around zero volts. In practice these sinusoids may well be superimposed on d.c. bias voltages. But, as explained in Chapter 1, small-signal parameters are only concerned with variations around a d.c. operating point.

Self-assessment question

5 The loss in signal strength in some of the older undersea cables across the Atlantic Ocean is about 1 dB per km. Amplifiers—usually called 'repeaters'—are inserted in the cable at a spacing of 50 km. How much gain is needed in each amplifier to overcome the loss in the cable? What is the total gain required in a 3000 km cable to counteract the loss in the cable? Express your answer in dB and as a ratio.

4.2 THE OPERATIONAL AMPLIFIER

A very important type of amplifier, judging by how widely it is used, is the **operational amplifier**. It differs from the kind of amplifier discussed so far in that it has a **differential input** and that its output is always centred around zero volts. It is represented in diagrams by the triangular symbol shown in Fig. 4.3. This diagram also shows the differential input terminals, to which an input signal voltage v_{in} is applied, and the output terminal whose output signal voltage v_{out} is referred to the zero voltage line. The d.c. supplies are shown by dotted lines as a reminder that they are needed. Typically they might be $+15$ V and -15 V relative to the 0 V line and they obviously set limits to the possible voltage swing at the output. With these supplies, a peak-to-peak output swing of perhaps 27 V might be possible. Normally these d.c. connections are not shown in circuit diagrams; it is understood that they are connected in the real circuit.

With this device the input voltage is applied between two input terminals *neither* of which is the common or zero voltage line. Thus the input voltage to the amplifier is the *difference* between the two instantaneous input voltages v_a and v_b. It does not matter (ideally) what their actual d.c. voltage is with respect to the zero voltage line; it is the voltage difference between them that is

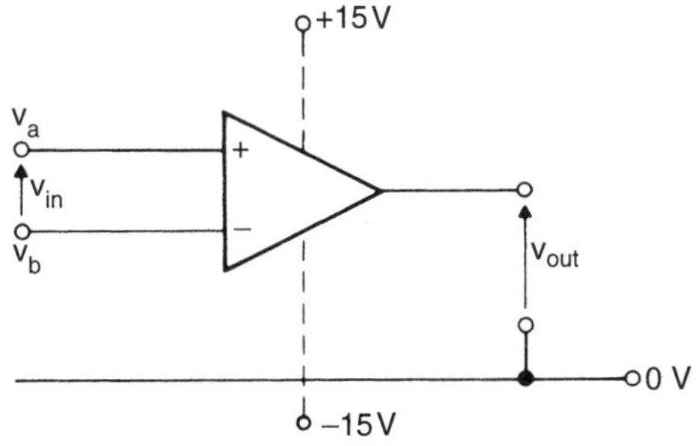

Fig. 4.3 The symbol for an operational amplifier, showing the differential input. The connections to the d.c. supplies are shown by dashed lines; they are usually omitted from circuit diagrams.

amplified by the amplifier. This is what is meant by a differential input. The voltage gain of the amplifier is therefore given by

$$A_v = \frac{\text{output signal voltage}}{\text{differential input signal voltage}}$$

Hence, when the output is open circuit, $v_{out} = A_v(v_a - v_b)$.

The frequency response of a typical op amp is shown in Fig. 4.4.

An input voltage applied simultaneously to *both* input terminals is called a **common-mode input voltage**. Ideally this would produce no output signal at all, but in practice such an input does result in a small proportional voltage appearing at the output. The ratio of output voltage to the common-mode input voltage is called the **common-mode gain**. Typically in an operational amplifier the low-frequency differential gain is 100 000 or more, whilst the common-mode gain might be between 1 and 10. In other words the differential gain greatly exceeds the common-mode gain. So for now we can ignore common-mode gain as being quite negligible compared with the differential gain.

Since the output is referred to the zero voltage line, the output should be 0 V when the differential input voltage is zero (e.g. if the two input terminals are connected together). In practice this may require an initial adjustment—called the *offset adjustment*—but as the design of operational amplifiers improves this is becoming less and less necessary.

Initially, we can ignore these departures from ideal performance and concentrate on the op amp's most important properties at low frequencies (below the corner frequency of Fig. 4.4) which are:

1. It has a pair of differential input terminals. One input is called the **inverting input** and is *labelled with a minus sign*. The other input is called the **non-inverting input** and is *labelled with a plus sign*. The output voltage is positive when the non-inverting (+) terminal is positive with respect to the inverting (−) terminal.
2. The output voltage is referred to 0 V, so the d.c. output voltage is zero when the input terminals are connected together or have the same signal input.

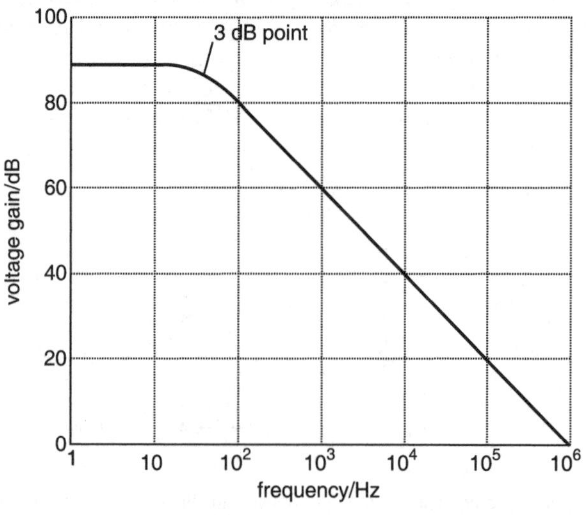

Fig. 4.4 The gain expressed in dB.

3. It has a very large differential voltage gain (usually greater than 10^5) at low frequencies.
4. It has a high input resistance between the two input terminals (usually greater than 200 kΩ), so they draw very little current.
5. It has a small output resistance (usually less than 1 kΩ) for small signals (although the output current is limited to, typically, less than 20 mA).

These properties (apart from the frequency response) are all represented in the equivalent circuit of Fig. 4.5. Notice that, as compared with the equivalent circuit of Fig. 4.1b there is now no direct common connection between the input part of the circuit and the output part, because of the differential input. Actual operational amplifiers contain many transistors and other components but they are designed so that this simple equivalent circuit, which includes a Thévenin equivalent circuit at the output, is an accurate representation of their low-frequency performance.

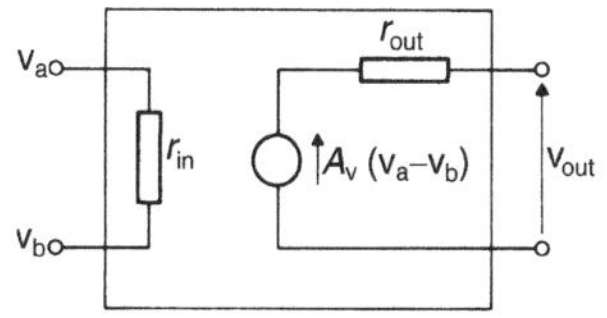

Fig. 4.5 The equivalent circuit of an operational amplifier. Note that there is no common connection between the input and the output. Typically $r_{in} > 100$ kΩ, $r_{out} <$ 1 kΩ and $A_v > 100\ 000$ at low frequencies.

4.3 FEEDBACK AND OPERATIONAL AMPLIFIERS

One reason why operational amplifiers are so widely used is that they can easily be converted into different kinds of amplifiers by the use of **feedback**, namely the connection of the output of an amplifier to its own input by some kind of circuit. The term 'feedback' has now become quite a common word in the general vocabulary, so that it has lost some of the more precise meaning that engineers gave it when the term was first coined. In engineering applications such as control systems, the output of the system is compared with the desired output and if there is a difference between the two, feedback is used to ensure that this information is connected into a point in the circuit in such a way that the output is brought back very close to its intended value. This is usually called **negative feedback** because the correction is in the opposite sense to the cause of the error.

The 'system' containing the feedback is not necessarily an amplifier, it can be almost anything. For instance, the process of driving a car down the road is a feedback process. The output (the position and velocity of the car) is sensed by the driver and compared with his or her expectations (the input). Corrections are then made to the settings of steering wheel, the accelerator, the brake, etc., to ensure that the output conforms with the intentions of the driver. The process is one of comparing the output with the input and then adjusting the system to correct any errors.

In amplifier circuits the output should be a multiple of the input, so in a 'feedback amplifier' the input is compared, instant by instant, with an attenuated version of the output. The difference between the two is then applied to the amplifier in such a way that the error is reduced. For example, if the intended gain is 100, the input is compared with 1/100th of the output and if there is a difference, a correcting signal is applied to the amplifier. How this can be done will be illustrated by several examples in the next sections.

In practice, in designing negative-feedback amplifiers, an amplifier with far too much gain, such as an op amp, is used, and feedback in the form of resistors is used to reduce this gain to the required value. You might think that this would be a pretty stupid thing to do. Why have the extra gain in the first place if you are only going to reduce it? Harold S. Black, who attempted to patent negative feedback in 1928, was greeted with the same response. In his words, 'Our patent application was treated in the same manner as one for a perpetual motion machine.' (see *IEEE Spectrum* December 1977.) True, it does lower the gain, but in exchange it also improves other characteristics, most notably freedom from distortion and noise, linearity, flatness of frequency response and predictability.

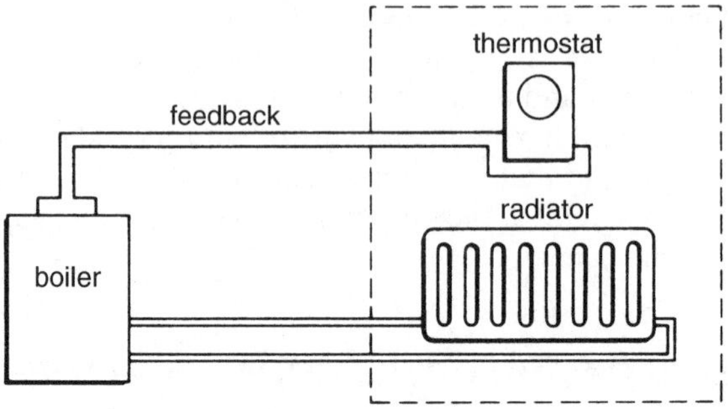

Fig. 4.6 A central heating system showing the feedback from the thermostat to the boiler. The thermostat compares the actual temperature with the intended temperature and causes the boiler to be switched on or off appropriately.

You can see in principle how these beneficial properties come about by considering one of the commonest negative feedback systems around today, the thermostatically-controlled room heating system, illustrated in Fig. 4.6. In such a system the thermostat is set to the desired output temperature (the input), and a boiler is installed which is capable of providing enough heat to achieve all reasonable room temperature settings, even in mid-winter. Obviously this is a greater heating capability than is needed most of the time. When the temperature in the room gets too hot the boiler is switched off, and when it gets too cold in the room the boiler is switched on again. The 'error' in the output temperature is controlling the boiler. So, as you would expect, the system keeps the room at approximately the intended temperature. But the point is that it does this despite fluctuations in the outside temperature, despite variations in the use of the room, despite the addition of extra heating such as an open fire, despite variations in the gas or oil pressure supplying the boiler and even despite some loss of efficiency in the boiler. In other words the system gives a 'predictable and flat response' despite 'distortion' produced by variations in the boiler output, despite noise or interference within the feedback loop, and despite variations in the load on the system caused by usage and the weather. You would expect this of your heating system, so you can expect a similar behaviour of properly designed feedback amplifiers too.

It is also possible to produce systems that are often referred to as 'positive feedback'. In such systems any error is *reinforced* rather than reduced by the feedback signal. Vicious circles are such systems. A central heating system connected the wrong way round, so that too high a temperature turns the boiler on, is such a system. We are not concerned with such systems in this chapter, even though they have their uses in such circuits as oscillators and electronic memory elements (see later chapters).

Now let us see how feedback can be applied to operational amplifiers.

4.3.1 A basic non-inverting feedback amplifier (series feedback connection)

An approximate analysis

Figure 4.7 shows the basic design of a non-inverting feedback amplifier. It consists of an op amp with a potential divider, consisting of R_2 and R_1, connected across the op amp's output. We assume that the potential divider's resistance is much greater than r_{out} of the amplifier, so it has negligible loading effect. The output from this potential divider is connected back to the *inverting* input as shown. We also assume that r_{in} of the amplifier is much greater than R_1, so it has negligible effect on the potential divider. The input signal voltage v_{in} is connected between the *non-inverting* input and 0 V line. To understand how this circuit works it is first necessary to understand that the differential voltage $v_a - v_b$ must be extremely small.

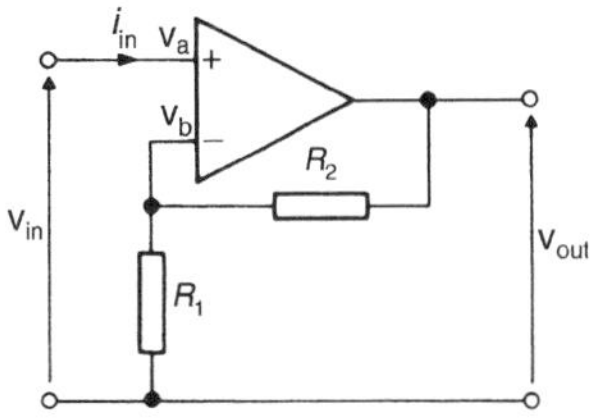

Fig. 4.7 The circuit of a basic non-inverting feedback amplifier.

The open loop voltage gain A_v of the op amp is very large, perhaps even a million or more. This means that if the maximum voltage swing of the output voltage is to be no more than a few volts, then the *differential* input voltage to the op amp, namely $v_a - v_b$, must be no more than a few *microvolts*. In other words v_b and v_a must be at *very nearly the same signal voltage*.

Suppose, for example, the overall feedback amplifier is to have a voltage gain of 100 and an output signal amplitude of, say, 5 V. Then the input signal amplitude v_{in}—between the non-inverting input and 0 V—must be 5 V/100 = 50 mV, of which $v_a - v_b$ contributes only a *few microvolts*. So the voltage drop across R_1 is very nearly equal to the input voltage. The output voltage is obviously the voltage drop across R_2 and R_1 in series, so the voltage gain G of the whole circuit, ignoring $v_a - v_b$, is given by

$$G = \frac{v_{out}}{v_{in}} = \frac{R_2 + R_1}{R_1}$$

Approximate voltage gain of non-inverting op amp feedback circuit

Use your analogue circuit package to build a non-inverting amplifier with series feedback, using resistor feedback components. Choose an op amp model from the package's component library. Check the circuit's frequency response, first without feedback and then with it, and confirm the predicted closed-loop gain. With the feedback connected, monitor the voltage between the inverting and non-inverting inputs of the op amp.

G is called the **closed-loop voltage gain** of the circuit, and is determined almost entirely by the feedback circuit. It is hardly affected by small changes in A_v.

Closed-loop voltage gain

You can see how the circuit adjusts itself to maintain this value of G as follows. Suppose the output increases a little for some reason; this will cause the voltage drop across R_1 to increase a little too. But a small increase in the voltage drop across R_1 results in a proportionately much larger reduction in $v_a - v_b$, causing a corresponding rapid decrease in the output voltage, bringing it back to where it should be.

Worked example

What difference to the above argument would a change in A_v from 500 000 to 200 000 make, when $G = 100$ and when V_{out}, the amplitude of the output, is 5 V?

The validity of the argument that $v_a \approx v_b$ depends on the fact that if A_v is large then $v_a - v_b$ must be small, because, no matter what feedback is connected, the ratio of $v_{out}/(v_a - v_b)$ must be equal to A_v.

If the output amplitude $V_{out} = 5$ V, and the gain G is 100, $V_{in} = 50$ mV. Now if $A_v = 500\ 000$, then the maximum value of $v_a - v_b$ must be 10 μV. On the other hand if

$A_v = 200\ 000$ then $v_a - v_b$ will not exceed 25 μV. The difference the change in A_v makes is the difference between ignoring 10 μV in comparison with 50 mV and ignoring 25 μV in comparison with 50 mV—the difference between a 0.02% error and a 0.05% error.

The feedback not only determines the voltage gain of the circuit, it also affects the *input resistance* R_{in} of the amplifier. We expect this to be very high, because the input current must be small. (The input resistance is high and the differential input voltage $v_a - v_b$ very small.)

We know that the input resistance r_{in} of the op amp alone—without feedback—is $(v_a - v_b)/i_{in}$, and that the input resistance of the whole feedback amplifier is $R_{in} = v_{in}/i_{in}$. So to see the effect of the feedback resistors on the input resistance we have to compare v_{in} with $v_a - v_b$.

Now v_{in} in the feedback circuit is approximately equal to the voltage drop across R_1 because $v_a \approx v_b$. But the voltage drop across R_1 is clearly $R_1/(R_1 + R_2)$ times the output voltage v_{out}. But v_{out} is simply the input voltage to the op amp $(v_a - v_b)$ times A_v. Thus,

$$v_{in} \approx (\text{the voltage drop across } R_1) = \frac{v_{out} \times R_1}{R_1 + R_2} = \frac{A_v(v_a - v_b)R_1}{R_1 + R_2}$$

So, rearranging this, the change in input resistance is given by

$$\frac{R_{in}}{r_{in}} = \frac{v_{in}}{v_a - v_b} \approx \frac{A_v \times R_1}{R_1 + R_2} = \frac{A_v}{G}$$

or

$$R_{in} = \frac{A_v}{G} r_{in}$$

Input resistance of non-inverting op amp feedback circuit

The input resistance is increased as compared with that of the op amp by quite a large factor. For example if $A_v = 200\ 000$ and $(R_1 + R_2)/R_1 = 100$ (giving $G = 100$) then the input resistance is increased by a factor of 2000. The larger the gain G, the smaller the increase in input resistance.

Self-assessment question

6 In the circuit of Fig. 4.7 the input resistance r_{in} of the op amp on its own is 100 kΩ and the voltage gain A_v of the op amp is 200 000. What are the overall gain, G, and the input resistance R_{in} of the circuit as a whole when (a) $R_1 = 1$ kΩ and $R_2 = 49$ kΩ, (b) $R_2 = 0$.

The effect on the output resistance of the feedback circuit is discussed shortly.

A more exact analysis: the feedback equation

In the following analysis, the 'virtual zero volts' approximation is no longer made, so that the gain can be predicted at frequencies where the open-loop gain A_v is not much greater than one.

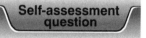

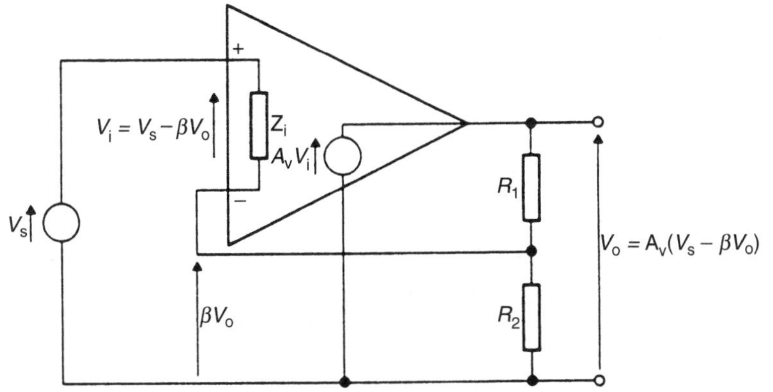

Fig. 4.8 Simplified equivalent circuit. The feedback fraction $\beta = R_2/(R_1 + R_2)$.

At such frequencies, we must also take account of all phase shifts, so phasor quantities must be used, and we must use input and output *impedances* instead of resistances.

We can assume that the resistance of the voltage divider, $(R_1 + R_2)$ in Fig. 4.7, is much greater than the magnitude of the output impedance of the amplifier, $|\mathbf{z}_o|$. This ensures that the amplifier output voltage appearing across the voltage-divider chain is very close to its open-circuit value. Similarly, at the input, we can assume that the amplifier input impedance $\mathbf{z}_i$ is sufficiently large to avoid loading either the voltage source or the voltage divider.

When these conditions are met we can use the simplified equivalent circuit in Fig. 4.8, in which the signal source is shown as an ideal voltage generator with negligible source impedance. The voltage fed back to the inverting input is expressed as βV_o, where β is known as the **feedback fraction**. In this case, β is given by $R_2/(R_1 + R_2)$.

Feedback fraction

In the input circuit, the source voltage V_S and the feedback voltage βV_o are effectively connected *in series* with the amplifier input impedance $\mathbf{z}_i$. This type of feedback is therefore known as **series feedback**. The feedback is also **voltage-derived** because it is derived from and proportional to the amplifier output voltage, V_o.

The feedback equation

We can use the simplified equivalent circuit to find an expression for the closed-loop gain of the amplifier, **G**. Applying Kirchhoff's voltage law to the input circuit in Fig. 4.8, the amplifier input voltage V_i is given by the phasor difference

$$V_i = V_S - \beta V_o$$

Hence, using the relationship $V_o = \mathbf{A}_v \mathbf{V}_i$, we obtain

$$V_o = \mathbf{A}_v(V_S - \beta V_o)$$

This expression can be rearranged as follows:

$$V_o = \mathbf{A}_v V_S - \beta \mathbf{A}_v V_o$$

$$V_o(1 + \beta \mathbf{A}_v) = \mathbf{A}_v V_S$$

so that

$$\mathbf{G} = \frac{V_o}{V_S} \approx \frac{\mathbf{A}_v}{(1 + \beta \mathbf{A}_v)}$$

Feedback equation

Feedback factor

This is known as the **feedback equation** and the term $(1 + \beta \mathbf{A}_v)$ which appears in the denominator is called the **feedback factor**. If the op amp gain $|\mathbf{A}_v|$ is very much greater than $1/\beta$, so that $|\beta \mathbf{A}_v| \gg 1$, then the feedback factor approximates to $(1 + \beta \mathbf{A}_v) \approx \beta \mathbf{A}_v$. Under these conditions, the closed-loop gain becomes

$$\mathbf{G} = \frac{V_o}{V_S} \approx \frac{\mathbf{A}_v}{\beta \mathbf{A}_v} = 1/\beta$$

where $1/\beta = (R_1 + R_2)/R_2$.

This is the result which you should have used in the low-frequency analysis of Self-assessment question 6. The feedback equation is a more exact expression which may be used both at low frequencies and at higher frequencies where $|\mathbf{A}_v|$ is reduced in value and $|\beta \mathbf{A}_v|$ is no longer much greater than 1.

The loop gain of a feedback amplifier

In the previous section you saw that the approximation $\mathbf{G} = 1/\beta$ depended on the value of the product $\beta \mathbf{A}_v$, rather than on the value of $\mathbf{A}_v$ alone. This now leads to an important feedback parameter known as the **loop gain**. If you imagine that the feedback loop is broken, then the **loop gain** is, effectively, the gain *around the feedback loop*, taking into account any inversion provided by the amplifier. For the circuit being considered it is given by

Loop gain

$$\text{Loop gain} = -1 \times (\text{feedback fraction}) \times (\text{open-loop gain}) = -\beta \mathbf{A}_v$$

Remember that the factor of −1 represents the effect of connecting the feedback signal to the *inverting* input of the op amp.

Worked example

How does the loop gain enter the feedback equation?

The loop gain is $-\beta \mathbf{A}$, so that the feedback factor $(1 + \beta \mathbf{A}_v)$ can be expressed as (1−loop gain). Since $\mathbf{A}_v$ is the open-loop gain of the op amp, the feedback equation gives the closed-loop gain:

$$\mathbf{G} = \frac{V_o}{V_S} = \frac{\mathbf{A}_v}{(1 + \beta \mathbf{A}_v)} = \frac{\textit{open-loop gain}}{(1 - \textit{loop gain})}$$

This is, in fact, a universal form of the feedback equation which can be applied to any feedback amplifier. Now, at low frequencies, the op amp gain is simply A_0, and the loop gain is given by

$$\text{Loop gain} = -\beta A_0 = \beta A_0 \angle 180°$$

The low-frequency loop gain is thus real and *negative*. This means that when the feedback loop is closed, the feedback signal will be effectively *subtracted* from the input signal. We then have a clear case of **negative feedback**, causing an overall *reduction* in the op amp gain.

Self-assessment question

7 Suppose that the op amp in the circuit of Fig. 4.8 has a single-lag frequency response, with $\mathbf{A}_v = A_0/(1 + jf/f_A)$, where $A_0 = 10^5$ and $f_A = 10$ Hz. If $R_1 = 180$ kΩ and $R_2 = 20$ kΩ, find:
(a) The loop gain and phase shift at very low frequencies and at the frequency f_A.
(b) The loop phase shift at very high frequencies.

At higher frequencies, there will be an extra phase shift incurred in the op amp so that the phase shift around the feedback loop is no longer precisely 180°. In addition, the loop gain magnitude $|\beta\mathbf{A}_v|$ will be reduced since $|\mathbf{A}_v|$ 'rolls-off' with increasing frequency. The closed-loop gain magnitude must then be found using the feedback equation:

$$|\mathbf{G}| = \frac{|V_O|}{|V_S|} = \frac{|\mathbf{A}_v|}{|(1 + \beta\mathbf{A}_v)|}$$

The feedback is usually considered to be *negative* in any frequency range where the magnitude of the feedback factor $|(1 + \beta\mathbf{A}_v)|$ is greater than 1. According to the feedback equation, negative feedback will always *reduce* the open-loop gain magnitude by the ratio $1/|(1 + \beta\mathbf{A}_v)|$.

At frequencies where the loop gain magnitude $|\beta\mathbf{A}_v|$ is sufficiently large, we can make the approximation $|(1 + \beta\mathbf{A}_v)| \approx |\beta\mathbf{A}_v|$ and the closed-loop gain magnitude becomes $|\mathbf{G}| = 1/\beta$. In this case, then, negative feedback gives a closed-loop gain which is almost independent of the op amp gain and is dependent only on the feedback fraction β. So negative feedback can be used to hold the voltage gain constant in spite of changes in $\mathbf{A}_v$.

Some of the terms used in feedback analysis are quite similar and yet have very different meanings. For this reason you should attempt the following self-assessment question.

Self-assessment question

8 Explain the difference between:
(a) open-loop gain, closed-loop gain and loop gain;
(b) feedback fraction and feedback factor.

4.3.2 Stability in feedback amplifiers

Suppose that at a particular frequency the phase shift of the amplifier and feedback network becomes exactly 180°. This has the effect of cancelling out the 180° phase-shift provided by connecting the feedback voltage to the inverting input of the amplifier. The overall phase shift around the feedback loop is then zero, resulting in a positive value for the loop gain.

When this happens, the feedback voltage is in exactly the right phase relationship to *reinforce* the input voltage and we have a particular case of **positive feedback** at this particular frequency. If the loop gain magnitude becomes equal to 1 while the loop phase shift is zero, then the feedback voltage will have exactly the amplitude required to maintain the output waveform at this frequency without any further assistance from the signal source.

The feedback system is then said to be in a state of **oscillation**. This effect can be exploited to design and build sinusoidal oscillators.

Worked example

What does the feedback equation tell us about this condition?

The closed-loop gain is

$$G = \frac{\textit{open-loop gain}}{(1 - \textit{loop gain})}$$

If the loop gain becomes equal to 1, the denominator becomes equal to zero, giving an infinitely large value for the closed-loop gain.

The fact that the gain is then indeterminate means that the feedback equation cannot be used to relate the input and output voltages in these circumstances.

The situation where the loop gain becomes real, positive and *greater* than 1 is more difficult to analyse. However, it is likely that the feedback system will be **unstable**, that is it is capable of bursting into oscillation in an unpredictable and intermittent fashion depending on, for instance, the temperature and slight variations in the circuit parameters.

Figure 4.9a shows that an *uncompensated* op amp is liable to become unstable when resistive feedback is used to give low values of closed-loop gain.

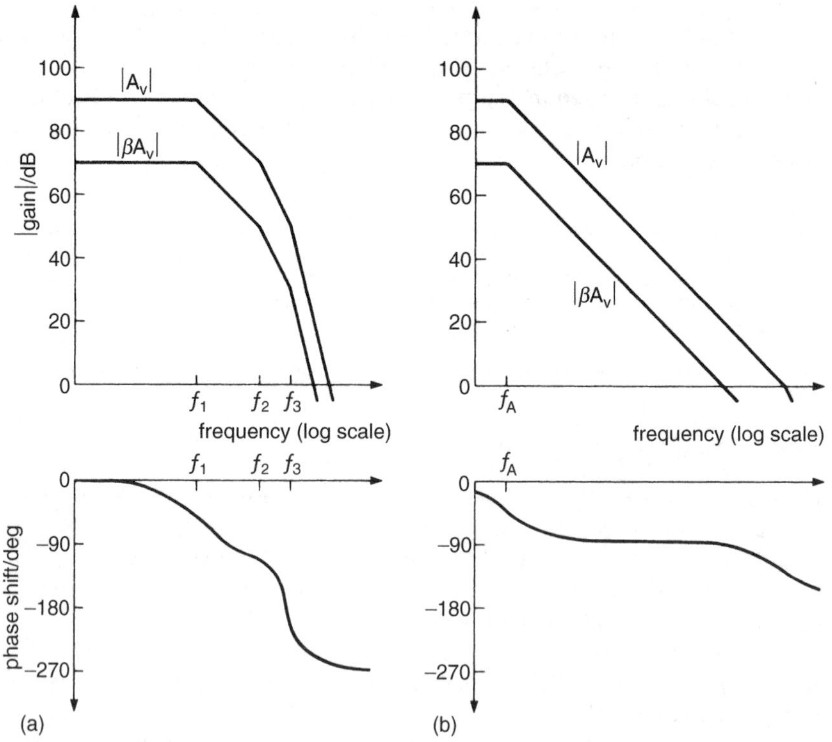

Fig. 4.9 (a) Uncompensated amplifier with three break points. (b) Compensated amplifier. The feedback fraction $\beta = 1/10$ in both cases.

Even though the feedback network contributes no phase shift, the phase shift of the amplifier alone can reach 180° at a frequency approaching f_3 which is less than the unity-gain frequency. As Fig. 4.9a shows, the loop-gain magnitude $|\beta \mathbf{A}_v|$ can be substantially greater than 1 at this frequency. Since the net phase shift around the feedback loop is then zero, spurious oscillations are able to start up, rendering the amplifier useless as a linear device.

In contrast, *frequency-compensated* op amps are designed to give stable operation when used as simple voltage amplifiers with resistive feedback. Figure 4.9b shows that, in this case, the amplifier phase shift is able to reach 180° only for frequencies well above the unity-gain frequency—where $|\mathbf{A}_v|$ takes very low values. Even in the very worst case, the corresponding value of the loop gain magnitude $|\beta \mathbf{A}_v|$ will always be less than 1, so that stable operation is maintained for *any value* of the closed-loop gain.

Worked example

When does this 'worst case' condition occur?

When the entire output voltage is fed back to the input, giving a feedback fraction of 1. The series voltage feedback amplifier then operates as a unity-gain buffer with a low-frequency gain of $G = 1/\beta = 1$.

In summary, the stability of a feedback amplifier depends on the magnitude and phase angle of the loop gain $-\beta \mathbf{A}_v$. To avoid unwanted oscillations and instability, the loop gain magnitude must be less than 1 at any frequency where the phase shift around the feedback loop is zero. If a frequency-compensated op amp is used, then stable operation will always be maintained with simple resistive feedback for *any* value of closed-loop gain.

For a stable feedback amplifier, the closed-loop gain is equal to the open-loop gain divided by the feedback factor $(1 + \beta \mathbf{A}_v)$. Under conditions of stable *negative feedback*, the magnitude of the feedback factor $|(1 + \beta \mathbf{A}_v)|$ is greater than 1. You will see in the following sections that the properties of feedback amplifiers are governed almost entirely by the feedback factor.

4.3.3 Frequency response and gain–bandwidth product

Consider the open-loop frequency response of the frequency-compensated amplifier of Fig. 4.9b. For frequencies up to the unity-gain frequency, this has the same form as the frequency response of the low-pass networks of Chapter 3. These low-pass networks have the transfer function:

$$\frac{V_o}{V_i} = \frac{1}{1 + jf/f_c}$$

This has a low-frequency value of 1, and a cut-off frequency f_c. For the amplifier, we can use the same transfer function if we introduce a factor A_0, the value of the gain at very low frequencies, and change the cut-off frequency to f_A of Fig. 4.9b:

$$\mathbf{A}_v = \frac{A_0}{1 + jf/f_A}$$

We will assume throughout this section that the feedback fraction is a real ratio β. The feedback is then said to be *frequency independent* and the feedback equation is

$$\mathbf{G} = \frac{\mathbf{A}_v}{(1 + \beta\mathbf{A}_v)}$$

Substituting for $\mathbf{A}_v$ in the feedback equation we obtain

$$\mathbf{G} = \frac{A_0/(1 + \mathrm{j}f/f_A)}{[1 + \beta A_0/(1 + \mathrm{j}f/f_A)]}$$

If both the numerator and the denominator are multiplied by $(1 + \mathrm{j}f/f_A)$, then $\mathbf{G}$ becomes

$$\mathbf{G} = \frac{A_0}{(1 + \mathrm{j}f/f_A) + \beta A_0}$$
$$= \frac{A_0}{(1 + \beta A_0) + \mathrm{j}f/f_A}$$

Finally, dividing the numerator and denominator by $(1 + \beta A_0)$ gives the result

$$\mathbf{G} = \frac{A_0/(1 + \beta A_0)}{1 + \mathrm{j}f/f_A(1 + \beta A_0)}$$

The expression for the closed-loop gain has the same form as the open-loop gain, except that the closed-loop gain has a low-frequency value

$$G_0 = A_0/(1 + \beta A_0)$$
$$\approx 1/\beta \qquad \text{when } \beta A_0 \gg 1$$

and 3 dB cut-off frequency

$$f_B = f_A(1 + \beta A_0)$$
$$\approx \beta A_0 f_A \qquad \text{when } \beta A_0 \gg 1$$

Notice that the factor $(1 + \beta A_0)$ is equal to the low-frequency value of the feedback factor $(1 + \beta\mathbf{A}_v)$.

The effect of applying negative feedback is summarized in Fig. 4.10. You can see that:

1. The low-frequency gain is *reduced* by the factor $(1 + \beta A_0) \approx \beta A_0$.
2. The 3 dB frequency is *increased* by the same factor.

The **gain–bandwidth product** is defined as

$$\text{(Low-frequency voltage gain)} \times \text{(3 dB bandwidth)}$$
$$= A_0/(1 + \beta A_0) \times f_A(1 + \beta A_0) = A_0 f_A$$

which is the same as that of the open-loop amplifier. In other words, for an amplifier with a single-lag response, the gain–bandwidth product with frequency-independent feedback remains constant, whatever the value of β. Increasing the feedback fraction β reduces the gain while increasing

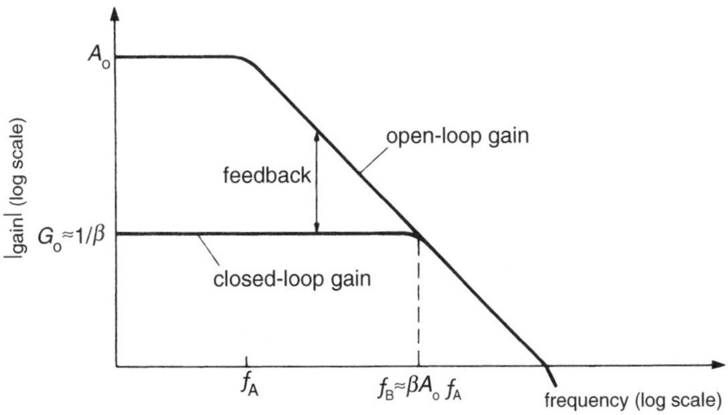

Fig. 4.10 Gain–bandwidth relationship for a compensated op amp (approximate values have been used).

the bandwidth. Reducing the feedback fraction increases the gain while reducing the bandwidth.

Self-assessment question

9 Negative feedback is applied to an amplifier having gain $\mathbf{A}_v = 10^5/(1 + jf/2)$. Construct the Bode gain plot for this amplifier. Calculate the low-frequency closed-loop gain and bandwidth for a feedback fraction β of (a) 1/10, (b) 1/200. Sketch the Bode plots for the closed-loop gain in each case on the same graph as the amplifier response.

Self-assessment question

10 An amplifier has a low-frequency voltage gain of 90 dB and a 3 dB frequency of 100 Hz. The high-frequency roll-off is 20 dB/decade up to the frequency where the gain is 0 dB. Frequency-independent negative feedback is applied to bring the low-frequency gain down to 50 dB. Find the new 3 dB frequency, and the gain–bandwidth product under both open-loop and closed-loop conditions.

The gain–bandwidth relationship applies to a.c. amplifiers as well as to d.c. amplifiers. You will see from Fig. 4.11 that while the mid-band gain is reduced with negative feedback, the lower cut-off frequency f_1 is *reduced*

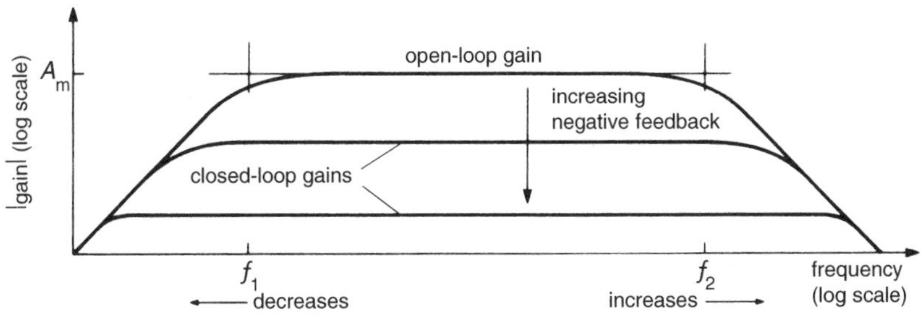

Fig. 4.11 Gain–bandwidth relationship for an a.c. amplifier.

and the upper cut-off frequency is *increased* in the same proportion. The effect overall is an increase in the 3 dB bandwidth, governed by the factor, $(1 + \beta A_m)$, where A_m is the mid-band gain in the absence of feedback.

4.3.4 Input and output impedance

This section shows that negative feedback not only increases bandwidth but, depending on the type of feedback (series or shunt, voltage-derived or current-derived) also affects the input and output impedances.

To be consistent with conventions used earlier, the small-signal input and output impedances of an amplifier are denoted by the lower-case symbols, $\mathbf{z}_i$ and $\mathbf{z}_o$. The *closed-loop* values, that is the values obtained when negative feedback is applied, are written as capital symbols, $\mathbf{Z}_i$ and $\mathbf{Z}_o$.

Closed-loop input impedance

In a series-applied negative feedback circuit, the signal fed back opposes the input voltage signal, thereby reducing the current drawn from the signal source. As explained in Section 4.3.1, the effect of negative feedback is therefore to increase the input impedance of the feedback amplifier above that of the op amp.

In order to find the general equation that applies to this circuit, we can assume at the outset that the closed-loop input impedance is very large. We can then simplify the analysis by supposing that the entire source voltage V_s appears at the non-inverting input as shown in the circuit of Fig. 4.12. Notice that now the feedback fraction $\boldsymbol{\beta}$ is in bold-face type to indicate that it is a frequency-dependent complex quantity.

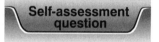

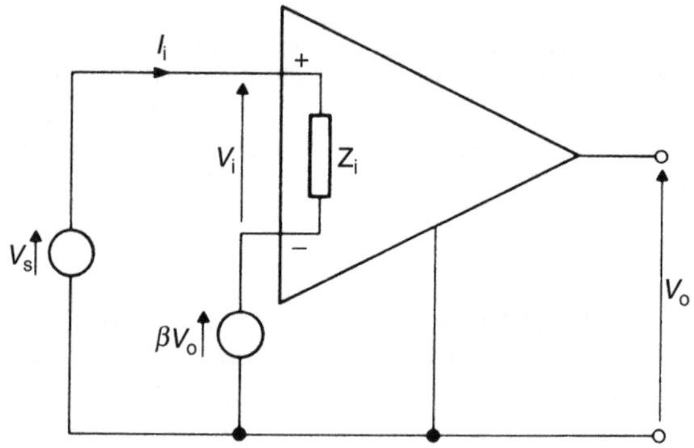

Fig. 4.12

Self-assessment question

11 Show that in the voltage-derived series feedback amplifier, the amplifier input voltage $\mathbf{V}_i$ is equal to the source voltage $\mathbf{V}_S$ divided by the feedback factor.

Now, the phasor voltage $\boldsymbol{V}_i$ and current $\boldsymbol{I}_i$ are related by $\boldsymbol{V}_i/\mathbf{I}_i = \mathbf{z}_i$, where $\mathbf{z}_i$ is the input impedance of the amplifier alone. Hence

$$\boldsymbol{I}_i = \boldsymbol{V}_i/\mathbf{z}_i$$

The overall input impedance is

$$\mathbf{Z}_i = V_S / I_i$$

substituting for I_i

$$\mathbf{Z}_i = V_S \mathbf{z}_i / V_i$$

From Self-assessment question 11 we have the relationship $V_i = V_S/(1 + \beta\mathbf{A}_v)$, so that $V_S = V_i(1 + \beta\mathbf{A}_v)$. The input impedance of the overall circuit, as seen by the signal source, is therefore

$$\mathbf{Z}_i = (1 + \beta\mathbf{A}_v)\mathbf{z}_i$$

This result shows that series negative feedback increases the input impedance of the amplifier by the feedback factor. This is true for *any* series-applied feedback amplifier and not only to the non-inverting case considered here.

Closed-loop output impedance

The effect of voltage-derived feedback on the op amp output impedance can be found by considering the equivalent circuit of the output end of the amplifier shown in Fig. 4.13.

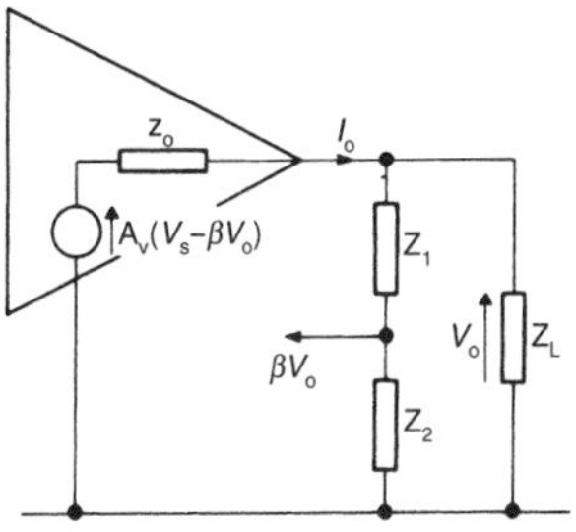

Fig. 4.13

Intuitively, the likely effect of the feedback can be deduced by imagining what happens if the load is reduced, causing an increase in output current. The increased current through $\mathbf{z}_o$ tends to cause V_o to fall, but any fall in V_o means an increase in input voltage to the amplifier (i.e. $V_S - \beta V_o$ increases). The output generator voltage $\mathbf{A}_v(V_S - \beta V_o)$ increases by a much larger amount, allowing the current through $\mathbf{z}_o$ to change significantly, with only a small change in V_o. Such a large output current change with only a small output voltage change implies a low output impedance.

To find the output impedance with negative feedback, we can follow the method used in Chapter 1 and find the open-circuit output voltage and the short-circuit output current, on the assumption of perfect linearity. These values will then give the output impedance of the Thévenin (or Norton) linear equivalent circuit.

The analysis can be simplified by assuming that the magnitude of the impedance $(\mathbf{Z}_1 + \mathbf{Z}_2)$ is much larger than the magnitude of the amplifier output impedance $\mathbf{z}_o$. The open-circuit output voltage for an input voltage V_S can then be obtained directly from the feedback equation:

$$V_{oc} = \frac{V_S \mathbf{A}_v}{(1 + \beta\mathbf{A}_v)}$$

When the output is short-circuited, $V_o = 0$ and the short-circuit output current is given by

$$I_{sc} = V_S \mathbf{A}_v / \mathbf{z}_o$$

Hence, on taking the ratio of these two quantities, we obtain the closed-loop output impedance:

$$\mathbf{Z}_o = V_{oc}/I_{sc} = \mathbf{z}_o/(1 + \beta\mathbf{A}_v)$$

The effect of applying voltage-derived negative feedback is therefore to *reduce* the output impedance of the op amp by the feedback factor $(1 + \beta\mathbf{A}_v)$.

Self-assessment question

12 An op amp has the following open-loop parameters; a low-frequency voltage gain of not less than 100 dB, an input impedance equivalent to a resistor of not less than 1 MΩ shunted by a capacitor of not more than 50 pF and a resistive output impedance of not more than 100 Ω. Frequency-independent voltage-derived series feedback is applied to give a closed-loop gain of 100.

(a) Calculate the low-frequency values of the closed-loop output impedance and input impedance.

(b) What will be the values of the closed-loop impedances $\mathbf{Z}_i$ and $\mathbf{Z}_o$ for frequencies much greater than the closed-loop bandwidth?

4.3.5 Frequency limitations of negative feedback

When you make use of negative feedback to increase the bandwidth of an amplifier, it is important to realize that the degree of feedback becomes progressively less as the frequency is increased beyond the open-loop bandwidth. As you saw earlier in Fig. 4.10, the feedback ceases to have any effect when the frequency exceeds the closed-loop bandwidth, so that in this region the closed-loop gain is equal to the open-loop gain.

It follows that all other improvements conferred by negative feedback, such as reduction in sensitivity to gain variations, control of input and output impedance, etc., apply only within the closed-loop bandwidth and are not maintained beyond this frequency.

4.3.6 The inverting feedback amplifier (parallel or shunt feedback)

A second 'building block' circuit is shown in Fig. 4.14. This example of **shunt feedback** is the basic *inverting amplifier*.

At low frequencies, the gain of the op amp in Fig. 4.14 is real and we have almost perfect inversion at the inverting input. The resistor R_F carries a current i_F which appears in parallel, or *in shunt*, with the input signal current i_S.

Provided the op amp gain magnitude is large enough, the operation of the circuit is almost independent of the op amp properties and is dependent only on the feedback network formed by R_F and R_S. Under these conditions, it can be shown that the current fed back depends only on the output voltage v_o. This means that we are dealing with a case of **voltage-derived shunt feedback**.

The exact analysis of this circuit, for all real and complex values of amplifier gain, is rather complicated. Fortunately, we can make a number of simplifying assumptions which enable approximate results to be found with only a few lines of working.

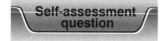

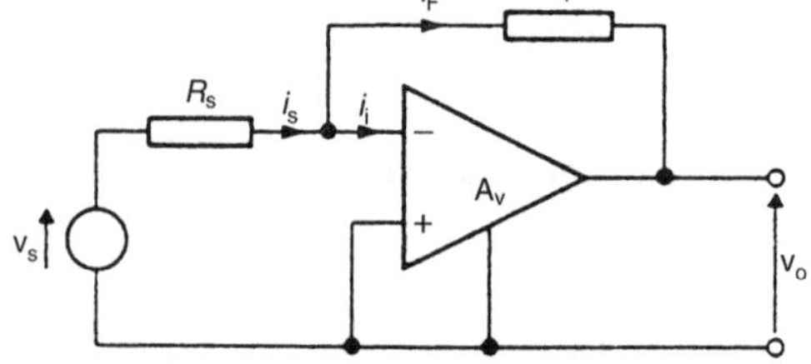

Fig. 4.14

The virtual-earth (ground) approximation

Following the argument of Section 4.3.1, the voltage between the input terminals of the op amp must be very small. The maximum output voltage swing can only be a few volts and, with a low-frequency open-loop voltage gain A_v of up to one million, the differential input voltage cannot be more than a few microvolts. In the shunt feedback circuit of Fig. 4.14, the non-inverting input is earthed (grounded), so the inverting input must have a voltage very close to zero. For this reason, it is commonly called a **virtual earth**, or **virtual ground**, because the feedback holds this point 'virtually' at zero volts. (It obviously cannot be *actually* earthed because then the circuit would not work.)

This 'virtually zero volts' concept is extremely useful in the analysis of feedback circuits, as you will see shortly.

In the more general a.c. circuit of Fig. 4.15, the feedback connection is made from the output of the amplifier through an impedance $\mathbf{Z}_F$ to the inverting input. The input signal V_S is also connected to the inverting input by the impedance $\mathbf{Z}_S$. In a practical circuit, $\mathbf{Z}_S$ may be the Thévenin equivalent impedance of a more complicated input circuit.

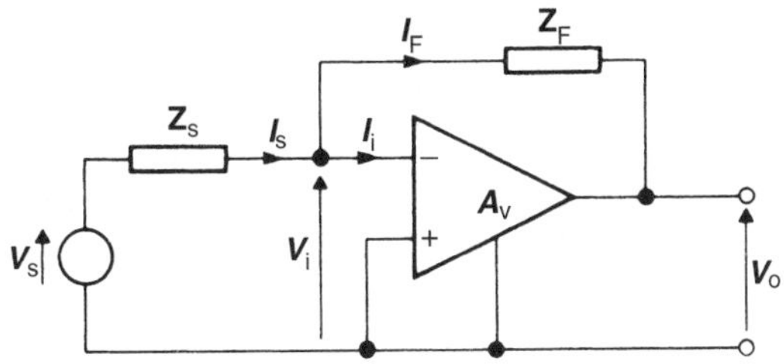

Fig. 4.15

For a typical op amp, the virtual earth assumption will usually be valid at low frequencies, where the open-loop gain magnitude $|\mathbf{A}_v|$ takes very high values. Now, if $V_i \approx 0$, this will also be true of the current I_i into the inverting input, which must be effectively zero. Thus, $I_S \approx I_F$. Assuming that $V_i \approx 0$, we have

$$I_S \approx \mathbf{V}_S/\mathbf{Z}_S \quad \text{and} \quad I_F \approx -V_o/\mathbf{Z}_F$$

Using the approximation $I_S = I_F$ gives

$$\frac{V_S}{\mathbf{Z}_S} = \frac{-V_o}{\mathbf{Z}_F}$$

The closed-loop voltage gain is therefore,

$$\mathbf{G} = \frac{V_o}{V_S} = \frac{-\mathbf{Z}_F}{\mathbf{Z}_S}$$

Under d.c. conditions, a positive input voltage results in a negative output voltage and vice versa.

The minus sign in this expression indicates a phase reversal of 180°, so the gain has a phase angle equal to the phase of $\mathbf{Z}_F/\mathbf{Z}_S$ plus a phase angle of 180°. Explicitly:

$$\mathbf{G} = \frac{V_o}{V_S} = \frac{\mathbf{Z}_F}{\mathbf{Z}_S}\angle 180°$$

Self-assessment question

13 In the feedback circuit of Fig. 4.15, $\mathbf{Z}_S = 10\ \text{k}\Omega \angle 0$ and $\mathbf{Z}_F = 33\ \text{k}\Omega \angle -10°$ at a particular frequency. Assuming that the virtual earth approximation is valid, what will be the output voltage for an input expressed as the phasor $0.3 \angle 30°$ V?

If $|\mathbf{A}_v|$ has the value of 20 000 what will be the corresponding magnitude of the voltage at the amplifier input?

Using the same op amp model as before, build an inverting amplifier with shunt feedback, using resistor feedback components. Check the circuit's frequency response, and confirm the predicted closed-loop gain. Also monitor the 'virtual earth' voltage at the inverting input of the op amp.

To identify the limitations of the virtual-earth approach it is necessary to carry out a fairly detailed analysis using a voltage amplifier with finite frequency-dependent gain. The important result to emerge can be summarized as the following 'rule of thumb':

If the virtual-earth approximation predicts a closed-loop gain **G**, then this result will be valid in any frequency range where $|\mathbf{G}|$ is much less than the open-loop gain magnitude $|\mathbf{A}_v|$.

In engineering terms, 'much less than' implies a factor of 10 or more.

Self-assessment question

14 Voltage-derived shunt feedback with $\mathbf{Z}_S = R_S$ and $\mathbf{Z}_F = R_F$ is applied to an op amp having a single-lag frequency response with $A_0 = 2 \times 10^5$ and $f_A = 5$ Hz.

If $R_S = 1\ \text{k}\Omega$ and $R_F = 100\ \text{k}\Omega$, over what frequency range will the closed-loop voltage gain be approximately -100?

Input and output impedance with shunt feedback

The calculation of the output impedance of the shunt feedback inverting amplifier follows the same lines as that given for the series feedback case, but is altogether more cumbersome. The result supports the general conclusion that *voltage-derived* feedback always gives rise to a circuit with low output impedance. Thus, if we sample the output voltage to generate a feedback signal, we always tend to produce a voltage source which is relatively insensitive to loading effects. Thus the output impedance of this amplifier is reduced, again by the feedback factor $1 + \boldsymbol{A}_v\boldsymbol{\beta}$.

No matter how the feedback signal is derived, the effect on the *input impedance* depends on whether the feedback is applied in series or in parallel with the input signal. As you have seen, the application of series negative feedback tends to *increase* the input impedance. However, in a shunt feedback amplifier, the input signal source must not only supply the input current to the amplifier, it must also provide a current to offset the current in the feedback impedance. This increase in input current for a given value of applied signal voltage is equivalent to a *reduction* in input impedance compared with the input impedance of the amplifier without feedback.

Look now at the shunt feedback circuit in Fig. 4.16. In any frequency range where the virtual-earth approximation is valid, we have $V_i \approx 0$. The impedance at the inverting input is reduced almost to zero by the feedback action and the input current supplied from the voltage source is simply $\boldsymbol{I}_S = \boldsymbol{V}_S/\boldsymbol{Z}_S$.

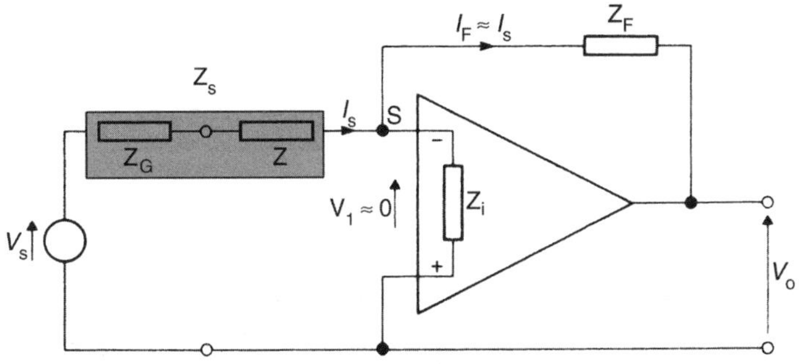

Fig. 4.16

You will see that $\mathbf{Z}_S$ consists of the source internal impedance $\mathbf{Z}_G$ in series with an impedance **Z**. In practice, the value of $\mathbf{Z}_G$ will usually be fixed, while **Z** can be chosen to set the overall gain and to limit the current supplied from the source. Clearly, if **Z** is set to zero, the voltage source will be required to work under virtually short-circuit conditions with the input current limited solely by its internal impedance, $\mathbf{Z}_G$.

For this circuit, therefore, the input impedance 'seen' by the signal source is the impedance **Z** which, within reason, can be any value that we wish to make it.

4.3.7 Current-derived feedback

All the examples of feedback dealt with so far have used voltage-derived feedback, where the feedback signal was proportional to the output *voltage* of the amplifier. It is also possible to generate a feedback signal proportional to the output *current* of the amplifier, and in that case the feedback is called **current derived.**

Most of the effects of negative feedback described so far are not dependent on whether the feedback is voltage or current derived. The reduction in gain and increase in bandwidth still occur with constant gain–bandwidth product. The input impedance is high for series-applied feedback, and is under the control of the circuit designer for shunt feedback.

The difference occurs in the closed-loop *output impedance* of the amplifier. With voltage-derived feedback the closed-loop output impedance is always very low, given by $\mathbf{z}_o/(1+\beta A_v)$. The amplifier output becomes an almost ideal voltage source, where the output voltage is independent of the load on the source. With current-derived feedback, the closed-loop output impedance is generally very high, so that the amplifier output behaves like an *ideal current source*, where the output current is independent of the value of the load.

Figure 4.17 shows one configuration of a series-applied, current-derived feedback amplifier. Notice that the load $\mathbf{Z}_L$ is 'floating', in that one end is no longer connected to the common line. The voltage fed back to the inverting input is equal to I_oR_F, where I_o is the current flowing through the load impedance. In other words, the signal fed back is proportional to the output current, hence the term current derived.

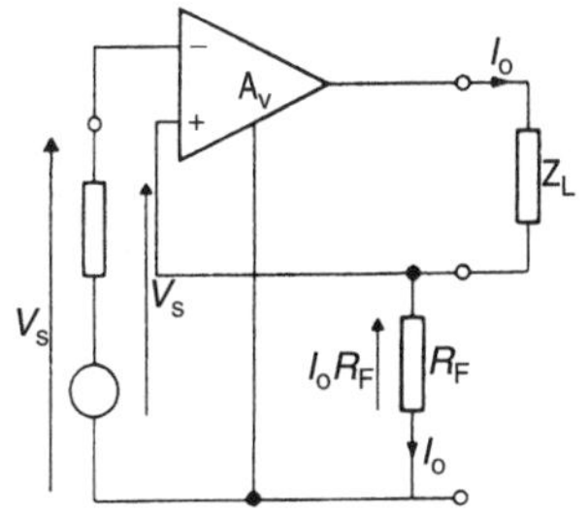

Fig. 4.17 Current-derived series-applied feedback.

Using the approximation that the voltages on the inverting and non-inverting inputs are very nearly equal, we can write

$$V_S = \boldsymbol{I}_o R_F$$

so

$$\boldsymbol{I}_o = V_S / R_F$$

This result shows that the output current of the amplifier is proportional to the input voltage V_i, and is independent of the value of the load impedance $\mathbf{Z}_L$. Since the feedback depends *only* on the output current, $\boldsymbol{I}_o$ remains proportional to V_i for both linear (ohmic) and non-linear load impedances.

A more complete analysis yields the result

$$\frac{\boldsymbol{I}_o}{V_i} = \frac{\mathbf{A}_v}{\mathbf{Z}_L + (1 + \mathbf{A}_v)R_F}$$

showing that $\boldsymbol{I}_o$ is affected by the value of $\mathbf{Z}_L$, but only to a small degree if the product $(1 + \mathbf{A}_v)R_F$ is much greater than $|\mathbf{Z}_L|$.

4.4 SUMMARY OF FEEDBACK PRINCIPLES

The properties of a stable feedback amplifier are governed by the *feedback factor* $(1 + \boldsymbol{\beta}\mathbf{A}_v)$, which enters almost all calculations. The following points apply in the case of *negative feedback*.

(a) When the *feedback fraction* β is frequency independent, feedback reduces gain, improves gain stability and increases bandwidth in single lag (or lead) amplifiers.

(b) Voltage-derived feedback leads to a more constant output voltage amplitude for a constant input signal amplitude and, hence, to low output impedance. Current feedback leads to a more constant output current amplitude and, hence, to a relatively high output impedance.

(c) When the feedback is applied in series with the input, it increases the input impedance. Shunt feedback decreases it.

In all of these instances, negative feedback is effective only for frequencies less than the closed-loop bandwidth.

It can also be shown that negative feedback will reduce the output magnitude (expressed as an r.m.s. voltage) of both internally generated distortion and noise by the feedback factor, provided the distortion and noise are sufficiently small in the first place. This is a fact worth remembering, although it will not be proved here. High-fidelity audio amplifiers, as well as other types, use negative feedback to reduce distortion.

4.5 FURTHER EXAMPLES OF FEEDBACK CIRCUITS

All the feedback circuits dealt with here have only two elements; an amplifier and a β-network. This is sufficient to introduce you to the basic ideas of feedback. However, feedback loops can include more than two elements and a system can have more than one loop. What is more, a circuit can have more than one input. Although the analysis of such circuits may be, at first sight, formidable it turns out that a good understanding of the operating principles can almost always be obtained by assuming that the amplifiers are ideal. This means that you take the gain of each amplifier to be very high and frequency independent, the input impedance to be high, the output impedance to be low, and so on. You may further assume at the outset that the voltage across the input of any amplifier is very small. In the case of op amps, the voltages at the non-inverting and inverting inputs will be very nearly the same; if one of the inputs is connected to the zero-volt line, then you can use the virtual-earth approximation and take advantage of the simplified expressions which follow.

These ideas are put into practice in the following sections which describe some of the applications of feedback circuits. In each case the simplest possible approach is used to arrive at an understanding of the circuit operation. However, since the loop gain of any feedback circuit falls off at high frequencies, there will always be a frequency beyond which simplifying assumptions no longer apply. It is in this regime that CAD packages prove to be invaluable. With computer assistance you can quickly set up circuit models and evaluate their performance over a wide frequency range without resorting to tedious calculations. However, do remember that the use of the CAD software only makes sense when it is guided by a basic understanding of the underlying principles of the circuit under investigation.

4.5.1 Summing amplifiers

Figures 4.19 and 4.20 show two amplifier circuits, each with two inputs. These circuits can be modified to have any number of inputs simply by adding more resistors, one for each input. When the resistors are chosen correctly, the first circuit has an output voltage which is the *sum* of its input voltages. Alternatively, the output can be the sum multiplied by a gain factor, or the output can be the *weighted sum* of its inputs (some inputs are amplified more than others).

The same remarks apply to the second circuit, but its output voltage is *inverted* too.

The resistive summing network

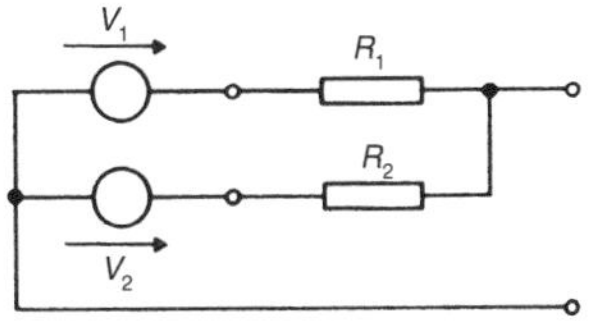

Fig. 4.18 A resistive summing network.

To understand how these circuits work, first consider the resistive summing network of Fig. 4.18. We need to know its Thévenin and Norton equivalent circuits:

(i) *Short-circuit current.* If the output terminals are shorted, the current through the short-circuit is simply

$I_{sc} = V_1/R_1 + V_2/R_2 = V_1G_1 + V_2G_2$

This represents a weighted sum of the input voltages.

(ii) *Output resistance (and conductance).* This is found by looking into the network's output, with the voltage sources replaced by short-circuits. Thus

$R_o = R_1 /\!/ R_2$

or $G_o = G_1 + G_2$

(iii) *Open-circuit output voltage*. This is simply

$$V_{oc} = I_{sc}R_o = I_{sc}/G_o$$
$$= (V_1G_1 + V_2G_2)/G_o$$

This network is used in both the summing amplifiers. Let us look at the first one.

The non-inverting summing amplifier

In Fig. 4.19 the resistive summing network is connected to the non-inverting input of an op amp. The amplifier's input resistance is very high, due to the series feedback, so we can assume that its loading effect on the summing network is negligible. Thus the output voltage is simply the open-circuit voltage of the summing network multiplied by the voltage gain of the amplifier:

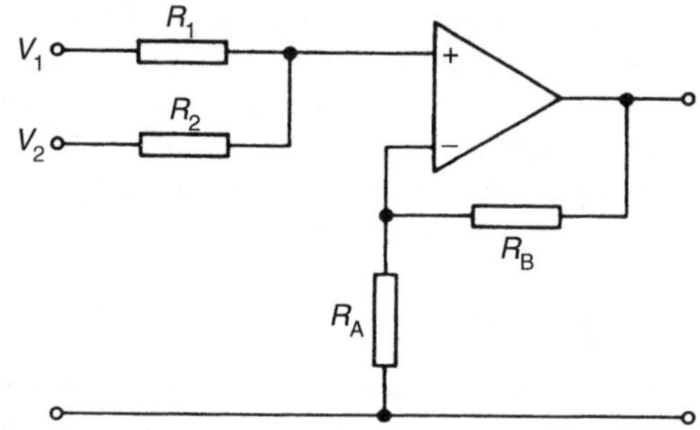

Fig. 4.19 A non-inverting summing amplifier.

$$V_o = GV_{oc} = \frac{1}{\beta} \times \frac{V_1G_1 + V_2G_2}{G_o}$$

where β is the feedback ratio $R_A/(R_A + R_B)$.

If an unweighted sum is required, then we make $G_2 = G_1$; so $G_0 = 2G_1$. Then

$$V_o = \frac{1}{\beta} \times \frac{V_1 + V_2}{2}$$

For a true sum, we make $\beta = 1/2$, so

$$V_o = V_1 + V_2$$

For a summing amplifier with N inputs, a true sum is obtained if all the summing network resistors are equal (so $G_o = NG_1$), and if $\beta = 1/N$.

The inverting summing amplifier

In Fig. 4.20 the resistive summing network is connected to the inverting input of the op amp. Feedback is taken from the output via R_F, and the inverting input is a virtual earth. That is, the negative feedback operates to keep the voltage at the inverting input very close to zero volts. So the summing network is working as if its output was short-circuited. The current which flows from it (its short-circuit output current) virtually all flows through R_F, so the amplifier output voltage is

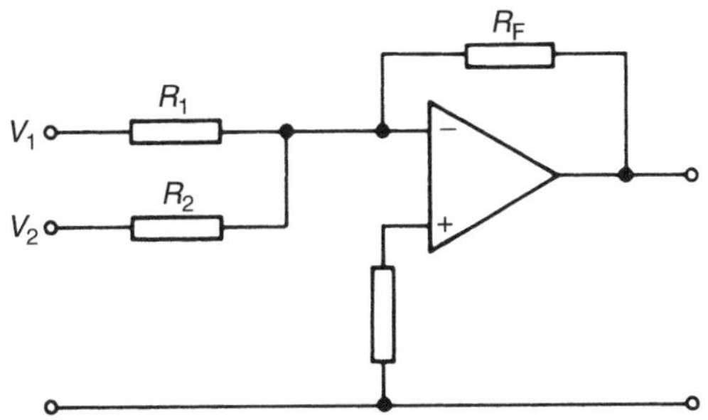

Fig. 4.20 An inverting summing amplifier.

$$V_o = -I_{SC}R_F = -(V_1G_1 + V_2G_2)R_F$$

As with the previous circuit, if an (inverted) unweighted sum is required, then we make $G_2 = G_1$. Then

$$V_o = -(V_1 + V_2)G_1R_F$$

For an inverted true sum, we make $R_1 = 1/G_1$ equal to R_F, so

$$V_o = -(V_1 + V_2)$$

For an inverting summing amplifier with N inputs, a true inverted sum is obtained if all the summing network resistors are equal, and equal to R_F.

Because the inverting input (the virtual earth) is the node where all the currents from the inputs add together to provide the current through R_F, this node is sometimes called the **summing junction**.

Summing junction.

Self-assessment question

15 In a weighted non-inverting summing amplifier, suppose $R_1 = 10$ kΩ, $R_2 = 20$ kΩ, $R_3 = 40$ kΩ and $\beta = 1/7$.

(i) Derive an expression for V_o in terms of the input voltages V_1, V_2 and V_3.
(ii) From this result, calculate V_o when $V_1 = 1$ V, $V_2 = 2$ V and $V_3 = 3$ V, all d.c. voltages.
(iii) What should be the ratio of the feedback resistors?

4.5.2 *CR*-active filters

We can define a filter to be a circuit or a network, with well-defined voltage transfer properties, that is used to suppress unwanted signal components lying in a particular frequency range. So far we have dealt only with *passive* filters containing resistors, capacitors and inductors. The next step is to introduce some elementary ***CR*-active filters** constructed using op amps with resistive and capacitive feedback elements.

Filters are used also for other applications, such as phase-changing, beyond the scope of this book.

CR-active op amp filters find extensive use in the audiofrequency range (up to about 20 kHz) and, with some effort in the choice of appropriate op amp, at frequencies up to 1 MHz and higher. Particular applications are in communications systems and in instrumentation systems.

You will notice that inductors are not included in these active filters. The reason is that by using appropriate combinations of resistors, capacitors and op amps it is possible to eliminate the need for inductors in many applications.

Worked example

Why should this be desirable?

Good quality inductors for low-frequency applications tend to be bulky and expensive compared with resistors and capacitors. Also, they are not so readily available in a wide range of standard values and tolerances. At higher frequencies, inductors are widely used in filters.

The first example will help you to understand why, despite their relative complexity, *CR*-active filters are often chosen in preference to their passive counterparts.

A *CR*-active low-pass filter

This is the shunt feedback circuit illustrated in Fig. 4.21. To see how this circuit works (and to show that it is, indeed, a low-pass filter) consider first of all its operation at very low frequencies. In this regime, the capacitive reactance $1/\omega C_F$ is much greater than the feedback resistance R_F. We can then ignore the capacitor and treat the circuit as a simple inverter with closed-loop gain $\mathbf{G} = -R_F/R_i$.

At higher frequencies we must look at the feedback impedance $\mathbf{Z}_F$ consisting of C_F in parallel with R_F and write the gain of the circuit as $\mathbf{G} = -\mathbf{Z}_F/R_i$. Now, as the frequency is increased, the capacitor impedance decreases until it becomes less than R_F, when $\mathbf{Z}_F \approx 1/j\omega C_F$. The gain magnitude then becomes $|\mathbf{G}| \approx 1/\omega C_F R_1$ which rolls off at a rate of 6 dB/octave. The behaviour over all frequencies is therefore similar to that of a passive *CR* low-pass filter. The essential difference in this case is that the low- frequency gain magnitude is no longer fixed at a value of unity.

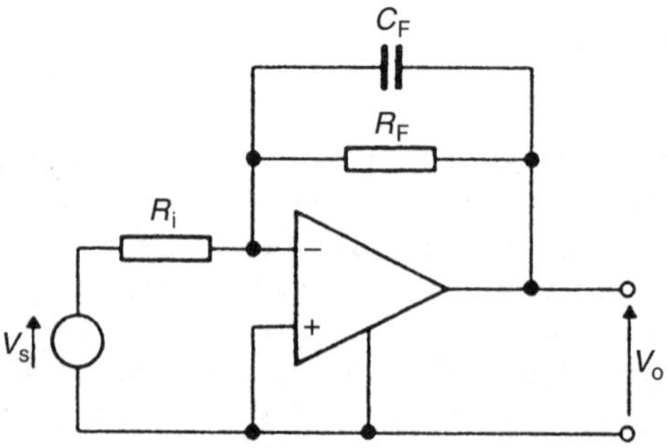

Fig. 4.21 A *CR*-active low-pass filter.

Self-assessment question

16 (a) Using the virtual-earth approximation, derive an expression for the closed-loop voltage gain of the circuit in Fig. 4.21.

(b) Identify the low-frequency gain G_0 and the closed-loop cut-off frequency f_c. Hence, show that the closed-loop gain can be written in the single-lag form $\mathbf{G} = G_0/(1 + jf/f_c)$ and state clearly how the phase shift varies with frequency.

(c) If $R_i = 10\ \mathrm{k\Omega}$, find the values of R_F and C_F to give a low-frequency gain magnitude of 12 dB and a cut-off frequency of 500 Hz.

Remember that the results of Self-assessment question 16 will be valid, provided that the magnitude of the op amp gain $|\mathbf{A}_v|$ is sufficiently large compared with the predicted value of the closed-loop gain $|\mathbf{G}|$, at all frequencies of interest.

Figure 4.22 shows two Bode gain plots. The upper plot is for the open-loop gain of the op amp, $\mathbf{A}_v$, which is assumed to be single lag. The lower plot is for the shunt feedback circuit of Fig. 4.21.

In practice, the low-frequency gain A_0 and bandwidth f_A of the op amp will be ill-defined. However when *CR* feedback is applied, both the low-frequency gain G_0 and the cut-off frequency become well-defined, since these depend only on the values of the feedback components, R_i, R_F and C_F. Comparing this graph with that given for purely resistive feedback in Fig. 4.10, you will see that capacitive feedback reduces the gain–bandwidth product. This is the price to be paid for having the low-frequency gain *and* the closed-loop bandwidth under the control of the designer.

More detailed Bode plots can be obtained using CAD software, showing quite clearly the influence of the op amp gain $\mathbf{A}_v$ at high frequencies, where the feedback becomes ineffective.

> Construct a *CR*-active low-pass filter as in Fig. 4.21. Run a frequency response to check the cutoff frequency. Also investigate performance at much higher frequencies, where the op amp's open-loop gain falls.

Self-assessment question

17 Contrast and compare (a) the passive *CR* circuit and (b) the *CR*-active circuit of Fig. 4.21, having regard to the following properties:

Voltage gain magnitude at low frequencies
Phase shift
Sensitivity to loading
Input impedance

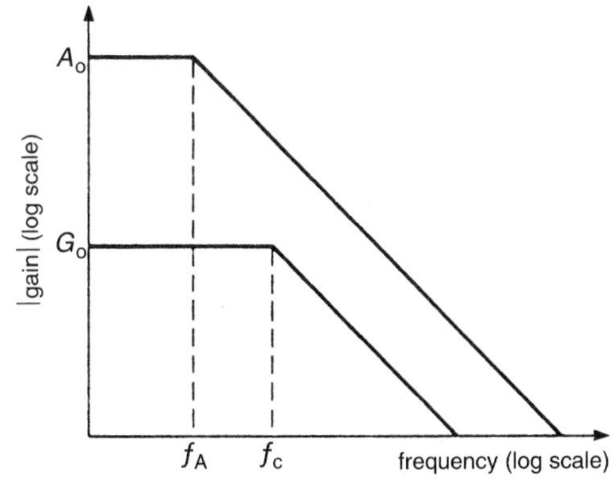

Fig. 4.22

A *CR*-active bandpass circuit

Figure 4.23 shows a feedback circuit with bandpass properties. Capacitor C_i offers a high impedance to signal currents at low frequencies, causing a progressive reduction in gain as the frequency is reduced. At higher frequencies, the response is governed primarily by the reduction in the impedance of the shunt feedback capacitor C_F. This results in an increase in the amount of negative feedback applied, so that the overall gain is reduced as the frequency is increased.

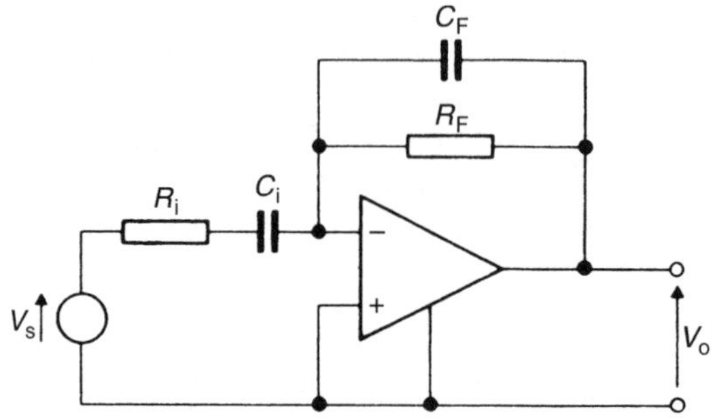

Fig. 4.23 A *CR*-active bandpass circuit.

Between these two extremes there is a range of frequencies in which the gain magnitude attains a maximum mid-band value, so that the operation overall is that of a *bandpass* filter.

You should see that this circuit is simply a variation of the standard shunt feedback circuit. Its behaviour is rather like two *CR* circuits—one high-pass, the other low-pass—separated by an ideal buffer amplifier.

4.6 OUTPUT OFFSET AND EQUIVALENT INPUT OFFSET SOURCES

An ideal differential d.c. amplifier, such as an op amp, would have an output of 0 *V* when its differential input voltage was zero. Real op amps are likely to have a d.c. output voltage even when the input is zero. This is because it is impossible to make a differential amplifier with perfect symmetry. The d.c. voltage at the output of an amplifier, when both inputs are grounded, is called the **output offset voltage**. If the amplifier is an op amp, the output offset depends on certain properties of the op amp itself, and on the particular feedback configuration chosen. It is usual, therefore, to specify the three *input* properties described below.

4.6.1 Bias current and input offset voltage and current

The circuit at the input of an op amp is usually a symmetrical arrangement of transistors. At this stage, you do not need to know anything about transistors, except that they need a small d.c. input current to 'bias' them for correct operation. (Transistors are explained in Chapter 9.) In an op amp, these bias currents must flow into each of the two input terminals. In many of the op amp circuits you have seen so far, the bias currents flow directly from ground, or through resistors connected to the signal source or to the output terminal.

This section develops a formula for calculating the output offset voltage caused by the bias currents flowing through resistors, and caused also by the effects of two other properties called the input offset current and the input offset voltage.

Figure 4.24a shows an operational amplifier in a test circuit. The two current generators supply bias currents I_B to the input terminals. For instance, if a type 741 op amp was used, the bias currents would be the base currents of its two input transistors. The op amp of Fig. 4.24a is a hypothetical one which is perfectly symmetrical; its input transistors are perfectly matched, the input terminals are at the same voltage, other transistors are perfectly matched, and there is zero output offset voltage.

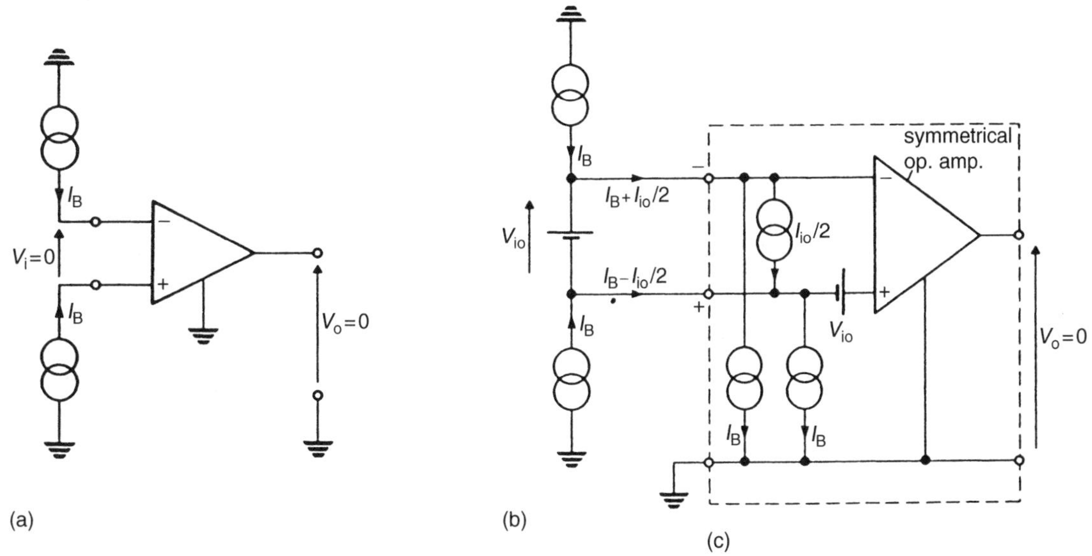

Fig. 4.24 Biasing, offsets, and equivalent circuit of a real op amp. (a) Biasing a perfectly symmetrical op amp, (b) biasing a real op amp, (c) equivalent circuit of a real op amp.

In Fig. 4.24b, an adjustable d.c. source is connected across the input terminals of a real op amp which is not perfectly symmetrical. This source is adjusted until the amplifier's output voltage is brought to zero. The resultant source voltage is defined as the *input offset voltage,* V_{IO}. The **input offset voltage,** V_{IO} is the voltage which much exist across the two input terminals of an operational amplifier in order to bring its output voltage to zero. It is caused by manufacturing tolerances.

Because the real op amp is asymmetrical, its two input bias currents must differ slightly too if the output voltage is to be zero: their difference is defined as the **input offset current,** I_{IO}. I_{IO} is the difference which exists between the two input bias currents of an operational amplifier when its output voltage is brought to zero by the application of V_{IO} across its input terminals.

The **bias current** I_B in a real op amp is defined as the average of the two input bias currents. So these have the values $I_B + I_{IO}/2$ and $I_B - I_{IO}/2$. Bias currents are typically 5 to 10 times greater than offset currents.

Figure 4.24c shows the equivalent circuit of a real op amp modelled by an ideal op amp together with an input offset voltage source V_{IO}, and input current sources which model I_B and I_{IO}. You can see that the external test circuit supplies these input currents. The test circuit also provides a voltage source which opposes V_{IO} in the model. The net effect is that no current flows into the ideal amplifier of the model, and no voltage exists across its input terminals, so its output voltage is zero. Thus the equivalent circuit of Fig. 4.24c models correctly the input effects and bias current.

I will use this model shortly to calculate the output offset voltage, but first let us look at some typical values of I_B, I_{IO}, and V_{IO}, to give you a 'feel' for them. Manufacturers' data sheets for op amps show values for the quantities V_{IO}, I_B and I_{IO}. For instance, the 741 has a quoted V_{IO} of 6 mV maximum, 1 mV typical at 25° C. This means that V_{IO} is guaranteed not to exceed 6 mV (it could be either polarity) in the worst case. 'Typically' means that 'most' 741s will have a V_{IO} which does not exceed 1 mV. Most types of op amp have similar specifications for V_{IO}: many are poorer at about 10 mV typical; some are much better at about 10 μV. For the 741, I_B is quoted as 500 nA maximum and typically 80 nA at 25° C. Other bipolar op amps have bias current ranging from a few microamps down to less than 100 pA. The 741 has a quoted maximum input offset current I_{IO} of 200 nA and a typical value of 20 nA at 25° C. Other bipolars have a typical I_{IO} ranging from a few microamps down to about 1 pA.

Op amps with FET inputs have much smaller input currents, with I_B ranging typically from 1 pA to 30 pA. The input offset current I_{IO} is correspondingly smaller too, with typical values ranging from 0.5 pA to 10 pA. (The FET—field effect transistor—and the bipolar type are explained in Chapter 9).

These values are summarized in Table 4.1.

Table 4.1 Op amp bias currents and offsets at 25° C

Type of op amp	*741*		*Other bipolar types*	*FET input types*
	Max.	*Typical*	*Typical*	*Typical*
Input offset voltage, V_{IO}	6 mV	1 mV	30 μV to 10 mV	30 μV to 10 mV
Input bias current, I_B	500 nA	80 nA	10 pA to a few μA	1 pA to a few nA
Input offset current, I_{IO}	200 nA	20 nA	1 pA to a few μA	0.5 pA to 10 pA

4.6.2 Calculating the output offset voltage

As far as d.c. conditions are concerned, Fig. 4.25 represents both non-inverting and inverting circuit configurations for an operational amplifier with feedback. R_3 represents the equivalent resistance of the bias current path from the non-inverting input to the 0 V rail. In the non-inverting case, the signal is fed in here, so this bias path will include the source resistance if the signal is directly coupled.

Both R_F and R_1 provide paths for the bias current to the non-inverting input. In the inverting case, the signal is fed in between R_1 and 0 V, and R_1 includes the source resistance if the signal is directly coupled. If the path from

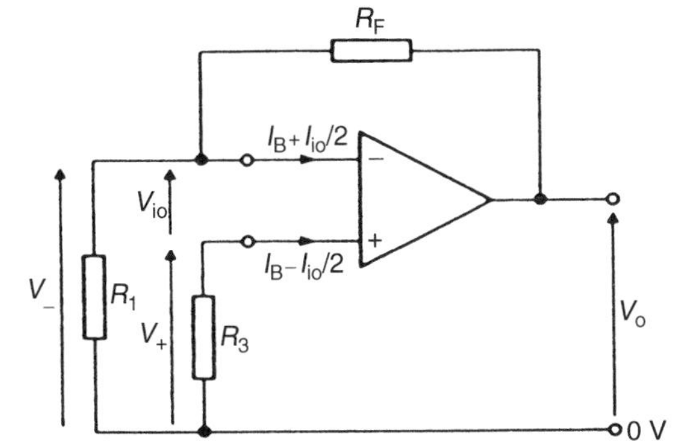

Fig. 4.25 D.C. bias paths for a real op amp.

the non-inverting input to the 0 V rail includes a series capacitor (in either inverting or non-inverting cases) then R_1 is infinite as a bias current path, and all the bias current must flow through R_F.

Effect of input offset voltage, V_{IO}

The principle of superposition allows us to calculate the effects of the input offset sources separately, so we shall start with the input offset voltage. Figure 4.26 shows the general configuration of Fig. 4.25 again, but with the real op amp replaced by an equivalent. This equivalent is the circuit of Fig. 4.24c, but with only the input offset voltage source V_{IO}. The current sources are switched off, and replaced by open circuits.

You should be able to see that this is the circuit of a non-inverting amplifier, with an input voltage of V_{IO}. So the output voltage is

$$V_o = (1 + R_F/R_1)V_{IO}$$

due to V_{IO} alone.

So for instance, if $V_{IO}(\max) = 6$ mV and $R_F/R_1 = 100$, then $V_o(\max) =$ 606 mV due to the input offset voltage alone. However, if the resistor R_1 is returned to 0 V via a capacitor, then the d.c. resistance of this path becomes infinite, and $V_o(\max) = V_{IO}(\max) = 6$ mV.

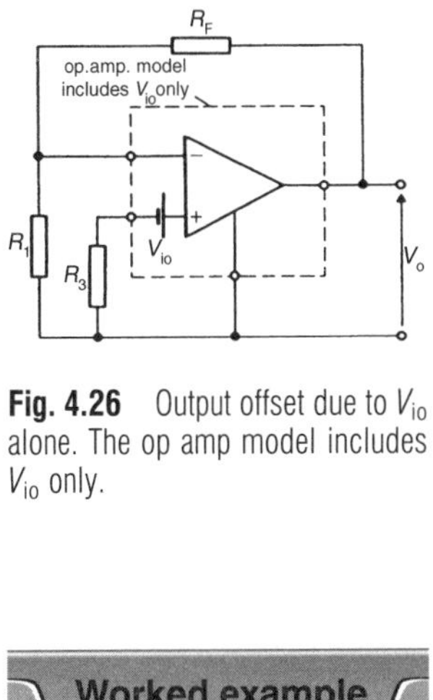

Fig. 4.26 Output offset due to V_{io} alone. The op amp model includes V_{io} only.

Worked example

What do you think would be the effect of putting a capacitor in series with R_F (with R_1 connected directly to 0 V)?

The resistance of this path would be infinite, so the formula predicts an infinite output voltage. The output would actually saturate at nearly the positive or negative supply voltage. There must always be a d.c. path from output to non-inverting input to avoid this.

Effect of input currents, I_B and I_{IO}

Figure 4.27 uses the configuration of Fig. 4.25 again, but this time the op amp model includes only the input current sources of Fig. 4.24c. The input offset voltage is switched off, and replaced by a short circuit.

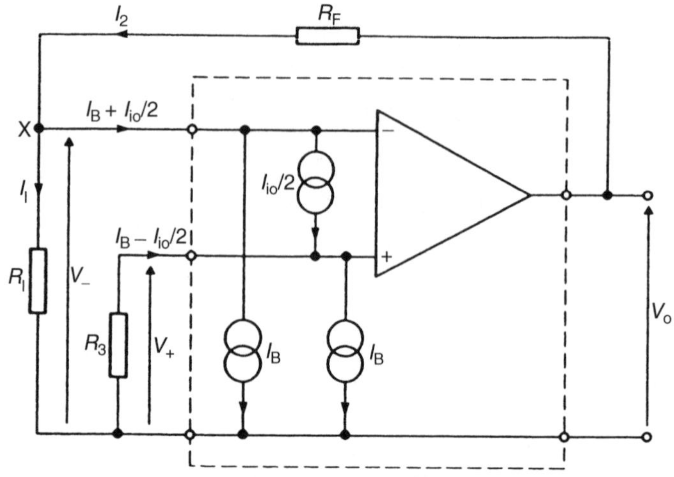

Fig. 4.27 Output offset due to input currents. The op amp model includes I_{io} and I_B, but not V_{io}.

At the node X,

$$I_2 = I_1 + (I_B + I_{IO}/2)$$

Substituting for I_2 and I_1:

$$\frac{V_o - V_-}{R_F} = \frac{V_-}{R_1} + I_B + I_{IO}/2$$

Therefore,

$$V_o = V_-(1 + R_F/R_1) + (I_B + I_{IO}/2)R_F$$

Because of the very high d.c. open-loop gain of the op amp, $V_- \approx V_+$ Thus

$$V_- \approx V_+ = -(I_B - I_{IO}/2)R_3$$

So, substituting for V_- in the expression for V_o:

$$V_o = -(I_B - I_{IO}/2)R_3(1 + R_F/R_1) + (I_B + I_{IO}/2)R_F$$

Collecting terms gives:

$$V_o = I_B[R_F - R_3(1 + R_F/R_1)] + I_{IO}[R_F + R_3(1 + R_F/R_1)]/2$$

This complicated expression can be made independent of I_B if we make $R_3(1 + R_F/R_1) = R_F$. That is, if

$$R_3 = \frac{R_F}{(1 + R_F/R_1)} = \frac{R_1 R_F}{R_1 + R_F}$$
$$= R_1 \;//\; R_F \qquad (R_1 \text{and}\, R_F \text{ in parallel})$$

If this is done, then $V_o = I_{IO}R_F$ due to input currents alone.

It is good practice to make $R_3 = R_1//R_F$. Then the bias current I_B has no effect on the output. This is the best strategy to minimize the output offset voltage. What is more, the output is then easier to calculate!

Total output offset voltage

The net effect of V_{IO}, I_B and I_{IO} is simply the sum of the appropriate expressions for V_o involving these quantities. The complete expression is:

$$V_o = (1 + R_F/R_1)V_{IO} + I_B[R_F - R_3(1 + R_F/R_1)] + I_{IO}[R_F + R_3(1 + R_F/R_1)]/2 \quad (4.1)$$

When $R_3 = R_1//R_F$:

$$V_o = (1 + R_F/R_1)V_{IO} + I_{IO}R_F \quad (4.2)$$

Self-assessment question

18 Suppose a 741 op amp has the maximum value of input bias current $I_B = 500$ nA but negligible values of input offset current and voltage. Suppose $R_F = 1\ \text{M}\Omega$, and $R_1 = 10\ \text{k}\Omega$ (to provide a gain of -100), and that the non-inverting input is connected directly to 0 V.

Calculate the output offset voltage.

The answer to this self-assessment question shows the appreciable output offset possible, even when the input offsets are negligible, if the d.c. bias paths are not made equal.

Self-assessment question

19 Calculate the maximum output offset voltage of the circuit of Fig. 4.28 with (a) the capacitor in circuit and (b) the capacitor replaced by a short-circuit.

Assume in both cases that R_3 is chosen to equal the d.c. value of R_1 in parallel with R_F. What should be the value of R_3 in each case?

Op amp is a 741. $R_S = 1\ \text{k}\Omega$, $R_i = 10\ \text{k}\Omega$, $R_F = 1\ \text{M}\Omega$.

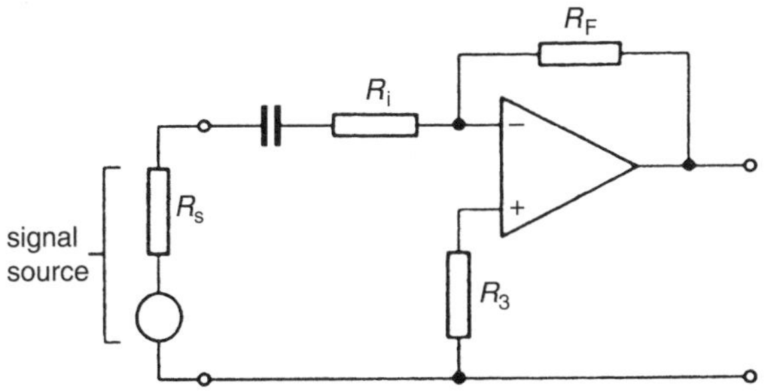

Fig. 4.28 The amplifier of Self-assessment question 19.

4.7 NOISE AND EQUIVALENT NOISE SOURCES

4.7.1 Sources of internal noise

Chapter 2 discusses the fact that if the volume control is turned right up on a radio or 'hi-fi', and there is little or no intentional signal or interference, it is often possible to hear a hiss from the loudspeaker. The electrical phenomenon causing this kind of hiss is called internal noise.

Where does the noise come from? There are three important sources of electrical noise: *thermal or Johnson noise*, which occurs in resistors and other resistive material, and is temperature dependent; *shot noise*, which occurs in active devices and is current dependent; and *flicker* or *1/f noise*, which occurs in active devices and resistors carrying direct current.

Thermal noise (Johnson noise)

In the absence of an applied electric field, the electrons in a conductor are in random motion, their agitation increasing as the temperature is raised. The random motion leads to a randomly varying potential difference between the ends of the conductor.

The uniform distribution of thermal noise throughout the frequency spectrum leads to its being called **white noise** (by analogy with white light, which is thought of as containing all visible frequencies equally).

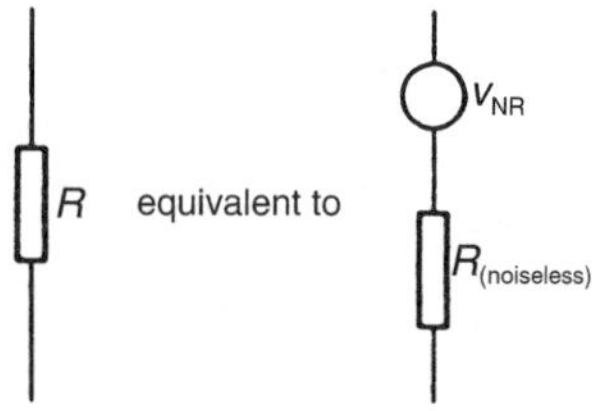

Fig. 4.29 A representation of thermal noise.

A real resistor, generating noise, can be *represented* by a noise voltage generator in series with a noiseless resistor of the same resistance as the real resistor. This is shown in Fig. 4.29. The mean-square value of the equivalent noise voltage generator associated with a resistance R is

$$\overline{v^2{}_{NR}} = 4RkT\Delta f$$

where k is Boltzmann's constant, T is the absolute temperature and Δf is the bandwidth over which the measurement is made. The square root of this is, of course, the r.m.s. value

$$V_{NR} = 2\sqrt{(RkT\Delta f)}$$

Consider, for example, the thermal noise generated by a 10 kΩ resistor at a temperature of 25°C (298 K) over the audiofrequency range, say from 100 Hz to 20 kHz. In this case Δf is equal to $(20 \times 10^3 - 100)$ Hz $\approx 20 \times 10^3$ Hz. Since Bolzmann's constant is equal to 1.38×10^{-23} J K^{-1}, the resistor will produce a fluctuating voltage with mean-square value

$$\overline{v^2{}_{NR}} = 4 \times 10 \times 10^3 \times 1.38 \times 10^{-23} \times 298 \times 20 \times 10^3 \text{ V}^2 \approx 3.3 \times 10^{-12} \text{ V}^2$$

This is equivalent to an r.m.s. value of $\sqrt{3.3} \times 10^{-6}$ V or about 1.8 μV.

Worked example

What would be the r.m.s. noise voltage produced by two resistors, R_1 and R_F, connected in series?

To answer this question we must assume that the noise voltages generated by the two resistors are quite independent of each other. The total mean-square voltage can then be obtained by adding together the mean-square contribution from each resistor taken separately. Thus, if

$\overline{v^2_{NRI}} = 4R_1kT\Delta f$

and

$\overline{v^2_{NRF}} = 4R_FkT\Delta f$

then the total mean-square voltage is

$\overline{v^2_{NT}} = \overline{v^2_{NRI}} + \overline{v^2_{NRF}} = 4(R_1 + R_F)kT\Delta f$

The total fluctation has an r.m.s. value equal to the square root of this quantity. This means that if $R_1 = R_F$, the mean-square noise voltage is doubled, while the r.m.s. noise voltage increases by a factor of $\sqrt{2}$. For the case $R_1 = R_F = 10\ \text{k}\Omega$, with the same conditions as before, we have

$V_{NT} = \sqrt{2} \times 1.8\ \mu\text{V} \approx 2.5\ \mu\text{V}$

Shot noise

Shot noise is also white noise and it results from the discrete nature of the charge carriers that constitute a flow of direct current. At any instant the number of carriers crossing a plane is different from the number crossing at any other instant. So noise will be generated by the electrons crossing the emitter–base or collector–base junctions within a transistor, for example.

Providing the carriers do not interact, so that their motions are independent, the r.m.s. shot noise current can be calculated using the Schottky equation,

$$I_N = \sqrt{(2qI\,\Delta f)}$$

where q is the carrier charge and I is the d.c. component upon which the noise current is superimposed.

Flicker noise

Flicker noise, also called **excess noise** or **1/*f* noise**, is noise which predominates at low frequencies. Its root mean square value over a limited frequency range is proportional to $\sqrt{(\Delta f/f)}$. It is therefore *not* white noise, because the spectral density increases as frequency falls.

Flicker noise occurs in active devices such as bipolar transistors and FETs, and in resistors with applied d.c. voltages.

Self-assessment question

20 (a) In an amplifier with a bandwidth of 20 kHz, calculate the r.m.s. noise voltage contributed by (i) thermal noise in a 1 kΩ resistor at 25°C (Boltzmann's constant $k = 1.38 \times 10^{-23}\ \text{J K}^{-1}$), (ii) shot noise current with 0.1 mA flowing through the 1 kΩ resistor (the electronic charge $q \simeq 1.6 \times 10^{-19}\text{C}$).

(b) Calculate the total r.m.s. noise voltage across the resistor. (Assume that the flicker noise contribution is negligible at the frequency of operation.) Remember that r.m.s. voltages cannot be added together simply, when the voltages are at different frequencies or are random. Powers *can* be added, and power is proportional to mean-square voltage, so mean-square voltages *can* be added.

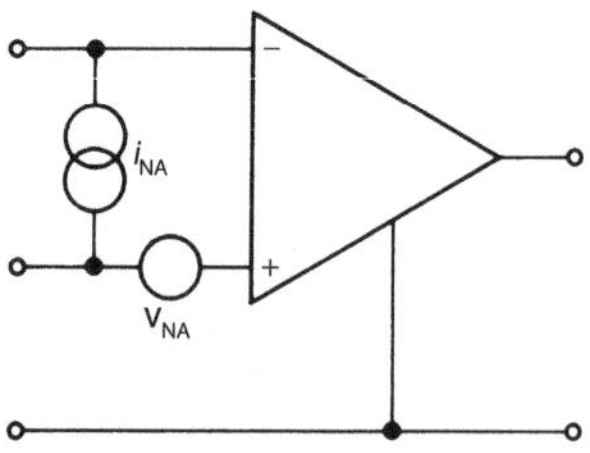

Fig. 4.30 Equivalent input noise sources.

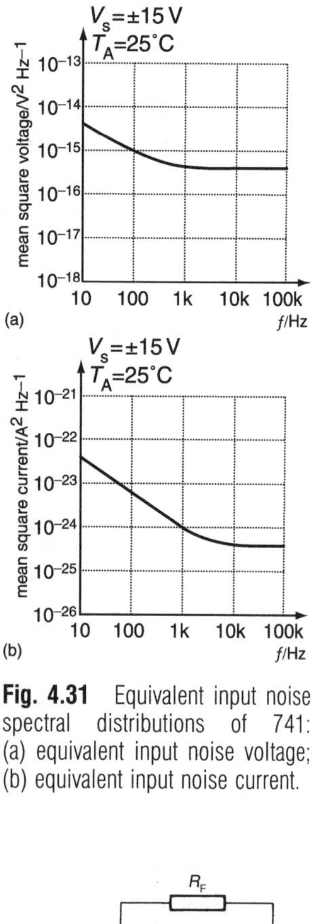

Fig. 4.31 Equivalent input noise spectral distributions of 741: (a) equivalent input noise voltage; (b) equivalent input noise current.

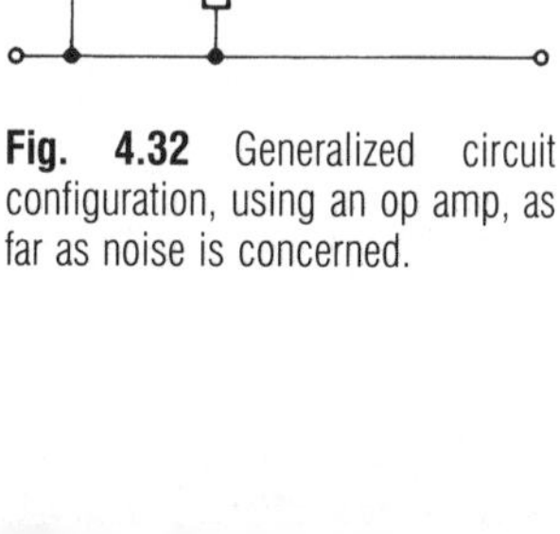

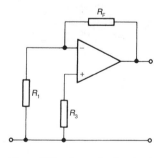

Fig. 4.32 Generalized circuit configuration, using an op amp, as far as noise is concerned.

4.7.2 Amplifier equivalent noise generators

Every amplifier contains resistors and active devices (usually transistors), so it generates thermal noise, shot noise and flicker noise, all of which appear at the output. Which of the three types has greatest effect will depend on the temperature, the types of active device, their operating current, and the frequency of operation.

The contribution of each resistor and transistor to the noise output depends also on its position in the circuit. A transistor in the output stage contributes directly to the output noise. But the noise from a transistor in the *input* stage is amplified by the succeeding stages of the amplifier, so the noise contribution to the output from the first stage usually far exceeds the noise from later stages.

The thermal noise generated in the internal resistance of the signal source is amplified by the whole amplifier, so it can be significant too. Thus the noise output of an amplifier is affected by the signal source resistance at its input.

When specifying an amplifier, the total noise output can be specified for the amplifier working with signal sources of specified source resistances. However, for design purposes it is much more convenient to specify the **equivalent input noise**, sometimes called 'noise referred to input'. The amplifier is *modelled* as a *noiseless* but otherwise similar amplifier together with noise sources at its input, as in Fig. 4.30. These provide noise at the input of the noiseless amplifier and make its output noise equal to that of the real, noisy, amplifier. We can think of these noise sources as analogous to the input offset voltage and current sources. In fact, we can think of the noise simply as *perturbations* of the offset voltage and current: in effect, a.c. components added to the d.c. offset values. Thus Fig. 4.30 is an 'a.c. only' noise equivalent of Fig. 4.24c. The bias currents I_B are removed. V_{IO} is replaced by its noise counterpart v_{NA}, and $I_{io}/2$ is replaced by its counterpart i_{NA}. These two noise sources are called:

the **equivalent input noise voltage generator,** v_{NA}
the **equivalent input noise current generator,** i_{NA}

As an example of these equivalent noise sources, we can look at the specification of the type 741 operational amplifier. Figure 4.31 is taken from the data sheet, and shows the way that both equivalent sources vary with frequency. Both plots are of spectral density against frequency. In Fig. 4.31a the mean-square voltage $\overline{v^2_{NA}}$ is plotted in V^2Hz^{-1}. In Fig. 4.31b the mean-square current $\overline{i^2_{NA}}$ is plotted in A^2Hz^{-1}. Both spectral densities increase at lower frequencies, showing the effect of flicker noise ($1/f$ noise).

These noise spectral density distributions can be used to calculate the output noise of an op amp in any circuit configuration. This is explained in the next section.

4.7.3 Calculating amplifier output noise

If an amplifier is built using an op amp, or uses an op amp for its first stage, it will have a circuit configuration equivalent to that of Fig. 4.32 as far as noise is concerned. The actual circuit might be inverting or non-inverting, with the signal fed into the appropriate input. Coupling capacitors might be used, chosen to have negligible impedance at all frequencies for which the circuit is intended. Thus, whatever actual configuration is used, R_1 and R_3 represent the equivalent resistances from the two input terminals to 0 V, including the effect of the signal source resistance where appropriate, and assuming all coupling capacitors act as short circuits.

Thus Fig. 4.32 represents the circuit configuration as far as noise is concerned, at frequencies of interest. To calculate the output noise, we must now include the noise equivalent of the real op amp, that is the noiseless amplifier together with the two input equivalent noise sources. We should also include the equivalent thermal (Johnson) noise voltage (or current) of each of the resistors R_1, R_F and R_3. But their effects are commonly negligible compared with v_{NA} and i_{NA} so, to keep things as simple as possible, they are left out in this analysis.

Figure 4.32 for the noise case is similar to Fig. 4.25 for the offset case, except that (i) R_1, R_F and R_3 now represent effective resistances of the a.c. paths instead of the d.c. paths and (ii) we must use the equivalent noise model of the op amp rather than the offsets model.

Thus, to find the output noise voltage, we can simply substitute appropriate values for the resistances, and noise sources for offset sources, in the expressions for the output offset voltage from Section 4.6.2 (Equations 4.1 and 4.2). As before, I_B is removed, V_{IO} becomes v_{NA} and $I_{IO}/2$ becomes i_{NA}. Thus, for the output offset voltage due to V_{IO}.

$$V_o = (1 + R_F/R_1)V_{IO}$$

This becomes, for output noise voltage due to v_{NA}:

$$v_{oNV} = (1 + R_F/R_1)v_{NA}$$

Because $(1 + R_F/R_1)$ is independent of frequency, we can substitute r.m.s. values of v_{oNV} and v_{NA}:

$$V_{oNV} = (1 + R_F/R_1)V_{NA}$$

The output offset voltage due to bias currents is

$$V_o = I_B[R_F - R_3(1 + R_F/R_1)] + I_{IO}[R_F + R_3(1 + R_F/R_1)]/2$$

Since I_B does not appear in the noise equivalent circuit, we set $I_B = 0$ and replace $I_{IO}/2$ by i_{NA}, giving the voltage noise due to i_{NA}:

$$v_{oNI} = i_{NA}[R_F + R_3(1 + R_F/R_1)]$$

The r.m.s. value of the voltage noise due to i_{NA} is then

$$V_{oNI} = I_{NA}[R_F + R_3(1 + R_F/R_1)]$$

where I_{NA} represents the r.m.s. value of i_{NA}.

At any instant, the total output noise voltage is equal to the sum of v_{oNV} and v_{oNI}:

$$v_{oN} = v_{oNV} + v_{oNI}$$

We can find the mean-square value of v_{oN} by assuming that the fluctuating voltages v_{oNV} and v_{oNI} vary independently of each other. We can then simply add their mean-square values, giving

$$\overline{v^2{}_{oN}} = \overline{v^2{}_{oNV}} + \overline{v^2{}_{oNI}}$$

The r.m.s. value of the noise output is the square root of this quantity. This can be written in terms of the r.m.s. values V_{oNV} and V_{oNI} defined earlier, giving the r.m.s. value

$$V_{oN} = \sqrt{(V^2_{oNV} + V^2_{oNI})}$$

Self-assessment question

21 If $R_1 = 10\ \text{k}\Omega$, $R_F = 1\ \text{M}\Omega$ and $R_3 = 10\ \text{k}\Omega$, estimate the output noise voltage of a type 741 op amp, using the curves of Fig. 4.31, within the band 100 Hz to 10 kHz.

4.8 VOLTAGE, CURRENT AND SPEED LIMITATIONS

There are limits to the amount of voltage and current which an amplifier can supply to a load, and there are limits to how quickly the output voltage can change. All these effects are aspects of *non-linear* behaviour. That is, if the output waveform is large enough to be limited in its voltage, or current, or in its rate of change, then the amplifier is behaving non-linearly. Usually, this is a state of affairs which should be avoided.

The following sections explain these effects and their terminology.

4.8.1 Voltage swing

When an amplifier's load is 'open-circuit', or has an impedance high enough to impose negligible loading on the output circuit, the output voltage swing is limited by the power supply and by inevitable voltage drops in the output transistors and associated components. Typically, in an op amp, the maximum available voltage (at low frequencies) is about 2 V less than the supply voltage. For instance, many op amps with supplies of ± 15 V will provide a maximum swing of ± 13 V, or 26 V peak-to-peak. Data sheets of op amps usually show a graph of available voltage swing (or 'maximum peak output voltage') versus supply voltage, with a relatively high-resistance load.

When the amplifier is driving a load, the available voltage is lower still, because the load current causes further voltage drops in the components of the output circuit. Part of the voltage drop can be attributed to the small-signal output resistance r_o, but this is not the whole story. With large signals, the effective output resistance is non-linear, and the available output voltage is likely to be lower than you might expect due to voltage drop in r_o alone. Data sheets of op amps commonly show a plot of available output voltage versus load resistance, with fixed supply voltages.

4.8.2 Current limit: short-circuit protection

Most op amps, hi-fi amplifiers, and many others, have short-circuit protection. The output circuit is able to monitor the output current, and limits it to a pre-set amount. Under normal load conditions, this current is not attained and the circuit works linearly. But if the load impedance is decreased sufficiently, or becomes short-circuited, the current will attempt to rise to a level which, in most amplifiers, would cause overheating and destruction of the output transistors. The current limit protects against low-value loads and accidental short-circuits by limiting the output current to a safe value. Op amp data sheets usually quote the value of the short-circuit output current with the amplifier operating on typical supply voltages.

4.8.3 Slew rate

At higher frequencies, large-signal operation of an amplifier is limited by how quickly the output can swing from one voltage to another. In the extreme, if a squarewave is being amplified, the output waveform's rising and falling edges become near-linear 'ramps' with a limited rate of change, dv_o/dt, known as the slew rate. Increasing the input voltage of the squarewave will not increase this slew rate: it is limited by the amplifier. So the output waveform's rise and fall times increase instead.

This fixed value of dv_o/dt limits linear operation to lower frequencies and/or small signals, where the demanded dv_o/dt is less than the slew rate.

Most op amp data sheets quote the slew rate, commonly in units of $\text{V}\ \mu\text{s}^{-1}$.

4.8.4 Full-power bandwidth

Many op amp specifications quote the full-power bandwidth. This is the range of frequencies over which the specified maximum available voltage swing is obtainable. Typically, the value specified at low frequencies is also available up to a frequency f_p (say) but, at higher frequencies, much smaller output voltage swings are available. This fall-off at higher frequencies is linked to the finite slew rate.

If the output signal is a sinewave, and the output swing is the maximum available low-frequency value $V_{o(max)}$, then the output voltage is

$$v_o = V_{o(max)} \sin \omega t$$

The rate of change of this output voltage is:

$$dv_o/dt = \omega V_{o(max)} \cos \omega t$$

The peak value of this is

$$[dv_o/dt]_{(max)} = \omega V_{o(max)}$$

At low frequencies, this peak value of dv_o/dt is less than the slew rate S, so the output waveform is unaffected. But at higher frequencies, the required dv_o/dt is higher than the slew rate, and cannot be attained. Thus, at higher frequencies, the output *amplitude* must be reduced to avoid distortion.

The full-power bandwidth is determined by the frequency f_p at which the peak value of dv_o/dt, that is $\omega V_{o(max)}$, is just equal to the slew rate S. Thus

$$\omega_p V_{o(max)} = S$$

or

$$f_p = \omega_p V_{o(max)}/2\pi = S/2\pi V_{o(max)}$$

4.9 A SUMMARY OF SOME OP AMP FEEDBACK CIRCUITS

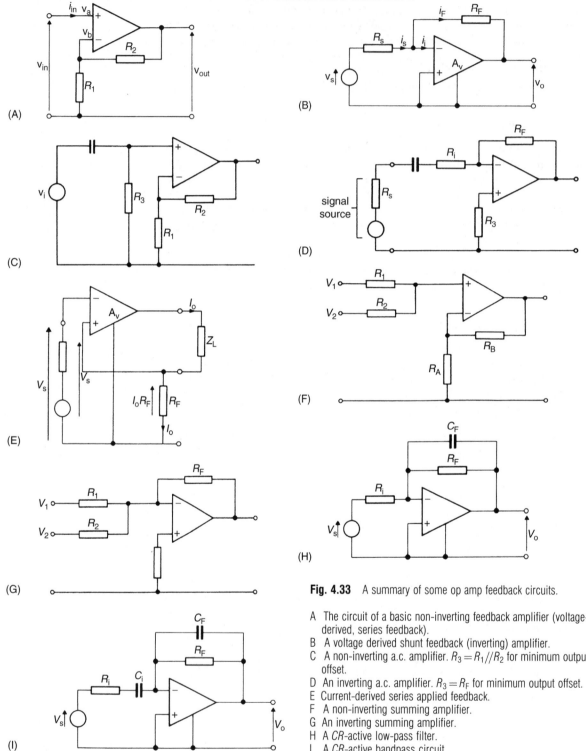

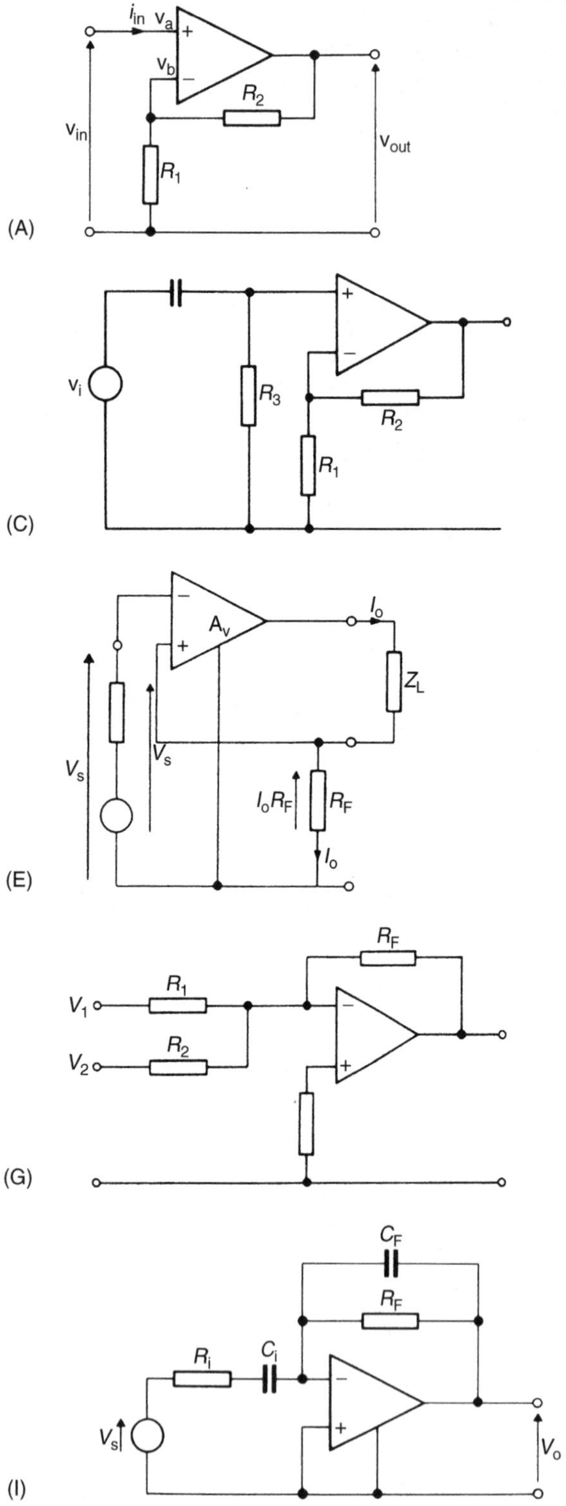

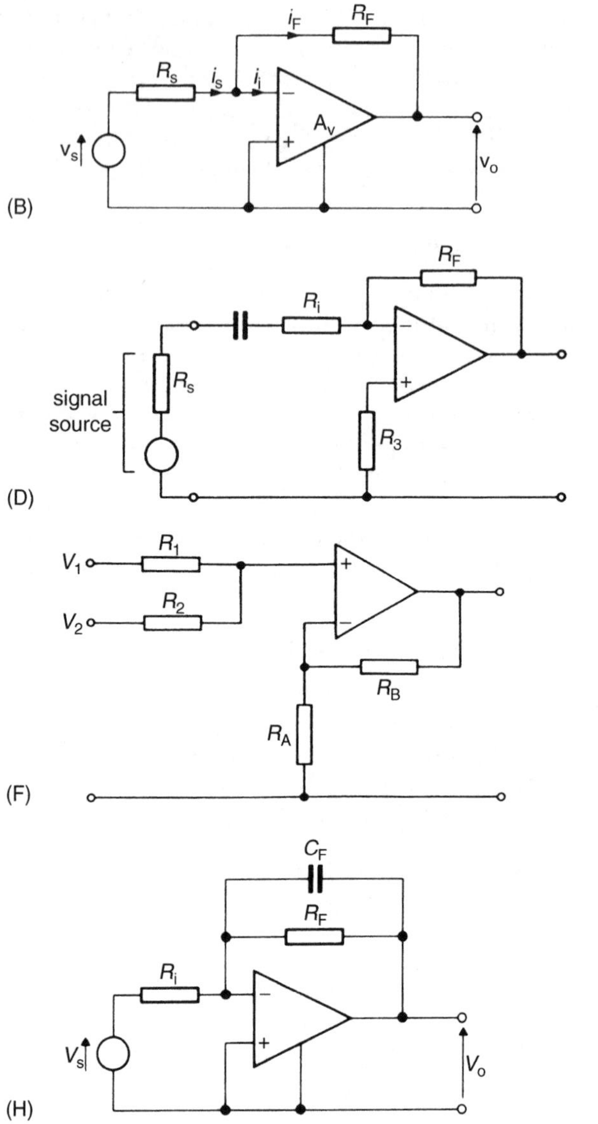

Fig. 4.33 A summary of some op amp feedback circuits.

A The circuit of a basic non-inverting feedback amplifier (voltage-derived, series feedback).
B A voltage derived shunt feedback (inverting) amplifier.
C A non-inverting a.c. amplifier. $R_3 = R_1 // R_2$ for minimum output offset.
D An inverting a.c. amplifier. $R_3 = R_F$ for minimum output offset.
E Current-derived series applied feedback.
F A non-inverting summing amplifier.
G An inverting summing amplifier.
H A *CR*-active low-pass filter.
I A *CR*-active bandpass circuit.

Answers to self-assessment questions

1.

Resistor:	passive; it only absorbs power.
Capacitor:	passive, it can store energy, but cannot increase signal power.
Inductor:	passive; as for capacitor.
Transformer:	passive; can only transfer power, but not increase it.
Battery:	passive; although it supplies d.c. power, it cannot increase *signal* power, so it is passive.
FET and bipolar transistor:	both active; they can both increase signal power (if they have a d.c. power source such as a battery).

2. Let V_{in} and V_{out} be the input and output voltages of the amplifier. Then

$$V_{in} = \frac{V_s \times r_{in}}{r_{in} + R_s} = \frac{0.002 \times 91000}{91000 + 9000} = 1.82\,\text{mV}$$

$$A_v \times V_{in} = 100 \times 1.82\,\text{mV} = 0.182\,\text{V}$$

$$V_L = A_v \times V_{in} \times \frac{R_L}{r_{out} + R_L} = 0.182 \times \frac{1000}{1000 + 100} = 0.165\,\text{V}$$

The voltage across R_L without the amplifier would be

$$V_S \times \frac{R_L}{R_L + R_S} = 2\,\text{mV} \times 1\,\text{k}\Omega/10\,\text{k}\Omega = 0.2\,\text{mV}$$

The amplifier has therefore increased the voltage across the load by a factor of 825!

3. Since $r_{in} = R_S$ it follows that $V_{in} = V_S/2$.
Since $A_v = 10$, it follows that $A_v V_{in} = 10V_{in} = 5V_S$.
Since $r_{out} = R_L$, it follows that $V_L = A_v V_{in}/2 = 2.5V_S$.
The overall voltage ratio is therefore 2.5

4. The amplifier has the equivalent circuit shown in Fig. 4. 1b. The problem is to evaluate the equivalent circuit elements by measuring the terminal voltages and currents

(i) r_{in} is the ratio v_{in}/i_{in}, so it is only necessary to measure v_{in} and i_{in}.
(ii) The voltage gain A_v is the ratio v_{out}/v_{in} when the output is open-circuit. So you have to connect a small signal voltage source to the input, as in Fig. 4.2b and measure both v_{in} and v_{out} when the output is an open circuit (i.e. when R_L is very large, or much larger than r_{out}).
(iii) To measure r_{out} you have to measure the voltage across it and the current flowing through it. In measuring A_v you have established a signal voltage of $A_v v_{in}$ on one side of r_{out}. To ensure that the voltage on the other side is zero you have to short circuit the output (i.e. make $R_L = 0$) and measure the output current i_{out}. Then $r_{out} = A_v v_{in}/i_{out}$.
Alternatively you can adjust R_L until v_{out} is half its open-circuit value. Then $R_L = r_{out}$.

5. With a spacing of 50 km, the cable attenuation between repeaters is 50 km × 1 dB/km = 50 dB. So each repeater should have a gain of 50 dB. In the 3000 km cable there will be 3000 km/50 km = 60 repeaters, with a total gain of 60 × 50 dB = 3000 dB. (Or 3000 km × 1 dB/km = 3000 dB total cable attenuation, to be counteracted by a total gain of 3000 dB.) The power gain is 10^{300}.

6. To answer this question you simply have to substitute the given values in the appropriate equations. Thus:

(a) $G = (R_2 + R_1)/R_1 = 50\,\text{k}\Omega/1\,\text{k}\Omega = 50$.
$R_{in} = r_{in} \times A_v R_1/(R_2 + R_1) = 100\,\text{k}\Omega \times 200\,000/50$
$= 400\,\text{M}\Omega$.
(b) Similarly $G = 1$ and $R_{in} = 20\,000\,\text{M}\Omega$!

7. The loop gain is given by $-\beta\mathbf{A}_v$, where β has the value $R_2/(R_1 + R_2) = 20\,\text{k}\Omega/200\,\text{k}\Omega = 0.1$.

(a) At very low frequencies, for $f \ll f_A$, $\mathbf{A}_v = A_o \angle 0°$ and the loop gain is

$$-\beta\mathbf{A}_v = -0.1 \times 10^5\angle 0° = 10^4\angle 180°.$$

The loop gain magnitude is thus 10^4 and the loop phase shift is 180°.

At the 3 dB frequency, $f = f_A$, the op amp gain magnitude falls to $1/\sqrt{2}$ of its low-frequency value and the op amp phase shift becomes −45°. Hence, for $f = f_A$,

$$\mathbf{A}_v = (1/\sqrt{2}) \times 10^5\angle -45° \approx 7 \times 10^4\angle -45°.$$

The loop gain at $f = f_A$ is then

$$-\beta\mathbf{A}_v = -0.1 \times 7 \times 10^4\angle -45°$$
$$= 7 \times 10^3\angle(180° - 45°).$$

The loop gain magnitude is 7×10^3 and the loop phase shift is now 135°.

(b) At very high frequencies, for $f \gg f_A$, the phase shift of the op amp approaches −90°. The loop phase shift therefore becomes equal to 180° − 90° = 90°.

8. (a) The open loop gain of an amplifier, denoted by $\mathbf{A}_v$, is the gain before feedback is applied.

The closed-loop gain, **G**, is the gain with feedback, i.e. when the feedback loop is 'closed'.

The loop gain depends on the feedback fraction β and the open-loop gain. When the feedback is connected to the inverting input of an op amp, the loop gain is $-\beta\mathbf{A}_v$.

(b) The feedback fraction, β, is the proportion of the output that is fed-back to the input.

The closed-loop gain is equal to the open-loop gain multiplied by $1/(1+\beta\mathbf{A}_v)$ where $(1+\beta\mathbf{A}_v)$ is the feedback factor. The feedback factor can also be expressed as (1 – loop gain).

9 The amplifier is single-lag with low-frequency gain of 10^5 (100 dB) and a 3 dB bandwidth of 2 Hz. Its Bode gain plot is shown in Fig. 4.34 (upper plot). When feedback is applied, the low-frequency gain becomes $A_o/(1+\beta A_o) \approx 1/\beta$, if $\beta A_o \gg 1$. The closed-loop bandwidth is increased to $f_A\,(1+\beta A_o) \approx \beta A_o f_A$, if $\beta A_o \gg 1$.

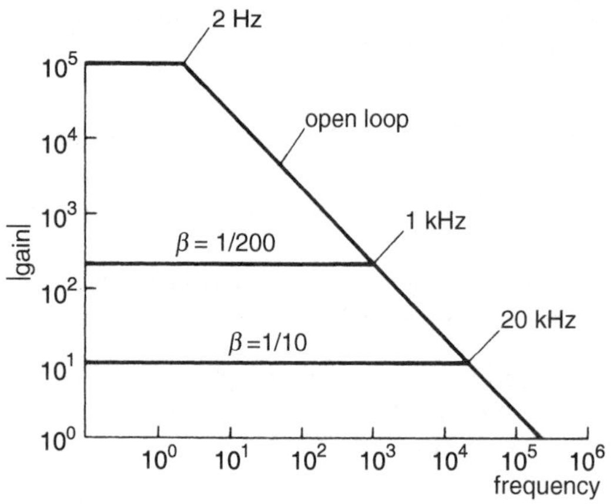

Fig. 4.34

(a) If $\beta = 1/10$, the low-frequency gain is 10 (20 dB) and the closed-loop bandwidth is $(1/10) \times 10^5 \times 2$ Hz = 20 kHz.

(b) If $\beta = 1/200$, the low-frequency gain is 200 (46 dB) and the closed-loop bandwidth is $(1/200) \times 10^5 \times 2$ Hz = 1 kHz.

The closed-loop Bode gain plots are shown in Fig. 4.34. Notice that the break point in the closed-loop gain always occurs on the 20 dB/decade slope of the open-loop plot.

10 The amplifier has a single-lag response up to the unity-gain (0 dB) frequency. The application of feedback reduces the low-frequency gain from 90 dB to 50 dB, which is equivalent to a reduction of 100 times. The 3 dB bandwidth will be increased in the same proportion, giving closed-loop bandwidth = 100 × 100 Hz = 10 kHz. 50 dB is equivalent to a voltage gain of 316. The closed-loop gain–bandwidth product is therefore 316 × 10 kHz = 3.16 MHz.

This is the same as the open-loop gain–bandwidth product of $31.6 \times 10^3 \times 100$ Hz = 3.16 MHz

11 With series applied voltage-derived feedback, the input to the amplifier is

$$\mathbf{V}_i = \mathbf{V}_S - \beta\mathbf{V}_o$$

Substituting $\mathbf{V}_o = \mathbf{A}_v\mathbf{V}_i$, we obtain

$$\mathbf{V}_i = \mathbf{V}_S - \beta\mathbf{A}_v V_i$$

Hence

$$\mathbf{V}_i(1+\beta\mathbf{A}_v) = \mathbf{V}_S, \quad \mathbf{V}_i = \frac{\mathbf{V}_S}{(1+\beta\mathbf{A}_v)}$$

showing that the input voltage is equal to the source voltage $\mathbf{V}_s$ divided by the feedback factor.

12 (a) With a closed-loop gain of 100, we have $1/\beta = 100$ and $\beta = 0.01$. A_o is at least 100 dB, that is a ratio of 10^5 or more, hence $\beta\mathbf{A}_o \geqslant 1000$. The feedback factor at low frequencies is $1+\beta\mathbf{A}_o \geqslant 1001$, say 1000. Hence, the low-frequency closed-loop output impedance is at most 100/1000 = 0.1 Ω.

The input impedance is equivalent to a resistor R shunted by a capacitor C, that is by a reactance of $1/\omega C$. Both the resistance and the reactance increase by a factor of at least 1000 at low frequencies. The equivalent resistance increases to at least 1000 MΩ and $1/\omega C$ increases at least 1000-fold, which is equivalent to a decrease in C to 50/1000 = 0.05 pF, at most.

(b) The feedback factor for a single lag amplifier is

$$(1+\beta\mathbf{A}_v) = 1 + \frac{\beta\mathbf{A}_o}{1+\mathrm{j}f/f_A}$$

As the frequency is increased beyond f_A, the feedback factor magnitude begins to roll off at 6 dB/octave as shown in the Bode magnitude plot of Fig. 4.35. This trend continues up to the closed-loop bandwidth, $f_B = \beta A_o f_A$, beyond which $|\beta\mathbf{A}_v|$ falls to less than 1. The high-frequency part of the Bode magnitude is thus described by

$$|1+\beta\mathbf{A}_v| \approx 1 \quad \text{for} \quad f \geqslant f_B$$

Since the feedback is no longer effective for $f \geqslant f_B$, the closed-loop input and output impedances are respectively reduced and increased to their open-loop values.

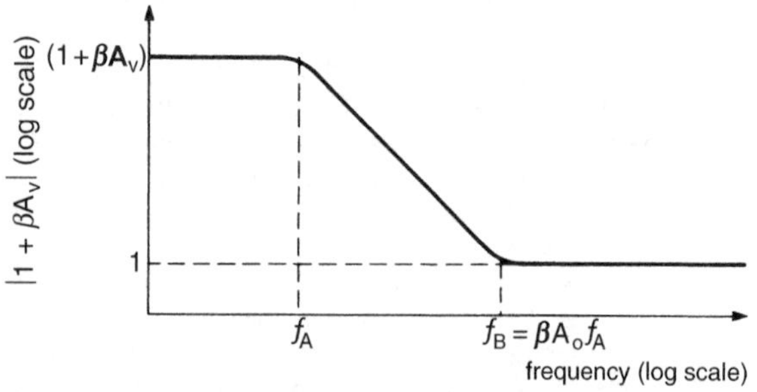

Fig. 4.35 Bode magnitude plot for the feedback factor $(1+\beta\mathbf{A}_v)$.

13 The voltage gain is given by

$$\mathbf{G} = \frac{\mathbf{Z}_F}{\mathbf{Z}_S}\angle 180^\circ = \frac{33\times 10^3\angle -10^\circ}{10\times 10^3\angle 0^\circ}\angle 180^\circ$$
$$= 3.3\angle(180^\circ - 10^\circ) = 3.3\angle 170^\circ$$

For a phasor input 0.3 ∠ 30° V, the output voltage will be

$\mathbf{V}_o = 0.3\angle 30^\circ \times 3.3\angle 170^\circ\ \text{V} \approx 1\angle 200^\circ\ \text{V}$

The output therefore *leads* the input by 200°. Alternatively, we can say that the output *lags* the input by 360° − 200° or 160°.

If $|\mathbf{A}_v| = 2\times 10^4$, the voltage across the amplifier input terminals will have magnitude

$|\mathbf{V}_i| = |\mathbf{V}_o|/|\mathbf{A}_v| = 1/(2\times 10^4)\ \text{V} = 50\ \mu\text{V}$

The input voltage will thus have a magnitude of 50 μV compared with $|\mathbf{V}_S| = 0.3$ V and $|\mathbf{V}_o| = 1$ V.

14 Using the virtual-earth approximation, the closed-loop voltage gain is $\mathbf{G} = -R_F/R_S = -100$. This result will be valid for all frequencies for which $|\mathbf{G}| \leqslant |\mathbf{A}_v|$. We therefore need to find the range of frequencies of which $|\mathbf{A}_v| \geqslant 100$. Now, $|\mathbf{A}_v| = 10^5$ for $f \leqslant f_A$ where $f_A = 5$ Hz. For frequencies $f \geqslant f_A$, we can approximate the single-lag response by $\mathbf{A}_v \approx -jA_of_A/f$. So, $|\mathbf{A}_v| = A_of_A/f$ and we require that

$A_of_A/f \geqslant 100 \quad \text{or} \quad 2\times 10^5\times 5/f \geqslant 100$

This will be true for frequencies *f*, such that

$f \leqslant 2\times 10^5\times 5/100\,\text{Hz} \leqslant 10^4\,\text{Hz} \quad \text{or} \quad 10\,\text{kHz}$

The closed-loop voltage gain will thus be approximately −100 for frequencies much less than 10 kHz, up to a maximum, say, of about 1 kHz.

15 (i) From the values given: $G_1 = 1/10\ \text{k}\Omega = 100\ \mu\text{S}$; $G_2 = 1/20\ \text{k}\Omega = 50\ \mu\text{S}$; $G_3 = 1/40\ \text{k}\Omega = 25\ \mu\text{S}$.

$G_o = G_1 + G_2 + G_3 = 175\,\mu\text{S}, \quad 1/\beta = 7$

From the text,

$$V_o = \frac{1}{\beta}\frac{V_1G_1 + V_2G_2 + V_3G_3}{G_o}$$

Expressing conductances in μS we have,

$$V_o = 7\times\frac{100V_1 + 50V_2 + 25V_3}{175} = 4V_1 + 2V_2 + V_3$$

(ii) $V_o = 4\times 1\ \text{V} + 2\times 2\ \text{V} + 3\ \text{V} = 11$ V.

(ii) $\beta = R_A/(R_A + R_B)$, so $1/\beta = (R_A + R_B)/R_A = 1 + R_B/R_A$. Thus $R_B/R_A = 1/\beta - 1 = 7 - 1 = 6$.

16 (a) Using the virtual-earth approximation, the closed-loop gain is $\mathbf{G} = -\mathbf{Z}_F/R_i$ where $\mathbf{Z}_F$ is the impedance of R_F shunted by C_F:

$$\mathbf{Z}_F = \frac{R_F}{1 + j\omega C_FR_F}$$

The closed–loop gain is therefore

$$\mathbf{G} = -\frac{Z_F}{R_i} = \frac{-R_F/R_i}{1 + j\omega C_FR_F}$$

(b) At very low frequencies, when $\omega C_FR_F \leqslant 1$, the gain is

$G_o = -R_F/R_i$

The gain is reduced to $1/\sqrt{2}$ of its low-frequency value at the angular cut-off frequency $\omega_c = 1/C_FR_R$. The gain can thus be expressed in the single-lag form

$$\mathbf{G} = \frac{G_o}{1 + jf/f_c}$$

where $G_o = -R_F/R_i$ and $f_c = 1/2\pi C_FR_F$.

A single-lag response on its own would normally imply a phase shift $\phi = -\tan^{-1}(f/f_c)$, which varies from 0° at low frequencies to −90° for $f \geqslant f_c$. However, we must include an additional phase shift of 180° due to the inversion provided by the amplifier. The phase shift is therefore

$\phi = 180^\circ - \tan^{-1}(f/f_c)$

where $\phi \approx 180^\circ$ for $f \leqslant f_c$, $\phi = 180^\circ - 45^\circ = 135^\circ$ for $f = f_c$ and $\phi = 180^\circ - 90^\circ = 90^\circ$ for $f \geqslant f_c$.

(c) We first choose the value of R_F to give a low-frequency gain magnitude of 12 dB (×4). Thus, $R_F/R_i = 4$, where $R_i = 10\ \text{k}\Omega$, giving $R_F = 40\ \text{k}\Omega$.

The cut-off frequency is to be 500 Hz, hence $1/(2\pi C_FR_F) = 500$ giving

$$C_F = \frac{1}{2\pi\times 500\times R_F} = \frac{1}{2\pi\times 500\times 40\times 10^3}\text{F}$$
$$\approx 8\,\text{nF}$$

17 *Voltage gain magnitude at low frequencies*. At low frequencies $|\mathbf{G}| = 1$ for the passive *CR* circuit. Using an active circuit. $|\mathbf{G}|$ can be made equal to or greater than I as required.

Phase shift. The phase shift of the passive circuit varies from 0° at low frequencies to −90° at high frequencies. The active circuit introduces an additional phase shift of 180° over all frequencies. At very low frequencies therefore, the output of the active circuit is always inverted with respect to the input.

Loading effects. If a load resistor is connected across the output of a passive *CR* circuit both the low-frequency gain magnitude and the cut-off frequency will be affected. The active circuit is a voltage-derived feedback circuit, which means that its output impedance will be relatively low over all frequencies of interest. This makes the active circuit relatively insensitive to loading effects.

The low-frequency gain and cut-off frequency will thus remain substantially constant for a wide variation in load resistance connected across the output.

Input impedance. The input impedance of a passive *CR* circuit is frequency-dependent, given by $\mathbf{Z}_{in} = R + 1/j\omega C$. For all frequencies for which the virtual-earth approximation is valid, the input impedance of the active circuit is fixed by the value of the input resistor R_i.

18 From Equation 4.1, with V_{IO} negligible and $R_3 = 0$,

$$V_o = I_B R_F \text{ (since } I_{IO} \text{ is negligible)}$$
$$= 500\,\text{nA} \times 1\,\text{M}\Omega$$
$$= 500\,\text{mV}$$

19 From Equation 4.2, output offset $V_o = (1 + R_F/R_1) \times V_{IO} + I_{IO}R_F$ (with equal bias paths).

(a) With *C* in circuit, no d.c. current flows in R_1 and the circuit behaves as if R_1 was infinite, giving

$$V_o = V_{IO} + I_{IO}R_F$$
$$= 6\,\text{mV} + 200\,\text{nA} \times 1\,\text{M}\Omega$$
$$= 6\,\text{mV} + 200\,\text{mV} = 206\,\text{mV}$$

(b) With *C* shorted.

$$V_o = (1 + 1\,\text{M}\Omega/11\,\text{k}\Omega)6\,\text{mV} + 200\,\text{nA} \times 1\,\text{M}\Omega$$
$$= 551\,\text{mV} + 200\,\text{mV} = 751\,\text{mV}$$

For equal bias current paths, $R_3 = R_1/R_F$.
In case (a), $R_1 = \infty$, so $R_3 = R_F = 1\,\text{M}\Omega$.
In case (b), $R_3 = 11\,\text{k}\Omega // 1\,\text{M}\Omega \approx 11\,\text{k}\Omega$.

20 (a) (i) $\overline{v^2_{NR}} = 4RkT\Delta f = 4 \times 1\,\text{k}\Omega \times 1.38 \times 10^{-23}\,\text{J K}^{-1} \times (273 + 25)\,\text{K} \times 20\,\text{kHz} \approx 3.29 \times 10^{-13}\,\text{V}^2$
So r.m.s. value is $V_{NR} \approx 0.6\,\mu\text{V}$.

(ii) $\overline{i^2_N} = 2qI\Delta f = 2 \times 1.6 \times 10^{-19}\,\text{C} \times 0.1\,\text{mA} \times 20\,\text{kHz} = 6.4 \times 10^{-19}\text{A}^2$
The r.m.s. value is $I_N = 0.8$ nA. The r.m.s. shot noise voltage dropped across a 1 kΩ resistor is therefore

$$I_N R = 0.8\,\text{nA} \times 1\,\text{k}\Omega = 0.8\mu\text{V}$$

(b) Total mean-square noise voltage across the resistor is

$$(0.6\,\mu\text{V})^2 + (0.8\,\mu\text{V})^2 = 1(\mu\text{V})^2$$

So total r.m.s. noise voltage is 1 μV.

21 In Fig. 4.31 the input noise generators are specified in terms of their *spectral densities*, plotted as functions of frequency. The total noise voltage or current over a particular frequency range is then found by calculating (or estimating) the *area* under these curves between the two frequencies of interest.

For simplicity, regard the voltage spectral density as being constant over the band 100 Hz to 10 kHz, with a value of $3 \times 10^{-16}\,\text{V}^2/\text{Hz}$. The total mean-square noise voltage over this frequency band of width 9.9 kHz is then approximately

$$v^2_{NA} \approx 3 \times 10^{-16}\,\text{V}^2/\text{Hz} \times 9.9\,\text{kHz} \approx 3 \times 10^{-12}\,\text{V}^2$$

The r.m.s. noise voltage is therefore $V_{NA} \approx 1.7\,\mu\text{V}$.

To calculate the current noise, assume an *average* spectral density of approximately $3 \times 10^{-24}\,\text{A}^2/\text{Hz}$ over the band 100 Hz to 1 kHz. The mean-square current in this frequency range is then approximately

$$\overline{i^2_{NA}} \approx 3 \times 10^{-24}\text{A}^2/\text{Hz} \times 0.9\,\text{kHz} \approx 3 \times 10^{-21}\,\text{A}^2$$

Over the band 1 kHz to 10 kHz we can assume an average spectral density of $5 \times 10^{-25}\,\text{A}^2/\text{Hz}$, giving a mean-square current

$$\overline{i^2_{NA}} \approx 5 \times 10^{-25}\text{A}^2/\text{Hz} \times 9\,\text{kHz} \approx 5 \times 10^{-21}\text{A}^2$$

The *total* mean-square current is the total area under the curve between 100 Hz and 10 kHz, given by the sum of these two contributions. Thus total $\overline{i^2_{NA}} \approx 8 \times 10^{-21}\text{A}^2$. The noise current is thus

$$\text{total}\,\overline{i^2_{NA}} \approx 8 \times 10^{-21}\,\text{A}^2$$

The noise current thus has an r.m.s. value $I_{NA} \approx 90 \times 10^{-12}$ A or 90 pA.

The mean-square output voltage is now

$$V^2_{oN} = (1 + R_F/R_1)^2 V^2_{NA} + i^2_{NA}[R_F + R_3(1 + R_F/R_1)]^2$$
$$= [(1 + 1\,\text{M}\Omega/10\,\text{k}\Omega) \times 1.7\,\mu\text{V}]^2 + (90\,\text{pA})^2$$
$$\times [1\,\text{M}\Omega + 10\,\text{k}\Omega(1 + 1\,\text{M}\Omega/10\,\text{k}\Omega)]^2$$
$$= 2.9 \times 10^{-8}\,\text{V}^2 + 3.3 \times 10^{-8}\,\text{V}^2 = 6.2 \times 10^{-8}\,\text{V}^2$$

So the total output r.m.s. noise voltage is about 250 μV.

Combinational logic circuits

5

AIMS

The main aims are to introduce basic techniques of combinational logic:

- To express the inputs and outputs of a system in binary form.
- To develop the relationship between these inputs and outputs as a truth table.
- To simplify the Boolean expression using algebra or Karnaugh maps.
- To select suitable electronic devices to implement the required function.

You will also be introduced to the terminology and Boolean notation for some logic functions and combinational devices.

GENERAL OBJECTIVES

Understand the meaning of the following terms:

- address
- AND
- AND gate
- application-specific integrated circuit (ASIC)
- ASCII
- associative law
- base
- BCD (binary coded decimal)
- bipolar
- bit
- Boolean algebra
- Boolean function
- Boolean variable
- buffer
- byte
- CMOS
- code
- code word
- combinational logic
- commutative law
- complement
- data
- deMorgan's theorem
- distributive law
- don't-care
- dual
- ECL
- error detecting
- EXCLUSIVE-NOR (XNOR)
- EXCLUSIVE-OR (XOR)
- fan-out
- full-custom
- function table
- gate array
- gate equivalence
- HCT CMOS
- hex
- hexadecimal
- INVERTER
- Karnaugh map
- location
- logic family
- low-power Schottky (LSTTL)
- LSB (least-significant bit)
- LSI
- mask-programmed ROM
- minterms
- MSB (most-significant bit)
- MSI

- NAND
- natural binary number
- *n*-bit code
- negative-logic
- noise margin
- NOR
- octal
- OR
- parity bit
- positive-logic
- power dissipation capacitance
- power-delay product
- product terms
- program
- programmable array logic (PAL) device
- programmable logic device (PLD)
- programmable read only memory (PROM)
- programmer
- propagation delay
- quad
- RAM
- read
- ROM
- Schottky (STTL)
- semi-custom
- sequential
- seven-segment display
- source and sink current
- SSI
- standard cell
- sum-of-minterms
- sum-of-products
- tri-state output
- truth table
- TTL (transistor–transistor logic)
- user-specific integrated circuit (USIC)
- VLSI, ULSI
- word
- write

SPECIFIC OBJECTIVES

1. Formulate the description of a simple interlock, covering binary inputs and outputs, truth table, electrical logic levels, and electronic components.
2. Determine the number of individual items or symbols which can be coded by *n* bits.
3. Translate numbers from denary to binary and vice versa. Express binary numbers as octal or hexadecimal numbers.
4. Write down the truth table, algebraic notation and symbols for some common logic functions.
5. Starting with a description of a problem in words, write down a truth table of the problem and express it in a sum-of-minterms form.
6. Implement the function from its sum-of-minterms expression using logic gates.
7. To be able to manipulate Boolean expressions.
8. To use Karnaugh maps to simplify a Boolean expression with up to four variables.
9. To implement any Boolean expression using the sum-of-products circuit.
10. To implement the sum-of-products circuit using only NAND gates.
11. To be able to implement a given logic function using gates or programmable logic devices.

12. To appreciate the differences between logic families and to understand their limitations.
13. To calculate power-delay product and fan-out, given appropriate data.

In the analogue circuits covered in earlier chapters, signals are represented by levels of voltage, current or charge which vary continuously between the limits of the circuit. In a digital circuit, however, these electrical levels are not measured on a continuous scale, but are simply classified as either 'high' or 'low' to represent only the two possible binary values 0 and 1. Digital integrated circuits are relatively cheap to make and extremely complex devices that can be made to operate accurately, reliably and with low power consumption. It is this ability of digital electronics to carry out complex tasks cheaply and reliably that has led to many of the dramatic developments in modern electronics.

In a **combinational logic** circuit, the subject of this chapter, the digital outputs are defined only by the present input values. In contrast, the outputs of a **sequential** circuit, to be described in the next chapter, depend upon present *and* previous inputs.

The term **logic** is often used to describe digital devices and circuitry.

5.1 A SIMPLE INTERLOCK EXAMPLE

To see a little more clearly what combinational logic is, consider the example of a protective interlock designed to guard against misuse of an industrial machine. Suppose that there are three conditions to be satisfied before operation is possible:

Conditions
The workpiece is in position.
Adequate lubricant is available.
A safety guard is in position.

The function of the interlock is to check the conditions to see whether they are satisfied, and if they are *all* satisfied the power will be connected to the machine. The three conditions can therefore be regarded as the inputs to the interlock. The interlock's output is its ability to select whether the power is connected to the machine or not.

If the input conditions and the output requirement are to be expressed in binary form, they must be phrased so that there are only two alternative possibilities in each case. These alternatives are then labelled 0 and 1.

There is usually more than one way of phrasing the conditions to give clear alternative possibilities. For example, a condition required of the workpiece can be expressed as the answer to any of the following questions:

'Is the workpiece in position?' Answer: YES or NO.

or

'Where is the workpiece? Is it in or out of position?' Answer: IN or OUT.

or as the statement

'The workpiece is in position' Answer: TRUE or FALSE.

The alternatives YES, IN or TRUE describe the same state of affairs, and could be represented by 1. The alternatives NO, OUT or FALSE would then correspondingly be represented by 0. However, this is a matter of choice—the important things are that the condition should be carefully phrased so that there are only two possible choices, and that the decision as to which shall be represented by 1 and which by 0 should be clearly understood.

To be able to describe the inputs and outputs more concisely, they can each be represented by a binary variable. For example, the three input conditions and the output requirement can be written A, B, C and P, respectively, where the variables A, B, C and P can each take on one of the two binary values 0 or 1 according to the following table:

Variable	*Value*	*[represents]*	*Situation*
A	1		Workpiece in position
	0		Workpiece not in position
B	1		Adequate lubricant available
	0		Adequate lubricant not available
C	1		Safety guard in position
	0		Safety guard not in position
P	1		Mains switch turns machine on
	0		Mains switch disconnected

We can then say that when $A=1$, for example, the workpiece is in position. This is the notation of **Boolean algebra**, named after the nineteenth century mathematician George Boole. A, B, C and P are described as **Boolean variables**, which can take only two values, 0 and 1.

5.1.1 Truth table

The required relationship between the inputs A, B and C and the output P can now be stated in a table which will form the specification for the required interlock. Each possible combination of input conditions must be considered in turn. Thus, if the workpiece is *not* in position ($A=0$), there is *not* adequate lubricant ($B=0$) and the safety guard is *not* in position ($C=0$), we require that the mains switch should *not* turn the machine on ($P=0$). This can be written

A	B	C	P
0	0	0	0

If only one or two of the input conditions are satisfied, we still require $P=0$. Thus, if the workpiece is in position and there is adequate lubricant, but the safety guard is not in position, this can be written

A	B	C	P
1	1	0	0

The mains switch should only turn the machine on when *all three* input conditions are suitable, that is when A, B and C all have the value of 1. Thus,

A	B	C	P
1	1	1	1

A *complete* description of the required interlock involves a complete list of all possible combinations of inputs with the required output. This is called the **truth table**, and it summarizes the relation between the binary variables.

A	B	C	P
0	0	0	0
0	0	1	0
0	1	0	0
0	1	1	0
1	0	0	0
1	0	1	0
1	1	0	0
1	1	1	1

Three inputs, as here, can have eight different combinations of their values, each row of the truth table corresponding to one of the input combinations. This ability to combine the inputs in different ways is the basis for calling the type of circuits that ultimately implement the truth table *combinational circuits*.

Each combination of inputs in the truth table is written here in order starting from 000 at the top to 111 at the bottom. This is the same order as the sequence of natural binary numbers (covered in Section 5.2.2). There would be no change in the interpretation of the truth table if the rows were rearranged. It is, however, usual to write truth tables with the possible input combinations arranged in the systematic way shown here to avoid missing any.

5.1.2 Implementation

The next step is to specify an electronic circuit that will act according to the specification given by the truth table. The electrical representation of 0 and 1 depends on having an electronic device that can be in one of two conditions. The two conditions must be distinguishable, and are often represented by ranges of voltage or ranges of current.

There are various agreed standards, one of the most common being that used in **TTL** devices, described further in Section 5.7 (TTL stands for transistor–transistor logic and reflects the internal structure of the TTL family of logic devices). In TTL the existence of a binary value 0 or 1 at any terminal depends on whether the voltage with respect to a common ground terminal is LOW or HIGH in comparison to a reference voltage, and the ranges used are 2.0 to 5.0 V for HIGH, and 0.0 to 0.8 V for LOW. Hence 3.5 V would be considered HIGH, whereas 0.2 V would be a LOW voltage, but there can be quite a spread of allowed values without affecting the decision about the logic value intended. This is one of the advantages of digital methods: that exact

Often a HIGH voltage is said to represent TRUE or 1 and a LOW voltage is said to represent FALSE or 0. This particular interpretation of voltages is a convention that is referred to as the **positive-logic** convention. In the alternative **negative-logic** convention a HIGH voltage represents FALSE or 0 and a LOW voltage represents TRUE or 1.

binary decisions between 0 and 1, on which everything depends, do not require exact voltages.

Voltages outside the two ranges cannot be interpreted as representing either HIGH or LOW and should therefore be avoided. If a voltage does occur outside the two ranges then the result of its application cannot be predicted.

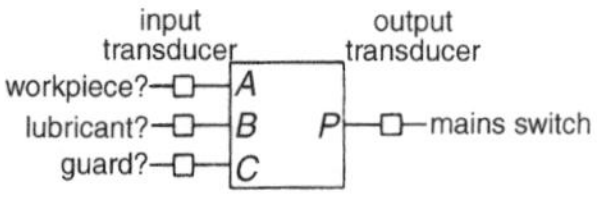

Fig. 5.1 Block diagram for the interlock system.

Now that there is a way of representing 0s and 1s with voltages, a block diagram for the interlock system can be drawn which assumes that electrical signals are present (Fig. 5.1).

It will also be necessary to find or devise input transducers which produce HIGH or LOW logic levels appropriate to the input conditions. An example of this might be a microswitch which produces an output of 5 V when open and 0 V when closed, as shown in Fig. 5.2. Similarly, output transducers are needed which convert logic levels into appropriate signals to control the interruption of the power circuit to the machine. An example of this might be a relay which closes a circuit when a voltage corresponding to a logic 1 is applied and is open when a voltage corresponding to a logic 0 is applied.

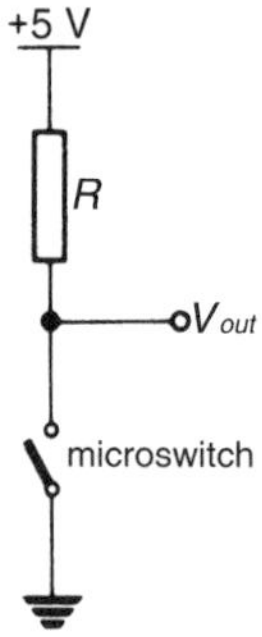

Fig. 5.2 Microswitch connected to give 0 V and 5 V logic levels.

There are available a great many electronic devices, mostly integrated circuits, that enforce relations between voltages which can be described in truth tables. There are not always devices that meet the precise requirements of an application, but there are techniques for 'customizing' those that are available or building up a required function from several interconnected devices.

For this particular application it is worth checking to see whether a single device is available. It must have (at least) three inputs and one output terminal, which can be labelled *A, B, C* and *P*, corresponding to the logical variables with these names discussed previously. The above truth table provides a specification for the required logic device. Thus if the device is to implement this truth table the application of input voltages corresponding to 1 at terminals *A, B*, and *C*, respectively, must produce an output voltage corresponding to 1 on *P*. If the input is changed to any other combination the output must change to 0.

This particular form of truth table, where $P = 1$ *only* when *A and B and C* are *all* equal to 1, describes the **Boolean function** called **AND**. Various circuits have been devised which provide this relation between input and output. These are called **AND gates** and are made with various numbers of inputs, but have only one output. For any AND gate *all* inputs must represent a 1 to give an output corresponding to a 1; otherwise the output that is generated corresponds to a 0. A three-input AND gate can be represented by the symbol shown in Fig. 5.3. The system resulting from using a single three-input AND gate to solve the interlock problem is shown in Fig. 5.4.

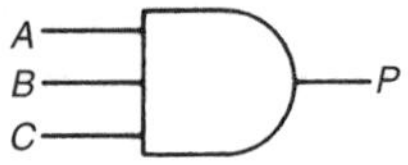

Fig. 5.3 AND gate.

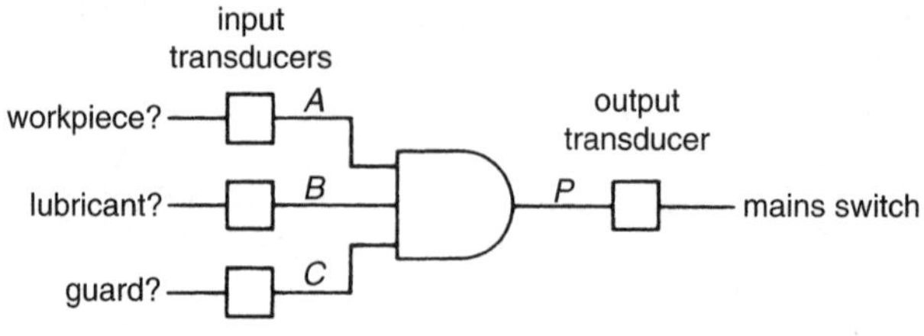

Fig. 5.4 The interlock system.

5.1.3 Summary of Section 5.1

In this section you were shown how a designer might approach a particular design problem in electronics. This can be broken down into four identifiable steps, as follows:

1. The input and output conditions are expressed in binary form and 0 and 1 assigned to the two possible values of each.
2. A truth table is constructed which consists of all the possible combinations of inputs and the corresponding value of the output represented by 0s and 1s. This completely defines the problem.
3. Suitable transducers are found which convert the input conditions into two voltage levels.
4. A digital electronic device is found which has the same truth table as the problem.

5.2 BINARY INPUTS AND OUTPUTS

The interlock example in Section 5.1 has shown one way in which inputs and outputs can be expressed in the binary form required for a combinational logic device. However there are several other ways binary inputs and outputs can arise. Some inputs and outputs may be classified as 'intrinsically binary' (as in the interlock); but there are others which are not, requiring to be 'coded' into binary form in some way if they are to be processed by electronic digital circuits. In this second group are the binary codes which represent numbers (as needed for calculators), other symbols like letters, or any other set of distinct items. The important case where analogue signals are converted to digital signals is the subject of Chapter 7.

5.2.1 Coding

In the interlock case considered in Section 5.1, the inputs and outputs could be described in binary form because they were expressed as questions with clear YES/NO answers or statements which were clearly TRUE or FALSE, that is, they each had only two distinct values. Every combination of input conditions was then represented by a group of 0s and 1s in the truth table with the corresponding output, also a 0 or a 1.

A single binary input or variable can have only two possible values, 0 or 1. When you have two inputs there are four possible combinations of 0s and 1s that can be made. For three inputs there are eight possible combinations of 0s and 1s and so on, each additional input doubling the number of combinations. This can be generalized so that if there are n inputs then there are 2^n possible combinations.

Worked example

How many combinations are possible with 10 inputs?

There would be $2^{10} = 1024$ combinations.

The value of 2^{10} is an important number in the world of digital systems and as such has been given its own symbol, capital or upper-case K, so 1 K = 1024, 2 K = 2048, and so on. This should not be confused with the standard metric (SI) prefix, lower-case k, which represents a multiple of 1000.

For reference, Table 5.1 lists values of 2^n up to $n = 20$.

The 2^n different combinations that are possible with n variables can be made to represent some other set of distinct items, for example the letters of the alphabet or products on sale in a supermarket. This is called coding: the 2^n combinations and their interpretations represent the **code** whereas the individual combinations represent the **code words**. Since each of the code words contains a combination of n binary digits or **bits**, the code is referred to as an n-**bit code**.

Any group of bits is called a binary **word**. A word consisting of 8 bits is generally known as a **byte**.

Self-assessment question

1 What is the minimum number of bits needed to represent the letters 'a' to 'z' as a binary code?

5.2.2 Number representation

A very common requirement is the conversion of familiar denary numbers (such as 173, 35 700, 8.26, 0.0044—often loosely called decimal numbers) to a binary form so that they can be handled by a computer, and several methods are available for this.

Table 5.1

n	2^n
0	1
1	2
2	4
3	8
4	16
5	32
6	64
7	128
8	256
9	512
10	1024 = 1 K
11	2048 = 2 K
12	4096 = 4 K
13	8192 = 8 K
14	16384 = 16 K
15	32768 = 32 K
16	65536 = 64 K
17	131072 = 128 K
18	262144 = 256 K
19	524288 = 512 K
20	1048576 = 1 M (1 Meg)

Natural binary numbers

The normal denary number system uses ten symbols 0, 1, 2, ..., 8, 9. Symbols are arranged side by side with a 'positional weighting' system which changes their significance (or weight) according to their position. Thus, 397 means 3 hundreds with 9 tens and 7 ones, or

$$397 = 3 \times 10^2 + 9 \times 10^1 + 7 \times 10^0$$

Binary numbers use only two symbols, 0 and 1. The **natural binary number** system uses the same principal of changing the significance of the digits but in this case each digit, or bit, corresponds to an increasing power of 2. The interpretation of '101' using the binary number system is

$$\begin{aligned} 101(\text{natural binary}) &= 1 \times 2^2 + 0 \times 2^1 + 1 \times 2^0 \\ &= 1 \times 4 + 0 \times 2 + 1 \times 1 \\ &= 5 \text{ (denary)} \end{aligned}$$

The standard pattern of input conditions used in truth tables simply corresponds to numbering the rows in the natural binary number system (including leading 0s to make up a fixed number of bits).

This positional notation can be extended indefinitely, just like the denary system, by adding digits to the left representing 'eights' (2^3), 'sixteens' (2^4), and so on. The bit at the extreme right (representing units 2^0) is known as the least-significant bit (**LSB**), and that at the extreme left (representing 2^{n-1} for n bits) the most-significant bit (**MSB**). Where it is necessary to avoid confusion, the subscripts 2 and 10 are introduced to indicate natural binary or denary respectively, for example 110_2 and 61_{10}.

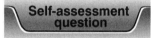

A natural binary number can be converted into denary simply by multiplying each bit by the appropriate power of 2 and adding them together to get the result. For instance:

$$\begin{aligned}1011011_2 &= 1\times 2^6 + 0\times 2^5 + 1\times 2^4 + 1\times 2^3 + 0\times 2^2 + 1\times 2^1 + 1\times 2^0\\ &= 64 \quad + \quad 0 \quad + \quad 16 \quad + \quad 8 \quad + \quad 0 \quad + \quad 2 \quad + \quad 1\\ &= 91_{10}\end{aligned}$$

A denary number can be converted into natural binary by looking for powers of 2.

$$\begin{aligned}397_{10} &= 256 \; + \; 128 \; + \; 0 \; + \; 0 \; + \; 0 \; + \; 8 \; + \; 4\\ &\qquad + \; 0 \; + \; 1\\ &= 1\times 2^8 + 1\times 2^7 + 0\times 2^6 + 0\times 2^5 + 0\times 2^4 + 1\times 2^3 + 1\times 2^2\\ &\qquad + 0\times 2^1 + 1\times 2^0\\ &= 1 \quad 1 \quad 0 \quad 0 \quad 0 \quad 1 \quad 1\\ &\qquad 0 \quad 1_2\end{aligned}$$

Another way of arriving at this is to keep dividing by 2 and list the remainders, r, in the order shown below by the arrows.

$$\begin{array}{rll}397 \div 2 = 198 & \text{remainder } r = 1 & \text{LSB}\\ 198 \div 2 = 99 & \text{remainder } r = 0 & \uparrow\\ 99 \div 2 = 49 & \text{remainder } r = 1 & \uparrow\\ 49 \div 2 = 24 & \text{remainder } r = 1 & \uparrow\\ 24 \div 2 = 12 & \text{remainder } r = 0 & \uparrow\\ 12 \div 2 = 6 & \text{remainder } r = 0 & \uparrow\\ 6 \div 2 = 3 & \text{remainder } r = 0 & \uparrow\\ 3 \div 2 = 1 & \text{remainder } r = 1 & \uparrow\\ 1 \div 2 = 0 & \text{remainder } r = 1 & \text{MSB}\end{array}$$

Table 5.2 BCD code

Denary	Natural binary	8:4:2:1 BCD Tens	Units
00	0000	0000	0000
01	0001	0000	0001
02	0010	0000	0010
03	0011	0000	0011
04	0100	0000	0100
05	0101	0000	0101
06	0110	0000	0110
07	0111	0000	0111
08	1000	0000	1000
09	1001	0000	1001
10	1010	0001	0000
11	1011	0001	0001
12	1100	0001	0010
13	1101	0001	0011
14	1110	0001	0100
15	1111	0001	0101

Binary-coded decimal (BCD) codes

Denary numbers of any size can be coded into natural binary form provided enough bits are used. However, there is no clear link between the separate digits of the denary number and the digits of the resulting binary code. You have to do some arithmetic, as shown in the previous section, to convert from one to the other.

An alternative number system is the **BCD code** (binary-coded decimal) shown in Table 5.2, where *each denary digit* is replaced by 4-bit binary code. The reason that a 4-bit code is used is that there are ten different denary numerals, and four is the smallest number of bits, n, that make 2^n greater than 10. This means that 6 of the possible 16 values of the 4-bit code are redundant.

Each of the examples in Table 5.2 has two denary digits (including leading zeros below 10), and so the BCD code, which requires 4 bits per digit, is 8 bits long. The principle can be extended to any number of digits. So, for example, to find the BCD code for 397_{10} each digit is separately converted to a 4-bit natural binary number, then the bits are strung together:

> Normally, when we deal with denary numbers, we do not write down 'leading zeros'. Hence an 8 digit calculator will display 2167 rather than 00002167 (they both, of course, represent the same number). When recording binary codes in logic circuits, however, it is normal to include leading zeros up to the number of bits that the code or system is using. Thus counting in a 4-bit binary system would be written 0001, 0010, 0011, ..., etc., rather than 1, 10, 11 ...

Denary	Hundreds	Tens	Units	BCD
397	0011	1001	0111	001110010111

Self-assessment question

2 What are the natural binary and 8 : 4 : 2 : 1 BCD numbers for denary 420? What denary numbers would 01101001 represent if it were (a) in natural binary, (b) in 8 : 4 : 2 : 1 BCD?

You will have noticed that the BCD equivalent of a large denary number is longer than the corresponding number in natural binary code, but that it is much easier (for denary-educated people) to translate quickly to and from BCD. Computers in their internal operations use codes like natural binary but, close to the points of contact with people, such as in keyboards and in displays, they commonly use BCD codes.

Octal and hexadecimal

Numbers in the denary and binary system use positional weighting, where the significance of each digit is a multiple of powers of 10 for denary and 2 for binary. In **octal** and **hexadecimal**, positional weighting is also used, but the significance of the digits is a multiple of power of 8 and 16, respectively. These representations provide two alternative ways of quickly reducing a long string of binary digits to a more manageable form, as follows.

Table 5.3 Octal and hex

Natural binary	*Denary*	*Octal*	*Hexa-decimal*
0000	0	0	0
0001	1	1	1
0010	2	2	2
0011	3	3	3
0100	4	4	4
0101	5	5	5
0110	6	6	6
0111	7	7	7
1000	8	10	8
1001	9	11	9
1010	10	12	A
1011	11	13	B
1100	12	14	C
1101	13	15	D
1110	14	16	E
1111	15	17	F
10000	16	20	10
10001	17	21	11

Suppose the initial binary number is 11010110 (which is 214_{10}). To express this in octal form, the binary number is divided into groups of three bits, counting from the right. For this example the groups are 11, 010, 110. The natural binary code is then used to replace each group by the corresponding single digit, 0 to 7. That is

$$\begin{matrix}(0)11, & 010, & 110\\ 3 & 2 & 6\end{matrix}$$

326_8 is then the octal equivalent to 11010110_2. It can be translated into denary using powers of 8.

$$\begin{aligned}326_8 &= 3\times 8^2 + 2\times 8^1 + 6\times 8^0\\ &= 192 + 16 + 6\\ &= 214_{10}\end{aligned}$$

Hexadecimal (abbreviated to 'hex') follows the same idea, but it splits the binary number up into groups of four bits at a time, that is 1101, 0110 in the present example. Each group of four bits is replaced by a single symbol. 0000 to 0101 can use 0 to 9 as usual, but for the remaining 4-bit groups extra symbols are needed: it is normal to use the letters A to F, as in Table 5.3.

So, for example, 214_{10}, which is equivalent to 11010110_2, can be written as $D6_{16}$.

$$\begin{aligned}D6_{16} &= 13\times 16^1 + 6\times 16^0\\ &= 208 + 6\\ &= 214_{10}\end{aligned}$$

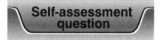

Conversions like these to base 8 and, more commonly, to base 16 are used widely as a convenient shorthand for binary. It is much easier to recognize or recall short strings of octal or hex digits than long strings of 0s and 1s.

Self-assessment question

3 Express 110001011010_2 in octal, denary and hexadecimal.

5.2.3 Other symbols and distinct items

Another common requirement is to have a code for letters (as in Self-assessment question 1) and other characters as well as numbers, so that written messages can be sent in binary digital form. An important example is the **ASCII** code (American Standard Code for Information Interchange), used for feeding information into computers from keyboards, and operating alphanumeric displays, printers, and so on.

The number of distinct symbols used in any system of counting is called the **base** of the system. Thus denary numbers are to the base 10, binary to base 2, octal to base 8 and hexadecimal to base 16.

The ASCII code uses 7 bits to represent $2^7 = 128$ characters, consisting of upper- and lower-case letters, numbers, punctuation marks and various standard messages. Some of them are listed in Table 5.4.

In this table the seven bits are labelled b_1 to b_7, starting from the LSB. There are two more columns (with $b_7b_6b_5 = 000$ and 001) not included as they represent messages requiring detailed explanation. Note that numerals 0 to 9 are included in the column $b_7b_6b_5 = 011$ where the other four bits (b_1 to b_4) are the natural binary code for these numerals.

Table 5.4

$b_7b_6b_5$						$b_4b_3b_2b_1$
010	*011*	*100*	*101*	*110*	*111*	
SP	0	@	P	`	p	0 0 0 0
!	1	A	Q	a	q	0 0 0 1
"	2	B	R	b	r	0 0 1 0
#	3	C	S	c	s	0 0 1 1
$	4	D	T	d	t	0 1 0 0
%	5	E	U	e	u	0 1 0 1
&	6	F	V	f	v	0 1 1 0
'	7	G	W	g	w	0 1 1 1
(	8	H	X	h	x	1 0 0 0
)	9	I	Y	i	y	1 0 0 1
*	:	J	Z	j	z	1 0 1 0
+	;	K	[	k	{	1 0 1 1
,	<	L	\	l	|	1 1 0 0
–	=	M	]	m	}	1 1 0 1
.	>	N	^	n	~	1 1 1 0
/	?	O	–	o	DEL	1 1 1 1

Worked example

What is the ASCII code for 'G' and for 'g'?

1000111 and 1100111

In practice, an eighth **parity bit** b_8 may be appended at the MSB end. A parity bit is chosen to make the number of 1s in the eight bits odd (say) for all symbols. (It could alternatively be chosen to make the sum even, but the choice must be the same for all symbols.) Thus, the ASCII code for '3' (0110011) has four 1s, and so requires an additional 1 to give it odd parity (10110011). The code for 'R' (1010010) with three 1s has odd parity already, so the additional bit would be 0 (01010010). The reason for doing this is that if a single error occurs when the code is first produced, or during transmission from one place to another, a 1 is substituted for a 0 or 0 for a 1, and so the parity would appear to be even. If the receiver is expecting a code word with odd parity the error would be evident. Thus the inclusion of the parity bit is an example of an **error-detecting** code.

Self-assessment question

4 A four-letter word is encoded in ASCII with no parity check. In octal the code corresponding to the characters is
102, 111, 124, 123_8
What word does this represent? What would the sequence of numbers be in hexadecimal if the ASCII characters had even parity?

Obviously other codes than ASCII are possible but there is no difference in principle: all that is needed is a table listing the symbols and the corresponding code. A binary code can equally well be associated with names of individuals, objects in a catalogue, and so on. The only general limitation is the 2^n rule—with n bits only 2^n distinct code words can be formed, so this is an upper limit on the number of distinct items which can be distinguished by an n-bit code.

5.2.4 Summary of Section 5.2

In some problems the input conditions and output requirements are naturally expressed in binary form (ON/OFF conditions, YES/NO answers, TRUE/FALSE statements.) A combination of n binary input conditions defines 2^n possible input combinations. A problem can therefore be specified in a truth table, with 2^n rows and one output column.

If the input is not naturally in binary form, but consists of a choice between individuals or symbols, the choice can be specified by a suitable binary code. An n-bit code can describe up to 2^n distinguishable items. The n bits describing a particular input combination can be described as a code word.

Examples of codes for numbering systems and symbol representation were shown. These included natural binary, BCD, octal and hexadecimal for numbers and ASCII for symbols.

5.3 TRUTH TABLES AND BOOLEAN NOTATION

Section 5.1 introduced the truth table as the specification of an electronic device. This section considers the development of truth tables of increasing complexity, and introduces the terminology and notation for some standard logic functions.

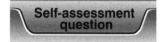

5.3.1 Problems with only one input

Fig. 5.5 A one-input one-output combinational logic device.

Starting with the very simplest case, consider a one-input, one-output combinational logic device, shown in Fig. 5.5. With one input there are only two possible input values, so there are two rows in the truth table. The output column can therefore be completed in four different ways as follows.

A	P	A	P	A	P	A	P
0	0	0	0	0	1	0	1
1	0	1	1	1	0	1	1

The first and fourth of these are of little practical significance. In Boolean notation they indicate $P=0$ and $P=1$, meaning that the output value P does not depend on the input at all, but is 'stuck at' 0 or 1, respectively.

The second truth table shows that the output is the same as the input, which can be written in Boolean notation as $P=A$. As far as logic levels are concerned, the terminals A and P might as well be connected together by a single wire. However, the electrical characteristics of this $P=A$ device might provide a **buffer** between a source of logic signals having limited power and a further logic device needing more power to drive it than the source can provide. Figure 5.6 shows the graphical symbol for the buffer, a triangle, which is the same as the standard symbol for an analogue amplifier.

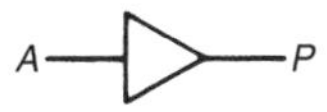

Fig. 5.6 Buffer.

The third truth table specifies the logic device called the **INVERTER**. Whatever the value of A (0 or 1), P is always the other binary value (1 or 0). The relationship is variously referred to as 'P is NOT-A' or 'P is the inverse of A' or 'P is the **complement** of A'. The Boolean notation for the relationship between the input and the output of an INVERTER is $P=\overline{A}$, where the bar above A indicates inversion. The relationship implies that the value of P is always the inverse of A, but does not say what that value is.

Worked example

What happens when you take the inverse of an already inverted variable?

Inverting twice restores the original value, that is, if $P=\overline{A}$*, then*

$$\overline{P}=\overline{\overline{A}}=A$$

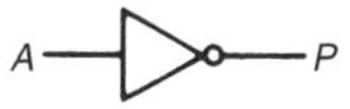

Fig. 5.7 Inverter.

The graphical symbol for an INVERTER is shown in Fig. 5.7, where it is the small circle that denotes inversion. The triangle is again the standard amplifier symbol, representing the fact that an INVERTER may also act as a buffer (but with inversion of the output signal).

5.3.2 Problems with two inputs

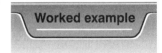

Fig. 5.8 A two-input one-output combinational logic device.

Consider a two-input one-output device, shown in Fig. 5.8. With two inputs there are $2^2=4$ input combinations or rows in the truth table. The output column can be completed in $2^{2^n}=16$ different ways. Once again, some of these will depend on fewer than two variables.

Self-assessment question

5 (a) Write out all 16 possible output columns for the truth table.
(b) What are the two truth tables where the output, *P*, is independent of the inputs, *A* and *B*, (*P* does not change if the inputs change)?
(c) What are the four truth tables where *P* depends on only one input? Describe them all as Boolean expressions.

Removing the six simple cases found in Self-assessment question 5 leaves ten more interesting ones where the output depends on some *combination* of the inputs. The first is the AND function (column P_8 in the solution to Self-assessment question 5), taking practical form in an AND gate. You have met the three-input version once before in the interlock problem in Section 5.1. In the truth table shown below, the output is 1 only when both input *A* AND input *B* are 1:

A	*B*	*P*
0	0	0
0	1	0
1	0	0
1	1	1

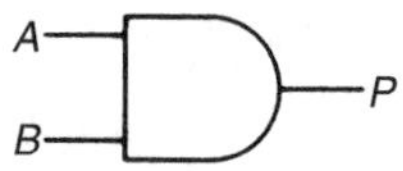

Fig. 5.9 AND gate.

The AND gate is specified in Boolean notation by the expression

$$P = A \cdot B$$

where the dot '·' represents the AND relationship. It is graphically symbolized by Fig. 5.9.

Another important function is the **OR** function (P_{14}), taking practical form in an **OR gate**. It is defined by the truth table shown below:

A	*B*	*P*
0	0	0
0	1	1
1	0	1
1	1	1

The output is 1 when *any* or *all* of *A* OR *B* (OR *C*..., for more inputs) are 1. It is specified in Boolean notation by the expression.

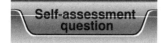

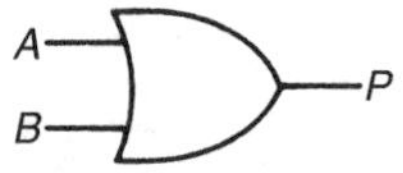

Fig. 5.10 OR gate.

$$P = A + B$$

where the sign '+' denotes the OR relationship. It is graphically symbolized by Fig. 5.10.

Two points that often cause confusion are worth mentioning here. When *A* and *B* are Boolean variables:

- Be careful not to read $A+B$ as *A and B*, or *A plus B*. In this Boolean notation it means *A* OR *B*, which is not the same as addition. Similarly, $A \cdot B$, the notation for *A* AND *B*, is not the same as multiplication.
- Some English sentences containing 'and' in fact imply the OR function. The statement 'students and teachers will find this book useful' means that a reader will find the book useful (*P*) if the reader is a student (*A*) OR a teacher (*B*) OR both, in the logic sense of OR.

Returning to the truth tables, there are six more of the possible output columns which can be described by a combination of AND or OR with inversion. Two of them are shown below:

A	B	P	A	B	P
0	0	1	0	0	1
0	1	1	0	1	0
1	0	1	1	0	0
1	1	0	1	1	0

The first is the inverse of the truth table which was described as the AND relationship, and is called **NAND** which is short for NOT-AND. The Boolean expression for P in this case is written as

$$P = \overline{A \cdot B}$$

The second is the inverse of the truth table which we described as the OR relationship, and is called **NOR** which is short for NOT-OR. The Boolean expression for P in this second case is written as

$$P = \overline{A + B}$$

The graphical symbols for NAND and NOR gates are shown in Fig. 5.11. The symbols are basically the same as the ones for AND and OR gates, but with the small ring on the outputs denoting inversion.

If appropriate, a 'bubble' like the one on the output of the NAND and NOR gates can be applied to represent inversion on the inputs as well as on the outputs of any type of logic symbol.

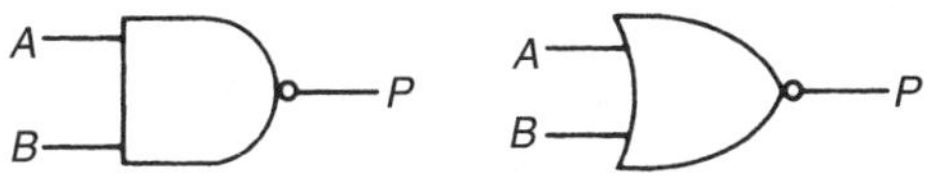

Fig. 5.11 NAND and NOR gates.

Inverting just one of the inputs provides a description for some of the remaining possibilities. If you examine the following truth table you will see that P is 1 when A is 0 AND B is 1. However, if we include the intermediate column labelled $\overline{A}$, then you can see that P is 1 when $\overline{A}$ AND B are both 1. The output can therefore be described by the Boolean expression $P = \overline{A} \cdot B$.

A	B	$\overline{A}$	P
0	0	1	0
0	1	1	1
1	0	0	0
1	1	0	0

Similarly, P is 1 in the following truth table when $\overline{A}$ OR B is 1. The output can therefore be described by the Boolean expression $P = \overline{A} + B$.

A	B	$\overline{A}$	P
0	0	1	1
0	1	1	1
1	0	0	0
1	1	0	1

Then, using the same ideas the following truth tables can be described by $P = A \cdot \overline{B}$ and $P = A + \overline{B}$.

A	B	P
0	0	0
0	1	0
1	0	1
1	1	0

A	B	P
0	0	1
0	1	0
1	0	1
1	1	1

We are now left with just two more possibilities shown below. The first is called **EXCLUSIVE-OR** or **XOR**, conveying that the output is 1 if A or B is 1, but *not if both are* 1. (Another way of saying this is that P is 1 when A and B are different from one another.)

$A\ B$	P
0 0	0
0 1	1
1 0	1
1 1	0

$A\ B$	P
0 0	1
0 1	0
1 0	0
1 1	1

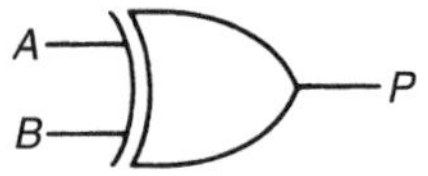

Fig. 5.12 XOR gate.

The XOR operation between two variables is written as:

$$P = A \oplus B$$

The graphical symbol for the XOR gate is shown in Fig. 5.12.

The second of the truth tables shown above is the inverse of the EXCLUSIVE-OR, called the **EXCLUSIVE-NOR** or **XNOR**, and it can be written as

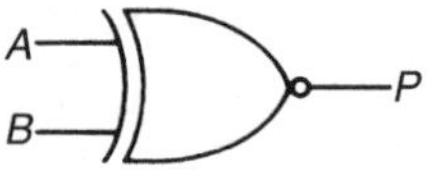

Fig. 5.13 XNOR gate.

$$P = \overline{A \oplus B} \quad \text{or} \quad A \overline{\oplus} B$$

In this case the output P is 1 whenever both inputs are the same. The graphical symbol for the XNOR gate is shown in Fig. 5.13.

This completes the preliminary study of all of the sixteen possible two-input functions. The most important ones to remember are the AND, OR and INVERTER operations, since these can always be combined to form any of the others including the EXCLUSIVE-OR and EXCLUSIVE-NOR. This can be seen by writing the EXCLUSIVE-OR operation in the following way.

$$P = 1 \text{ only if either } A = 1 \text{ and } B = 0 \text{ or if } A = 0 \text{ and } B = 1$$

You can think of this as the output of one AND gate whose inputs are A and $\overline{B}$ *OR* the output of another AND gate whose inputs are $\overline{A}$ and B. This is shown in Fig. 5.14.

The Boolean expression for the output of an XOR gate can be written as:

$$P = A \cdot \overline{B} + \overline{A} \cdot B$$

The behaviour of digital circuits is normally simulated with programs in which the circuit is represented by suitably characterized blocks of logic such as gates, rather than by individual transistors. Analogue simulation packages working at the transistor level are usually far too slow to produce useful results from moderately complicated logic circuits, but are sometimes used to simulate in more detail small, critical parts of a digital circuit.

If you have access to a digital simulator package, then you might find it helpful to simulate the behaviour of some circuits in this chapter and, more especially, in Chapter 6. Some recommended computer work is indicated in the marginal notes.

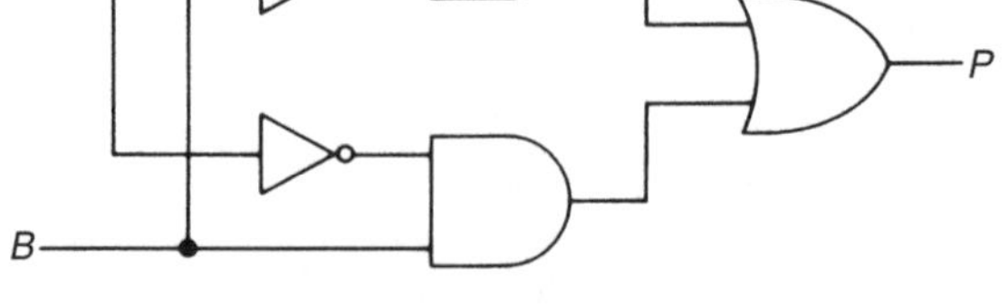

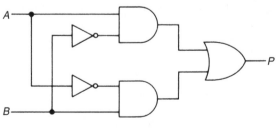

Fig. 5.14 XOR circuit.

which as you can see, reflects the gate structure of the XOR circuit shown in Fig. 5.14.

Self-assessment question

6 The logical statement for an XNOR gate can be written:

$P = 1$ only if either $A = 0$ and $B = 0$ or if $A = 1$ and $B = 1$

What is the equivalent Boolean expression for this statement? How would you construct a circuit that performs the XNOR using AND, OR, and INVERTERs?

(If you have a suitable computer package, then simulate your solution. Apply waveforms so that all four input combinations are reached and check that the output follows the required truth table.)

Finally, the AND and OR functions can also be shown to be related to each other through the following equations:

$$A \cdot B = \overline{\overline{A} + \overline{B}}$$
$$\overline{A \cdot B} = \overline{A} + \overline{B}$$
$$\overline{\overline{A} \cdot \overline{B}} = A + B$$
$$\overline{A} \cdot \overline{B} = \overline{A + B}$$

(These can be checked by substituting each possible combination of A and B values in both sides of each equality.) The first of these expressions, $A \cdot B$, might represent, for example, a simple interlock, allowing a car to start only if (A) the doors are locked AND (B) seat belts of occupied seats are fastened. Then the second expression could represent the condition for a warning light to come on when the conditions are not met. You can look at this in two ways, either as:

> $\overline{A \cdot B}$, the light comes on if it is NOT the case that both conditions A AND B are satisfied,

or as

> $\overline{A} + \overline{B}$, the light comes on if either condition A is NOT satisfied OR condition B is NOT satisfied.

The substitution of values and the verbal descriptions both indicate that these amount to the same thing: the NAND function. NAND can be seen either as AND followed by NOT, or as NOT for each input followed by OR.

The two alternatives are shown in Fig. 5.15, with circles for inversion 'before' or 'after' (that is, at the inputs or outputs of) the gate symbols.

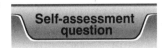

Fig. 5.15 NAND.

Worked example

How would you use a two-input AND gate and two INVERTERs to perform the same function as a two-input NOR gate?

The last of the expressions given above, namely $\overline{A}\cdot\overline{B}=\overline{A+B}$, *can be used to show the relation between a two-input NOR gate and a two-input AND gate with both inputs inverted. This is shown in Fig. 5.16.*

Fig. 5.16 NOR.

These relationships between AND and OR are commonly called **deMorgan's theorem,** named after the nineteenth-century mathematician Augustus deMorgan. This can be stated as:

> NAND can be obtained either by inverting the output of AND or by inverting ***both*** the inputs of OR, that is $\overline{A\cdot B}=\overline{A}+\overline{B}$.
> NOR can be obtained either by inverting the output of OR, or by inverting ***both*** the inputs of AND, that is $\overline{A+B}=\overline{A}\cdot\overline{B}$.

5.3.3 Problems with more than two inputs

Some three-input functions simply extend the definitions of AND, OR and their inverses which you have already met. For example, the AND of three inputs is written:

$$P=A\cdot B\cdot C$$

The truth table for a three-input AND, shown in Section 5.1.1, has a single 1 at the bottom of the output column. Any other truth table with a single 1 can be expressed as AND with one or more of the inputs inverted. For example, let the following three conditions describe the condition of two telephones:

$A=1$ when the caller's handset is on the rest.
$B=1$ when the correct number is dialled.
$C=1$ when the receiver's handset is on the rest.

For the initial connection to be made (P), the right combination of conditions must be satisfied. This requirement is that $A=0$ AND $B=1$ AND $C=1$, as shown in the following truth table:

$A\,B\,C$	$\overline{A}$	P
0 0 0	1	0
0 0 1	1	0
0 1 0	1	0
0 1 1	1	1
1 0 0	0	0
1 0 1	0	0
1 1 0	0	0
1 1 1	0	0

The additional column called $\overline{A}$ shows clearly that P is 1 when $\overline{A}$ AND B AND C are all 1. The expression for P can therefore be written

$$P = \overline{A} \cdot B \cdot C.$$

The relationship between a single 1 in the truth table and its Boolean (AND) expression can be summarized as:

> Where the output is 1, if an input variable is 0, then that variable appears inverted in the Boolean expression for the output.
> Where the output is 1, if an input variable is 1, then that variable appears non-inverted in the Boolean expression for the output.

The OR of three inputs is written:

$$P = A + B + C$$

The truth table has a single 0 in the top line. Hence, any truth table with a single 0 in the output column can be written as OR, possibly with one or more of the inputs inverted, as in the following Self-assessment question.

Self-assessment question

7 In a chemical process, an alarm buzzer sounds (P) if the temperature rises above a specified level (A), the pressure rises above a specified level (B) or the supply of raw materials is *not* above a specified minimum (C).

Write down the truth table and a Boolean expression for the required conditions.

Finally, there are many three-input functions which have 2, 3 or 4 0s or 1s in the output column, not covered by any of the above. For example, a car seat belt interlock might require that:

The car should only start (P)
if the driver's seat belt is fastened ($A = 1$)
and either (i) the front passenger seat is occupied ($B = 1$) and the passenger seat belt is fastened ($C = 1$)
or (ii) the front passenger seat is unoccupied ($B = 0$).

(We can assume that the driver's seat will always be occupied if the seat belt is fastened and not have a separate variable to test for this.)

The truth table is

$A\ B\ C$	P	
0 0 0	0	
0 0 1	0	
0 1 0	0	
0 1 1	0	
1 0 0	1	← Option (ii): if $B = 0$ (passenger seat unoccupied),
1 0 1	1	← it does not matter whether C is 0 or 1
1 1 0	0	
1 1 1	1	← Option (i)

Option (i) alone could be written $P = A \cdot B \cdot C$. Option (ii) could be written $P = A \cdot \overline{B}$, which is independent of C and covers both 1s. Because the requirement is $P = 1$ if either (i) OR (ii) is satisfied, they may be combined using the notation for OR:

$$P = A \cdot B \cdot C + A \cdot \overline{B}$$

You will have seen by now that each 1 in the output column of three-input truth tables corresponds to a term like $A \cdot B \cdot C$ in the expression for P. These terms with or without inversion bars ($A \cdot B \cdot C$, $\overline{A} \cdot B \cdot C$, $\overline{A} \cdot \overline{B} \cdot C$, and so on) are called **minterms**. So any truth table can be expressed in **sum-of-minterms** form by writing the minterms corresponding to each of the 1s in the output column and 'summing' them (in the logic sense, that is combining them by the OR function).

The following truth table can be used to demonstrate this. The four 1s in the output column, P, are also shown as single 1s in four separate output columns labelled Q, R, S and T. Each of the single 1s corresponds to an AND of the inputs with some of the inputs inverted. (The columns should, of course, contain 0s but these have been omitted so that the diagram is less cluttered.) The output, P, can then be found by taking the OR of the four separate AND functions.

A B C D	P	Q	R	S	T
0 0 0 0	0				
0 0 0 1	1	1			
0 0 1 0	1		1		
0 0 1 1	0				
0 1 0 0	0				
0 1 0 1	1			1	
0 1 1 0	0				
0 1 1 1	0				
1 0 0 0	1				1
1 0 0 1	0				
1 0 1 0	0				
1 0 1 1	0				
1 1 0 0	0				
1 1 0 1	0				
1 1 1 0	0				
1 1 1 1	0				

This gives : $P = Q + R + S + T$

$$P = \overline{A} \cdot \overline{B} \cdot \overline{C} \cdot D + \overline{A} \cdot \overline{B} \cdot C \cdot \overline{D} + \overline{A} \cdot B \cdot \overline{C} \cdot D + A \cdot \overline{B} \cdot \overline{C} \cdot \overline{D}$$

Thus the Boolean expression corresponding to any truth table can be written down in a sum of minterms form using these rules:

> Inputs which are 0 when the output is 1 appear inverted in the minterm that corresponds to that 1
> Inputs which are 1 when the output is 1 appear non-inverted in the minterm that corresponds to that 1.

This way of writing a Boolean expression does not necessarily give the most economical description, but it is one certain and comprehensive way of doing it.

5.3.4 Implementation using gates

The Boolean expression of *any* truth table can be stated as the sum-of-minterms. Each minterm is an AND function with some of the inputs inverted, and can therefore be directly implemented using an AND gate with appropriate INVERTERs. For example, a minterm of the form $A \cdot \overline{B} \cdot C \cdot \overline{D}$ can be implemented with a four-input AND gate and two INVERTERs, as shown in Fig. 5.17.

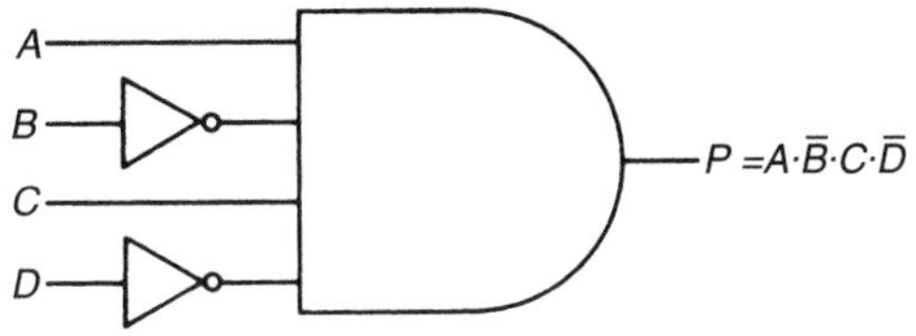

Fig. 5.17 An implementation of $P = A \cdot \overline{B} \cdot C \cdot \overline{D}$.

The 'sum' part of the sum-of-minterms is the OR function, and can therefore be implemented by an OR gate with the correct number of inputs. The circuit, which can be described by the expression

$$P = \overline{A} \cdot \overline{B} \cdot \overline{C} \cdot D + A \cdot \overline{B} \cdot C \cdot \overline{D} + \overline{A} \cdot B \cdot \overline{C} \cdot \overline{D} + A \cdot \overline{B} \cdot \overline{C} \cdot \overline{D}$$

can be constructed using four four-input AND gates, a four-input OR gate, and a number of INVERTERs. This is shown in Fig. 5.18.

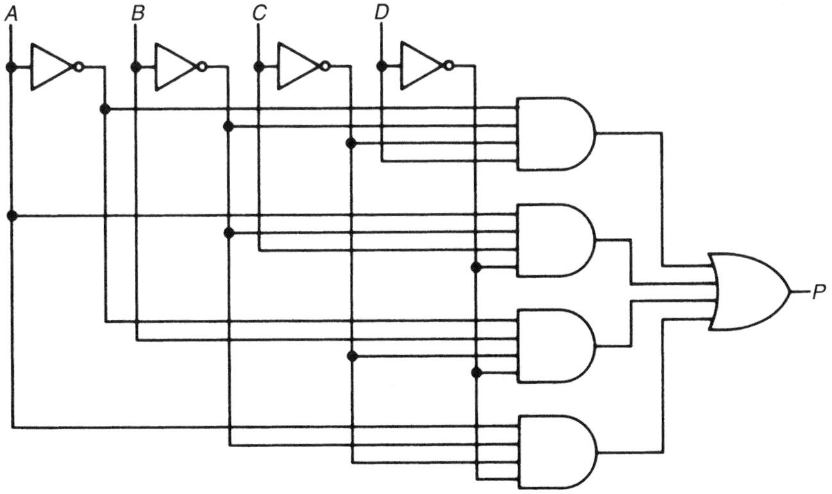

Fig. 5.18

In principle you now have a systematic method for designing a digital circuit. This circuit will probably not be the most economical, meaning that it could probably have been implemented using fewer gates. Procedures for finding more economical designs are described in Sections 5.4 and 5.5.

Self-assessment question

8 An insurance company classifies requests for motor insurance according to whether (*A*) applicant is under 25 years old; (*B*) age of vehicle is under 5 years; (*C*) engine capacity is under 2 litres; (*D*) previous insurance claims have been made. The maximum premium is charged (*P*) for

(i) applicants under 25 with vehicles over 5 years old;
(ii) applicants under 25 with vehicles under 5 years old who have made previous claims;
(iii) all applicants with vehicles under 5 years old and capacity over 2 litres with previous claims; or
(iv) applicants of 25 or over with vehicles under 5 years old and under 2 litres capacity with previous claims.

Complete the truth table in the margin and write down a Boolean expression for *P*.

A	*B*	*C*	*D*	*P*
0	0	0	0	
0	0	0	1	
0	0	1	0	
0	0	1	1	
0	1	0	0	
0	1	0	1	
0	1	1	0	
0	1	1	1	
1	0	0	0	
1	0	0	1	
1	0	1	0	
1	0	1	1	
1	1	0	0	
1	1	0	1	
1	1	1	0	
1	1	1	1	

5.3.5 Summary of Section 5.3

In this section you have been introduced to the common logic gates and their Boolean expressions. The INVERTER, AND, NAND, OR, NOR, XOR and XNOR are widely used as a basis for the implementation of combinational logic circuitry. The truth tables, the Boolean notation and graphical symbols used to describe these 2-input gates are as follows.

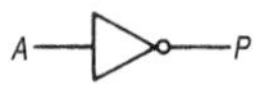

INVERTER

A	*P*
0	1
1	0

$P = \overline{A}$

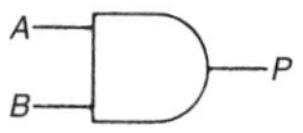

AND

A B	*P*
0 0	0
0 1	0
1 0	0
1 1	1

$P = A \cdot B$

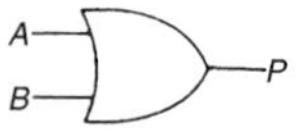

OR

A B	*P*
0 0	0
0 1	1
1 0	1
1 1	1

$P = A + B$

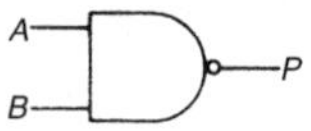

NAND

A B	*P*
0 0	1
0 1	1
1 0	1
1 1	0

$P = \overline{A \cdot B}$

NOR

A B	P
0 0	1
0 1	0
1 0	0
1 1	0

$$P = \overline{A + B}$$

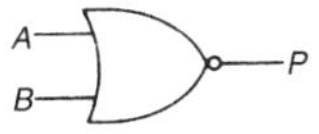

XOR

A B	P
0 0	0
0 1	1
1 0	1
1 1	0

$$P = A \oplus B$$

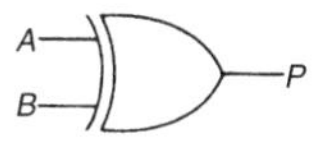

XNOR

A B	P
0 0	1
0 1	0
1 0	0
1 1	1

$$P = \overline{A \oplus B}$$

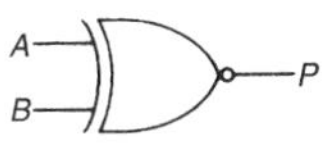

The AND and OR functions are linked by deMorgan's theorem, which shows that $\overline{A \cdot B} = \overline{A} + \overline{B}$ or $\overline{A + B} = \overline{A} \cdot \overline{B}$.

All of these gates were shown as two-input devices, and it was shown how some of them could be extended to any number of inputs. Each 1 in the output column of a truth table can be described by an AND expression with some of the inputs inverted. This expression is called a minterm, and the Boolean expression for the whole output column can be written as the sum-of-minterms, which is the OR of all the conditions for 1 in the output column. We therefore have a systematic way of getting a Boolean expression from any truth table.

5.4 BASIC RULES OF BOOLEAN ALGEBRA

The answer to Self-assessment question 8 showed that sometimes alternative, simpler expressions can be found which can make it much easier to implement truth tables. To be able to simplify any Boolean expression it is necessary to use algebraic manipulation, and a number of important rules to do this can be deduced from the AND and OR truth tables. These rules are stated without proof in Table 5.5, but to check them, substitute 0 and 1 for X in each case where X is a Boolean variable and use the appropriate truth table to find each result.

Table 5.5

AND	*OR*
$X \cdot X = X$	$X + X = X$
$X \cdot \overline{X} = \overline{X} \cdot X = 0$	$X + \overline{X} = \overline{X} + X = 1$
$X \cdot 1 = 1 \cdot X = X$	$X + 1 = 1 + X = 1$
$X \cdot 0 = 0 \cdot X = 0$	$X + 0 = 0 + X = X$

These rules also apply to functions where there are more than two inputs, for example, $P = X \cdot X \cdot X$ reduces to $P = X$.

Self-assessment question

9 What does the expression $P = X + \overline{X} + X$ reduce to?

You can see from the table above that both the AND and OR operations are independent of the ordering of the inputs, for example, $X + 1 = 1 + X$

whatever the values of X. This property can be summarized by the following **commutative law**:

> Commutative law: $A \cdot B = B \cdot A$ and $A + B = B + A$

Similarly, it does not matter which variables are grouped together or in which order. The same is true for the OR function, and so we can summarize the **associative law** as:

> Associative law: $A \cdot (B \cdot C) = (A \cdot B) \cdot C$ and $A + (B + C) = (A + B) + C$

Finally, the **distributive law** relates to the use of the AND and the OR functions.

> Distributive law: $A \cdot (B + C) = A \cdot B + A \cdot C$

These three laws—commutative, associative and distributive—are the same laws that apply to ordinary arithmetic, where '+' and '·' are interpreted as addition and multiplication. It is therefore not difficult to apply these laws even if you cannot remember them explicitly.

Self-assessment question

10 Simplify the following expressions and name the laws being used:

(a) $P = A \cdot (B + C) + A \cdot (\overline{C} + \overline{B})$

(b) $P = A \cdot B \cdot C + \overline{A} \cdot C \cdot B$

5.4.1 Simplifying Boolean expressions

You saw earlier that any truth table could be expressed in Boolean notation as the sum-of-minterms, where each minterm corresponded to a logic 1 in the output column of the truth table. Terms which are of the form $A \cdot B \cdot C$, that is the AND of several variables and their inverses but *not necessarily all* of the input variables, are called **product terms**. Minterms, which *do contain all* of the input variables in inverted or non-inverted form, are a special kind of product term.

The aim of this section is to show you how the sum-of-minterms expression can be simplified. This means that the total number of product terms in the expression, and the number of variables in each product term, are fewer than in the original sum-of-minterms. The final Boolean expression is then said to be in a **sum-of-products** form, where the '+' and '·' symbols represent OR and AND operators, respectively.

When a Boolean expression contains minterms which have variables in common, it is possible to use brackets using the distributive law. If the expression inside the brackets contains only a single variable in inverted and non-inverted form, it is possible to simplify the expression. For example, if

$$P = A \cdot \overline{B} \cdot C \cdot D + A \cdot \overline{B} \cdot C \cdot \overline{D}$$

then the two minterms have $A \cdot \overline{B} \cdot C$ in common, and so the expression becomes

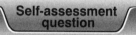

$$P = A \cdot \overline{B} \cdot C \cdot (D + \overline{D})$$

From Table 5.5 we know that $X + \overline{X} = 1$, so

$$P = A \cdot \overline{B} \cdot C \cdot 1$$

Also, from the table, we know that $X \cdot 1 = X$, so

$$A \cdot \overline{B} \cdot C \cdot 1 = A \cdot \overline{B} \cdot C$$

making

$$P = A \cdot \overline{B} \cdot C$$

We have succeeded in simplifying the expression from two minterms to one product term which only has three input variables instead of all four.

Self-assessment question

11 A car seat belt interlock has the following truth table:

A	B	C	P
0	0	0	0
0	0	1	0
0	1	0	0
0	1	1	0
1	0	0	1
1	0	1	1
1	1	0	0
1	1	1	1

Write down the Boolean expression for this problem and use the method shown above to simplify this expression to

$P = A \cdot \overline{B} + A \cdot B \cdot C$.

The expression in Self-assessment question 11 could just as easily have been simplified to $P = A \cdot \overline{B} \cdot \overline{C} + A \cdot C$. It is even possible to simplify it further to $P = A \cdot \overline{B} + A \cdot C$. It is sometimes difficult to spot ways of simplifying expressions. Fortunately there are techniques to make the task easier, one of which will be discussed in the next section.

5.4.2 Summary of Section 5.4

- In designing a logic circuit it is sometimes necessary to simplify the Boolean expression of a particular problem so that it can be implemented using fewer gates. Some of the rules which can be used to manipulate a Boolean expression are:

AND	OR
$X \cdot X = X$	$X + X = X$
$X \cdot \overline{X} = \overline{X} \cdot X = 0$	$X + \overline{X} = \overline{X} + X = 1$
$X \cdot 1 = 1 \cdot X = X$	$X + 1 = 1 + X = 1$
$X \cdot 0 = 0 \cdot X = 0$	$X + 0 = 0 + X = X$

A	B	P
0	0	a
0	1	b
1	0	c
1	1	d

Truth table

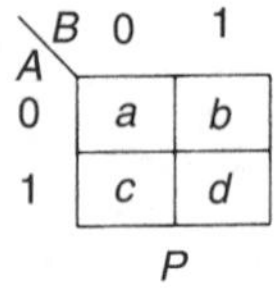

Fig. 5.19 Truth table and Karnaugh map for a two-input problem.

A	B	P
0	0	0
0	1	1
1	0	1
1	1	1

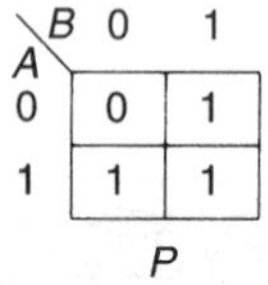

Fig. 5.20 Truth table and Karnaugh map for $P = A + B$.

(a)

A	B	P
0	0	1
0	1	0
1	0	1
1	1	1

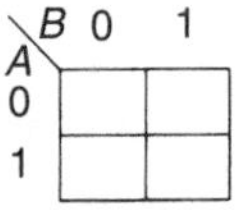

(b)

A	B	P
0	0	
0	1	
1	0	
1	1	

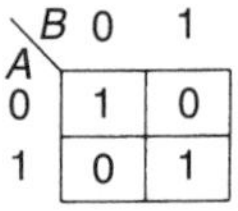

Fig. 5.21 Figure for Self-assessment question 12.

- The commutative law: $A \cdot B = B \cdot A$ and $A + B = B + A$.
- The associative law: $A \cdot (B \cdot C) = (A \cdot B) \cdot C$ and $A + (B + C) = (A + B) + C$.
- The distributive law: $A \cdot (B + C) = A \cdot B + A \cdot C$.
- When a sum-of-minterms expression is simplified, some or all of the minterms become product terms which do not contain all the input variables.

5.5 KARNAUGH MAPS

Karnaugh maps are a pictorial way of simplifying a design for a combinational circuit. They are named after Maurice Karnaugh, an American engineer who first suggested their use in 1953. They are an alternative way of writing down the information that is contained in a truth table, and in this form it is easy to identify groups of minterms that can be reduced to smaller and fewer product terms.

5.5.1 Obtaining product terms

Two inputs

Figure 5.19 shows the correspondence between the Karnaugh map and the truth table for the general case of a two-input problem. The values inside the squares are copied from the output column of the truth table, so there is one square in the map for every line in the truth table. Around the edge of the Karnaugh map are the values of the two input variables: B along the top and A down the left-hand side. Figure 5.20 shows a particular example where $P = A + B$.

The values around the edge of the map can be thought of as coordinates. So, for example, the square on the top right-hand corner of the map in Fig. 5.20 has coordinates $A = 0$ and $B = 1$. This square corresponds to the line in the truth table where $A = 0$, $B = 1$, and in this example $P = 1$.

Self-assessment question

12 Transfer the information from the truth table to the Karnaugh map or vice versa in the examples shown in Fig. 5.21.

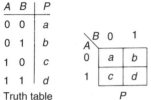

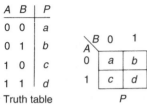

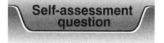

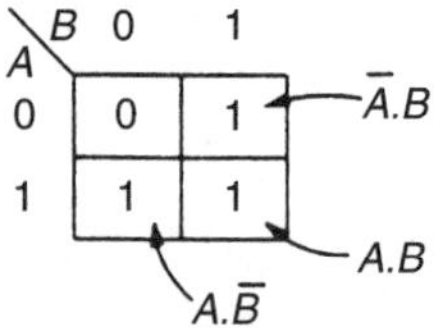

Fig. 5.22 Minterms for $P = A + B$.

Each square in a Karnaugh map containing a 1 corresponds to a minterm in the Boolean expression of the problem. For example, the 1 that can be located by the condition that $A = 1$ and $B = 0$ in Fig. 5.20 has the corresponding minterm $A \cdot \overline{B}$, as shown in Fig. 5.22.

The way the Karnaugh map is used to simplify a Boolean expression for a particular problem is to group the 1s together into rectangular clusters of two or four. Figure 5.23 shows two groups that can be formed for the example given above. In this example, the largest rectangular clusters that can be made consist of two 1s. Notice that a 1 can belong to more than one group.

The first group, labelled I, consists of two 1s which correspond to $A = 1$ and $B = 0$, and $A = 1$ and $B = 1$. Put another way, all of the squares in this example that correspond to the area of the map where $A = 1$ contain 1s, independently of the value of B. So when A is 1 the output is 1, and the expression for the output will contain the term A. Similarly, the second group, labelled II, corresponds to the area of the map where $B = 1$. The group can therefore be defined by the single term B. This implies that when B is 1 the output is 1. The output is therefore 1 whenever A is 1 or B is 1, so the Boolean expression for the output is obtained by finding the OR of the two terms.

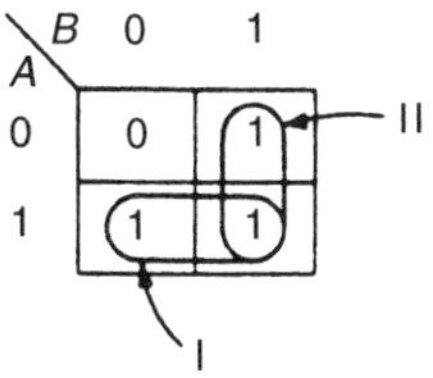

Fig. 5.23

$$P = \underset{\text{group I}}{A} + \underset{\text{group II}}{B}$$

To summarize, there are three general points to remember about Karnaugh maps:

- Each square in the Karnaugh map corresponds to a line in the truth table.
- The coordinates of a square that contains a 1 corresponds to a minterm in a Boolean expression.
- A rectangular cluster of squares corresponds to a product term in a simplified Boolean expression.

Three inputs

Figure 5.24 shows an example of three input variables, where $P = A \cdot \overline{B} \cdot \overline{C} + A \cdot \overline{B} \cdot C + A \cdot B \cdot C$ (this is the example used in Self-assessment question 11). Notice how the values of B and C are written along the top of the Karnaugh map. They are written in such a way that the value of only one of the coordinates changes as you move from column to column. *It is important that you remember the order in which the coordinates are written.*

To find the simplified Boolean expression, we proceed as we did for the two-variable case. First group the 1s together into rectangular clusters of two or four, then find the terms that correspond to these groups, and finally OR all of these terms.

In this example there are two groups, labelled I and II. The first is independent of the value of C and corresponds to the area of the map where $A = 1$ and $B = 0$, as shown in Fig. 5.25a. This means that it has a corresponding product term of $A \cdot \overline{B}$ in the Boolean expression for P. The second group can

A	B	C	P
0	0	0	0
0	0	1	0
0	1	0	0
0	1	1	0
1	0	0	1
1	0	1	1
1	1	0	0
1	1	1	1

A \ BC	00	01	11	10
0	0	0	0	0
1	1	1	1	0

Fig. 5.24 Truth table and Karnaugh map for a three-input problem.

be found in the area where $A=1$ and $C=1$, shown in Fig. 5.25b, and therefore has the corresponding product term of $A \cdot C$. The expression for P is then found by taking the OR of the two product terms:

$$P = A \cdot \overline{B} + A \cdot C$$

This solution could not have been obtained in Self-assessment question 11 just using the manipulation rules that you were given then.

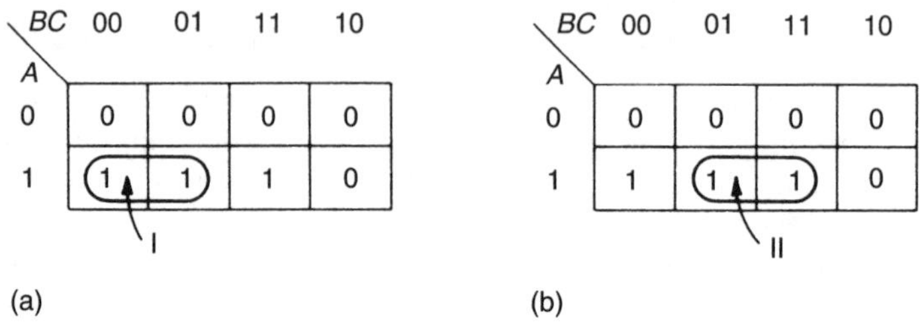

Fig. 5.25

Self-assessment question

13 Write down the terms that define the areas shown in Fig. 5.26.

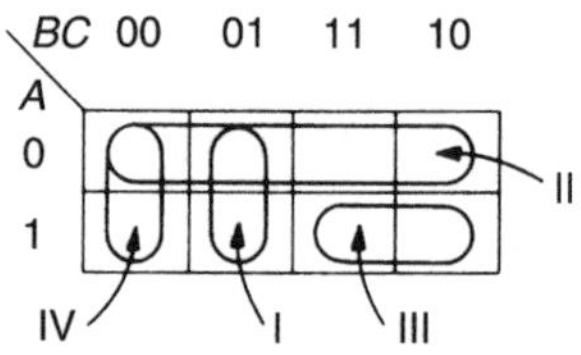

Fig. 5.26 Figure for Self-assessment question 13.

Four inputs

Karnaugh maps can be used for problems with more variables, but four will be the limit in this book. Figure 5.27 shows the truth table and Karnaugh map of a four-input problem. It is the insurance problem again that was used in Self-assessment question 8. Notice that the values along the left-hand column are labelled in the same special way as those along the top. This again is a convention that you have to observe.

It has been emphasized that groups should be formed which are rectangular and which contain two, four, eight (or sixteen, ...) 1s, and that 1s can belong to more than one group. As a designer, your task might be to find the simplest Boolean expression and hence implement the design using the smallest number of gates. You would do this by forming the largest possible groups, and the smallest number of groups.

In this particular problem, the two groups selected to find a reduced Boolean expression are shown in Fig. 5.28. Group I is independent of the variables A and C and can be identified by the area where $B=1$ and $D=1$. The product term $B \cdot D$ represents this group in the Boolean expression for P. The second group, II, is the single row identified with the area where $A=1$ and $B=0$. It is therefore represented in the expression for P by the product term $A \cdot \overline{B}$. The Boolean expression for P is found by taking the OR of these two terms:

$$P = A \cdot \overline{B} + B \cdot D$$

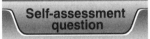

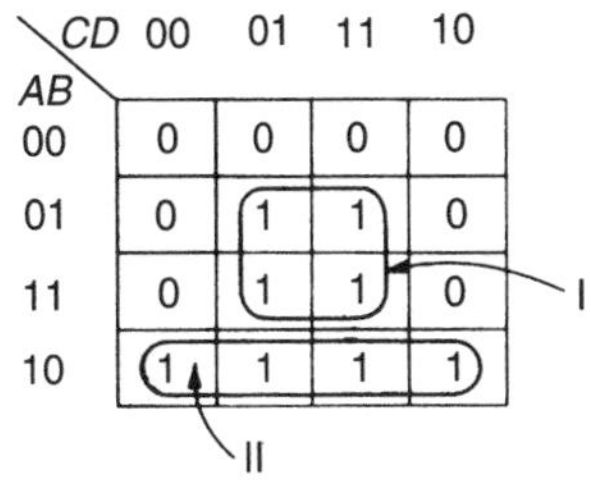

A	B	C	D	P
0	0	0	0	0
0	0	0	1	0
0	0	1	0	0
0	0	1	1	0
0	1	0	0	0
0	1	0	1	1
0	1	1	0	0
0	1	1	1	1
1	0	0	0	1
1	0	0	1	1
1	0	1	0	1
1	0	1	1	1
1	1	0	0	0
1	1	0	1	1
1	1	1	0	0
1	1	1	1	1

CD\AB	00	01	11	10
00	0	0	0	0
01	0	1	1	0
11	0	1	1	0
10	1	1	1	1

P

Fig. 5.27 Truth table and Karnaugh map for a four-input problem.

CD\AB	00	01	11	10
00	0	0	0	0
01	0	1	1	0
11	0	1	1	0
10	1	1	1	1

I

II

Fig. 5.28

There are many other groups that could have been used but this would have resulted in a more complicated Boolean expression.

---

14 (a) Find the three groups of four 1s in the Karnaugh map shown in Fig. 5.29.
  (b) What is the corresponding Boolean expression for the function using the groups that you found in (a)?

Using the boundaries

There is one more property of Karnaugh maps that you need to know. Figure 5.30 shows the truth table for a three-input-variable problem and its corresponding Karnaugh map.

You should be able to see from the truth table that the output, P, is the inverse of C, so the Boolean expression for P is $P = \overline{C}$. In the Karnaugh map, there appear to be two groups, each having a pair of 1s. The first can be identified with the area where $B = 0$ and $C = 0$, and therefore by the product term $\overline{B} \cdot \overline{C}$. The second group can be identified by the area where $B = 1$

CD\AB	00	01	11	10
00	1	1	0	0
01	1	1	0	0
11	0	1	0	0
10	1	1	1	1

P

Fig. 5.29

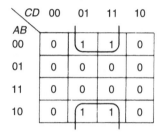

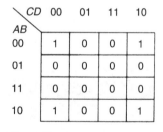

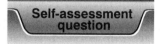

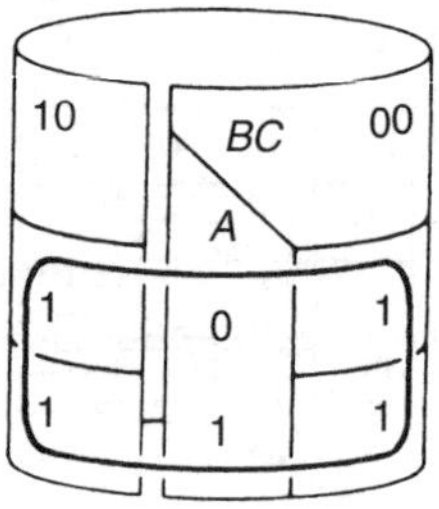

Fig. 5.31 Group of four found by joining the sides of a Karnaugh map.

A	B	C	P
0	0	0	1
0	0	1	0
0	1	0	1
0	1	1	0
1	0	0	1
1	0	1	0
1	1	0	1
1	1	1	0

A \ BC	00	01	11	10
0	1	0	0	1
1	1	0	0	1

P

Fig. 5.30

and $C = 0$ and therefore by the product term $B \cdot \overline{C}$. The full expression for the output is then found as the OR of the two product terms:

$$P = \overline{B} \cdot \overline{C} + B \cdot \overline{C}$$

We know from the previous discussion on Boolean algebra that this expression reduces to

$$P = (\overline{B} + B) \cdot \overline{C} = \overline{C}$$

but why did this not come directly out of the Karnaugh map?

The answer can be seen if the ends of the Karnaugh map are imagined as being joined together. In the example given above, if you join the sides together, the two groups of 1s can be combined to form a square as shown in Fig. 5.31. The four-variable Karnaugh map also has the top and bottom edges, which in theory are joined together. For example, the Karnaugh map in Fig. 5.32 has a single group of four 1s. Note the way that the group is represented, by drawing a box around the four 1s which wraps around at the top and bottom edges. The group can be identified by the area where $B = 0$ and $D = 1$ and therefore has the corresponding product term $\overline{B} \cdot D$.

There is one more group which sometimes appears as a result of the edges being joined, consisting of all four 1s in Fig. 5.33. This group has the coordinates $B = 0, D = 0$ and therefore has the product term $\overline{B} \cdot \overline{D}$. Figure 5.34 shows how the four 1s can still be made to form a rectangular group by joining the corners.

AB \ CD	00	01	11	10
00	0	1	1	0
01	0	0	0	0
11	0	0	0	0
10	0	1	1	0

Fig. 5.32 Group of four found by joining the top and bottom of a Karnaugh map.

AB \ CD	00	01	11	10
00	1	0	0	1
01	0	0	0	0
11	0	0	0	0
10	1	0	0	1

Fig. 5.33 Group of four in the corners of a Karnaugh map.

Self-assessment question

15 (a) Place the following groups of 1s into the Karnaugh map given in Fig. 5.35

Group I: $\overline{A} \cdot \overline{B} \cdot \overline{C} \cdot D$

Group II: $B \cdot \overline{C} \cdot \overline{D}$

Group III: $A \cdot B$

(b) There is a single square in the Karnaugh map in part (a) which can be used to reduce group II to $B \cdot \overline{C}$ by changing its contents from a 0 to a 1. Which square is it, how is it identified and how can it be used to simplify group 1?

5.5.2 Designing a seven-segment decoder using Karnaugh maps

The previous section described how Karnaugh maps can be used to simplify Boolean expressions. The reason we want to simplify the expression is because there is a direct link between the number of product terms in the expression and the number of electronic logic gates needed to implement the corresponding logic function.

There are many different ways in which a logic function can be finally implemented. If individual logic gates are used, one of the most straightforward ways of arranging the gates is to use an AND gate to represent each of the minterms, and the outputs of each AND gate connected to the input of a single OR gate. This form of circuit is often referred to as the *sum-of-minterms circuit*. Reducing the Boolean expression means that fewer AND gates are needed, and the number of inputs to each AND gate is also reduced. In this form, the circuit is known as the *sum-of-products* circuit.

To illustrate the general techniques of logic design using Karnaugh maps, a circuit is now described that will carry out the conversion from 8 : 4 : 2 : 1 BCD to the code needed to generate readable numbers on a **seven-segment display**. This is the familiar pattern used on calculators, voltmeters, clocks and so on. Seven line segments which can be separately illuminated can form recognizable denary numbers when they are arranged as in Fig. 5.36. Thus, to form '1', it is only necessary to illuminate segments b and c, but not a, d, e, f and g. To form '8' it is necessary to illuminate all seven segments, and so on.

Table 5.6 is a truth table where 1 indicates ON and 0 indicates OFF for the segments. The new notation, x, in the output columns of this truth table from row 10 to 15 represents the **don't-care** condition of the output, which indicates that the output can be 0 or 1 for that particular input combination. We can assign any value that we want to these don't-cares because they will never actually be required as there is no individual BCD code for the numbers 10 to 15. If we want to display the numbers 10 to 15 (and up to 99) we would need two

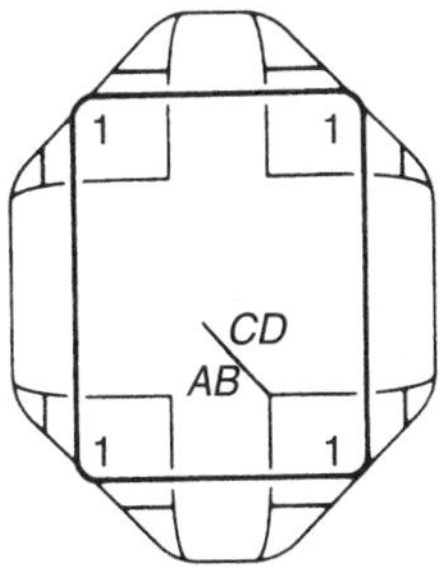

Fig. 5.34 The four corners of a Karnaugh map wrapped around.

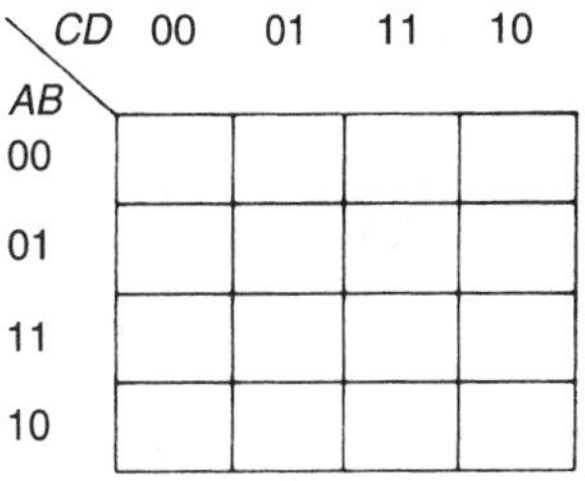

Fig. 5.35 Figure for Self-assessment question 15.

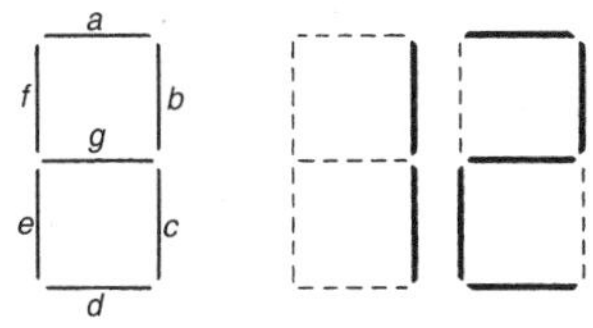

Fig. 5.36 Seven-segment display.

Table 5.6 Truth table for a BCD to seven-segment decoder

Denary digit	*Binary code*	*Segment (on = 1)* a b c d e f g
0	0 0 0 0	1 1 1 1 1 1 0
1	0 0 0 1	0 1 1 0 0 0 0
2	0 0 1 0	1 1 0 1 1 0 1
3	0 0 1 1	1 1 1 1 0 0 1
4	0 1 0 0	0 1 1 0 0 1 1
5	0 1 0 1	1 0 1 1 0 1 1
6	0 1 1 0	0 0 1 1 1 1 1
7	0 1 1 1	1 1 1 0 0 0 0
8	1 0 0 0	1 1 1 1 1 1 1
9	1 0 0 1	1 1 1 0 0 1 1
No meaning	1 0 1 0	× × × × × × ×
for 8:4:2:1	1 0 1 1	× × × × × × ×
BCD	1 1 0 0	× × × × × × ×
	1 1 0 1	× × × × × × ×
	1 1 1 0	× × × × × × ×
	1 1 1 1	× × × × × × ×

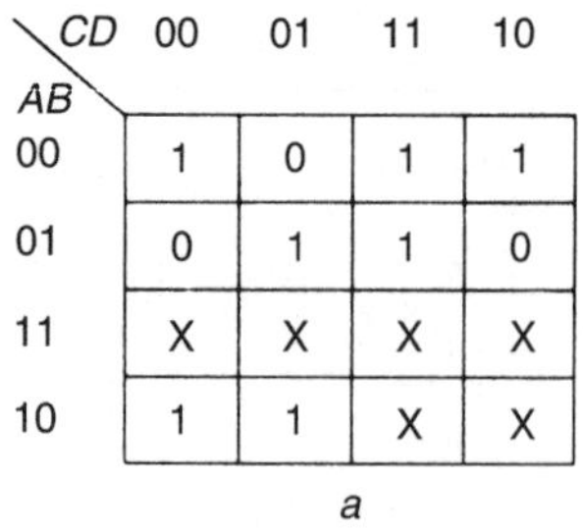

Fig. 5.37 Output *a*

separate seven-segment digits, each of which will only ever be required to display the numerals 0 to 9.

A don't-care can be assigned as a 0 *or* a 1, so that don't-cares can be used in a Karnaugh map to reduce a Boolean expression by assigning 1s to any square which would help to make up a rectangular group of 1s, while setting the remaining don't-cares to 0.

There are several ways to design the seven-segment decoder, each with its own merits. As the decoder is being used as an example to show how don't-cares are handled, a different circuit for each of the outputs will be illustrated. The first step is to construct separate Karnaugh maps for each output. The first Karnaugh map, for the output a, is shown in Fig. 5.37.

The aim is to reduce the Boolean expression for the output a, which can be achieved by grouping the 1s together into rectangular arrangements of 2, 4, 8 or 16. Bear in mind that:

- The larger the group the fewer input variables needed to locate that group, and
- We want to end up with the smallest number of groups.

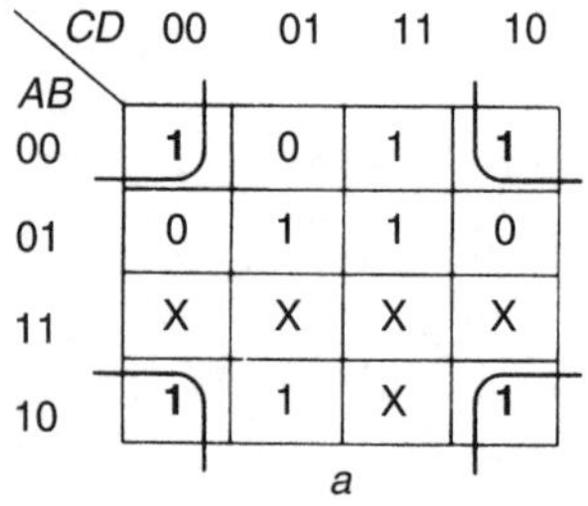

Fig. 5.38 Product term $\overline{B} \cdot \overline{D}$.

Groups can be formed anywhere, so start with the 1 in the top left-hand corner. This 1 can be made part of a group of four by including all of the four corners. Three of the corners are already 1, only the bottom right-hand corner is a don't-care. This don't-care can be assigned to a 1, as shown in Fig. 5.38.

Worked example

What is the product term that identifies the four corners?

You can see by inspection that this is $\overline{B} \cdot \overline{D}$.

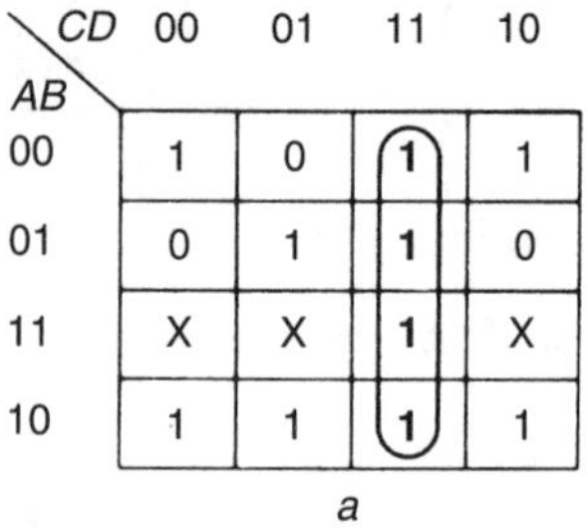

Fig. 5.39 Product term $C \cdot D$.

One of the don't-care terms has now been set to 1. *Having been set, this don't-care term has to stay at that value for as long as we are in the process of finding the Boolean expression for a*. The other don't-cares in this output have yet to be assigned values. The circuits that implement all of the other segments, b to g, are going to be designed separately, so the don't-cares in their respective output columns are unaffected.

Next, look at the 1 corresponding to the term $\overline{A} \cdot \overline{B} \cdot C \cdot D$. It can be placed in a group of four 1s which consists of the column under the coordinates $C = 1$ and $D = 1$, which therefore has the corresponding product term $C \cdot D$. To do this the two don't-care terms can be assigned to a value of 1, as shown in Fig. 5.39.

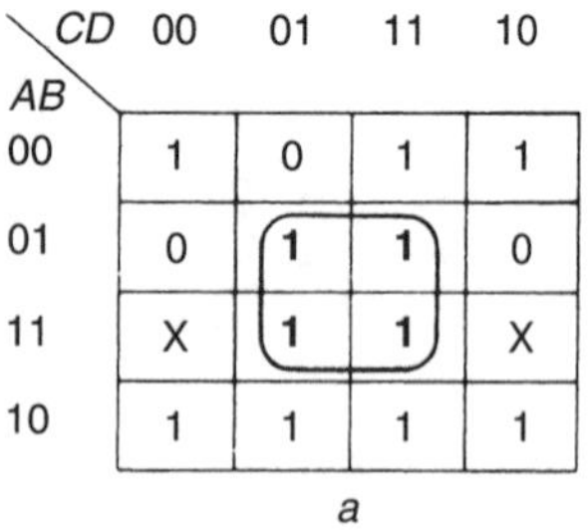

Fig. 5.40 Product term $B \cdot D$.

Next, the 1 at coordinates $\overline{A} \cdot B \cdot \overline{C} \cdot D$ can be made part of a group of four 1s as shown in Fig. 5.40. This requires one more of the don't-care terms to be assigned the value of 1. Remember that only rectangular groups of 2, 4, 8 or 16 are permitted. The product term that corresponds to this group is $B \cdot D$, since the middle two rows are selected when $B = 1$ and the middle two columns are selected when $D = 1$. The combination of $B = 1$ and $D = 1$ therefore isolates the middle four 1s.

Finally, the last row of the Karnaugh map still contains one of the original 1s from the truth table (at $A \cdot \overline{B} \cdot \overline{C} \cdot D$) which has not yet been incorporated into a group. This can be included in a group of eight 1s by assigning the final two don't-care terms to a value of 1, and making a group out of the bottom two rows, as shown in Fig. 5.41. This group has a corresponding product term which consists of the single variable A. This says that the output a is 1 whenever the input A is 1.

The complete Boolean expression is found by taking the OR of all these product terms:

$$a = A + B \cdot D + \overline{B} \cdot \overline{D} + C \cdot D$$

In deriving this expression I have assigned a value of 1 to the don't-care terms but in other cases it could have been a 0. There are several possible solutions to this problem, but this Boolean expression is probably the simplest one that could have been derived. If we were to implement this expression using AND, OR and INVERTER gates, the result would look like Fig. 5.42.

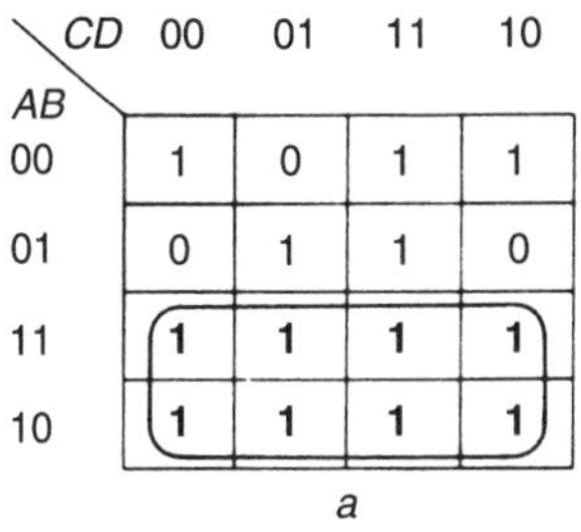

Fig. 5.41 Product term A.

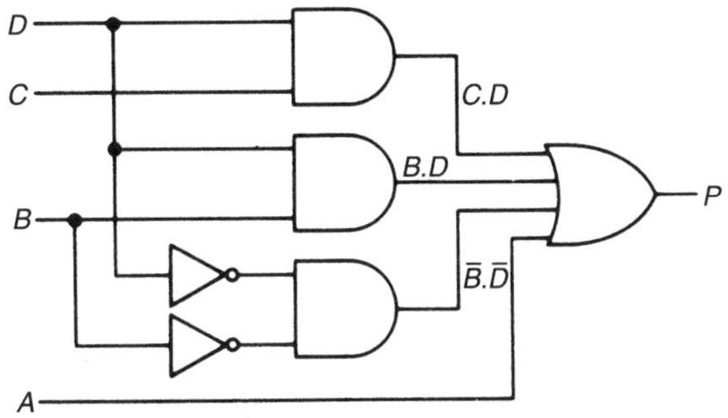

Fig. 5.42 An implementation for segment a.

Designing circuits for the other sections provides useful practice in these techniques of using a Karnaugh map. One result is as follows.

$$a = A + B \cdot D + \overline{B} \cdot \overline{D} + C \cdot D$$
$$b = \overline{B} + C \cdot D + \overline{C} \cdot \overline{D}$$
$$c = B + \overline{C} + D$$
$$d = \overline{B} \cdot C + \overline{B} \cdot \overline{D} + B \cdot \overline{C} \cdot D + C \cdot \overline{D}$$
$$e = C \cdot \overline{D} + \overline{B} \cdot \overline{D}$$
$$f = A + B \cdot \overline{C} + B \cdot \overline{D} + \overline{C} \cdot \overline{D}$$
$$g = A \cdot \overline{B} + B \cdot \overline{C} + \overline{B} \cdot C + C \cdot \overline{D}$$

If possible, use computer simulation to check your circuits against the truth table, Table 5.6.

Having seen how the circuit for one of the segments was designed, you should now be able to tackle the other six in Table 5.6. Remember that each one is being designed separately, so the fact that values have been assigned to the don't-cares for segment a does not affect the don't-cares for the other segments: they can be assigned to any value that you choose.

Self-assessment question

16 The Boolean expression for the output c in the BCD-to-seven-segment display converter can be reduced to

$$c = B + \overline{C} + D$$

This can be achieved by setting all except one of the don't-care terms to 1 in the Karnaugh map. Construct the Karnaugh map, and find which don't-care term to set to 0.

5.5.3 The inverted Karnaugh map

The Karnaugh map provides a good way of simplifying a Boolean expression by grouping the 1s together. In some cases, further simplification can be achieved if the inverse of the output is considered, particularly when the number and pattern of 0s in the map are simpler than the 1s. In the case of the BCD-to-seven-segment decoder there are fewer 0s in the output columns of the truth table than there are 1s. It is therefore possible that the Boolean expressions for the inverse of the outputs might be simpler than the expressions for the actual outputs.

For example, in Self-assessment question 16, the simplest Boolean expression for c was found by setting all except one of the don't-care terms to 1. If this time we look at the Karnaugh map of $\overline{c}$, the task becomes much simpler. The truth table and Karnaugh map for $\overline{c}$ are shown in Fig. 5.43. Initially there is only one 1 in the map which corresponds to the term $\overline{A} \cdot \overline{B} \cdot C \cdot \overline{D}$. As you can see, the expression for this single 1 has all four input variables. However, by setting the don't-care term in the bottom right-hand corner to 1, a group of two can be formed which has the corresponding product term $\overline{B} \cdot C \cdot \overline{D}$. Assuming all other don't-cares are 0, the expression for $\overline{c}$ can be reduced to

$$\overline{c} = \overline{B} \cdot C \cdot \overline{D}$$

A	B	C	D	$\overline{c}$
0	0	0	0	0
0	0	0	1	0
0	0	1	0	1
0	0	1	1	0
0	1	0	0	0
0	1	0	1	0
0	1	1	0	0
0	1	1	1	0
1	0	0	0	0
1	0	0	1	0
1	0	1	0	X
1	0	1	1	X
1	1	0	0	X
1	1	0	1	X
1	1	1	0	X
1	1	1	1	X

AB \ CD	00	01	11	10
00	0	0	0	1
01	0	0	0	0
11	X	X	X	X
10	0	0	X	X

$\overline{c}$

Fig. 5.43 Truth table and Karnaugh map for $\overline{c}$.

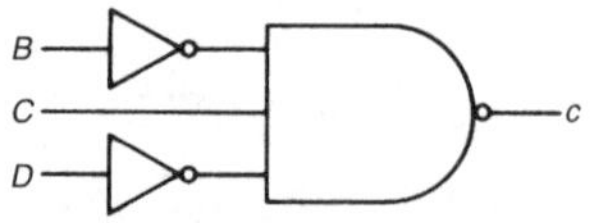

Fig. 5.44 An implementation for segment c.

To implement c, you would use a three-input AND gate to produce $\overline{c}$, and then invert the output. In other words, you would use a three-input NAND gate, as shown in Fig. 5.44.

In order to get a Boolean expression for c, the expression for $\overline{c}$ is inverted on both sides. Applying deMorgan's theorem to the above expression for $\overline{c}$ gives

$$c = \overline{\overline{B} \cdot C \cdot \overline{D}} = B + \overline{C} + D$$

which was the expression found earlier in Self-assessment question 16.

There are two points to note from this example. The first is that the simplified Boolean expression that we derived for $\overline{c}$ is the inverse of the simplified Boolean expression that we found for c. This may seem to be inevitable, but in fact it is not. The choice of values for the don't-cares for an output and for its inverse can sometimes be different.

The second point is that the final implementation of the inverted output is not simpler than the original non-inverted output (you end up using a three-input NAND gate instead of a three-input OR gate). The derivation, however, was certainly a lot easier since there were only two 1s in the Karnaugh map.

In this example we inverted the output and then grouped the 1s together. An alternative would have been to have used the Karnaugh map of the non-inverted output which we already had and grouped the 0s together. The Boolean expression for the groups of 0s is the same as the expression for the inverted output, so this is a quicker way than actually inverting the truth table. The Karnaugh map is then known as an inverted Karnaugh map since the resulting Boolean expression represents the inverted output.

To illustrate this, consider the truth table and Karnaugh map for g shown in Fig. 5.45. The first group of 0s that can be formed is the pair in the top left-hand corner. This group corresponds to the product term $\overline{A} \cdot \overline{B} \cdot \overline{C}$. The remaining 0 can be made into another group of two by making the don't-care term immediately below it a 0. Thus the group corresponding to the product term $B \cdot C \cdot D$ is formed. All the other don't-cares are set to 1. The final expression for $\overline{g}$ is therefore

$$\overline{g} = \overline{A} \cdot \overline{B} \cdot \overline{C} + B \cdot C \cdot D$$

Again, it is not difficult to show that the distribution of 0s and 1s in the resulting Karnaugh map is identical to the one used to get the Boolean expression for g in the previous section. Consequently, the simplified expression for $\overline{g}$ is the inverse of the simplified expression for g. As stated earlier, this is not generally true where don't-cares are involved but does often happen.

This new expression for $\overline{g}$ has only two product terms instead of the four that were present in the expression for g, so we have succeeded in simplifying our expression. If we were to implement this function using the sum-of-products circuit then we would have to include an INVERTER on the output, since our expression is for $\overline{g}$ and we want g. This is shown in Fig. 5.46.

One set of simplified expressions for the inverse of each of the segments is

$$\overline{a} = \overline{A} \cdot \overline{B} \cdot \overline{C} \cdot D + B \cdot \overline{D}$$

$$\overline{b} = B \cdot C \cdot \overline{D} + B \cdot \overline{C} \cdot D$$

$$\overline{c} = \overline{B} \cdot C \cdot \overline{D}$$

$$\overline{d} = B \cdot C \cdot D + \overline{B} \cdot \overline{C} \cdot D + B \cdot \overline{C} \cdot \overline{D}$$

$$\overline{e} = B \cdot \overline{C} + D$$

$$\overline{f} = C \cdot D + \overline{A} \cdot \overline{B} \cdot D + \overline{B} \cdot C$$
$$\overline{g} = B \cdot C \cdot D + \overline{A} \cdot \overline{B} \cdot \overline{C}$$

A	*B*	*C*	*D*	*g*
0	0	0	0	0
0	0	0	1	0
0	0	1	0	1
0	0	1	1	1
0	1	0	0	1
0	1	0	1	1
0	1	1	0	1
0	1	1	1	0
1	0	0	0	1
1	0	0	1	1
1	0	1	0	X
1	0	1	1	X
1	1	0	0	X
1	1	0	1	X
1	1	1	0	X
1	1	1	1	X

AB \ *CD*	00	01	11	10
00	0	0	1	1
01	1	1	0	1
11	X	X	X	X
10	1	1	X	X

g

Fig. 5.45

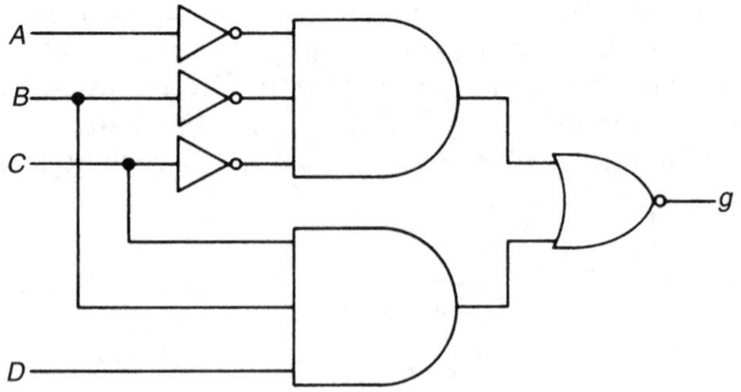

Fig. 5.46 An implementation for segment *g*.

The two methods for grouping 1s or 0s can be summarised as:

1. The don't-cares in the Karnaugh map of an output, P, can be assigned values of 0 or 1. Usually the choice is made to create groups of 1s that give the simplest Boolean expression for P.
2. Alternatively, the don't-cares can be assigned values of 0 or 1 in order to make groups of 0s and hence produce the simplest Boolean expression for the inverse of the output, $\overline{P}$.

Self-assessment question

17 Use the 0s in the Karnaugh maps of b and f shown in Fig. 5.47 to derive the expressions for $\overline{b}$ and $\overline{f}$ given above.

AB \ *CD*	00	01	11	10
00	1	1	1	1
01	1	0	1	0
11	X	X	X	X
10	1	1	X	X

b

AB \ *CD*	00	01	11	10
00	1	0	0	0
01	1	1	0	1
11	X	X	X	X
10	1	1	X	X

f

Fig. 5.47 Figure for Self-assessment question 17.

5.5.4 NAND gate representation

The Boolean expressions for the outputs of the BCD-to-seven-segment decoder could be implemented using the standard sum-of-products circuit. The implementation therefore requires several AND, OR and INVERTER gates. It is possible, however, to implement the same sum-of-products circuits using only NAND gates. To be more general, *all* Boolean expressions can be implemented using *only* NAND gates. To be such a basic building block, the NAND gate has to be able to replace all the AND gates, INVERTERs and OR gates in a sum-of-products circuit.

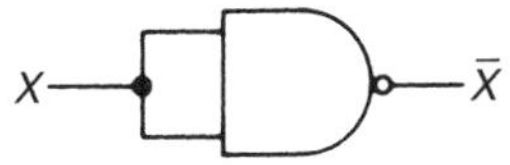

Fig. 5.48 NAND gate connected as an inverter.

Worked example

How can a NAND gate be used as an INVERTER?

If you connect all the inputs of a NAND gate together so that there is a single input, the NAND gate will act as an INVERTER. Figure 5.48 shows an example of a two-input NAND gate connected this way.

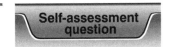

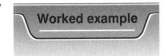

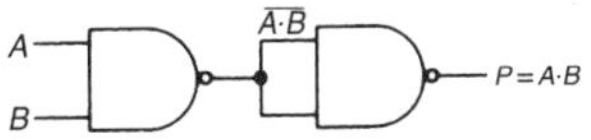

Fig. 5.49 AND gate formed from two NAND gates.

An AND gate can be constructed from a NAND gate by inverting its output. Since we have shown that a NAND gate can act as an INVERTER, an AND gate can be made using two NAND gates, as shown in Fig. 5.49.

Next, the OR gate. As we saw earlier, deMorgan's theorem gives the relationship between the OR and the AND functions. It can be summarized by the expression

$$\overline{A+B} = \overline{A} \cdot \overline{B}$$

Inverting both sides gives

$$A+B = \overline{\overline{A} \cdot \overline{B}}$$

In other words, it is possible to exchange AND gates and OR gates with suitable use of INVERTERs. An OR gate can be replaced by a NAND gate with INVERTERs on each of its inputs. This is shown in Fig. 5.50

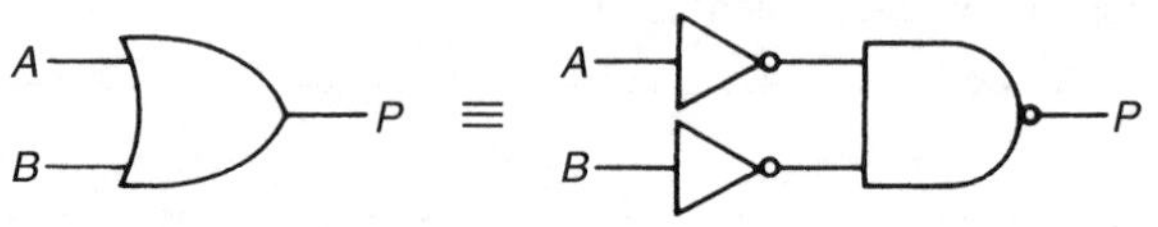

Fig. 5.50 OR gate formed from a NAND gate and two inverters.

We know that any Boolean expression can be implemented using the sum-of-products circuit, so all that remains is to show that the sum-of-products circuit can be implemented using only NAND gates. Figure 5.51a shows the sum-of-products implementation of a particular Boolean expression. Figure 5.51b shows

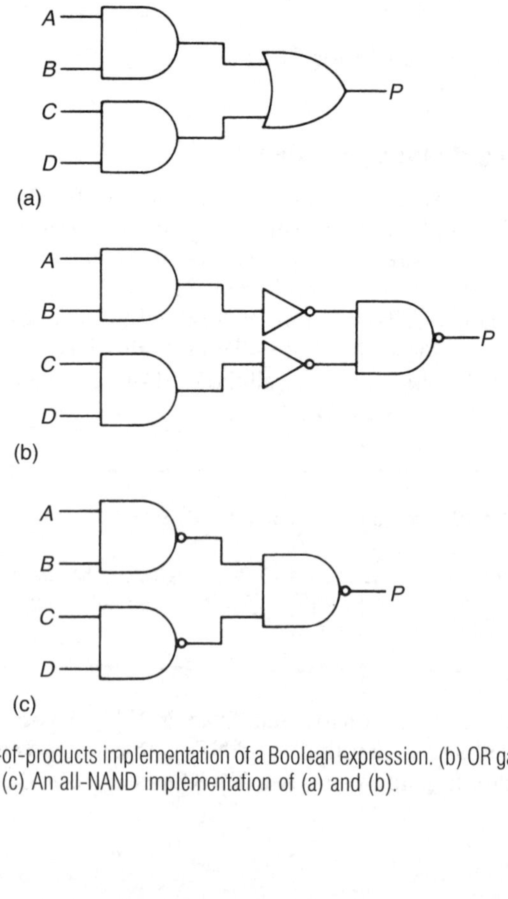

Fig. 5.51 (a) Sum-of-products implementation of a Boolean expression. (b) OR gate in (a) replaced by NAND and inverters. (c) An all-NAND implementation of (a) and (b).

the same circuit with the OR gate replaced with a NAND gate and two INVERTERs. Finally, Fig. 5.51c shows that the combination of INVERTERs and AND gates can replaced by NAND gates.

It should be noted that all Boolean expressions can also be implemented just by NOR gates. However, this is not as common as NAND gate implementation.

---

18 Implement the EXCLUSIVE-OR function for two variables using only NAND gates.

Self-assessment question

5.5.5 Summary of Section 5.5

1. The Karnaugh map is a convenient method of representing a logic problem that enables simplifications to be more easily spotted. Some of the important points to remember are:
   - Each square in the Karnaugh map corresponds to a line in the truth table.
   - The coordinates of a square that contains a 1 correspond to a minterm in a Boolean expression.
   - A rectangular cluster of squares corresponds to a product term in a simplified Boolean expression.
   - Coordinates around a Karnaugh map are arranged so that only one variable at a time changes between adjacent columns.
   - Edges and corners 'wrap around' to behave as if they were complete groups.
   - Don't-care terms can be assigned a value to make up a complete group.
   - An inverted map in which the 0s are grouped may give a simpler answer than the conventional map in which 1s are grouped.
2. Any logic design can be implemented either completely from NAND gates or completely from NOR gates. The NAND gate representation is much the more common.

5.6 ELECTRONIC COMBINATIONAL LOGIC

Previous sections in this chapter have looked at the way combinational problems can be described in binary digital form and specified by either truth tables or Boolean expressions. Some electronic integrated circuits (ICs) are now introduced that can be used to provide practical solutions to such problems. The number of different integrated circuits to be found in manufactures' catalogues is extensive, and new types of device are continually being produced. The aim here is to sample the information that manufacturers provide and help you to interpret it.

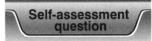

Logic circuits can be based on bipolar transistors or field-effect transistors. The most common group of **bipolar** logic integrated circuits is known as **TTL** (transistor–transistor logic), a term which reflects the basic circuit configurations within them. The biggest group that uses field-effect transistors is known as **CMOS** (from complementary metal-oxide semiconductor (MOS)

transistor). A great variety of simple and complex logic functions is available in both these technologies.

All digital integrated circuit logic components could be built using NAND gates. Hence a useful measure of the relative complexity of a digital integrated circuit is its **gate-equivalence**, which is the number of individual logic gates that would be interconnected to perform the same function provided by the IC. Common abbreviations for integrated circuits of different complexity are:

SSI (small scale integration)	up to about 20 gate equivalents
MSI (medium scale integration)	about 20 to 200 gate equivalents
LSI (large scale integration)	about 200 to 100 000 gate equivalents

Above this level the two terms **VLSI** (very large scale integration) and **ULSI** (ultra large scale integration) are rather more vaguely defined, and are often based on the number of transistors rather than gates. In 2002 the most complex ICs (microprocessors and memories) can have more than 100 million transistors in them.

5.6.1 Some SSI devices

Many manufacturers produce ranges of CMOS and bipolar logic devices in standard configurations known as the 74-series, and some examples of SSI devices in this series are outlined in Fig. 5.52. The logical function of each

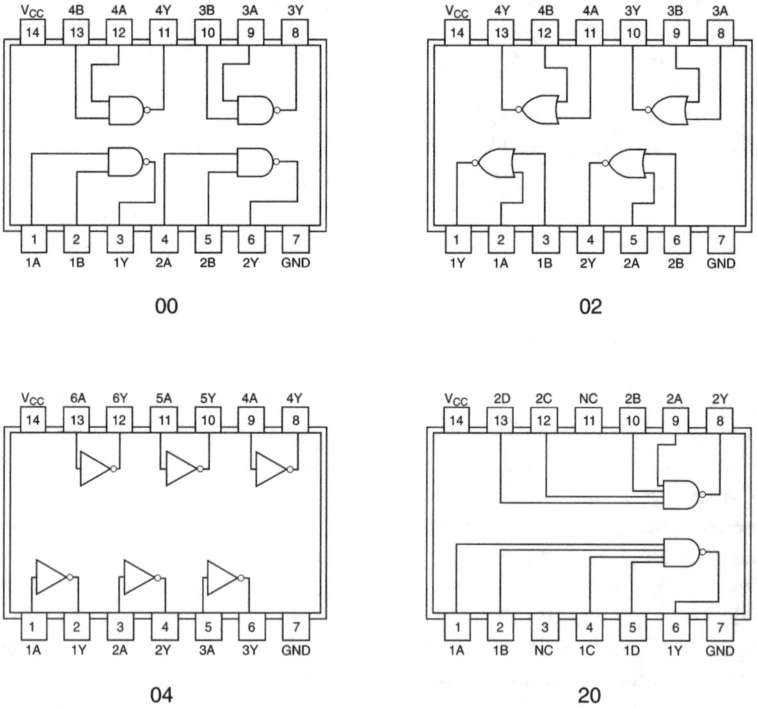

Fig. 5.52 Schematic diagrams of some SSI devices from the 74-series, showing the logic functions and their connections to the pins of a standard 14-pin dual-in-line package.

device is represented by a numerical code. Thus '00' devices contain four two-input NAND gates, '02' devices contain four two-input NOR gates, '04' devices contain six INVERTERs, and '20' devices contain two four-input NAND gates. The figure shows these functional arrangements of gates drawn within a standard 14-pin 'dual-in-line' package. There are over one thousand other functional codes in the 74-series, covering a wide range of SSI and MSI functions, and these are packaged in a variety of sizes and styles.

The full type number for a particular 00 device might be SN74LS00, DM74HC00 etc. where each group of letters or numbers gives more information about the origin and performance of the device. Thus the first two letters (SN, DM etc.) usually represent the manufacturer, and '74' confirms that its function is from the 74-series. (Note that there is another 54-series, that differs from the 74-series mainly in the temperature range over which it is specified to operate correctly.) The next group of letters are the 'technology' or 'family', for example LS for 'low power Schottky bipolar', and HC for 'high-performance CMOS'. The family is a very important aspect which indicates the overall electrical characteristics of the device, such as operating speed and power dissipation. After the function code ('00' in this case, meaning a combination of four two-input NAND gates) there may be a final letter defining the physical package, which is any one of a number of closely defined standard shapes and sizes.

The full type number is rarely needed until specific devices are to be purchased. For design purposes a simplified number such as 74LS04 or 74HC20 conveys all the important information. When describing just the function, the number can be further abbreviated to 74x04 etc.

Notice from the diagrams in Fig. 5.52 that, as well as connections to the gates, each integrated circuit (IC) has two power connections: GND for 0 V and V_{CC} for the positive 5 V power supply. Details of the performance characteristics of some typical CMOS and bipolar logic families are given in Section 5.7.

The following terms are commonly used to describe multiple devices in one IC package. Two devices, **dual**. Three devices, triple. Four devices, quadruple or **quad**. Six devices, **hex**. Eight devices, octal. Thus the 74LS00 is known as a 'quad NAND gate' package, and the 74LS04 as a 'hex inverter' chip.

Self-assessment question

19 Figure 5.53 shows a circuit that implements the relationship

$P = A \cdot B + \overline{A} \cdot C$

Show how one of the 74x00 packages in Fig. 5.52 could be connected to produce this function.

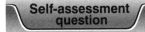

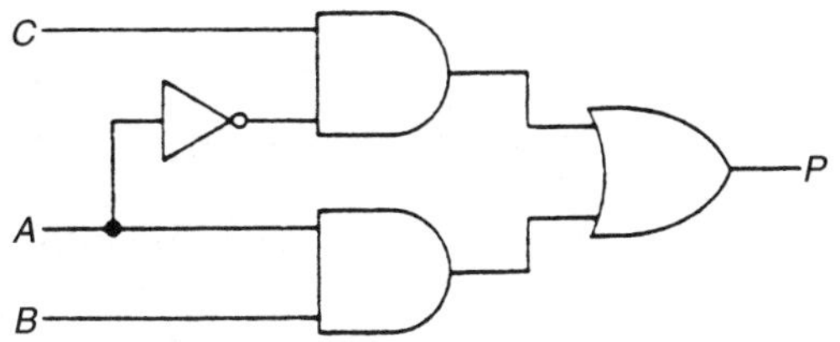

Fig. 5.53 Self-assessment question 19.

Self-assessment question 19 shows one of the advantages of designing using only NAND gates, which is that sometimes fewer standard 74-series packages can be used. We could have implemented *P* using two AND gates, an OR gate and an INVERTER, which exist on different packages. We would therefore require three separate packages (and several wasted gates) instead of the one package needed for the NAND gate design. However, note that some gates are also available in one-gate, two-gate and three-gate packages for applications which have limited requirements, and the use of such devices might be a good alternative in some cases.

5.6.2 MSI devices

The BCD-to-seven-segment decoder

Whenever the point is reached in the design process where a problem is ready to be implemented, it is always worth checking to see whether an integrated circuit already exists that can perform to the required specification. The BCD-to-seven-segment decoder in Section 5.5.2 was implemented by connecting individual gates together. However, a quick look at the manufacturers' literature reveals several integrated circuits that are available for this task, and one suitable device is referred to as the 74LS47. The 74 again indicates that it is in the standard 74-series numbering scheme, and the 47 is the reference number to its particular function, BCD-to-seven-segment code conversion. As before, the letter code LS refers to the low power Schottky family within the 74-series.

Some excerpts from the manufacturer's data sheet for the 74LS47 are shown in Fig. 5.54. Note the following points.

(a) What you can actually measure on the inputs and outputs of the device are voltage levels which are either high or low. In the positive logic convention these high and low levels are defined to be 1 and 0, respectively. However, the positive logic convention is not always used, so to be perfectly general the letters H for high and L for low are normally used in descriptions of device function.
(b) The 74LS47 is designed for displays which require a low logic level input to turn a segment on. Hence the segment outputs from this IC are connected through inverters, and an L in an output column means 'on'. The position of the seven segments labelled a to g are as shown in Fig. 5.36.
(c) The 74LS47 logic diagram (in Fig. 5.54) shows the arrangement of AND and NOR gates used to decode the BCD input into the correct segments. (If you cannot find the NOR gates, then look back to Fig. 5.16 for a hint!) There is also some NAND gate logic to deal with 'ripple-blanking' inputs and outputs, which automatically turn off 'leading zeros' in a multi-digit display. It is clear from the definitions given in the introduction to this section that the BCD-to-seven-segment decoder would be classified as an MSI device, having between 20 and 200 gate equivalents.
(d) The truth table and the 'numerical designations - resultant displays' diagram describe the symbols that would be displayed for inputs corresponding to the numbers 0–15. The codes 10–15, of course, should never arise with BCD inputs, and hence the form of these extra symbols is not important in most applications.

LOGIC DIAGRAM

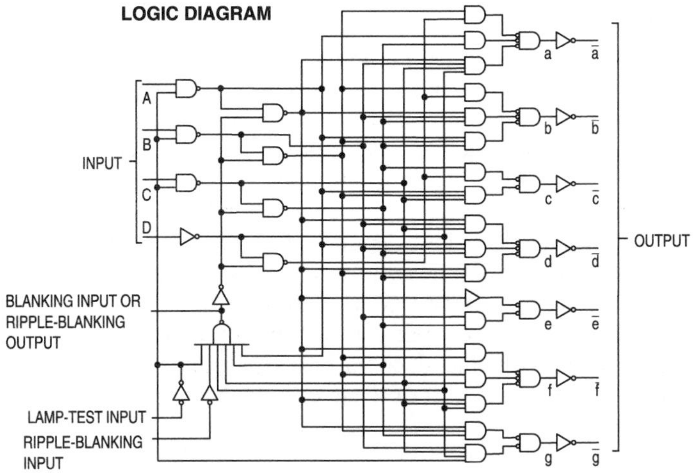

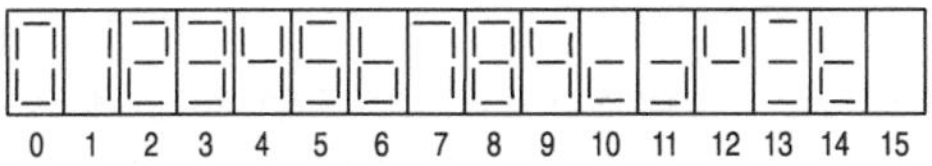

NUMERICAL DESIGNATIONS Ð RESUL TANT DISPLAYS

TRUTH TABLE

DECIMAL OR FUNCTION	INPUTS							OUTPUTS							NOTE
	$\overline{LT}$	$\overline{RBI}$	D	C	B	A	$\overline{BI/RBO}$	$\bar{a}$	$\bar{b}$	$\bar{c}$	$\bar{d}$	$\bar{e}$	$\bar{f}$	$\bar{g}$	
0	H	H	L	L	L	L	H	L	L	L	L	L	L	H	A
1	H	X	L	L	L	H	H	H	L	L	H	H	H	H	A
2	H	X	L	L	H	L	H	L	L	H	L	L	H	L	
3	H	X	L	L	H	H	H	L	L	L	L	H	H	L	
4	H	X	L	H	L	L	H	H	L	L	H	H	L	L	
5	H	X	L	H	L	H	H	L	H	L	L	H	L	L	
6	H	X	L	H	H	L	H	H	H	L	L	L	L	L	
7	H	X	L	H	H	H	H	L	L	L	H	H	H	H	
8	H	X	H	L	L	L	H	L	L	L	L	L	L	L	
9	H	X	H	L	L	H	H	L	L	L	H	H	L	L	
10	H	X	H	L	H	L	H	H	H	H	L	L	H	L	
11	H	X	H	L	H	H	H	H	H	L	L	H	H	L	
12	H	X	H	H	L	L	H	H	L	H	H	H	L	L	
13	H	X	H	H	L	H	H	L	H	H	L	H	L	L	
14	H	X	H	H	H	L	H	H	H	H	L	L	L	L	
15	H	X	H	H	H	H	H	H	H	H	H	H	H	H	
$\overline{BI}$	X	X	X	X	X	X	L	H	H	H	H	H	H	H	B
$\overline{RBI}$	H	L	L	L	L	L	L	H	H	H	H	H	H	H	C
$\overline{LT}$	L	X	X	X	X	X	H	L	L	L	L	L	L	L	D

H = HIGH Voltage Level
L = LOW Voltage Level
X = Immaterial

NOTES:

(A) $\overline{BI/RBO}$ is wire-AND logic serving as blanking Input (BI) and/or ripple-blanking output ($\overline{RBO}$). The blanking out (BI) must be open or held at a HIGH level when output functions 0 through 15 are desired, and ripple-blanking input ($\overline{RBI}$) must be open or at a HIGH level if blanking of a decimal 0 is not desired. X = input may be HIGH or LOW.

(B) When a LOW level is applied to the blanking input (forced condition) all segment outputs go to a LOW level regardless of the state of any other input condition.

(C) When ripple-blanking input ($\overline{RBI}$) and inputs A, B, C, and D are at LOW level, with the lamp test input at HIGH level, all segment outputs go to a HIGH level and the ripple-blanking output (RBO) goes to a LOW level (response condition).

(D) When the blanking input/ripple-blanking output ($\overline{BI/RBO}$) is open or held at a HIGH level, and a LOW level is applied to lamp test input, all segment outputs go to a LOW level.

Fig. 5.54 Function table and functional diagram for a 74LS47 BCD-to-seven-segment decoder. Copyright of Semiconductor Components Industries, LLC. Used by permission.

Binary decoders

Probably the most common MSI combinational device is the binary decoder, and a **function table** for one of the simplest is shown in Table 5.7. Here there are three inputs and four outputs. Input $\overline{G}$ is called an active-low enable input, simply because it has to be set low in order for the circuit to carry out its decoding function. When $\overline{G}$ is high, then all the outputs are high, as represented in line 1 of Table 5.7. However, when $\overline{G}$ is low, then one and only one of the outputs becomes low, depending on the binary code applied to the 'select' inputs A and B. As shown in lines 2 to 5 of Table 5.7, when both these inputs are low then output $Y0$ is low, when B is low and A is high then $Y1$ is low, and so on through the four possible combinations of B and A. Thus the function of this 2-line to 4-line decoder is to separate the four possible cases of a 2-bit binary input code. Each of these outputs corresponds to a minterm of the input code.

Table 5.7 Function table for a 2-line to 4-line binary decoder

Inputs						
	Select		*Outputs*			
$\overline{G}$	B	A	Y0	Y1	Y2	Y3
H	X	X	H	H	H	H
L	L	L	L	H	H	H
L	L	H	H	L	H	H
L	H	L	H	H	L	H
L	H	H	H	H	H	L

The 2-line to 4-line decoder in Table 5.7 requires a total of seven data connections. The 74-series device 74x139 consists of two of these decoders in a standard 16-pin package (with two pins used, of course, for the power supplies GND and V_{CC}).

If it is required to decode 3 select inputs, then there should be a total of eight decoded outputs. A common 74-series device which does this is the 3-line to 8-line decoder 74x138, which fits three select inputs (C, B and A), eight outputs ($Y0$ to $Y7$), three enable inputs, and the power connections into a 16-pin package. Devices like this have very wide applications whenever specific values of a binary code need to be selected for a particular purpose.

5.6.3 Programmable logic devices

Large combinational logic problems can be solved by connecting many SSI and MSI devices together, but cheaper and smaller solutions can usually be found by using some form of programmable device. The term *programmable* in this context means that initially the devices have no specified logical function. You, as the user, define the function and then alter the device in some way so that it conforms to your specification. There are very many forms of programmable logic device (**PLD**), and these are generally divided into simple PLDs (SPLDs) and complex PLDs (CPLDs). This section will cover two types of SPLD, namely **programmable array logic (PAL) devices** and **programmable read-only memory (PROM)**. As you will see, the PAL devices and PROM take very different approaches to the representation of combinational problems.

Programmable array logic (PAL) devices

The principle of the PAL device can be best illustrated by a simple example. Figure 5.55 shows a two-input single-output device consisting of two AND gates, an OR gate and eight fusible links F1 to F8. (Notice that on each input there is a symbol which indicates that the input and its inverse are available and can be connected to either of the two AND gates.) The device is programmed by applying electrical signals to the pins to 'blow' those fusible links which are not required.

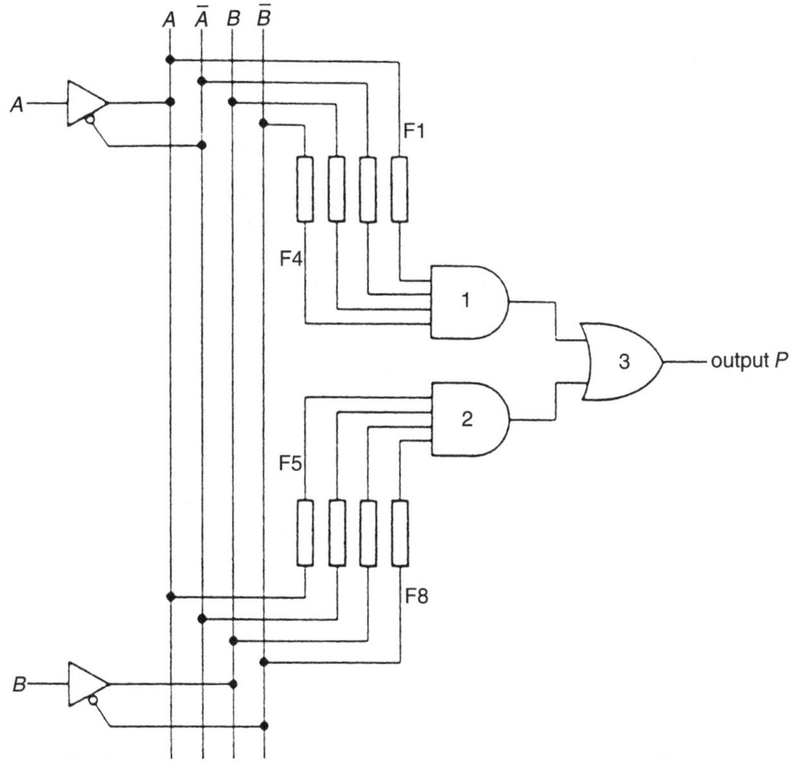

Fig. 5.55 Logic diagram for a simple PAL device.

Let us consider the problem of how to use this device to implement the EXCLUSIVE-OR relationship

$$P = A \cdot \overline{B} + \overline{A} \cdot B$$

The first thing that we need to know is whether the links (F1 to F8) are to be open or closed. Initially, before we blow any of the links, we can assume that they are all closed. We can use AND gate 1 to provide the first product term of P and gate 2 to provide the second. If we open or blow links F2 and F3 then the output of gate 1 will be $A \cdot \overline{B}$. Likewise if we open links F5 and F8 the output of gate 2 will be $\overline{A} \cdot B$.

Notice that the disconnected AND gate inputs of this device can be assumed to be HIGH, that is at a logic 1. This is sometimes referred to as 'floating high', and is a consequence of the circuitry used inside PAL chips.

The output of the OR gate is now the desired function P and is equivalent to the circuit shown in Fig. 5.56.

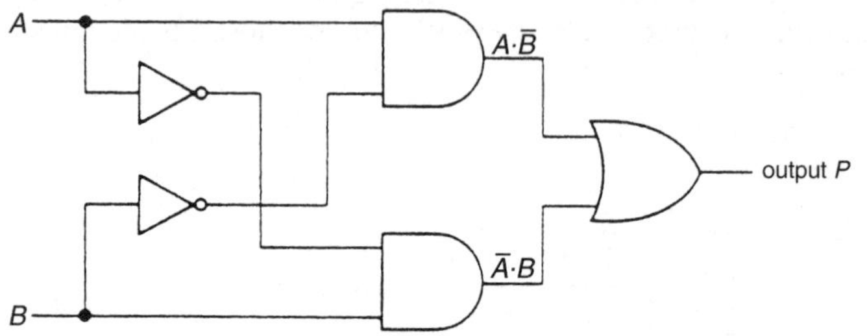

Fig. 5.56 An implementation of the XOR function.

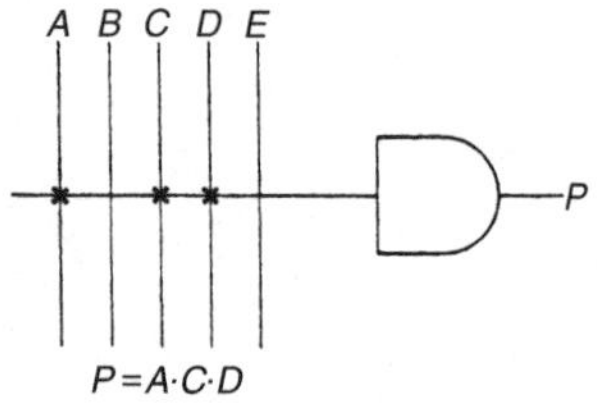

Fig. 5.57 Shorthand notation for internal PAL device connections.

PAL devices follow a basic structure like this, but have more inputs and outputs and represent many more product terms. Hence, a shorthand notation has been developed to help the designer when it comes to remembering which fuses have to be blown. Each AND gate is shown with only a single input, which is meant to represent all of its inputs. Only the fuses that are not blown are shown on the diagram as a cross, as in Fig. 5.57, which shows a five-input gate that implements the function

$$P = A \cdot C \cdot D$$

Each cross on the diagram represents one closed link used to connect a circuit input to one of the AND gate inputs. The other inputs on this line are disconnected from the gate once the fuses have been blown.

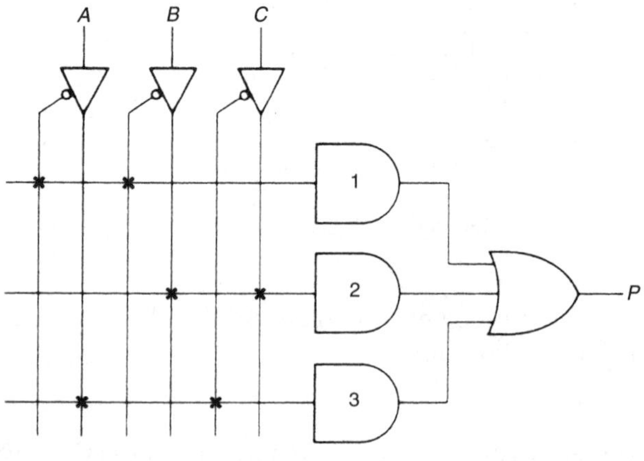

Fig. 5.58 Self-assessment question 20.

Self-assessment question

20 Derive the Boolean expression for the output *P* of the logic diagram shown in Fig. 5.58.

Figure 5.59 shows the internal structure of one particular PAL device, the PAL16L8. The code for the device indicates that there are up to 16 inputs and 8 outputs. Look first at the nine inputs (pins 1 to 9) shown entering on the left hand side of the diagram, and one more (pin 11) at the bottom right. Each of these ten inputs is connected directly to a special buffer so that it is available in both inverted and non-inverted form as we have seen before. Each one of these twenty input terms is then connected to one of the vertical tracks which go from the top to the bottom of the diagram.

The other six inputs are the pins marked 'I/O' down the right hand side of the diagram. I will explain in a moment how inputs and outputs can share the same pins, but for now notice that these six 'I/O' inputs are also connected through inverting and non-inverting buffers to twelve more of the vertical lines. Hence the 32 vertical lines represent the complete set of 16 inputs and their complements.

Next, identify the eight output OR gates that can be seen near the right hand side of the diagram, each one with a hat-shaped symbol which takes in seven inputs from the adjacent groups of small AND gate symbols. This is an AND–OR structure like that used in Fig. 5.58 that will generate a sum-of-products output.

The inputs to every one of the AND gates is shown as a horizontal line that cuts across all 32 vertical lines. This is the shorthand notation of Fig. 5.57, in this case showing that every AND gate has 32 inputs, each of which can be connected to any or all of the 16 inputs and their complements.

Each one of the OR gates is connected to seven of these AND gates. Hence an OR gate output is the sum of up to 7 product terms, where each product term can be a function of up to 16 input variables.

Now notice that the OR gate outputs are connected through special inverters which have an additional control input. With a logic H on the control input, the output of an inverter is simply the inverse of its input. With a logic L on the control input, the output is put into a high impedance mode, effectively disconnecting it from the rest of the circuit. This is called a **tri-state output**, as it can be in any of three states: H, L or Z, where Z represents the high impedance, disconnected state. The tri-state outputs are controlled by 32-input AND gates, and by blowing fuses correctly the output can be programmed to be always enabled (operating normally), always disabled (Z), or enabled/disabled by a product term from the inputs.

These tri-state outputs are the reason why the six 'I/O' connections can be used either as inputs or as outputs. To allow one of these to used as an input, the corresponding OR gate output has to be put permanently into the high impedance state by suitable fuse connections to its controlling AND gate. The benefit of having these dual purpose I/O pins is that the device can fit into a modest sized 20-pin package. The disadvantage is that there has to be a compromise over the total number of inputs and outputs used in an application. In practice, this disadvantage is not great, and chips like this are a very versatile solution to a wide range of logic problems.

logic diagram (positive logic)

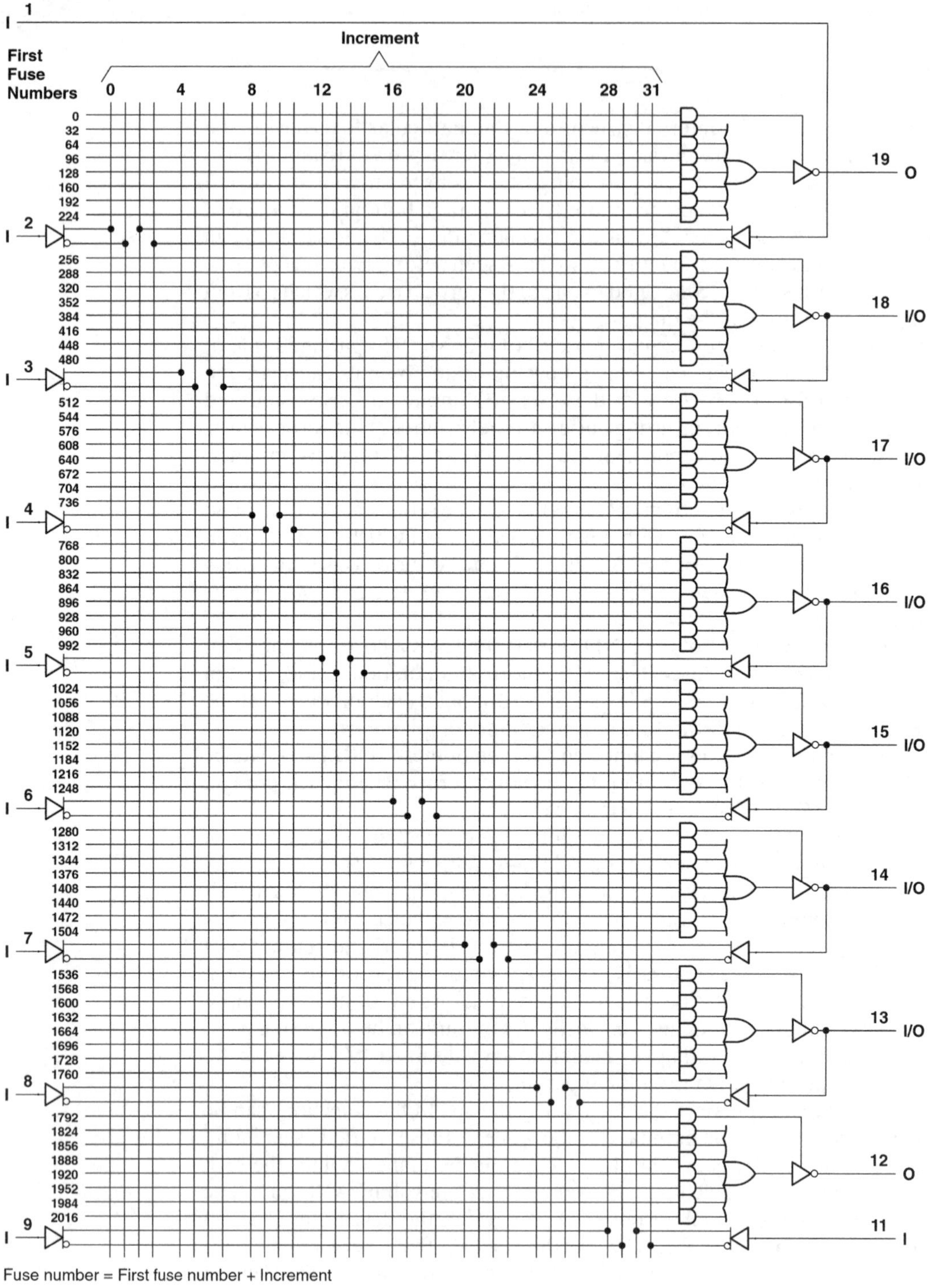

Fig. 5.59 Logic diagram for a TIBPAL16L8 device. Reproduced by permission of Texas Instruments.

To design a circuit using this device you have to decide which fuses to blow. As mentioned earlier, it is assumed at the start that all fuses are intact, so you blow the fuses on all of the input variables that are not in the product term. To do this you need a device called a **programmer**, a box which is normally attached to and controlled by a computer. The required logic configuration is usually input as a group of definitions and equations in a special programming language. The programmer then puts the chip into programming mode, and automatically generates the right sequence of signals on the device pins to blow the appropriate fuses.

PROM

Using a **read-only memory (ROM)** to implement a combinational logic problem has many benefits. The main one is that the design is finished once the truth table has been derived.

A memory can be thought of as a series of compartments or **locations** which each contain a binary word which can be changed. Each location has a unique **address**, which is fixed. If we want to find out what is in a particular location, we set the appropriate address at the input to the ROM, and the contents appear at the output. Figure 5.60 shows an example where the address is set to 2 at the input, and the content, in this case a single bit binary word with a value 1, is displayed at the output. This process is known as **reading** the memory, and the contents of the memory are referred to as **data**. Reading the memory does not alter the contents, it just produces a copy of the data at the output.

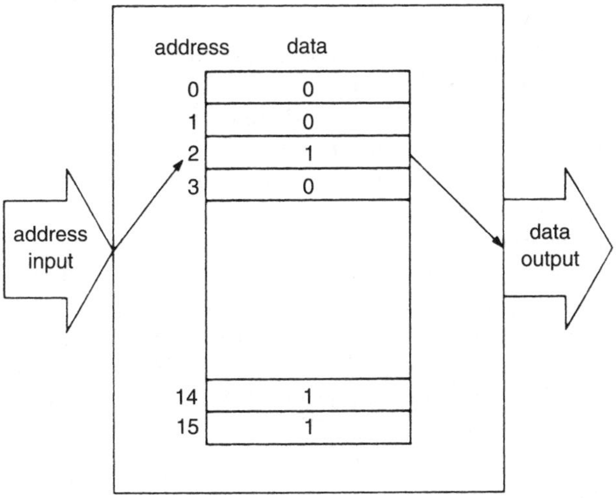

Fig. 5.60 Operation of a ROM.

The link between ROM and the truth table is that the input code in the truth table can be used as the address to the ROM and one or more outputs listed in the truth table can be stored as the corresponding data in the ROM, as shown in Fig. 5.61. This is a very different solution from that offered by the PAL device. In the PAL device you program the product terms from the Boolean expression, while in the PROM you program the 0s and 1s from its truth table.

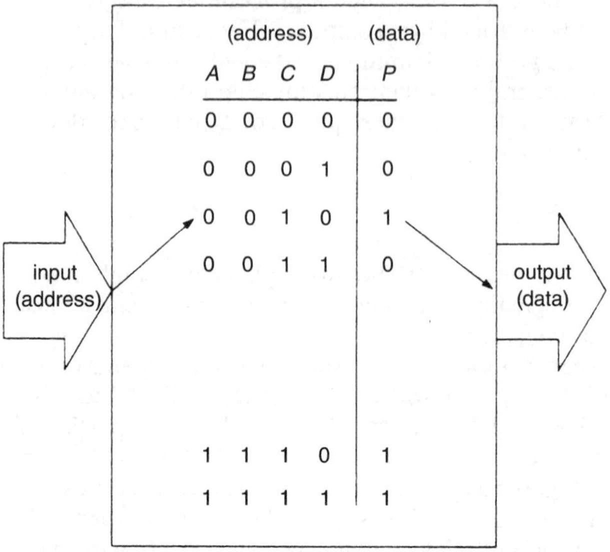

Fig. 5.61

So, how did the data get into the memory in the first place? Some ROMs are fabricated with the required truth table already built in. These are called **mask-programmed ROMs** because the manufacturer has to design appropriate masks (as part of the photographic processing) for the particular requirement. Designing and creating a mask is a relatively expensive procedure, and is the same irrespective of the number of ROMs that are produced. Consequently it is only appropriate to use mask-programmable ROMs when a design is settled and large numbers of identical ROMs are needed. The cost for each individual ROM is then relatively low.

Programmable ROMs or **PROMs** are more appropriate for the development of prototype systems, as they are supplied with an effectively blank truth table (all 0s or all 1s) which can be converted into the required truth table by the user. Some PROMs are programmed like the PAL devices by fusing internal links to fix the contents of each location. However, it is much more common to hold the data as stored packets of electric charge within the device. Some of this type of PROM are erasable, meaning that the contents of each location can be reset to their initial state by applying an electrical signal or, in other cases, ultraviolet light.

Like the PAL devices, a PROM is customized by a suitable **programmer** which is connected to a computer. (The equipment may actually be called a PROM programmer, a PLD programmer or a universal programmer, since most examples can handle a wide range of programmable devices.) After the required logical functions are defined in an appropriate programming language, the programmer selects each address in turn, and by applying suitable voltage levels to the pins, **writes** or **programs** the required data into each location.

An alternative form of memory which can be written to as easily as it can be read from is referred to as read–write memory. However, for historical reasons (and because the acronym RWM is hard to pronounce except in Welsh) it is more commonly called **RAM**, which stands for random-access memory. This is unsuitable for solving most combinational problems as it is generally volatile, meaning that every location loses its contents when the power is switched off. (But note that small capacity non-volatile RAM chips are available for special purposes.) However, RAM is used very extensively in computer systems, when data has to be continually modified during processing operations. Note that modern IC fabrication technology allows the mass production of both ROMs and RAMs that can store many millions of bits on a single chip.

5.6.4 Custom and semi-custom integrated circuits

Another option for the implementation of logic circuitry is to design an integrated circuit specifically for the required application. If an IC is designed from scratch then it is called a **full-custom chip**. Since IC design is a very expensive and time-consuming operation, this course of action is only practical when a very large number ($>10^5$) of identical devices is needed.

A more realistic option in many cases is to use some form of **semi-custom** chip otherwise known as an **application-specific integrated circuit (ASIC)**, or a user-specific integrated circuit **(USIC)**. One form of ASIC is the **standard cell** device, in which small, standard blocks of logic are automatically combined to form a layout of a silicon chip with the correct function. An alternative form of ASIC is a general-purpose array of gates that is dedicated to a particular function by means of interconnections made only during the final stages of processing. These devices are usually referred to as **gate arrays**. Both of these forms of ASIC are practical for much smaller production runs than a full-custom design, but use the area of the silicon wafer less efficiently and so are more expensive per device. Note that field programmable gate arrays **(FPGAs)** are also available which allow the chip to be programmed by the user in a similar way to PAL devices and PROMs.

5.7 ELECTRICAL CHARACTERISTICS OF LOGIC FAMILIES

> Ranges of integrated circuits operating to the same standards and designed to be used together to build up more complicated circuits are often referred to as **logic families**. Examples include 'LS' and 'ALS' in TTL, and 'HC' and 'HCT' in CMOS. Sometimes the word 'family' is used more generally to include, for example, the whole of TTL.

5.7.1 LS TTL

The most widely used bipolar TTL ICs are the LS family, standing for 'low-power Schottky' (see Section 5.7.2). The two-input NAND gate illustrated in Section 5.6.1 has electrical characteristics that are typical of the TTL 74LS family, and these are reproduced here as Fig. 5.62. The data shown actually covers the 54LS00 and 74LS00 devices, but the following description is based only on the 74LS00 column. The main points from this data sheet are as follows.

The first item in the section headed 'recommended operating conditions' is the positive supply voltage V_{CC}, which is nominally specified as 5 V, but can vary by up to 5 % higher or lower. The most common value for logic circuits is 5 V, and needs to be well stabilized to remain within the specified tolerance (see Section 8.4 on voltage regulation).

SN5400, SN54LS00, SN54S00
SN7400, SN74LS00, SN74S00
QUADRUPLE 2-INPUT POSITIVE-NAND GATES

SDLS025 – DECEMBER 1983 – REVISED MARCH 1988

recommended operating conditions

		SN54LS00			SN74LS00			UNIT
		MIN	NOM	MAX	MIN	NOM	MAX	
V_{CC}	Supply voltage	4.5	5	5.5	4.75	5	5.25	V
V_{IH}	High-level input voltage	2			2			V
V_{IL}	Low-level input voltage			0.7			0.8	V
I_{OH}	High-level output current			– 0.4			– 0.4	mA
I_{OL}	Low-level output current			4			8	mA
T_A	Operating free-air temperature	– 55		125	0		70	°C

electrical characteristics over recommended operating free-air temperature range (unless otherwise noted)

PARAMETER	TEST CONDITIONS †	SN54LS00			SN74LS00			UNIT
		MIN	TYP‡	MAX	MIN	TYP‡	MAX	
V_{IK}	V_{CC} = MIN, I_I = – 18 mA			– 1.5			– 1.5	V
V_{OH}	V_{CC} = MIN, V_{IL} = MAX, I_{OH} = – 0.4 mA	2.5	3.4		2.7	3.4		V
V_{OL}	V_{CC} = MIN, V_{IH} = 2 V, I_{OL} = 4 mA		0.25	0.4		0.25	0.4	V
	V_{CC} = MIN, V_{IH} = 2 V, I_{OL} = 8 mA					0.35	0.5	
I_I	V_{CC} = MAX, V_I = 7 V			0.1			0.1	mA
I_{IH}	V_{CC} = MAX, V_I = 2.7 V			20			20	μA
I_{IL}	V_{CC} = MAX, V_I = 0.4 V			– 0.4			– 0.4	mA
I_{OS}§	V_{CC} = MAX	– 20		– 100	– 20		– 100	mA
I_{CCH}	V_{CC} = MAX, V_I = 0 V		0.8	1.6		0.8	1.6	mA
I_{CCL}	V_{CC} = MAX, V_I = 4.5 V		2.4	4.4		2.4	4.4	mA

† For conditions shown as MIN or MAX, use the appropriate value specified under recommended operating conditions.
‡ All typical values are at V_{CC} = 5 V, T_A = 25°C
§ Not more than one output should be shorted at a time, and the duration of the short-circuit should not exceed one second.

switching characteristics, V_{CC} = 5 V, T_A = 25°C (see note 2)

PARAMETER	FROM (INPUT)	TO (OUTPUT)	TEST CONDITIONS	MIN	TYP	MAX	UNIT
t_{PLH}	A or B	Y	R_L = 2 kΩ, C_L = 15 pF		9	15	ns
t_{PHL}					10	15	ns

NOTE 2: Load circuits and voltage waveforms are shown in Section 1.

Fig. 5.62 Extract from the data sheet for a 74LS00 quad NAND gate. Reproduced by permission of Texas Instruments.

Below V_{CC} there are the voltage levels representing logic states. The subscript I represents input, and the additional subscripts H and L represent a HIGH and LOW voltage level. So V_{IH}, for example, refers to the input voltage corresponding to a high logic level. This must be at least 2 V. Similarly, V_{IL} must be less than 0.8 V for the 74LS00 device.

Below these are maximum recommended values of output current for logic high and low (I_{OH} and I_{OL}). These values are -0.4 mA for a high (the minus sign indicates that current flows out of the device), and 8 mA for a logic low.

Jumping down to the next section, headed 'electrical characteristics', values are given for the output voltages. The worst value of V_{OH}, the high level output voltage, is given as 2.7 V. This is specified at an output current equal to the maximum recommended, with a supply voltage of the minimum allowed and at any temperature over the full range. Under the heading TYP (= typical) is given a more optimistic figure assuming that $V_{CC} = 5$ V and the operating temperature is 25°C.

Two sets of figures for V_{OL}, the low level output voltage, are given, one for the maximum output current and one for a current of 4 mA. The worst-case output voltage is 0.5 V, although the 'typical' value is no more than 0.35 V.

These worst-case output values are drawn in Fig. 5.63 against the minimum specified input requirements. Thus in this specification, a high level output is guaranteed to be at least 0.7 V above the minimum requirement for an input, and low level output at least 0.3 V below the requirement for an input. These figures of 0.7 V and 0.3 V are called **noise margins**. They show that the logical output of one circuit can be accurately passed on to the input of the next even if the voltage levels are disturbed to that extent by noise or any other cause.

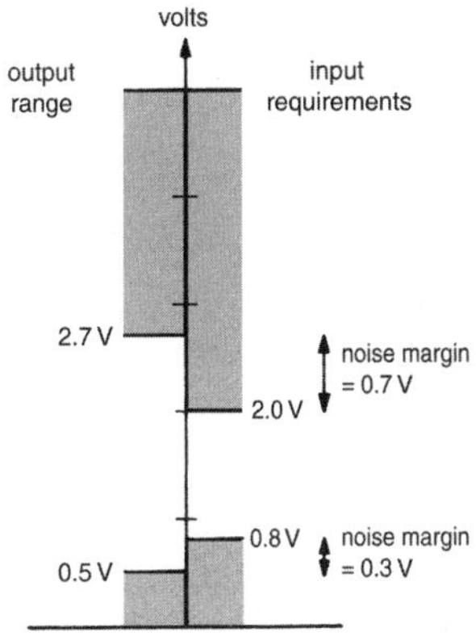

Fig. 5.63 Input and output voltage level limits for LSTTL.

Next look in more detail at the currents in the circuit. The supply current, I_{CC}, is the total current drawn from the power supply with the outputs open-circuit. Two values are quoted, one when the outputs are HIGH, I_{CCH}, and one when the outputs are LOW, I_{CCL}. The average value, usually just stated as I_{CC}, is quite often used instead of the two values. For this 74LS00 chip the average value at 25°C would be:

$$I_{CC} = \frac{I_{CCH} + I_{CCL}}{2} = \frac{0.8 + 2.4}{2} = 1.6\,\text{mA}$$

Note that this is the total current for the IC. The 74LS00 is a quad two-input device, which means that it has four separate NAND gates on it, and so the average current per gate is therefore only 0.4 mA.

We have already seen that the current supplied by a gate output at logic high, I_{OH}, has a maximum value of 0.4 mA, whereas a low logic output current, I_{OL}, has a maximum value of 8 mA. Turning to the currents on the input we see that I_{IH} can be as much as 20 μA and I_{IL} as much as 0.4 mA, again with a minus sign indicating that the current flows out of the device.

The values of these currents dictate how many gate inputs can be connected to the outputs of other gates, a measure known as the **fan-out** of a gate. Figure 5.64 shows the output of a NAND gate connected to the inputs of some other NAND gates when the output is a 1 and a 0.

For a logic 1 output with this 74LS family, $V_{out} \geqslant 2.7$ V and at least 400 μA is available to flow out of the gate. This current is often referred as a **source current**, because it comes from an effective voltage source within the gate. Each of the gates connected to this output could have an input current, I_{IH}, of up to 20 μA. Consequently, the maximum number of gates that can be reliably connected to the output of the first NAND gate is 400 μA/20 μA = 20.

Similarly when the output is low, V_{out} is guaranteed to be no more than 0.5 V when the gate is 'sinking' 8 mA, which is the maximum value of I_{OL}. The input current, I_{IL}, required by the other gates can be as much as 0.4 mA each.

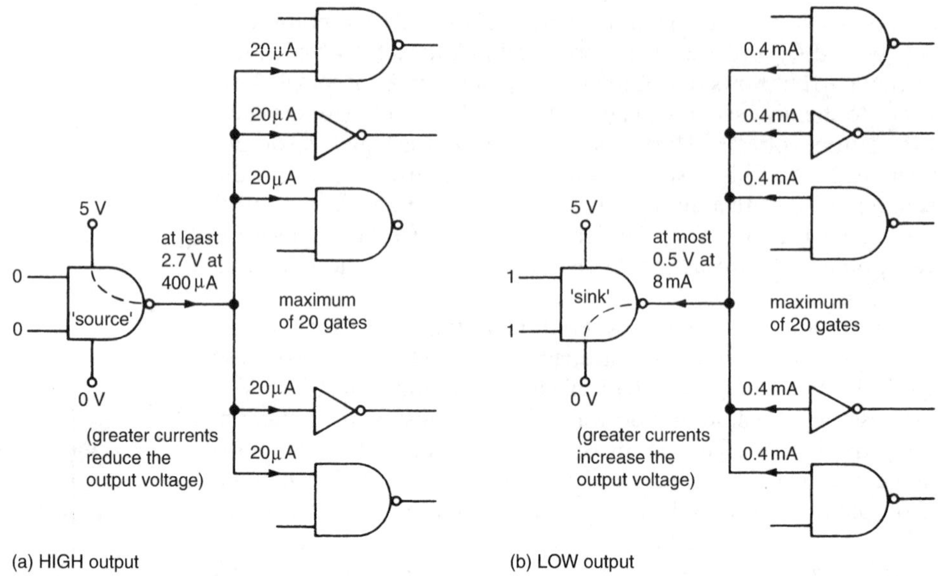

Fig. 5.64 Fan-out calculation for LSTTL.

In this case the current flows from the inputs of the gates into the output of the first NAND gate, and so the output is often referred to as a **sink current**. The maximum number of gates that can be supplied by the output of a single NAND gate when the output is low is 8 mA/0.4 mA = 20 again. Thus the fan-out for a 74LS family NAND gate is 20. (Usually the fan-out for low and high states is the same. If not, the lower of the two values is quoted.)

In situations where there are unused inputs, such as when you have a three-input NAND gate but a two-input problem, it is common practice to connect the unwanted input to one or other of the voltage rails. If the input is left unconnected the gate acts as if the input is a logic 1, but it is very prone to noise and can often produce spurious results. Manufacturers recommend that for 0 inputs you can connect the gate directly to the 0 V rail, whereas for 1 inputs the gate should be connected to the 5 V rail via a resistor known as a 'pull-up resistor'. A frequently used value is 1 kΩ.

An alternative is to connect the unused inputs to one of the other inputs on the NAND gate. The drawback of this method is that the limited fan-out of the gate providing the input might not be enough to ensure correct logic levels.

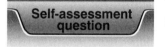

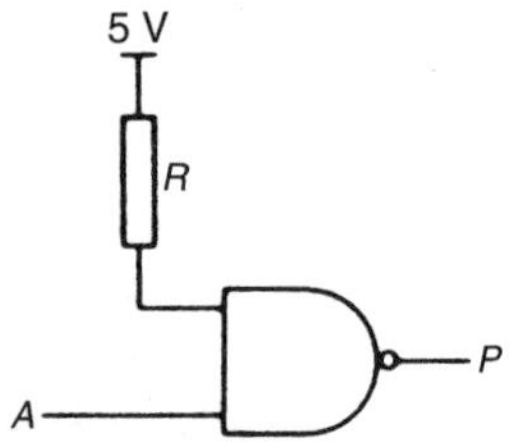

Fig. 5.65 Figure for Self-assessment question 21.

Self-assessment question

21 Figure 5.65 shows a two-input 74 series TTL NAND gate with one of its inputs connected to the 5 V supply rail via a resistor R which has the effect of making that input a permanent logic 1.

(a) What is the Boolean expression for the output, P?
(b) What is the largest value of R that can be used that would ensure that the input voltage is still a recognizable logic 1?
(c) If a 1 kΩ resistor is connected to the 5 V rail, how many logic 1 inputs to a NAND gate could be connected to it?

Another parameter which is of interest is the **propagation delay** time, t_{pd}, which is the delay between a change occurring on the input of the device and the resultant change appearing on the output. Under the heading 'switching characteristics' in Fig. 5.62 are shown two times. The delay between the input changing and the output going from low to high is t_{PLH} and the delay between the input changing and the output going from high to low is t_{PHL}, as shown in Fig. 5.66.

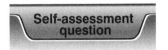

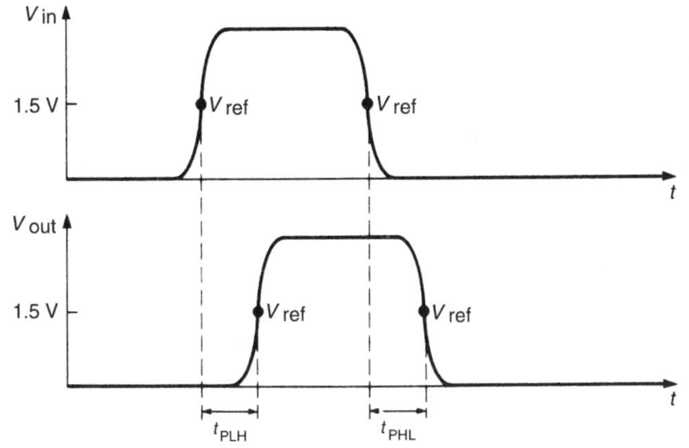

Fig. 5.66 Propagation delay.

More specifically, the delays are measured between the input passing through a fixed reference voltage and the corresponding output passing through the same fixed reference voltage. Figure 5.66 shows that the fixed reference voltages on the curves are 1.5 V for both input and output voltages. (This value is chosen as the reference voltage because when it is applied to the input of the gate the same voltage appears at the output. In other families a different reference voltage may be used.)

A characteristic propagation delay time can be taken to be the average of these two values. For the NAND gate under consideration this has a typical value of

$$t_{pd} = \frac{10 + 9}{2} = 9.5 \text{ ns}$$

The total delay along a path through a circuit is found by adding the delays of all of the gates in that path. The delay of the circuit is equal to the path with the longest delay, which is the worst case that can occur.

Self-assessment question

22 What are the shortest and longest propagation delays between any input and the output of the circuit shown in Fig. 5.67, assuming that NAND gates and INVERTERS are LSTTL devices with a typical delay of 9.5 ns? (If possible, simulate this circuit with a suitable package and examine carefully the relative timing of input and output changes.)

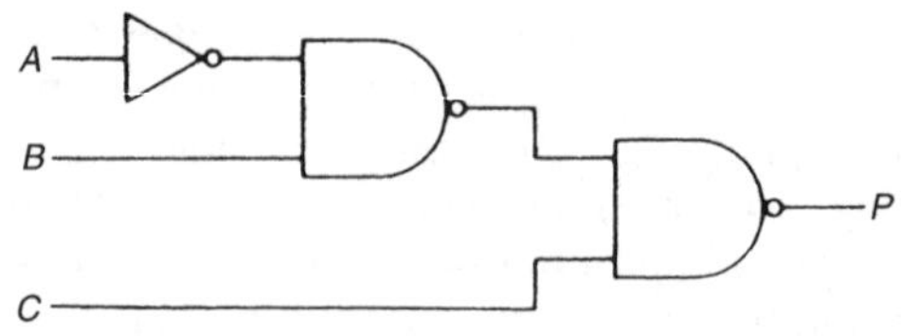

Fig. 5.67 Figure for Self-assessment question 22.

5.7.2 Other TTL families

The two parameters which are most widely used to characterize the performance of logic families are propagation delay time and power dissipation. They are often multiplied to form a 'figure of merit' with which to compare different logic families. This is known as the **power–delay product**, and the aim is to get its value as small as possible.

The total power dissipation, P_D, is calculated from the supply voltage and the supply current. Thus for the TTL 74LS00 NAND gate, a typical value for the power dissipation *per gate*, using the values for V_{CC} and I_{CC} found earlier, is

$$P_D = V_{CC} \times I_{CC} = 5 \times 0.4 \times 10^{-3} = 2\ \text{mW}$$

For the NAND gate, using the figure for delay time t_{pd} from the last section, the figure of merit is

$$\begin{aligned}\text{Power-delay product} &= 9.5 \times 10^{-9} \times 2 \times 10^{-3}\text{J}\\ &= 19 \times 10^{-12}\text{J}\\ &= 19\,\text{pJ}\end{aligned}$$

Notice that the product is expressed in joules because power in watts multiplied by time in seconds gives energy in joules. This value can therefore be interpreted as the energy required to bring about changes in the output.

Since they were introduced in the late 1960s, TTL logic devices have undergone many modifications in an effort to improve operating performance, and in particular most of the effort has been directed at increasing the switching speed and reducing the power dissipation of each gate. These two effects are interrelated, so that an increase in speed often introduces an increase in power dissipation. Improvements have been made largely by a combination of changes in the design of the basic gate circuit and by the reduction in size of the transistors and other integrated-circuit components.

The original 'standard' TTL configuration had a propagation delay of around 9 ns. Adding a Schottky diode to this basic circuit configuration produced single gates with propagation delays as short as 3 ns. However this **Schottky TTL** (**STTL**, with device codes such as 74S00) dissipates much more power than standard TTL. **Low-power Schottky TTL (LSTTL)** dissipates only one-fifth of the power of standard TTL and about a tenth of the power of STTL but still achieves a typical propagation delay less than 10 ns per gate. Further developments have led to other new TTL families, such as advanced low-power Schottky (ALS) with even better speed and power figures, but these are, at present, rather more expensive.

Self-assessment question

23 Table 5.8 shows the supply voltage, currents and propagation delay times for a quad input NAND gate in original TTL, STTL and ALSTTL. What are their power–delay products per gate?

Table 5.8

	7400	*74S00*	*74ALS00*
V_{CC}	5 V	5 V	5 V
I_{CCH}	4 mA	10 mA	0.5 mA
I_{CCL}	12 mA	20 mA	1.5 mA
t_{PLH}	11 ns	3 ns	4 ns
t_{PHL}	7 ns	3 ns	3 ns

5.7.3 Emitter-coupled logic

The second major logic family that uses bipolar transistors is called emitter-coupled logic (**ECL**). These devices use a totally different circuit design from TTL logic gates and offer very fast switching speeds with gate delays of less than 0.8 ns. However, the cost of this speed is power consumption approaching 40 mW per gate. ECL devices are used extensively in mainframe computer systems.

ECL devices operate from a negative voltage supply rail of −5.2 V. In the past this supply voltage difference has made it difficult for a digital design engineer to mix ECL and TTL devices in a single system, to gain the speed advantage of ECL in critical areas, with the lower power dissipation of TTL in non-critical areas. Semiconductor manufacturers have now addressed this particular problem and it is now possible to buy large-scale integrated circuit devices that use TTL input and output stages and ECL gates for all the internal logic. However, because of their high cost, ECL devices are only used when high speed is of very great importance.

5.7.4 CMOS

The main alternative to TTL is CMOS, which is based on MOS field-effect transistors instead of bipolar ones. A wide range of CMOS logic is available, in several 74-series families together with a much older 4000 series. Current usage of CMOS logic is about twice that of bipolar families, and the gap is steadily increasing.

Because of the fundamental difference in internal circuitry, the electrical characteristics of CMOS are very different from TTL, and typical values of some parameters for the **74HCT family** are shown in Table 5.9. This HCT family is designed to be compatible with LSTTL circuitry, and devices from both families can be mixed together in one circuit. Hence the maximum value for V_{IL} and the minimum value for V_{IH} are the same as for LSTTL. The maximum current output is 4 mA for either high or low level, implying that an HCT output can drive up to 10 LSTTL inputs. However, the input current is specified to be less than 1 μA for either high or low level, very much less than

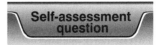

Table 5.9 Some basic parameters for the 74HCT CMOS family

Recommended operating conditions
V_{CC} supply voltage: 5 V ± 0.5 V
V_{IH} high level input voltage (minimum): 2 V
V_{IL} low level input voltage (maximum): 0.8 V
I_{OH} high level output current (maximum): −4 mA
I_{OL} low level output current (maximum): 4 mA
Electrical characteristics
I_{IH}, I_{IL} input current (maximum): ±1 μA
(typical): ±0.1 μA
With 20 HCT loads (±20 μA, with V_{CC} = 4.5 V)
Output voltage V_{OH} (minimum): 4.4 V
Output voltage V_{OL} (maximum): 0.1 V
With 10 LSTTL loads (±4 mA, with V_{CC} = 4.5 V)
Output voltage V_{OH} (minimum): 3.84 V
Output voltage V_{OL} (maximum): 0.33 V
I_{CC} supply current (V_{CC} = 5.5 V) (typical): 2 μA
t_{PLH} propagation delay (typical): 10 ns
t_{PHL} propagation delay (typical): 10 ns

that required by TTL. Hence, two sets of output voltages are quoted, covering the case of 20 HCT loads (a total of + or − 20 μA), and 10 LSTTL loads (a total of up to 4 mA). In the first case, the output voltages V_{OH} and V_{OL} are both within 0.1 V of the supply lines, leading to a very high noise margin. When LSTTL inputs are being driven, the output voltages are not as good, but still with a better noise margin than TTL.

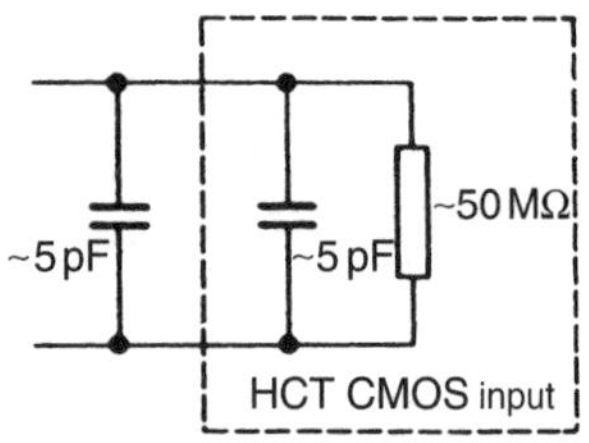

Fig. 5.68 Equivalent circuit for an HCT CMOS gate input for an applied logic high level.

This very low value of input current also implies that the maximum fan-out of an HCT gate driving other HCT inputs is 4000. In practice, however, the fan-out is limited by the total input capacitance of the devices connected to it long before that figure is reached. The equivalent circuit of an HCT gate input when a logic high is connected, outlined in Fig. 5.68, consists of a 5 pF capacitor in parallel with a resistance of about 50 MΩ. In addition, the physical connections to the gate add at least another 5 pF capacitance, and this is represented on the diagram by another capacitor across the input.

When the logic high input voltage is steady, then current flows through the 50 MΩ resistance, producing the 'typical' 0.1 μA input current. Whenever the input voltage is switched, however, this total capacitance of 10 pF has to be charged or discharged by the output that is driving it. An HCT device has a typical output resistance of around 100 Ω. If this output switches from logic low to logic high with a perfect step function, then the voltage across the capacitor rises exponentially with a time constant of about $100\ \Omega \times 10\ \text{pF} = 10^{-9}\ \text{s} = 1\ \text{ns}$. Therefore the time taken for this voltage to reach the threshold for switching the input to a logic high is slowed up by about this amount, meaning that a delay is effectively added to the circuit.

Another input connected in parallel across the output adds another 10 pF, and so doubles the rise time. As more are added, each one effectively adds another 1 ns to the circuit response time. Hence, the maximum number of inputs that can be connected to an output is limited by the additional delay that can be tolerated.

One other effect of charging and discharging capacitances is that the power dissipation is highly dependent upon frequency. When the inputs are not changing, there is only a small leakage current flowing through the device, and so the 'static' or 'd.c.' power consumption is almost negligible. Whenever the inputs change, then power is dissipated in the transistors in charging or discharging all the on-chip capacitances, together with the external ones like those shown in Fig. 5.68. All the internal effects are usually lumped together as one **power dissipation capacitance** C_{pd}, which is normally included as a parameter in the data sheet. If the total external load capacitance is C_{L}, then the total power dissipation P is given by the following approximate formula:

$$P = I_{\text{CC}}V_{\text{CC}} + C_{\text{PD}}V_{\text{CC}}^2 f_{\text{i}} + C_{\text{L}}V_{\text{CC}}^2 f_{\text{o}}$$

where f_{i} is the average frequency of changes at the input and f_{o} is the average frequency of changes at the output (these are not the same in most cases). The term $I_{\text{CC}}V_{\text{CC}}$ is, of course, the static or d.c. power dissipation. However, above a few hundred kilohertz, the 'a.c.' terms become larger, and the overall power dissipation increases with frequency.

For an SSI gate, an HCT chip dissipates much less power than the equivalent LSTTL version up to a switching frequency of a few megahertz. Above this frequency the power dissipation of the LSTTL increases as well, and the two stay much the same. In practice the average frequency of logic level changes in any one particular gate is likely to be much less than the maximum working frequency for a circuit. Hence HCT devices can usually be assumed to dissipate very much less power than their LSTTL equivalent except at the very highest speeds of operation.

An additional advantage of CMOS is its tolerance to a wide range of supply voltages. In order to be compatible with LSTTL, the HCT family is designed to be used with a supply of 5 V, and is usually specified over a range of 4.5 to 5.5 V (which is twice the range over which 74-series TTL is specified). However, the closely similar HC CMOS devices are suitable for use with power supplies from 2 V to 6 V. The problems of power supply stabilization are therefore removed (unless particular logic levels are required for interfacing with other circuitry), and together with the very low power consumption make such CMOS devices much more suitable for battery-operated equipment. In fact, many new CMOS families are designed to have optimum performance at much lower supply voltages, to take advantage of reduced power consumption and to be compatible with the supply voltages common with microprocessors and memory chips.

Self-assessment question

24 An HCT CMOS gate with the electrical characteristics shown in Table 5.9 has a power dissipation capacitance of 24 pF. It is used in a circuit where the total load capacitance is 40 pF and the average frequency of output changes is one quarter of the average frequency of input changes. Calculate the total power dissipation at (a) an input frequency of 4 kHz, and (b) an input frequency of 4 MHz.

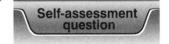

5.7.5 Summary of Sections 5.6 and 5.7

There are a great many integrated logic circuits available, mainly in various bipolar (TTL) and CMOS families, but only a few examples have been illustrated here. Some combinational problems can be solved by the use of a single chip, either a standard function (such as a decoder), a programmed PAL device or ROM, or a semi-custom chip. Other problems are solved by interconnecting ICs from the same or compatible families.

The data sheet specifies the functional, electrical, environmental and mechanical characteristics of a device. The most important electrical parameters are as follows:

- The worst-case output voltage when the maximum load current is being drawn. V_{OH} is the minimum value for a high logic level, and V_{OL} the maximum for a logic low.
- The maximum output current that can safely be used, I_{OH} and I_{OL}.
- The range of input voltages that represent logic levels. V_{IH} is the minimum value that will be interpreted as a logic high, and V_{IL} the maximum for a low.
- The maximum input current requirements I_{IH} and I_{IL}.
- The propagation delay t_{pd} is the time taken for a change in logic level at the input to appear as a change in level at the output.
- The noise margins are the differences between the worst-case output voltage values and the worst-case input voltage requirements.
- The fan-out is the maximum number of inputs that can be driven successfully by an output to either logic level. For TTL devices the fan-out is determined by the ability of outputs to supply steady input current. For CMOS devices this is not a problem, but each additional input connected to an output slows the rise time of the change in level and so affects the overall response speed of the circuit.

Logic families are often broadly characterized by typical values of propagation delay and of power dissipation per gate. The most widely used families of SSI and MSI logic devices are LSTTL (based upon bipolar transistors) and HC and HCT CMOS (based on MOSFETs). These families have a similar speed (about 10 ns delay per gate) but the CMOS ones have a much lower power dissipation except at the highest speeds of operation.

Answers to self-assessment questions

1. If we assign an arbitrary *n*-bit binary code to the letters 'a' to 'z' we will need 26 separate code words. The nearest number that is a power of 2 and that is greater than 26 is 32, or 2^5. We therefore need 5 bits at least, in which case 6 of the code words will be unused.

2. $420_{10} = 256 + 128 + 32 + 4$

$= 2^8 + 2^7 + 0\times2^6 + 2^5 + 0\times2^4 + 0\times2^3 + 2^2 + 0\times2^1 + 0\times2^0$

$= 110100100_2$

	4	2	0
=	0100	0010	0000 in BCD

(a) $01101001_2 = 2^6 + 2^5 + 2^3 + 2^0$

$= 64 + 32 + 8 + 1$

$= 105_{10}$

(b)

0110	1001	in BCD represents
6	9	in denary

3.

110 001 011 010	binary
6 1 3 2	octal
110001011010	binary
2048 + 1024 + 64 + 16 + 8 + 2 = 3162	denary
1100 0101 1010	binary
C 5 A	hexadecimal

4.

Octal	*ASCII*	*Binary (no parity)*	*Binary (even parity)*	*Hex (even* parity)
102	B	1000010	01000010	42
111	I	1001001	11001001	C9
124	T	1010100	11010100	D4
123	S	1010011	01010011	53

5. (a) The truth table can be completed in the following ways.

A	B	P_{15}	P_{14}	P_{13}	P_{12}	P_{11}	P_{10}	P_9	P_8	P_7	P_6	P_5	P_4	P_3	P_2	P_1	P_0
0	0	1	0	1	0	1	0	1	0	1	0	1	0	1	0	1	0
0	1	1	1	0	0	1	1	0	0	1	1	0	0	1	1	0	0
1	0	1	1	1	1	0	0	0	0	1	1	1	1	0	0	0	0
1	1	1	1	1	1	1	1	1	1	0	0	0	0	0	0	0	0

(b) The following columns in the truth table and Boolean expressions represent all of the outputs that can be generated that are independent of *A* and *B*:

$P_0 : P = 0$ $\quad$ $P_{15} : P = 1$

(c) Dependent on only one input:
Dependent on *A* only:

$P_{12} : \; P = A$ $\quad$ $P_3 : \; P = \overline{A}$

Dependent on *B* only:

$P_{10} : \; P = B$ $\quad$ $P_5 : \; P = \overline{B}$

6. $P = \overline{A}\cdot\overline{B} + A\cdot B$
A circuit that implements this is shown in Fig. 5.69.

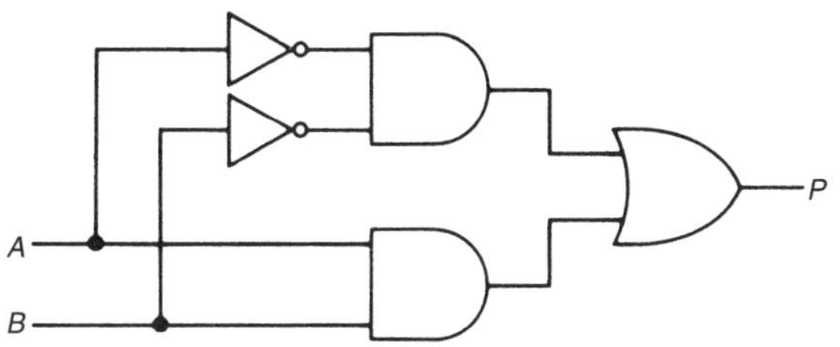

Fig. 5.69 Figure for Self-assessment question 6.

7.

A	*B*	*C*	*P*
0	0	0	1
0	0	1	0
0	1	0	1
0	1	1	1
1	0	0	1
1	0	1	1
1	1	0	1
1	1	1	1

$P = A + B + \overline{C}$

This truth table applies if *C* represents the statement 'supply of materials is *above* specified minimum'. If it had been 'supply *below* specified minimum', the truth table would have the normal OR form, with a single 0 at the top of the *P* column, and $P = A + B + C$.

8.

A	*B*	*C*	*D*	*P*	*Minterms*
0	0	0	0	0	
0	0	0	1	0	
0	0	1	0	0	
0	0	1	1	0	
0	1	0	0	0	
0	1	0	1	1	$\overline{A}\cdot B\cdot\overline{C}\cdot D$
0	1	1	0	0	
0	1	1	1	1	$\overline{A}\cdot B\cdot C\cdot D$
1	0	0	0	1	$A\cdot\overline{B}\cdot\overline{C}\cdot\overline{D}$
1	0	0	1	1	$A\cdot\overline{B}\cdot\overline{C}\cdot D$

A	*B*	*C*	*D*	*P*	*Minterms*
1	0	1	0	1	$A\cdot\overline{B}\cdot C\cdot\overline{D}$
1	0	1	1	1	$A\cdot\overline{B}\cdot C\cdot D$
1	1	0	0	0	
1	1	0	1	1	$A\cdot B\cdot\overline{C}\cdot D$
1	1	1	0	0	
1	1	1	1	1	$A\cdot B\cdot C\cdot D$

$$P = \overline{A}\cdot B\cdot\overline{C}\cdot D + \overline{A}\cdot B\cdot C\cdot D + A\cdot\overline{B}\cdot\overline{C}\cdot\overline{D} + A\cdot\overline{B}\cdot\overline{C}\cdot D + A\cdot\overline{B}\cdot C\cdot\overline{D} + A\cdot\overline{B}\cdot C\cdot D + A\cdot B\cdot\overline{C}\cdot D + A\cdot B\cdot C\cdot D$$

(sum-of-minterms)

It is possible, but not necessarily obvious, to derive simpler expressions from the description of the problem:

$P = A\cdot\overline{B} + A\cdot B\cdot D + B\cdot\overline{C}\cdot D + \overline{A}\cdot B\cdot C\cdot D$

(from the description of the problem in words)

$P = A\cdot\overline{B} + B\cdot D$

(from inspection of the truth table, or thinking about the description of the problem).

9. From the rules in Table 5.5, $P = X + \overline{X} + X = 1 + X = 1$. This can also be seen graphically as follows. Figure 5.70 shows a three-input OR gate with two of its inputs connected to the *X* and one of its inputs connected to $\overline{X}$. So when *X* is 0, the inputs to the OR gate will be 0, 1 and 0. The truth table for the three-input OR function is

A	B	C	P
0	0	0	0
0	0	1	1
0	1	0	1
0	1	1	1
1	0	0	1
1	0	1	1
1	1	0	1
1	1	1	1

When the inputs are 0, 1 and 0, the output is 1. Similarly, when X is 1, the inputs are 1, 0 and 1. From the truth table we can see that the corresponding output is again 1. So we can conclude that the expression $P = X + \overline{X} + X$ reduces to $P = 1$.

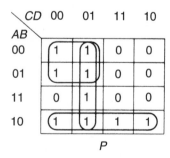

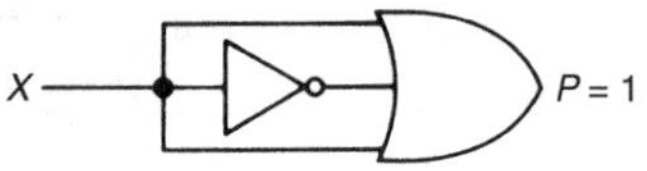

Fig. 5.70 Figure for Self-assessment question 9.

10. You may have decided to simplify the expressions in all sorts of different ways. My solutions show one way for each of the examples.

(a) $P = A \cdot (B + C) + A \cdot (\overline{C} + \overline{B})$

$= A \cdot B + A \cdot C + A \cdot \overline{C} + A \cdot \overline{B}$ Distributive law

$= A \cdot B + A \cdot \overline{B} + A \cdot C + A \cdot \overline{C}$ Commutative law

$= A \cdot (B + \overline{B}) + A \cdot (C + \overline{C})$ Distributive law

The expression can be further reduced using the rules for the AND and OR functions as follows:

$P = A \cdot 1 + A \cdot 1$ $(X + \overline{X} = 1)$

$= A + A$ $(X \cdot 1 = X)$

$= A$ $(X + X = X)$

(b) $P = A \cdot B \cdot C + \overline{A} \cdot C \cdot B$

$= A \cdot B \cdot C + \overline{A} \cdot B \cdot C$ Commutative law

$= (A + \overline{A}) \cdot B \cdot C$ Distributive law

Again the expression can be further simplified using the rules for the AND and OR functions:

$P = 1 \cdot B \cdot C$ $(X + \overline{X} = 1)$

$= B \cdot C$ $(1 \cdot X = X)$

11. $P = A \cdot \overline{B} \cdot \overline{C} + A \cdot \overline{B} \cdot C + A \cdot B \cdot C$
The first two terms have $A \cdot \overline{B}$ in common, so we can take these variables outside brackets:

$P = A \cdot \overline{B} \cdot (\overline{C} + C) + A \cdot B \cdot C$

$\overline{C} + C = 1$ and $A \cdot \overline{B} \cdot 1 = A \cdot \overline{B}$, so the expression reduces to

$P = A \cdot \overline{B} + A \cdot B \cdot C$

12. The truth tables and Karnaugh maps are shown in Fig. 5.71

(a)

A	B	P
0	0	1
0	1	0
1	0	1
1	1	1

A \ B	0	1
0	1	0
1	1	1

(b)

A	B	P
0	0	1
0	1	0
1	0	0
1	1	1

A \ B	0	1
0	1	0
1	0	1

Fig. 5.71 Figure for Self-assessment question 12.

13. The areas shown are defined by the following terms:

I $\overline{B} \cdot C$
II $\overline{A}$
III $A \cdot B$
IV $\overline{B} \cdot \overline{C}$

14. (a) The three groups are shown in Fig. 5.72.

(b) The first group, in the second column, correspond to the product terms $\overline{C} \cdot D$. The second group is along the bottom row and corresponds to the product term $A \cdot \overline{B}$. Finally, the third group is in the top left-hand corner and corresponds to the product term $\overline{A} \cdot \overline{C}$. The Boolean expression for this function is therefore

$P = A \cdot \overline{B} + \overline{A} \cdot \overline{C} + \overline{C} \cdot D$

AB \ CD	00	01	11	10
00	1	1	0	0
01	1	1	0	0
11	0	1	0	0
10	1	1	1	1

P

Fig. 5.72 Figure for Self-assessment question 14.

15. The three groups are shown in Fig. 5.73.

(a) Group I is defined by the term $\overline{A} \cdot \overline{B} \cdot \overline{C} \cdot D$, and therefore corresponds to a single 1. To place it in

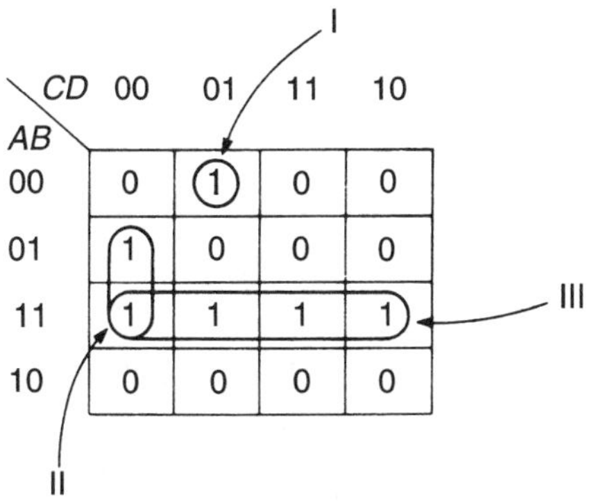

Fig. 5.73 Figure for Self-assessment question 15.

the map you look for the square that can be identified by the conditions that $A=0$, $B=0$, $C=0$ and finally $D=1$, which can be found in the top row, second from the left.

Group II is defined by the term $B \cdot \overline{C} \cdot \overline{D}$, and therefore corresponds to two 1s. These are found by locating the area of the map where $B=1$, $C=0$ and $D=0$. The middle two rows are located by $B=1$ and the left-hand column by the condition that both C and D are 0, so the area that we are looking for consists of the middle two squares of the left-hand column.

Finally, group III is defined by the term $A \cdot B$ and correspond to four 1s. The area of the map that we need to find is identified by the conditions $A=1$ and $B=1$, which is the second row from the bottom.

(b) Group II is currently $B \cdot \overline{C} \cdot \overline{D}$. We are told that this can be reduced to $B \cdot \overline{C}$, which is independent of D, by changing one of the values in a square to 1. Pointing out the area on the map that corresponds to $B \cdot \overline{C}$ should make it become clear which square needs to be changed.

The area corresponding to $B \cdot \overline{C}$ consists of four squares which are located by the conditions $B=1$ and $C=0$. The condition $B=1$ identifies the middle two rows, while $C=0$ identifies the two left-hand columns, so the area that we are looking for is the middle four squares in the two left-hand columns, as shown in Fig. 5.74 So for $B \cdot \overline{C}$ to be the

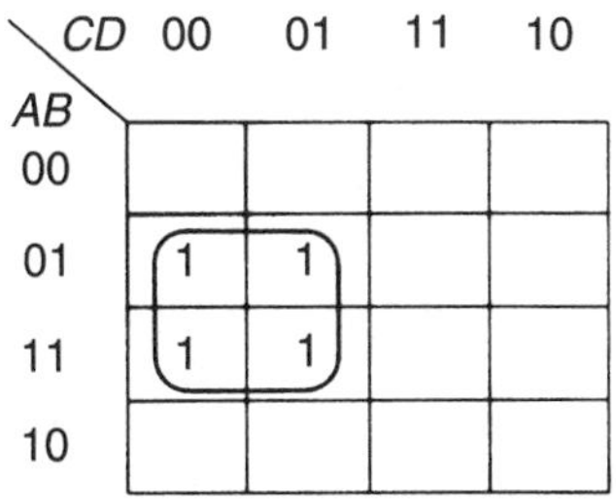

Fig. 5.74 Figure for Self-assessment question 15.

corresponding term for these four squares, there must be 1s in all of the squares. Comparing Fig. 5.74 with Fig 5.73, it is clear that the extra 1 has to be placed in the square that corresponds to the term $\overline{A} \cdot B \cdot \overline{C} \cdot D$.

The inclusion of this extra 1 means that group I, which was initially identified by the term $\overline{A} \cdot \overline{B} \cdot \overline{C} \cdot D$, can become a group of two 1s. This new group corresponds to the term $\overline{A} \cdot \overline{C} \cdot D$.

16. The Karnaugh map for c is shown in Fig. 5.75. The expression $c = B + \overline{C} + D$, when entered into a Karnaugh map, is shown in Fig. 5.76. A comparison of Fig. 5.75 and Fig. 5.76 shows that all the don't-cares are set to 1 except the one in the bottom right-hand corner, which corresponds to the product term $A \cdot \overline{B} \cdot \overline{C} \cdot D$.

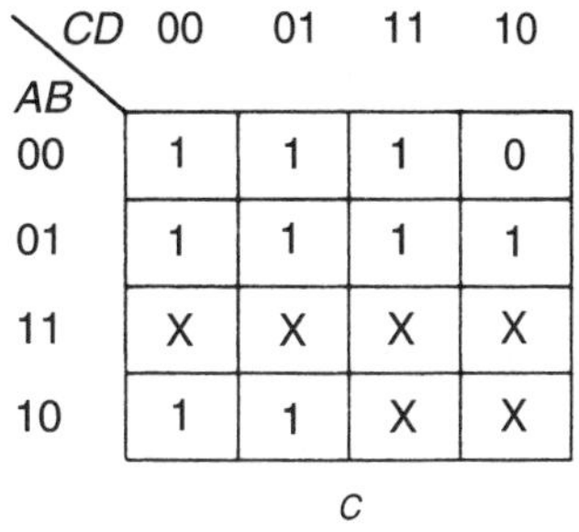

Fig. 5.75 Figure for Self-assessment question 16.

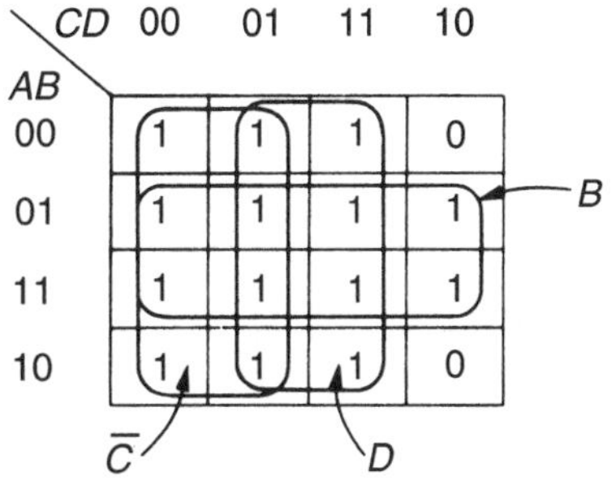

Fig. 5.76 Figure for Self-assessment question 16.

17. The Karnaugh maps for b and f are repeated in Fig. 5.77. Start with b and go through a step-by-step procedure:

1. Form a group of two 0s in the second column corresponding to the product term $B \cdot \overline{C} \cdot D$ as shown in Fig. 5.78.
2. Form a group of two 0s in the last column corresponding to the product term $B \cdot C \cdot \overline{D}$ as shown in Fig. 5.79.

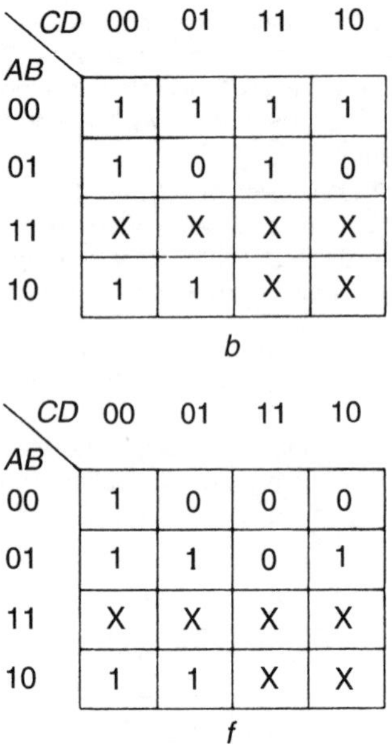

CD\AB	00	01	11	10
00	1	1	1	1
01	1	0	1	0
11	X	X	X	X
10	1	1	X	X

b

Fig. 5.77 Figure for Self-assessment question 17.

CD\AB	00	01	11	10
00	1	0	0	0
01	1	1	0	1
11	X	X	X	X
10	1	1	X	X

f

Fig. 5.77 Figure for Self-assessment question 17.

CD\AB	00	01	11	10
00	1	1	1	1
01	1	0	1	0
11	X	0	X	X
10	1	1	X	X

b

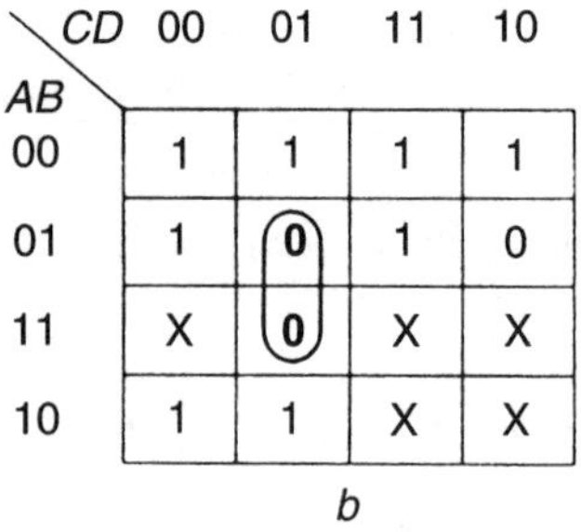

Fig. 5.78 Figure for Self-assessment question 17.

CD\AB	00	01	11	10
00	1	1	1	1
01	1	0	1	0
11	X	0	X	0
10	1	1	X	X

b

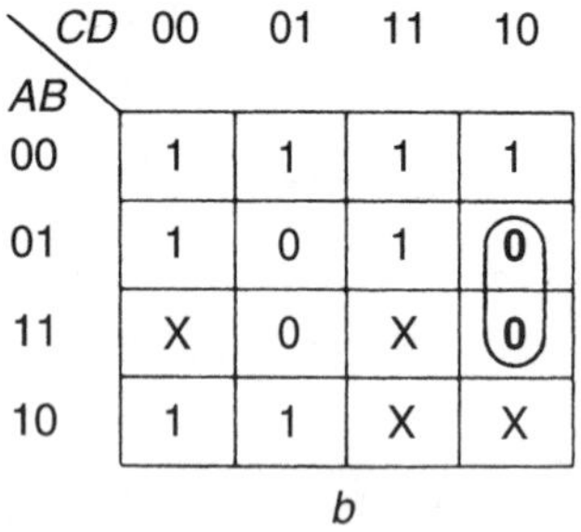

Fig. 5.79 Figure for Self-assessment question 17.

3. The remaining don't-cares are set to 1. The expression for $\overline{b}$ is then:

$$\overline{b} = B \cdot \overline{C} \cdot D + B \cdot C \cdot \overline{D}$$

Now I will go through the same procedure for f.

1. Form a group of four 0s in the third column which corresponds to the product term $C \cdot D$, as shown in Fig. 5.80.

CD\AB	00	01	11	10
00	1	0	0	0
01	1	1	0	1
11	X	X	0	X
10	1	1	0	X

f

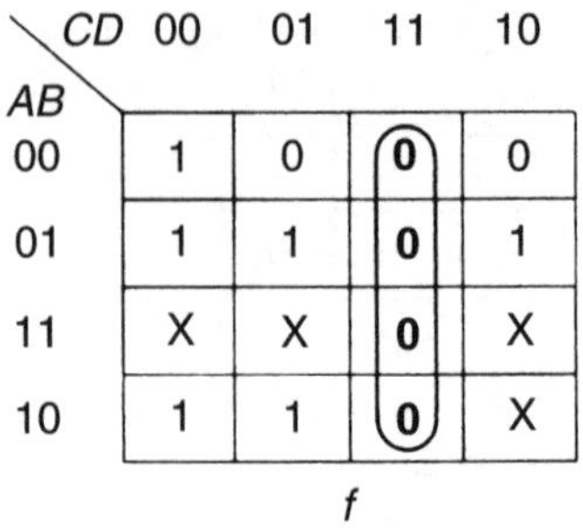

Fig. 5.80 Figure for Self-assessment question 17.

2. Form a group of four 0s in the last two columns using the top and bottom rows, corresponding to the product term $\overline{B} \cdot C$, as shown in Fig. 5.81.

CD\AB	00	01	11	10
00	1	0	0	0
01	1	1	0	1
11	X	X	0	X
10	1	1	0	0

f

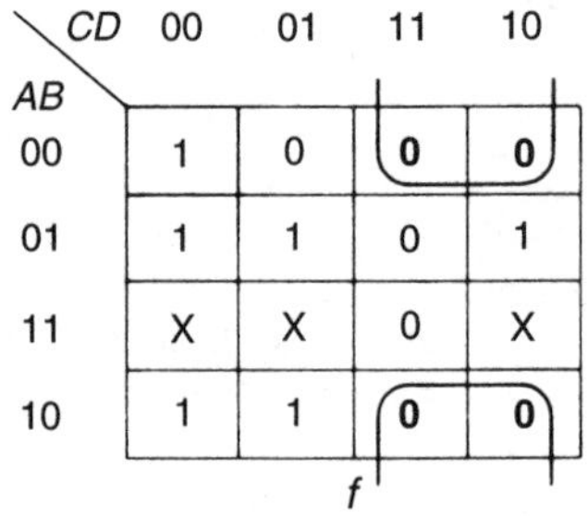

Fig. 5.81 Figure for Self-assessment question 17.

3. Form a group of two 0s in the top row corresponding to the product term $\overline{A} \cdot \overline{B} \cdot D$, as shown in Fig. 5.82.

CD\AB	00	01	11	10
00	1	0	0	0
01	1	1	0	1
11	X	X	0	X
10	1	1	0	0

f

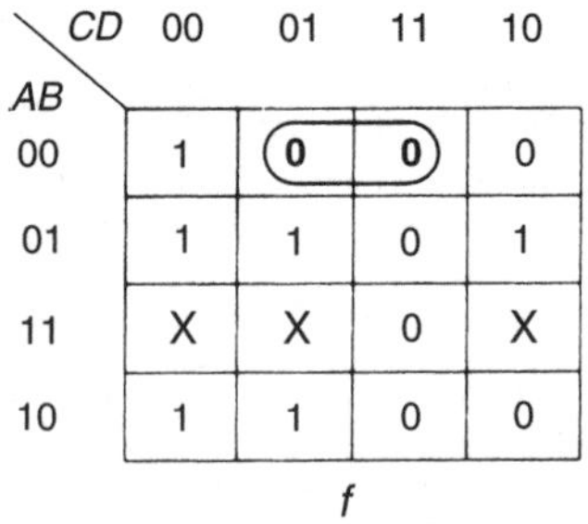

Fig. 5.82 Figure for Self-assessment question 17.

4. Set the remaining don't-care terms to 1. The expression for $\overline{f}$ becomes $\overline{f} = \overline{A} \cdot \overline{B} \cdot D + \overline{B} \cdot C + C \cdot D$.

18. The EXCLUSIVE-OR relationship can be expressed as a sum-of-products as follows:

$$P = A \cdot \overline{B} + \overline{A} \cdot B$$

The sum-of-products circuit for this function therefore consists of two AND gates connected to a single OR gate. We can replace the OR gate with a NAND gate if both of its inputs are inverted. These inversions are obtained by changing the AND gates into NAND gates. Finally, the INVERTERs on each of the inputs are replaced by two input NAND gates with the two inputs joined together. The result is shown in Fig. 5.83.

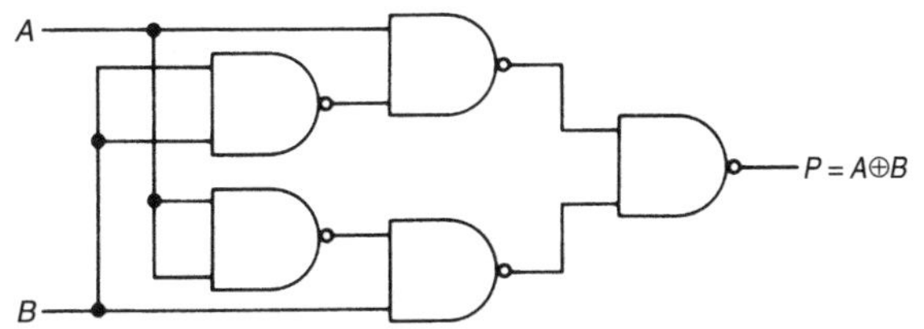

Fig. 5.83 Figure for Self-assessment question 18.

19. We know from deMorgan's theorem that the OR gate can be replaced with a NAND gate provided that its inputs are also inverted. This means converting the AND gates corresponding to the terms $A \cdot B$ and $\overline{A} \cdot C$ to NAND gates. The INVERTER can be made from the other NAND gate if both inputs are joined together. We therefore need a total of four NAND gates to implement P. One solution is shown in Fig. 5.84.

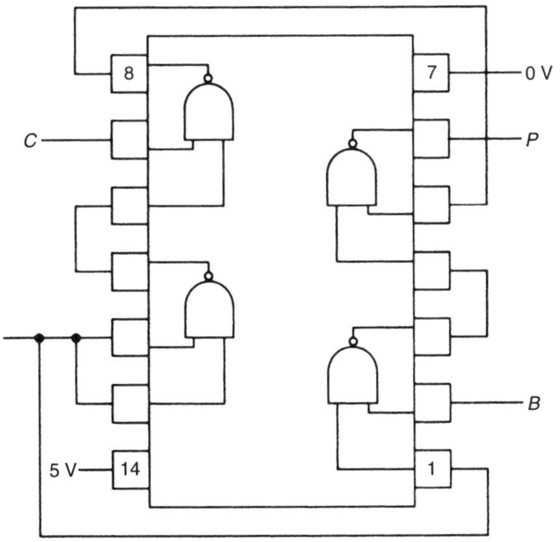

Fig. 5.84 Figure for Self-assessment question 19.

20. The crosses indicate where fuses are unblown and therefore where a connection is made. For gate 1 the connections are with $\overline{A}$ and $\overline{B}$, so the product term that this represents is $\overline{A} \cdot \overline{B}$. Gate 2 is connected to B and C and therefore represents the product term $B \cdot C$; and finally gate 3 is connected to A and $\overline{C}$ and so represents $A \cdot \overline{C}$. The output, P, is the OR of these three product terms:

$$P = \overline{A} \cdot \overline{B} + B \cdot C + A \cdot \overline{C}$$

21. (a) $P = \overline{A \cdot 1} = \overline{A}$

(b) If we say that the input voltage, V_{in}, has to be greater than 2 V to be a recognizable logic 1, then the voltage drop across R must be less than 3 V. The maximum input current for a logic 1 is 20 μA. From Ohm's relationship we have $3 = 20 \times 10^{-6} \times R$, so $R = 150$ kΩ

(c) Again, the voltage drop must be no more than 3 V, so if there are n gates and each gate takes a maximum input current, the total current drawn is $n \times 20$ μA. Using Ohm's relationship:

$$3 = n \times 20 \times 10^{-6} \times 1 \times 10^{3}$$

This gives $n = 150$. So, in theory, 150 gates could be connected to the single resistor. However, manufacturers usually suggest that no more than about 25 gates should be connected via a 1 kΩ resistor, to allow for fluctuations in the power supply and to provide a better noise immunity.

22. The shortest delay is from input C, which has only to propagate through one NAND gate. The delay in this case is 9.5 ns. The longest delay is from input A, which has to propagate through an INVERTER and two NAND gates. The delay in this case is 28.5 ns.

23. The power is calculated by multiplying the supply voltage by the average supply current per gate. The supply voltage is 5 V in all three cases. The average supply currents are 8 mA, 15 mA and 1.0 mA for standard TTL, STTL and ALSTTL, respectively. However, since these are quad gates, that is there are four gates per chip, the average supply current per gate is a quarter of these values: 2 mA, 3.75 mA and 0.25 mA. The power for each gate is therefore 10 mW, 18.75 mW and 1.25 mW, respectively. The propagation delay is the average of t_{PLH} and t_{PHL}, which are 9 ns, 3 ns and 3.5 ns. The power–delay products are therefore:

TTL Power-delay $= 10 \times 10^{-3} \times 9 \times 10^{-9} = 90$ pJ
STTL Power-delay $= 18.75 \times 10^{-3} \times 3 \times 10^{-9}$
$= 56.25$ pJ
ALSTTL Power-delay $= 1.25 \times 10^{-3} \times 3.5 \times 10^{-9}$
$= 4.4$ pJ

For comparison, LSTTL has a value of 19 pJ.

24. (a) All the necessary data can be found to use the formula in Section 5.7.4. From Table 5.9 $V_{cc} = 5$ V and $I_{CC} = 2$ μA. From the question, $C_{PD} = 24$ pF, $C_L = 40$ pF, $f_i = 4$ kHz and $f_o = 1$ kHz. So

$$P = 2 \times 10^{-6} \times 5 + 24 \times 10^{-12} \times 5^2 \times 4 \times 10^3$$
$$+ 40 \times 10^{-12} \times 5^2 \times 10^3$$
$$= 10^{-5} + 2.4 \times 10^{-6} + 10^{-6} = 1.34 \times 10^{-5}\text{W}$$
$$= 13.4\ \mu\text{W}$$

The biggest term in this case is the static dissipation, the first term.

(b) The only change here is that f_i and f_o are both larger by a factor of 10^3. Hence the corresponding terms in the formula are also increased by this factor and the total power disspation is given by

$$P = 10^{-5} + 2.4 \times 10^{-3} + 10^{-3} = 3.41 \times 10^{-3}\text{W}$$
$$= 3.41\ \text{mW}$$

This figure is close to the power expected from an LSTTL gate, and is dominated by the a.c. terms.

Sequential logic circuits

6

AIMS

The main aim of this chapter is to describe some of the basic building blocks of sequential logic, and to introduce the general sequential machine as a design technique for any synchronous circuit.

GENERAL OBJECTIVES

Understand the meaning of the following terms:

- asynchronous inputs
- asynchronous or ripple counter
- bistable logic circuit
- characteristic table
- clock or control input
- clocked *D* latch
- clocked logic
- clocked or gated latch
- *D* flip-flop
- disable
- down counter
- edge-triggered
- enable
- excitation table
- flip-flop
- frequency division
- function table
- general sequential machine
- hold time
- *JK* flip-flop
- latch
- master–slave
- memory
- memory device
- modulo-*n* counter
- negative and positive edge
- next state
- next-state logic circuit
- present state
- preset
- programmable logic sequencer (PLS)
- pulse-triggered
- register
- reset or clear state
- rising and falling edge
- sequential logic
- serial and parallel loading
- serial and parallel output
- set state
- set-up time
- shift register
- *SR* latch
- state
- state table
- state-assignment table
- state-transition diagram
- synchronous counter
- synchronous sequential circuit
- timing diagram
- toggle
- transparency
- up counter

SPECIFIC OBJECTIVES

1. To be able to differentiate between the various types of bistable circuits, and to know when it is appropriate to use one kind or another.
2. To describe the structure and operation of simple registers, shift registers and binary counters.

3. To sketch, and explain the features of, a timing diagram for an n-bit register made from clocked D latches or from edge-triggered D flip-flops.
4. To be able to connect an IC counter to create a modulo-n counter or to cascade several counters to extend the count range.
5. To generate a state transition diagram from the description of a problem, or to follow the flow of a given state transition diagram.
6. To apply the general sequential machine design method to sequential circuits such as counters and automatic processes.

6.1 INTRODUCTION

Chapter 5 introduced the idea of formulating simple logic problems in terms of a binary truth table. The truth table was shown to be a complete and systematic statement of such a problem, and also provided the basis for a practical solution using logic circuits. However, this discussion was restricted to a particular type of problem. It was understood that a certain combination of binary digits would appear on the output of a logic circuit whenever a particular combination of binary digits was applied to its inputs, irrespective of the past 'history' of its inputs. In other words, for combinational logic, the output depends only on the present input and is not influenced by what has gone before.

Digital simulation really comes into its own with sequential circuitry. Nearly every sequential circuit has potential problems with timing, and simulation is often the only practical method to identify and remove such problems.

As in Chapter 5, you might find it helpful to simulate the behaviours of some circuits in this chapter. Particularly suitable areas are indicated in the margin.

With this restriction, the solutions can be implemented using combinational logic devices, either by devising a suitable arrangement of standard gates and other elements or by using a programmable device such as a programmable array logic (PAL) device or read-only memory (ROM). The ROM can be regarded as a direct, practical realization of the truth table, giving a specific binary output for each valid combination of binary digits applied to its input or address terminals.

This chapter introduces you to the idea of **sequential logic** circuits in which the inputs and outputs have the form of sequences of binary patterns. The difference between combinational and sequential logic is that the behaviour of a sequential logic circuit is not only determined by its present inputs but also *by its past inputs*. As you will see, sequential logic circuits can be designed both to modify and to generate sequences of binary patterns. Application examples include the counting of events and the timing of operations for the sequential control of domestic and industrial processes.

In order to respond to past as well as present inputs, a fundamental requirement of sequential circuits is the ability to store binary data. Hence Section 6.2 describes the most basic circuit elements that have the property of memory. Sections 6.3 to 6.6 then describe how these elements can form the basis of practical devices. Finally Sections 6.7 and 6.8 introduce a systematic design approach to sequential circuitry.

6.2 LATCHES

The basic property of a **memory device** is that it should be able to hold its output at a fixed value until instructed to change it. For logical elements this output value should either be a 0 or a 1. This function is provided by **bistable logic circuits** which have two stable states, a set state and a reset or clear state. Using the positive logic convention, the **set state** is when the output is set to 1, and the **clear state** is when the output is **reset** or cleared to 0.

Figure 6.1 shows one of the most basic types of controllable bistables, constructed from two NOR gates. With the inputs S and R both equal to 0, it can be in one of two stable states at any time. You can use the NOR gate truth table to check that the two states are:

(a) Gate 1 output = 0, gate 2 output = 1.
(b) Gate 1 output = 1, gate 2 output = 0.

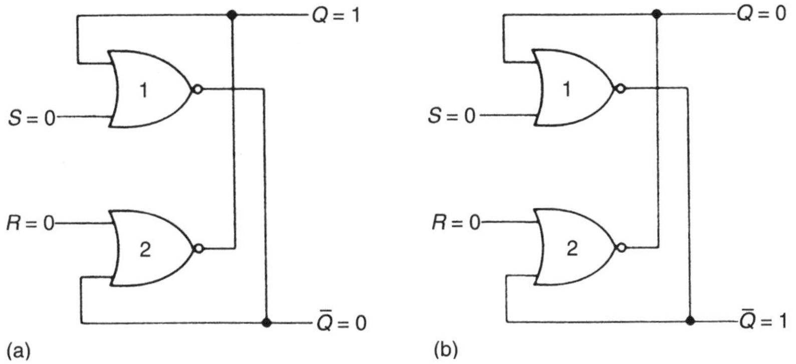

Fig. 6.1 A controllable bistable circuit.

Such a circuit is often called a **latch**, reminiscent of a mechanical device latching into a fixed position.

The circuit will remain fixed in one state or the other as long as S and R are both 0. However, the output may change to the other stable state according to what values are subsequently placed on the inputs S and R. Figure 6.2a shows what happens if S now goes to 1. The top NOR gate must produce an output of 0 regardless of the value of its second input, making $\overline{Q} = 0$. The inputs to the bottom NOR gate will be both 0, making $Q = 1$. Hence if the bistable had started off in the state in Fig. 6.1a, it would remain in this state. However, if it started off in the state of Fig. 6.1b, then it would have changed when S was connected to logic 1. If the input S is now returned to 0, the outputs will stay as they are, that is $Q = 1$ and $\overline{Q} = 0$. This is the set state.

Figure 6.2b shows what happens if R goes to 1 with $S = 0$. The bottom NOR gate must produce an output of 0 regardless of the value of its second input making $Q = 0$, and the inputs to the top NOR gate will be both 0, making

It is useful to test out the operation of this bistable circuit, either by experiment or by computer simulation.

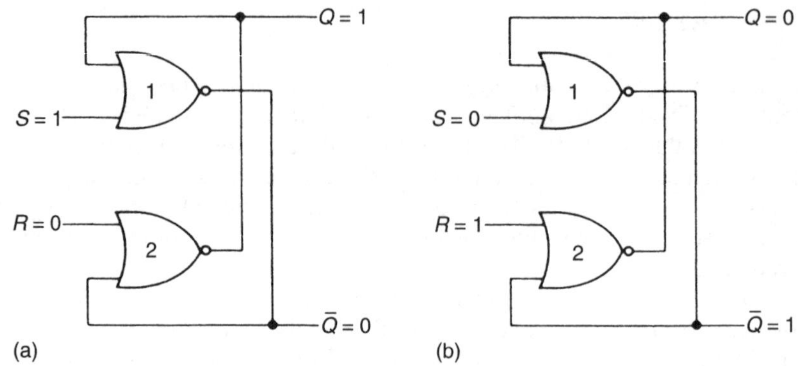

Fig. 6.2 Setting and resetting the controllable bistable.

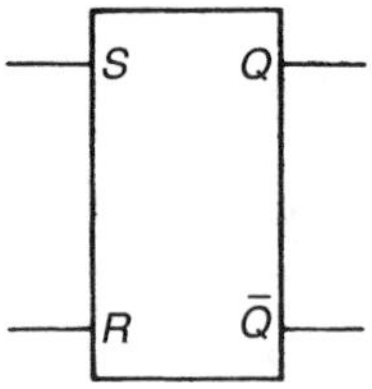

Fig. 6.3 A symbol for the *SR* latch.

$\overline{Q} = 1$. Even if the input R is returned to 0 the outputs stay as they are, in the reset or clear state.

For these input conditions the values of outputs Q and $\overline{Q}$ are the inverse or complement of one another, hence the reason for the labels. The inputs are labelled S and R because one sets and one resets the output—input S going to 1 causes the output Q to be set to 1, whereas the input R going to 1 causes the output Q to be reset to 0 or cleared. This circuit is known as *SR* latch, and its symbol is shown in Fig. 6.3.

Sometimes in circuit symbols the complementary output $\overline{Q}$ is shown with a small ring against it to represent inversion. When this is done, some authors label the output inside the symbol as Q (as in Fig. 6.4a), taking the logically correct view that subsequently passing through the 'inverter bubble' will turn it into $\overline{Q}$ outside the device. Others label it $\overline{Q}$ inside the device as in Fig. 6.4b. Strict conventions are not observed, so beware!

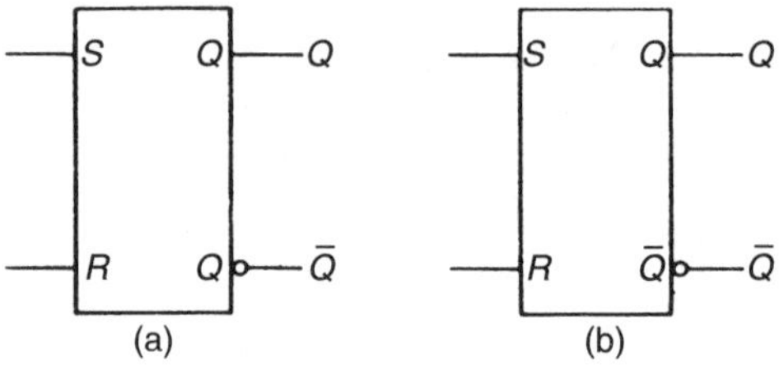

Fig. 6.4 Two alternative symbols for the *SR* latch.

Worked example

Assuming that initially Q is 1, $\overline{Q}$ is 0 and that R and S are both 0, what happens if both inputs go to 1?

Since both of the NOR gates will have at least one of their inputs set to 1, their outputs must both be 0. So when both inputs are 1, both outputs are 0 and the outputs are no longer the inverse of each other.

The case of $S = R = 1$ is a well-defined and stable state, and if either S or R is returned to 0 then the latch will enter the appropriate reset or set state. However, if the two inputs return to 0 together the state of the output is unpredictable—it could go either way. Because of the possibility that the output will become unpredictable, and because Q and $\overline{Q}$ are no longer complementary, it is generally undesirable that both inputs should be at 1 at the same time.

Thus, although the *SR* latch is one of the most important, fundamental methods of digital storage, in practice it is not often used as it stands but as the basis for the slightly more complex latches described in Sections 6.2.1 and 6.2.2.

6.2.1 The clocked *SR* latch

One way to introduce more control into the operation of the *SR* latch is to add a **control** or **clock input** via AND gates to the latch, as shown in Fig. 6.5. When the control input C is 0, then changes in the voltage levels applied to the S and R inputs cannot reach the latch because the outputs of the AND gates are both 0. When the control input is 1, the value of the S and R inputs appear on the AND gate outputs and so, in effect, are connected to the latch. If it is arranged that changes to S and R only occur when the control is 0, the timing problems referred to earlier can be avoided. A latch with gates attached to its inputs in this way is usually called a **clocked** or **gated latch**, and its symbol is also shown in Fig. 6.5.

The behaviour of a device like this cannot be summarized as a truth table, since the output may be latched into one state or the other depending on previous as well as present inputs. However, this time-dependent behaviour can be represented in a **function table**, and one form is shown in Table 6.1. (This is sometimes called a **characteristic table or excitation table**. There is no standard way of writing it down, but that shown is one of the most common and useful.)

The function table shows the changes that take place on the outputs, Q and $\overline{Q}$, for all possible changes on the inputs. The notation Q_n and Q_{n+1} is used to represent the present output and the next output, respectively. Generally, if the value of Q_{n+1} is given as Q_n, it means that, with those particular input conditions, the output is latched and does not change. More specifically, Q_n can be regarded as the value of the output, Q, after n clock pulses or active transitions have occurred, and so the value of the output after one more clock pulse, that is $n+1$ clock pulses, is then Q_{n+1}. The value of this notation will

Table 6.1 Function table for the clocked SR latch

Inputs			*Outputs*		*Comment*
S	R	C	Q_{n+1}	$\overline{Q}_{n+1}$	
x	x	0	Q_n	$\overline{Q}_n$	Hold
0	0	1	Q_n	$\overline{Q}_n$	Hold
0	1	1	0	1	Reset
1	0	1	1	0	Set
1	1	1	0	0	$Q = \overline{Q}$, normally unused

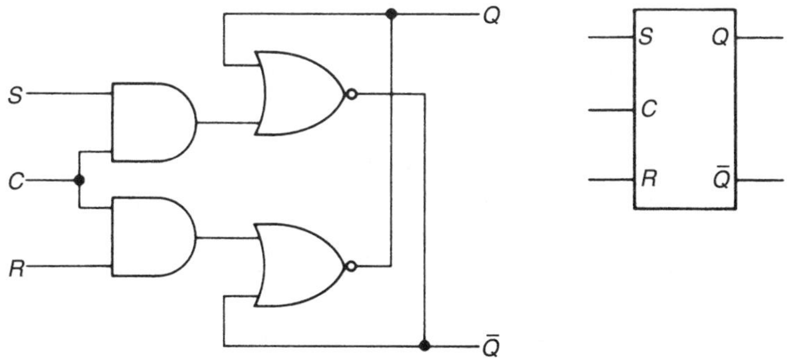

Fig. 6.5 A clocked latch and its symbol.

An alternative notation for the function table is to use Q for the next output (instead of Q_{n+1}) and Q_o for the previous one (instead of Q_n). An example of this is given in the data sheet in Fig. 6.24.

become more apparent when edge-triggered devices are considered in Section 6.3.2.

The table should be read as follows. First, when the control input is 0 the output Q is latched, staying the same as it was just before the control input changed from 1 to 0. The x in the S and R columns is the 'don't-care' symbol, meaning that it does not matter what values (0 or 1) are applied to these inputs at this time. The other rows in the table are all with the control input $C = 1$. When the inputs are both 0 the output, Q, stays at the same value as it was before the control input went to 1. When the inputs are both 1, Q and $\overline{Q}$ are both 0. Because Q and $\overline{Q}$ are no longer complementary, and because of the possibility of an unpredictable state if S and R return together, this state is normally unused. The other two entries in the table show that the output gets set when S is a 1 and reset when R is a 1.

The major advantage of the clock input to this latch is that it can be used to control when the circuit responds to a change of inputs. Thus output changes in different components can be synchronized with each other or with external factors. Such clocked logic is the basis of virtually all sequential circuit designs.

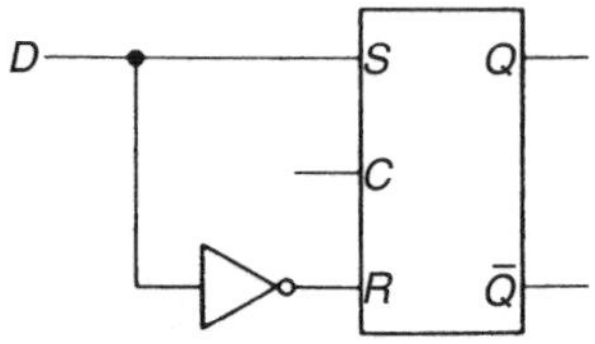

Fig. 6.6 Circuit for Self-assessment question 1.

Self-assessment question

1 Write out the function table for the circuit shown in Fig. 6.6, with inputs D and C, and outputs Q and $\overline{Q}$.

The circuit in Fig. 6.6, usually referred to as a gated or **clocked *D* latch** (or ***D*-type latch**), is widely used in logic circuits, and is described further in Section 6.2.2.

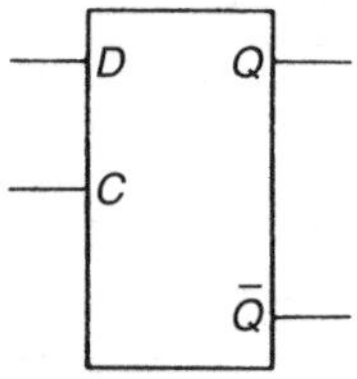

Fig. 6.7 A symbol for the *D* latch.

Table 6.2 Function table for a *D* latch

Inputs		*Outputs*		*Comment*
D	C	Q_{n+1}	$\overline{Q}_{n+1}$	
x	0	Q_n	$\overline{Q}_n$	Hold
0	1	0	1	Reset
1	1	1	0	Set

6.2.2 The clocked *D* latch

The simplest clocked latch of practical importance in logic design is the D-type latch previously shown in Fig. 6.6. Figure 6.7 shows a symbol that is often used to represent this circuit.

The D latch has a data input D, outputs Q and $\overline{Q}$, and the clock pulses are applied to the control input C. The function table for the D latch, shown in Table 6.2, is very simple. It shows that when the control input is 1, the output follows the value of the input. When the control input drops to the 0 level, then the output remains fixed in the state it was before the drop.

Figure 6.8 is another way to show how the output, Q, changes when a clock is applied to the control input and a sequence of binary values is applied to the data input D. Diagrams like Fig. 6.8 are called **timing diagrams** and are widely used in specification and design data to illustrate the relationship between the inputs of a sequential circuit and its outputs. From this figure you should be able to see that the output is the same as the input while the clock is 1. This behaviour is called **transparency**, as setting $C = 1$ to 'let the data through' is analogous to opening a shutter to let light through a window. When the clock goes to 0, the output stays at the value that it had just before the clock changed to 0. Any changes on the input when the clock is 0 are ignored until the clock rises again.

If possible use a computer to simulate the behaviour illustrated in Fig. 6.8, either using the circuitry of Figs. 6.5 and 6.6 or by using an appropriate D-type latch from the package library.

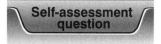

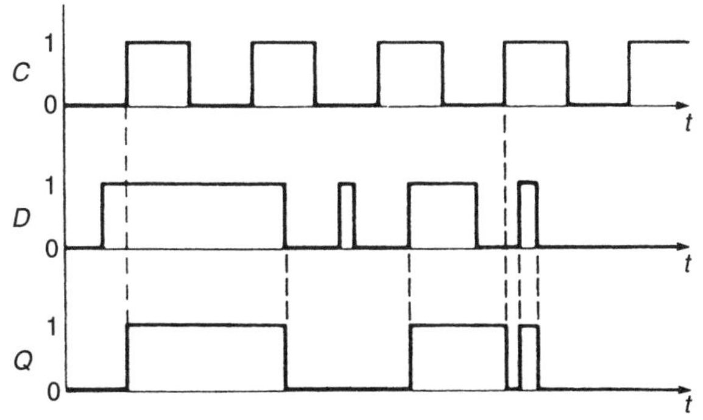

Fig. 6.8 A timing diagram for the D latch.

In circuit design the transparent D latch is often used as a basic memory element for the short-term storage of a binary digit applied to its data input. The binary input present at the time the control pulse is applied is transferred to the output and held there until another control pulse is applied.

6.2.3 Summary of Section 6.2

Two cross-coupled NOR gates form an SR (set and reset) latch, which is one of the basic memory elements of digital electronics. A clocked SR latch has an additional input which controls when setting and resetting can take place. A D latch has a single data input; the output is transparent with the clock input in one state, and the data is latched when the clock input switches to the other state. This behaviour can be represented by the following function table.

Inputs		*Outputs*		*Comment*
D	C	Q_{n+1}	$\overline{Q}_{n+1}$	
x	0	Q_n	$\overline{Q}_n$	Hold
0	1	0	1	Reset
1	1	1	0	Set

6.3 FLIP-FLOPS

The transparency of the clocked D latch can be a disadvantage in some applications: it is often necessary to be able to sample the input at a particular instant and hold this value internally. This can be achieved through the use of either **master–slave** or **edge-triggered flip-flops**. These two types of devices are described in the next two sections.

The term 'latch' is generally used to describe a transparent bistable circuit, and 'flip-flop' to describe an edge-triggered device. However, not everyone uses this terminology consistently, so do not rely on it to be an accurate indication of the characteristics of a device.

6.3.1 The master–slave flip-flop

The circuit diagram of a master–slave D-type flip-flop is shown in Fig. 6.9. It consists essentially of two clocked latches in cascade. The first, called the master, controls the operation of the slave. The clock waveform goes directly to

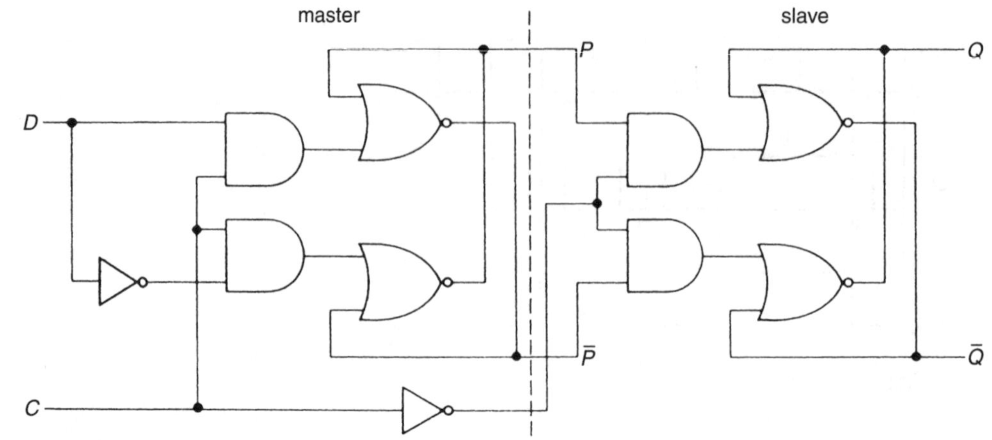

Fig. 6.9 A master–slave *D* flip-flop.

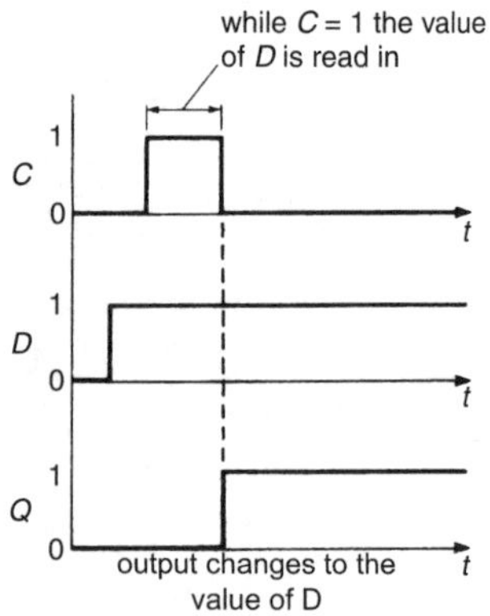

Fig. 6.10 Data storage in a master–slave flip-flop.

the control inputs of the master, but passes through an INVERTER before it reaches the control inputs of the slave. The circuit works as follows.

The D input can only influence the state of the master latch when $C = 1$. Similarly, because of the INVERTER in the control input of the slave circuit, the output of the master can only influence the outputs of the slave when $C = 0$. So, while $C = 1$, the new state of the master can be established using the D input without affecting the slave. Then when C falls to 0 the slave latch is able to change. Since the outputs of the master latch are always the inverse of one another, P and $\overline{P}$, the slave latch produces an output, Q, which is the same as the input, P. However, while $C = 0$, the master latch no longer responds to changes in the D input, and its state remains as it was when C fell to 0. Hence the slave latch also remains in this state.

The result of all this is that, although the master latch is transparent and follows the value of D as long as the clock is high, it is the value applied to D just before the falling edge of the clock which is transferred to the slave latch and hence to the output. Variations of D when the clock is high or low do not affect the output, nor does the 0 to 1 transition of the clock. This behaviour is called edge-triggering, in this case on the **negative edge** (1 to 0) of the clock waveform. The behaviour of such an **edge-triggered *D* flip-flop** is illustrated on a timing diagram in Fig. 6.10.

Self-assessment question

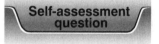

2 Figure 6.11 shows how the inputs to the circuit described in Fig. 6.9 change with time. Complete the diagram by sketching the values of P and Q, assuming that they are both 1 at the start. (If you have access to a computer package, then check your solution by simulation.)

6.3.2 The edge-triggered flip-flop

Although the master–slave configuration gives an edge-triggered characteristic to a D-type latch, there are other, simpler circuits which are inherently

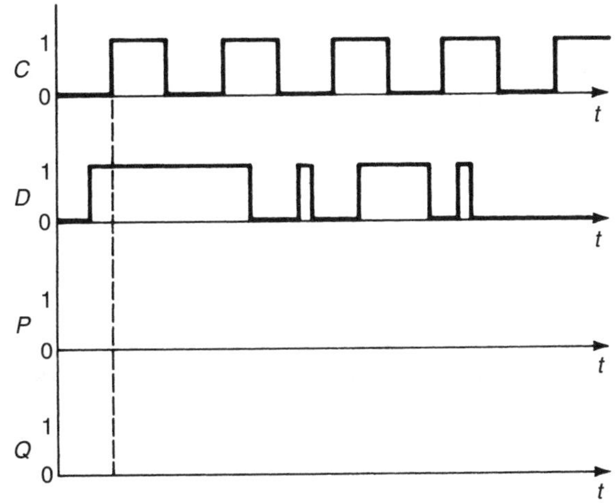

Fig. 6.11 Timing diagram for Self-assessment question 2.

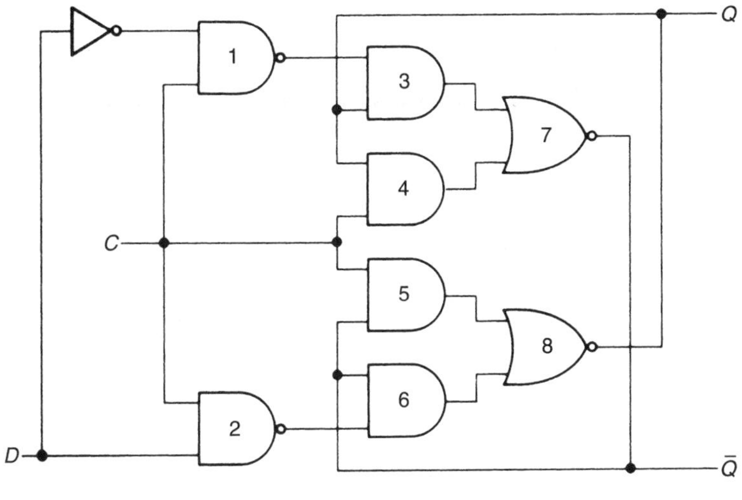

Fig. 6.12 A negative edge-triggered *D*-type flip-flop.

A good example of the effect of propagation delay is shown in Fig. 6.13. Substitution in the truth table shows that when the input *C* is held at a steady value of either 0 or 1, the output *Q* of the AND gate is 0. However, if *C* is rapidly switched from 0 to 1, input *A* changes immediately but there is a short interval before the new value is applied to input *B* due to propagation delay in the inverter. During this interval, both inputs of the AND gate are at logic 1, and so there is a brief pulse or 'glitch' on the output. It is simple and instructive to simulate the behaviour of this circuit with a suitable computer package or to examine the circuit experimentally. In either case, the glitch may be clearer if two or more inverters are placed in series to increase the delay time.

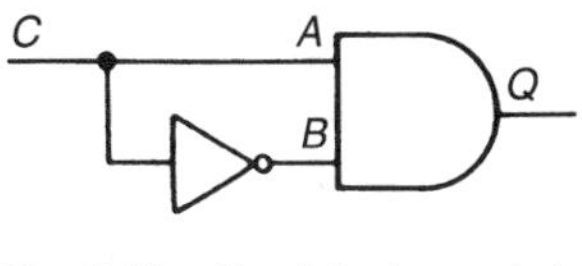

Fig. 6.13 Circuit to demonstrate propagation delay.

edge-triggered and which are used more widely. One form is shown in Fig. 6.12. In all logic gates there is always a delay between the input changing and the corresponding change appearing on the output (see Chapter 5), and the successful operation of this circuit relies on this propagation delay in the input pair of NAND gates.

Notice that this time there is no INVERTER in the control input line in Fig. 6.12 and that the input NAND gates are followed by AND gates. When $C = 1$ the outputs of the NAND gates 1 and 2 are the values of D and $\overline{D}$ respectively. The outputs of the AND gates 3 to 6 are then $Q \cdot D$, Q, $\overline{Q}$, and $\overline{Q} \cdot \overline{D}$ respectively which makes the outputs of the NOR gates equal to $\overline{Q}$ and Q.

Now consider what happens when C falls from 1 to 0. Ultimately, when everything has settled down after C goes to 0, the outputs of both NAND gates will be 1, and so the outputs of the AND gates will be Q, 0, 0, and $\overline{Q}$ making the outputs of the NOR gates equal to $\overline{Q}$ and Q again.

However, there is a short period after the clock changes to 0 when the values of the outputs of the NAND gates have not yet changed. The output of AND gate 4 will therefore change to 0 *before* the output of AND gate 3 changes to Q, causing the output of the NOR gate 7 to be temporarily equal to $\overline{D} + \overline{Q}$. If $\overline{D}$ is 1 (i.e. if D is 0) the flip-flop will be reset.

Similarly the output of AND gate 5 will change to 0 before the output of AND gate 6 changes to $\overline{Q}$, causing the output of the NOR gate 8 to be temporarily equal to $D + Q$. If D is 1 the flip-flop will be set.

You might like to check this by computer simulation.

Figure 6.14 shows the timing diagram for one of the paths through this edge-triggered flip-flop. Just as in the master–slave flip-flop, the value of the output will change on the falling edge of the clock to the values of the input D that existed just before the falling edge of the clock.

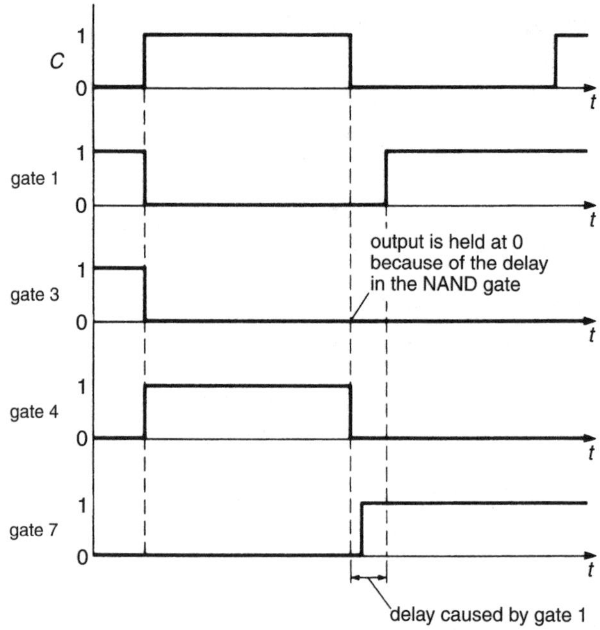

Fig. 6.14 Timing diagram for one path through the circuit of Fig. 6.12.

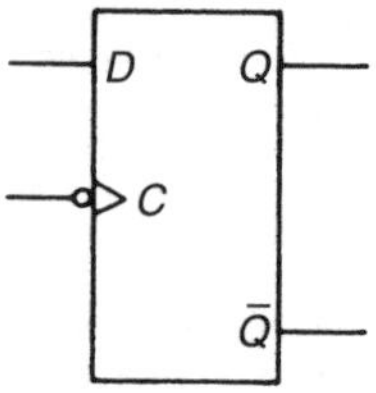

Fig. 6.15 A symbol for an edge-triggered D flip-flop.

The symbol for an edge-triggered D flip-flop is shown in Fig. 6.15. The $>$ symbol is used to indicate that the device is edge-triggered and the small circle on the control input shows that the device changes state when the pulse applied to it goes from 1 to 0, the **negative edge** (also called the **negative-going** or **falling** edge). Devices which trigger on the **positive** or **rising** edge (0 to 1) do not have the small circle.

The function table for a negative edge-triggered D flip-flop can be written in the form of Table 6.3. The $\downarrow$ symbol means a falling edge. The last line of the table, showing that the output is latched when the clock input is steady, is not always included in the function table for edge-triggered devices. However, it does emphasize that no change occurs to the output except at the clock edge.

Table 6.3 Function table for a D flip-flop

Inputs		Output	Comment
D	C	Q_{n+1}	
0	$\downarrow$	0	Reset
1	$\downarrow$	1	Set
X	X	Q_n	Hold

An important lesson to be drawn from this edge-triggered circuit is that relative timing is an important factor in the behaviour of sequential circuits. Propagation delays must always be taken into account to ensure that signals are present in the correct sequence. However, even with the relatively straightforward circuit of Fig. 6.12, the effects of propagation delay are by no means obvious on first inspection. Thus, in sequential design, a greater deal of use is made of circuit simulation to identify and correct timing problems.

Self-assessment question

3 Figure 6.16 shows a circuit commonly used for positive edge-triggered *D* flip-flops.

(a) The circuit is based on three pairs of NAND gates cross-coupled to form latches. Write out the function table for such a latch made from 2 two-input NAND gates. (Consider all four possible combinations of levels on the inputs $\overline{S}$ and $\overline{R}$. This circuit is normally called an $\overline{SR}$ latch, because the stable state is with $\overline{S} = 1$ and $\overline{R} = 1$ and each input achieves its function by going to logic 0.)

(b) Investigate the function of Fig. 6.16 by completing the following table line by line. This may be done by working through the circuit by hand or by using a digital simulation package on a computer.

	C	D	E	F	G	H	Q	$\overline{Q}$
Initial state	0	0	0	1	1	1	1	0
Positive clock edge	1	0						
Change *D*	1	1						
Negative clock edge	0	1						
Positive clock edge	1	1						
Negative clock edge	0	1						
Change *D*	0	0						
Positive clock edge	1	0						

(Notice that latching only occurs on a positive transition of the clock input, and that the outputs *Q* and $\overline{Q}$ are affected neither by changes in *D* when the clock is steady nor by a negative edge on the clock.)

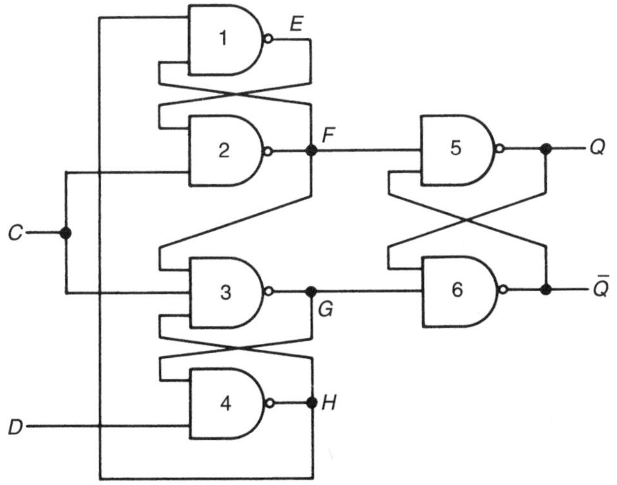

Fig. 6.16 A positive edge-triggered *D* flip-flop.

Self-assessment question

4 An edge-triggered *D* flip-flop is connected as shown as Fig. 6.17. The flip-flop is initially cleared to $Q = 0$. Write down the sequence of outputs which would be generated by a series of pulses applied to its control input. What would happen if the flip-flop was replaced by a transparent *D* latch?

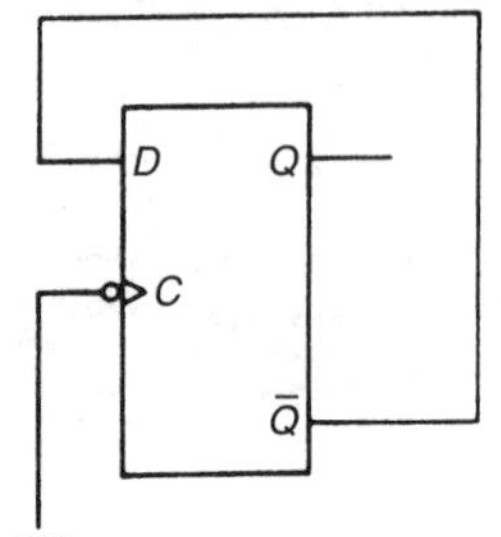

Fig. 6.17 Circuit for Self-assessment question 4.

6.3.3 The *JK* flip-flop

The most versatile form of flip-flop is the ***JK* flip-flop**. Its versatility results from the fact that it has two data inputs (*J* and *K*), and Table 6.4 shows the function table for all combinations of levels on these inputs.

The difference between this device and an *SR* flip-flip is the output value, Q_{n+1}, when the *J* and *K* inputs are both 1. In the *SR* flip-flop this condition results in $Q = \overline{Q}$, whereas in the *JK* flip-flop this combination of inputs causes the output to **toggle** on the next clock edge, which means that it changes to the opposite state.

Hence with $J = K = 1$, the output produces the sequence 0, 1, 0, 1, 0, 1, 0, . . . with a train of clock pulses.

Table 6.4 Function table for a positive edge-triggered *JK* flip-flop

Inputs *J K C*	Output Q_{n+1}	Comment
0 0 ↑	Q_n	Hold
0 1 ↑	0	Reset
1 0 ↑	1	Set
1 1 ↑	$\overline{Q_n}$	Toggle
x x x	Q_n	Hold

Early forms of integrated-circuit *JK* flip-flop were based on the master–slave principle, but these suffer from a particular problem. The *JK* master latch will change state in response to input signals occurring at any time while the clock input is a 1. Therefore additional pulses caused by interference or by switching problems in other parts of the circuit can cause the latch to end up in the wrong state when the clock finally makes the 1 to 0 transition. The best way to control these devices is by a very brief clock pulse, so that the opportunity for incorrect switching is minimized, and the input data must be steady throughout the duration of this pulse. Thus these master–slave *JK* flip-flops are often called **pulse-triggered** devices. Most recent designs of *JK* flip-flop are pure edge-triggered, in which, as in Fig. 6.12, the output state depends only upon the inputs that existed at the moment of transition. The symbol for a positive edge-triggered *JK* is shown in Fig. 6.18.

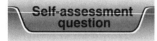

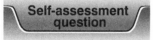

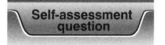

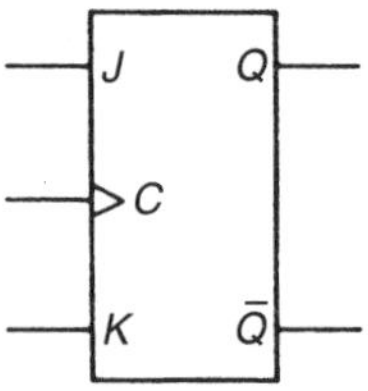

Fig. 6.18 A symbol for a positive edge-triggered *JK* flip-flop.

The *JK* flip-flop is popular because it is a kind of 'general-purpose' flip-flop that can be connected to replace all other types of flip-flop.

Self-assessment question

5 Indicate how you would use the *JK* flip-flop to act as a *D* flip-flop. (An additional component is necessary.)

Self-assessment question

6 (a) Write out the function table for the *T* flip-flop (or toggle flip-flop) shown in Fig. 6.19, obtained by connecting the *J* and *K* inputs together.
(b) Assuming this flip-flop is initially cleared to $Q = 0$, sketch the output waveform which would result when the clock input has equal 'high'

and 'low' periods and $T = 1$. Show clearly the timing relationship between the clock waveform and the output waveform.

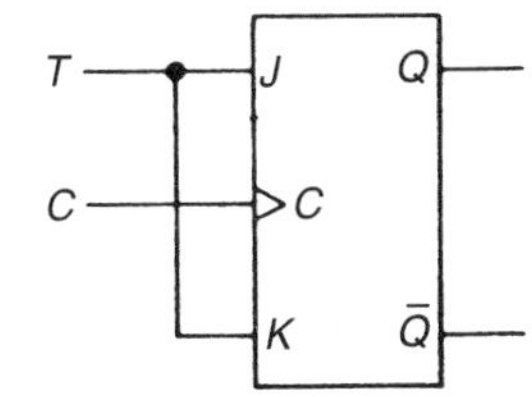

Fig. 6.19 Flip-flop connections for Self-assessment question 6.

6.3.4 Practical flip-flops

Flip-flops and latches are the basic elements of binary storage, and are available in a wide variety of forms in all logic families. For example, most 74-series TTL and CMOS families (see Chapter 5) contain the 74x74 dual, positive edge-triggered D flip-flop as outlined in Fig. 6.20a. The package contains two independent flip-flops, each with D and clock (CLK) inputs, and with Q and $\bar{Q}$ outputs. The internal logic circuit is similar to Fig. 6.17. In addition each flip-flop has two **asynchronous inputs**, PR (= preset) to set the flip-flop (so that $Q = 1$) and CLR to clear or reset it (so that $Q = 0$). These inputs are called asynchronous because they affect the state of the output immediately, regardless of the state of the clock input, and so do not need a clock transition to carry out their function. They are useful if you want to clear the outputs of all flip-flops when you switch on a circuit, for example, or to preset it to a given state at any time.

A common JK flip-flop, the 74LS112, is outlined in Fig. 6.20b. This has the same asynchronous inputs as the 74LS74, but is a negative edge-triggered device based on the circuit in Fig. 6.12. The package has two extra pins to allow for each flip-flop having J and K inputs instead of just a D input.

Although these examples are useful devices in many applications, integrated circuit technology allows much more complicated systems than dual flip-flops to be fabricated. Hence the most common storage devices consist of many bistable elements grouped together to form **registers** (which generally store a single binary word) or **memories** (which can store a great many binary words). Sections 6.4 and 6.5 will describe some common forms of register.

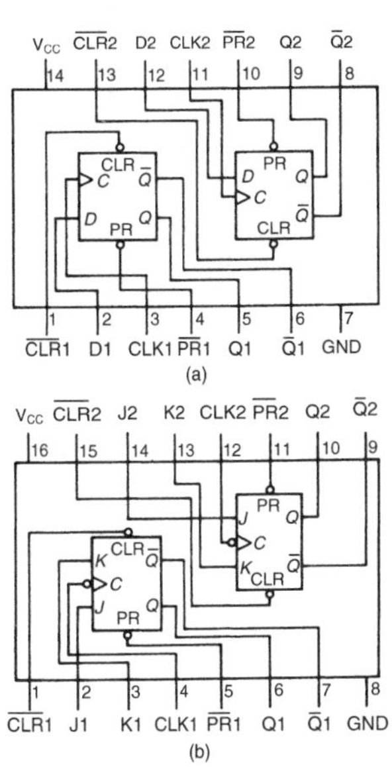

Fig. 6.20 Schematic package outlines of (a) 74LS74 dual positive edge-triggered D flip-flops. (b) 74LS112 dual negative edge-triggered JK flip-flops.

The PR and CLR inputs in Fig. 6.20 have a small ring against them to represent inversion because they are 'active low', so that it is a low logic level (0 in the positive logic convention) which causes the named action to take place. Often a bar is placed over the name of the active low function to emphasize this.

6.3.5 Summary of Section 6.3

The output of an edge-triggered flip-flop only changes when the clock input makes the appropriate transition. This characteristic can be achieved with two latches connected as a master and slave, or by special circuits which rely on the propagation delay in gates to trigger only on an edge. D flip-flops are widely used as a basic element of data storage devices, and the behaviour of a negative edge-triggered variety can be represented by the following function table.

Inputs		*Output*	*Comment*
D	*C*	Q_{n+1}	
0	↓	0	Reset
1	↓	1	Set

JK flip-flops are versatile components that can be configured to behave as an *SR*, toggle or *D*-type device. The *J* and *K* inputs behave as follows in a positive edge-triggered device:

Inputs JK	C	Output Q_{n+1}	Comment
00	↑	Q_n	Hold
01	↑	0	Reset
10	↑	1	Set
11	↑	$\overline{Q}_n$	Toggle

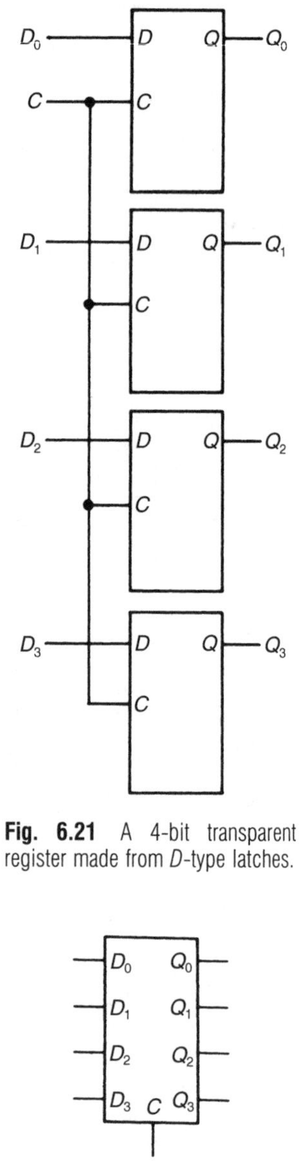

Fig. 6.21 A 4-bit transparent register made from *D*-type latches.

6.4 REGISTERS

Figure 6.21 shows how a group of clocked *D* latches can be combined to form a 4-bit register. Each latch has its own *D* input and *Q* output, but the clock input is fed to every latch so that each one stores one bit of the input data at the same instant. Depending on the use being made of the register, input *C* can be thought of as a *control* input or as a *clock* input.

Since the register is made up of clocked *D* latches, its behaviour is identical to that of these latches previously shown in Section 6.2.2. This means that when $C = 1$ it is transparent, and each output *Q* equals the corresponding input *D* (for instance, $Q_2 = D_2$). When $C = 0$ each output *Q* equals the value which was stored in the corresponding latch at the instant *C* went to 0.

Figure 6.22 shows the symbol generally used for a 4-bit transparent register made up of four gated *D* latches.

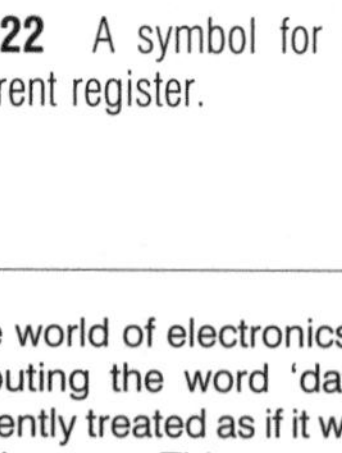

Fig. 6.22 A symbol for a 4-bit transparent register.

In the world of electronics and computing the word 'data' is frequently treated as if it were a singular noun. This convention is adopted in this text.

6.4.1 Timing characteristics of a transparent register

The behaviour of the edge-triggered circuit in Fig. 6.12 illustrated how important it is to consider the detailed timing of devices in the operation of sequential circuits. In the case of a transparent register made from clocked *D* latches, there are two main considerations. The first is that (when $C = 1$) the outputs do not change at the exact instant the inputs change: there is a small propagation delay t_{pd} between these two events. The second consideration is that the data on any *D* input should be steady around the time that the *C* input goes from 1 to 0—that is, when the latch is latching on to the data in order to store it. If the data is changing too closely to the instant of latching, the stored value is unpredictable.

Figure 6.23 illustrates the timing relationships that must be observed. Notice that for both *D* and *Q* two logic levels are shown, rather than just one. This is because the register has several inputs and outputs. Some may be 0 and some may be 1, and these two possibilities are indicated by the two levels on the timing diagram. The cross-hatching indicates when changes can be applied to the *D* input, or when the output is invalid.

The interval between points W and Y is called the **set-up time,** $t_{\text{set-up}}$ or t_{su}, and the data on the *D* inputs must not change for *at least* this time interval before the *C* input goes from 1 to 0. The interval between points Y and Z is called

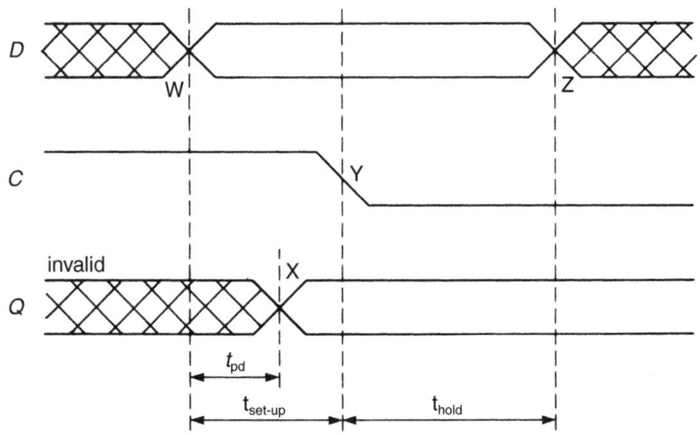

Fig. 6.23 Timing diagram for a transparent register.

the **hold time,** t_{hold} or t_h, and the data on the D inputs must be held steady for *at least* this time interval after the C input has gone from 1 to 0. Both the set-up time and the hold time are generally of the order of a few nanoseconds ($1\ \text{ns} = 10^{-9}\text{s}$). The user of the register must observe these two time intervals in a circuit design to ensure that the value on the D inputs stays steady while C changes from 1 to 0. If this is not done, the value stored is unpredictable.

Also shown on Fig. 6.23 is the propagation delay, t_{pd}, mentioned earlier. It is the interval between points W and X and is the delay between when the inputs change and when the outputs change. Finally, the slopes on the rises and falls in Fig. 6.23 serve to remind you that signal levels do not themselves change instantaneously, but rise and fall over a finite, though small, time interval.

6.4.2 A commercial 8-bit register

Figure 6.24 shows extracts from the data sheet for a typical 8-bit register, the SN74HCT573. (In fact, the data sheet also describes the SN54HCT573, which differs mainly in the temperature range over which it is specified.) The letters 'HCT' in the device codes indicate that these registers are implemented in TTL-compatible high speed CMOS (see Chapter 5). The number '573' describes the function of the device: an 8-bit register described by the manufacturer as 'octal transparent D-type latches with 3-state outputs'.

The term '3-state output' refers to a common feature of integrated circuits like this which helps their application in complex systems. It was used in the PAL device shown in Fig. 5.59. Two of the states that the outputs can be in correspond to the normal voltage levels of logic 1 and logic 0. In the third state the output is switched into a high-impedance condition, in which virtually no current can flow either to or from the output pin. In this state the outputs are effectively disconnected from the external circuit attached to them. Hence more than one device with 3-state outputs can share a common set of connecting wires, as long as only one output at a time is **enabled** and so has normal logic levels on it.

It is important not to confuse the high-impedance state of a 3-state output with the normal logic 0 to 1 levels. With these normal logic levels the output can source current (i.e. current can flow out if the external potential is lower) or sink current (i.e. current can flow in when the external potential is higher). In the high-impedance state, the output cannot source or sink current under any circumstances, and so cannot affect any inputs or outputs of other devices connected to it.

SN54HCT573, SN74HCT573
OCTAL TRANSPARENT D-TYPE LATCHES WITH 3-STATE OUTPUTS

SCLS176C – MARCH 1984 – REVISED FEBRUARY 1998

- **Inputs Are TTL-Voltage Compatible**
- **High-Current 3-State Outputs Drive Bus Lines Directly or Up to 15 LSTTL Loads**
- **Bus-Structured Pinout**
- **Package Options Include Plastic Small-Outline (DW) and Ceramic Flat (W) Packages, Ceramic Chip Carriers (FK), and Standard Plastic (N) and Ceramic (J) 300-mil DIPs**

SN54HCT573 . . . J OR W PACKAGE
SN74HCT573 . . . DW OR N PACKAGE
(TOP VIEW)

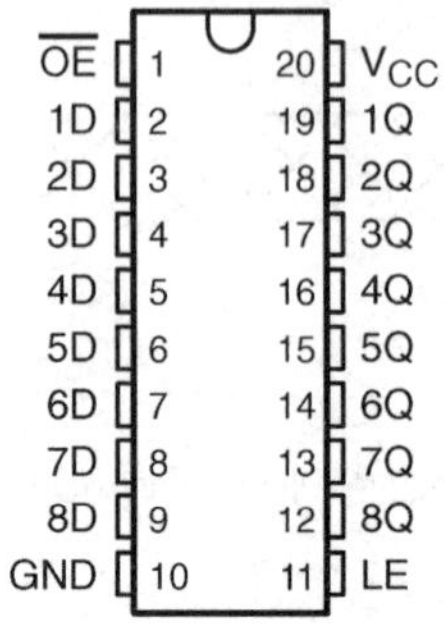

SN54HCT573 . . . FK PACKAGE
(TOP VIEW)

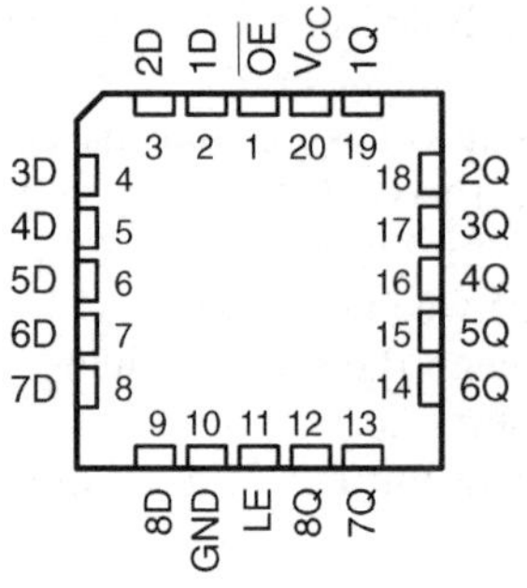

description

These octal transparent D-type latches feature 3-state outputs designed specifically for driving highly capacitive or relatively low-impedance loads. They are particularly suitable for implementing buffer registers, I/O ports, bidirectional bus drivers, and working registers.

While the latch-enable (LE) input is high, the Q outputs respond to the data (D) inputs. When LE is low, the outputs are latched to retain the data that was set up at the D inputs.

A buffered output-enable ($\overline{OE}$) input can be used to place the eight outputs in either a normal logic state (high or low logic levels) or the high-impedance state. In the high-impedance state, the outputs neither load nor drive the bus lines significantly. The high-impedance state and increased drive provide the capability to drive bus lines without interface or pullup components.

$\overline{OE}$ does not affect the internal operations of the latches. Old data can be retained or new data can be entered while the outputs are in the high-impedance state.

The SN54HCT573 is characterized for operation over the full military temperature range of –55°C to 125°C. The SN74HCT573 is characterized for operation from –40°C to 85°C.

FUNCTION TABLE
(each latch)

INPUTS			OUTPUT
$\overline{OE}$	LE	D	Q
L	H	H	H
L	H	L	L
L	L	X	Q_0
H	X	X	Z

Fig. 6.24 Extract from the data sheet for an 8-bit transparent register. Reproduced by permission of Texas Instruments.

logic diagram (positive logic)

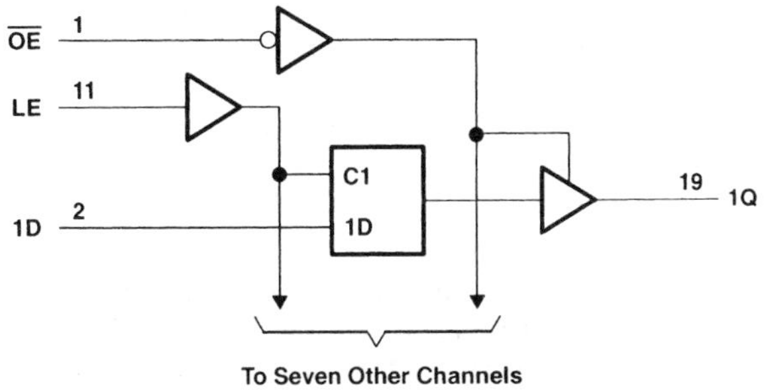

timing requirements over recommended operating free-air temperature range (unless otherwise noted)

		V_{CC}	T_A = 25°C		SN54HCT573		SN74HCT573		UNIT
			MIN	MAX	MIN	MAX	MIN	MAX	
t_w	Pulse duration, LE high	4.5 V	20		30		25		ns
		5.5 V	17		27		23		
t_{su}	Setup time, data before LE↓	4.5 V	10		15		13		ns
		5.5 V	9		14		12		
t_h	Hold time, data after LE↓	4.5 V	5		5		5		ns
		5.5 V	5		5		5		

switching characteristics over recommended operating free-air temperature range, C_L = 50 pF (unless otherwise noted) (see Figure 1)

PARAMETER	FROM (INPUT)	TO (OUTPUT)	V_{CC}	T_A = 25°C			SN54HCT573		SN74HCT573		UNIT
				MIN	TYP	MAX	MIN	MAX	MIN	MAX	
t_{pd}	D	Q	4.5 V		25	35		53		44	ns
			5.5 V		21	32		48		40	
	LE	Any Q	4.5 V		28	35		53		44	
			5.5 V		25	32		48		40	
t_{en}	$\overline{OE}$	Any Q	4.5 V		26	35		53		44	ns
			5.5 V		23	32		48		40	
t_{dis}	$\overline{OE}$	Any Q	4.5 V		23	35		53		44	ns
			5.5 V		22	32		48		40	
t_t		Any Q	4.5 V		9	12		18		15	ns
			5.5 V		9	11		16		14	

PRODUCT PREVIEW

PRODUCT PREVIEW

Fig. 6.24 continued

The function table in Fig. 6.24 is written in terms of logic levels L (= low) and H (= high) as they appear on the device. The L and H representation is a general description of device behaviour that can be used with either positive or negative logic conventions. ‘Z’ is the standard symbol used to indicate that the output is in its high-impedance state. Thus a low level on the $\overline{OE}$ (= active low output enable) pin causes a normal logic level to appear on each of the latch output pins. However, when a high level is applied to $\overline{OE}$, the outputs are **disabled** by being switched into the high-impedance state. Q_0 is used to denote 'the value the output had when LE last made the transition from 1 to 0', and is an alternative to the Q_{n+1}, Q_n notation used elsewhere in this chapter.

In the table of timing requirements in Fig. 6.24, notice that there is a requirement, t_w, on the minimum time that LE must be high (1 in positive logic terminology). This is an extra timing constraint over and above the set-up and hold times already mentioned in Section 6.4.1. '$LE\downarrow$' means 'when LE goes from 1 to 0 (high to low)'.

In the table of switching characteristics, notice that there are two propagation delays quoted, one measured from the instant D changes and the other from the instant LE changes. In this table, t_{en} (enable time) is the time delay between $\overline{OE}$ changing and the output changing from the high-impedance state to 0 or 1; t_{dis} (disable time) is the time delay between $\overline{OE}$ changing and the output changing from 0 or 1 to the high-impedance state; t_t (transition time) is the time for the output signal to change up (rise) from 10% to 90% of its final value or change down (fall) from 90% to 10% of its initial value.

Self-assessment question

7 Consider a SN74HCT573 device being used at an ambient temperature of 25°C and with a supply voltage of 4.5 V. (Use the column headed 'T_A = 25°C'.)

(a) While C is 1 and $\overline{OE}$ is 0, input 3D changes from 0 to 1. How long will it be before output 3Q also changes from 0 to 1?

(b) While 4D is 1, 4Q is 0 and $\overline{OE}$ is 0, LE changes from 0 to 1. How long will it be before 4Q changes from 0 to 1?

(c) How long before LE goes from 1 to 0 must the input data be present on the D inputs?

(d) What happens to the value of 2Q if 2D goes to 1 while LE = 1 and $\overline{OE}$ = 1? If OE then goes to 0, how long will it be before 2Q changes?

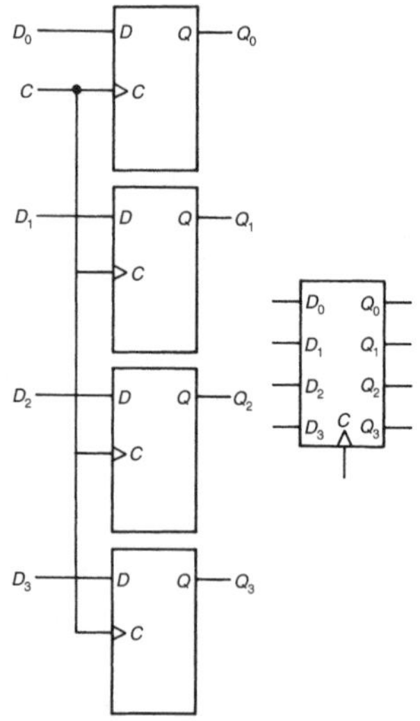

Fig. 6.25 A 4-bit edge-triggered register made from D-type flip-flops.

6.4.3 Timing characteristics for edge-triggered registers

As an alternative to the transparent latches used in the register of Fig. 6.21, it is also common to use edge-triggered flip-flops, and such an edge-triggered register is outlined in Fig. 6.25. In this device the output value only changes at the instant when C changes, and the behaviour is the same as that of a single flip-flop outlined in Section 6.3.2.

The timing considerations for an edge-triggered device are rather different from those of a transparent one and a timing diagram is shown in Fig. 6.26. Because the output only changes in response to a transition in the clock input, propagation delay is defined between the clock edge and output change. Similarly the set-up and hold times are the intervals over which the data must be held steady before and after the positive clock transition which stores the data.

6.5 SHIFT REGISTERS

Shift registers are an important class of device which allow stored data to be moved from one bit position to another. To illustrate the principle, Fig. 6.27

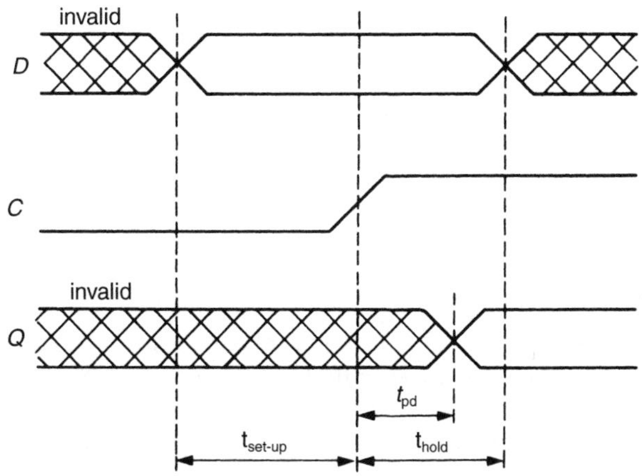

Fig. 6.26 Timing diagram for an edge-triggered register.

shows four D flip-flops connected to form one type of shift register. Suppose that each flip-flop has been reset so that the outputs Q_A, Q_B, Q_C and Q_D form the output word 0000. If D_A is set to logic 1, then the outputs after two clock pulses are as follows:

		Q_A	Q_B	Q_C	Q_D
$D_A = 1$:	initial values	0	0	0	0
	after clock 1	1	0	0	0
	after clock 2	1	1	0	0

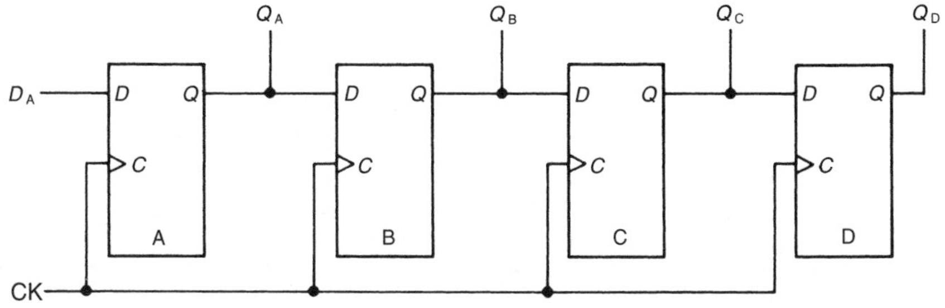

Fig. 6.27 A 4-bit shift register made from D-type flip-flops.

On the first clock edge a 1 was entered into the Q_A position. On the next pulse, this 1 was shifted to Q_B, and since D_A was still 1 a new 1 was loaded into Q_A.

Now suppose that with the present register contents D_A is connected to logic 0. As more clock pulses are applied the sequence continues:

		Q_A	Q_B	Q_C	Q_D
$D_A = 0$:	initial values	1	1	0	0
	after clock 3	0	1	1	0
	after clock 4	0	0	1	1
	after clock 5	0	0	0	1
	after clock 6	0	0	0	0

On each clock edge, each bit of data is shifted one place to the right. Data is lost from the Q_D position. Since the D_A input is held at 0, the new data added to Q_A is a succession of 0s.

It is useful to simulate this example on a computer, building up the shift register from *D*-type latches.

This example has illustrated a number of functions that shift registers can carry out:

- A sequence of values applied to the D_A input can be clocked in to fill the register. This is called **serial loading** of the register.
- The original contents of the register appear one by one on the Q_D output as each clock pulse is applied. This gives **serial output** of the register contents.
- If the values of Q_A to Q_D are available as external connections, then **parallel output** of the register contents is possible.
- If the state of each flip-flop can be set or reset directly then **parallel loading** of the register is possible.

A great many integrated circuit shift registers are available with various combinations of serial or parallel loading, serial or parallel output, and left or right shift. Their main use is in converting a parallel word into serial form, or vice versa. For example, a 16-bit word can be sent along a single wire by parallel loading it into a 16-bit shift register and then clocking it out serially along the wire. At the other end it can be converted back to parallel form with a serial load/parallel output shift register (making sure that the register is clocked in synchronization with the serial stream of data).

Notice that in Fig. 6.27 the data shifted serially into D_A appears 4 clock pulses later, in the same order, on output Q_D. Thus a shift register can be used to delay a serial stream of data by a number of clock pulses equal to the number of register stages. If the clock is controlled in a different way, then the shift register becomes a 'first in–first out' serial data store, with each stored bit clocked out when required.

Also, a number of important logical and arithmetic operations can be carried out by shifting, and shift registers form an important part of microprocessors. For example, multiplication by 2 in binary involves a left shift with a 0 loaded into the LSB (least significant bit) position, as in

Binary	(*Denary*)
$11 \times 10 = 110$	$(3 \times 2 = 6)$
$1011 \times 10 = 10110$	$(11 \times 2 = 22)$

This is very simply achieved with the appropriate form of shift register that can parallel load and parallel output, shift to the left and serially load a zero into the LSB position.

Self-assessment question

8 Figure 6.28 shows a 4-bit shift register, with the Q_D output connected back through an inverter to the D_A input. If the initial state of the register is 0000, draw timing diagrams for the outputs Q_A to Q_D for the next 10 clock pulses. Hence list the sequence of output states which are produced before repetition.

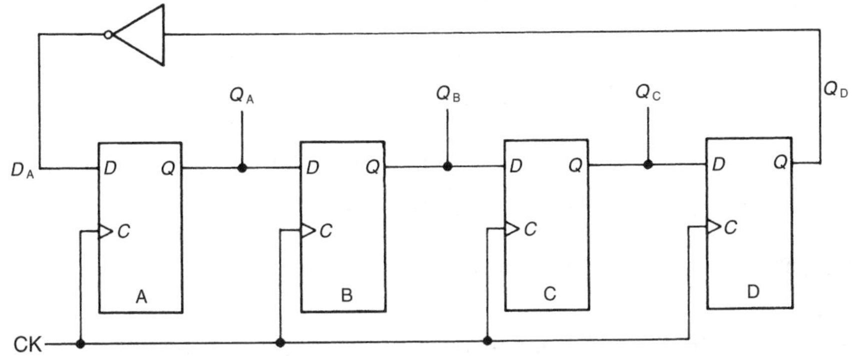

Fig. 6.28 Circuit for Self-assessment question 8.

6.6 COUNTERS

A very important aspect of sequential circuitry is the ability to count. Counting may be of regular clock pulses to determine a certain time duration, or of random pulses such as those arising from a sensor detecting objects passing on a conveyor belt in an industrial process.

The simplest and most common method of counting is by means of the natural binary sequence of codes. For three bits this sequence is 000, 001, 010, 011, 100, 101, 110, 111, then again 000, 001, ... and so on. A circuit which follows the sequence in this order is called an **up counter**. A **down counter** works with the reverse sequence 111, 110, 101, etc.

A simple circuit which functions as a 3-bit natural binary up counter is outlined in Fig. 6.29. It is based around three negative edge-triggered *JK*

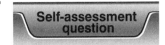

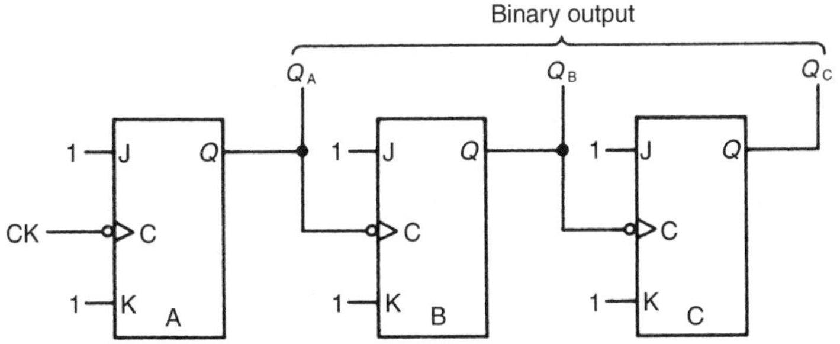

Fig. 6.29 A 3-bit ripple counter.

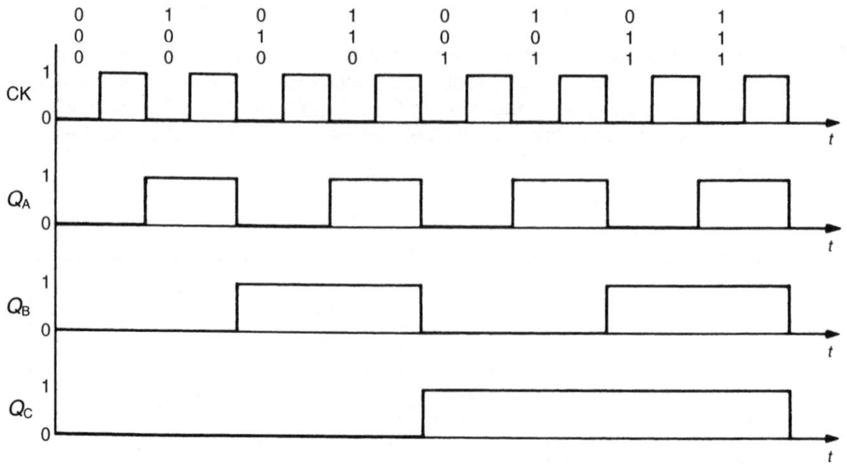

Fig. 6.30 Waveforms for a 3-bit binary counter.

flip-flops, with all J and K inputs connected to logic 1. This means that on every 1 to 0 transition of its C input, a flip-flop will change state. Flip-flop A is connected to a regular clock waveform CK, and the C inputs of the other two are connected to the output of the preceding flip-flop.

The waveforms that result from the regular clock applied to this circuit are shown in Fig. 6.30. On each 1 to 0 transition of the clock, the output Q_A toggles to the opposite state. On each 1 to 0 transition to Q_A, output Q_B toggles, and on each 1 to 0 transition of Q_B, output Q_C toggles. As you can see from these

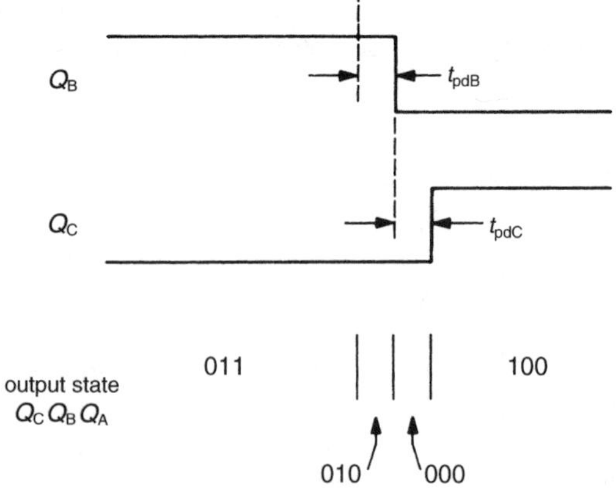

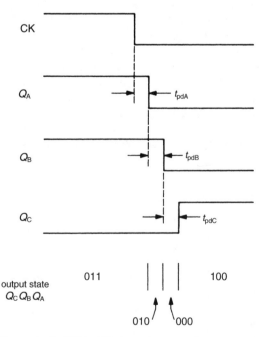

Fig. 6.31 Timing diagram for the 011 to 100 change in an asynchronous counter.

waveforms, the binary code formed by the outputs Q_C, Q_B and Q_A counts up through the natural binary sequence given above.

Although in principle this simple circuit achieves the aim of natural binary counting, there is a serious drawback which arises because of the propagation delay t_{pd} in each flip-flop. To explain the problem, Fig. 6.31 shows in more detail what happens at the change from 011 to 100. The output Q_A actually makes its transition after the propagation delay through the flip-flop of t_{pdA}. Similarly, Q_B only changes after an additional time t_{pdB}, and Q_C is later still by t_{pdC}. Thus the change from 011 to 100 actually involves the additional states of 010 and 000 existing for brief intervals of time. This effect of the time taken for the count to 'ripple' through from stage to stage can be a serious problem in many situations. Consequently, this form of counter, called an **asynchronous** or **ripple counter**, is not widely used.

The transient 000 state in the ripple counter can be seen if you build or simulate the circuit of Fig. 6.29 with a 3-input OR gate connected to outputs Q_A, Q_B and Q_C. If the delays are suitable, then this transient state will produce a brief glitch on the OR gate output.

6.6.1 Synchronous counters

A better solution in most cases is to use a **synchronous counter** in which each flip-flop changes simultaneously; a 2-bit synchronous counter is illustrated in Fig. 6.32. Notice that this circuit is also based on negative edge-triggered *JK* flip-flops, but this time the clock signal is applied in parallel to each one.

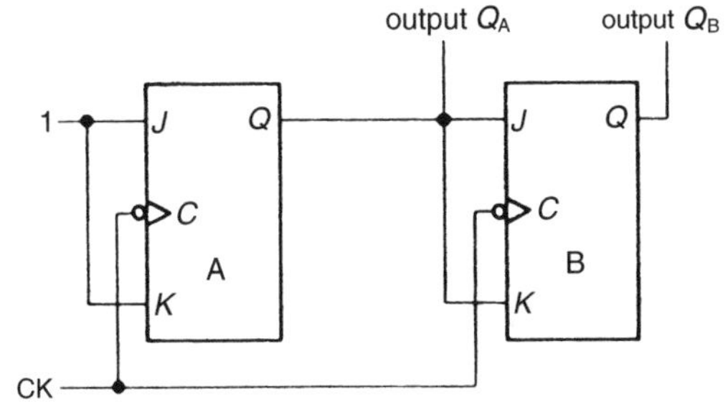

Fig. 6.32 A 2-bit synchronous counter.

The *J* and *K* inputs on flip-flop A are set to 1, so the output Q_A toggles on each negative edge of the clock signal. This behaviour is shown in the second trace in the timing diagram, Fig. 6.33. The *J* and *K* inputs to flip-flop B are connected to Q_A. Thus when Q_A is 0, the output Q_B is held without changing when the clock edge occurs. However, when Q_A is 1, output Q_B toggles on the clock edge in synchronization with Q_A changing. This behaviour is summarized in Fig. 6.33. The output state represented by Q_BQ_A then counts through the natural binary pattern 00, 01, 10, 11, 00,. . ., etc.

The waveforms required for a 3-bit counter are repeated in Fig. 6.34. Outputs Q_A and Q_B are the same as the 2-bit synchronous counter, and so could be produced by the same circuit as Fig. 6.32. The third waveform could be produced by another flip-flop, to be called device C.

At first sight it appears as if the relationship of waveform Q_C to Q_B is the same as the relationship between Q_B to Q_A. This could lead to the conclusion that all you have to do is to connect the output of flip-flop B to both inputs of flip-flop C to get the required output. However, this would not work at

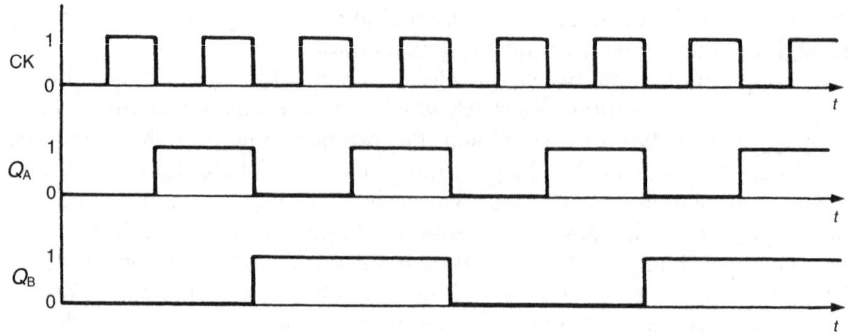

Fig. 6.33 Waveforms for a 2-bit binary counter.

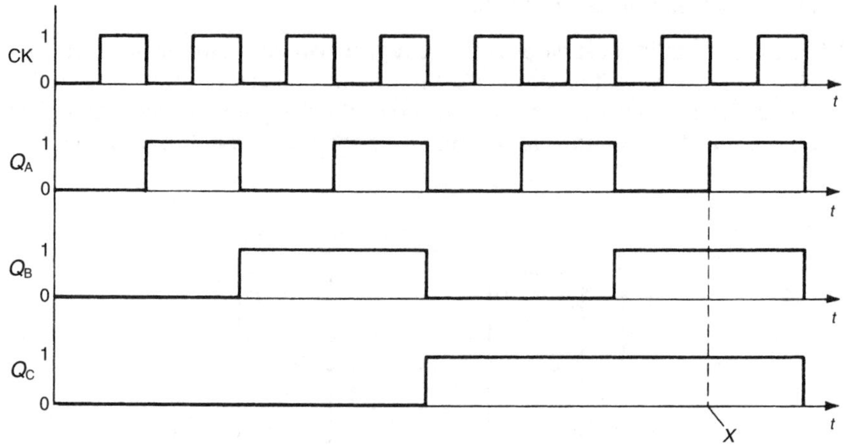

Fig. 6.34 Waveforms for a 3-bit binary counter.

the times marked X in Fig. 6.34. At these points there is a falling edge on the clock pulse, while waveform Q_B is at a logic 1. This would cause the output of flip-flop C to toggle if it were directly connected to Q_B, which is not wanted at this point in the sequence. A closer examination reveals that the condition to

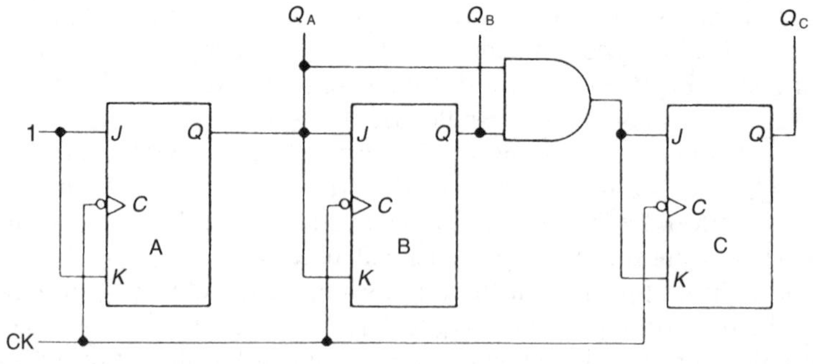

Fig. 6.35 A 3-bit synchronous counter.

satisfy is that the output of flip-flop C should toggle only when Q_B and Q_A are both already 1 at the falling edge of the clock. This is implemented in Fig. 6.35, using a two-input AND gate to select the required condition.

Self-assessment question

9 (a) Figure 6.36 shows the waveforms of a 4-bit synchronous counter. Assuming that the fourth waveform is to be produced from a *JK* flip-flop, what conditions have to be satisfied for the output of flip-flop D to toggle?
(b) Draw a circuit that implements this 4-bit counter. (*Hint*: the most compact solution uses 2 two-input AND gates with the four flip-flops.)
(c) If you have a suitable package, simulate the circuit to check your solution.

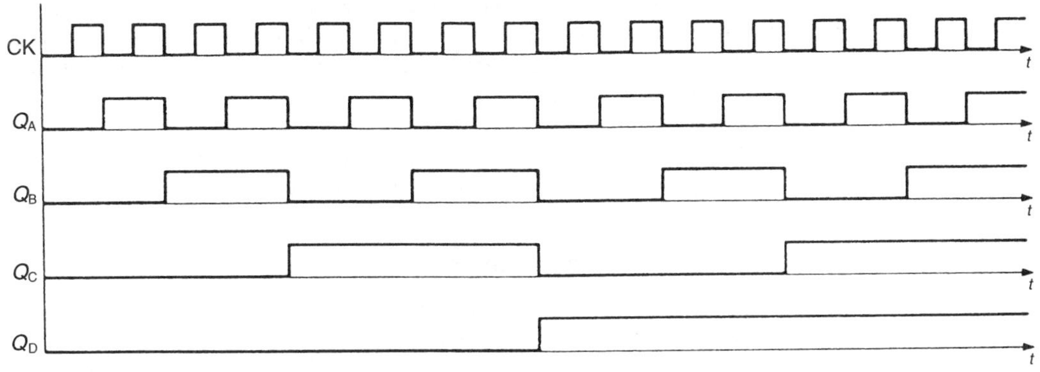

Fig. 6.36 Waveforms for a 4-bit binary counter (Self-assessment question 9).

6.6.2 An integrated circuit counter

Synchronous up and down counters for any number of bits can be designed from *JK* flip-flops and gates, and a great many forms are available as integrated circuits. Figure 6.37 is a schematic outline of one common example, the 74×163 4-bit synchronous binary up counter. In common with other general-purpose counters, this device has a number of extra functions built in which enable it to be tailored to suit a wide variety of applications with the minimum of external components. The nature of these functions can be seen from the external pin connections as follows.

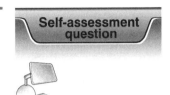

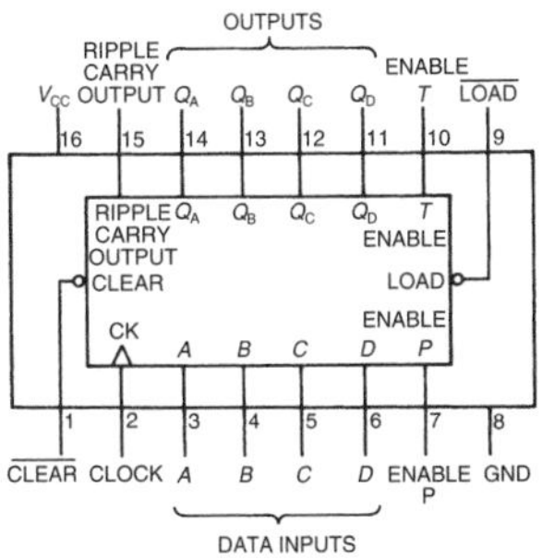

Fig. 6.37 Schematic package outline of a 74x163 4-bit synchronous binary counter.

INPUTS

Clock: The clock input. The count is advanced by 1 at each rising edge on this input.
$\overline{\textbf{Clear}}$: This is held at logic 1 for normal counting. If taken to 0 then the counter is cleared to the output state 0000 on the next edge of the clock. Hence this is a *synchronous clear* function, unlike the *asynchronous clear* illustrated on the flip-flops in Fig. 6.20. The inversion bubble on the input to the main function block in Fig. 6.37 and the bar over the input name both show that this is an active low input.

$\overline{\text{Load}}$: This is held at logic 1 for normal counting. If taken to 0 then the counter is set to the binary word connected to the **Data inputs A, B, C and D** on the next rising edge of the clock. Like the $\overline{\text{Clear}}$ input above, this is a synchronous, active low control.

Enable T and **Enable P**: Both must be held at logic 1 for normal counting. If either or both inputs are 0, then counting is inhibited. This ability to control when counting takes place is useful, for example, when joining these 4-bit devices together to form an 8-bit or 12-bit counter (see Section 6.6.3).

OUTPUTS

Q_A, Q_B, Q_C and Q_D: These are connections to the outputs of the flip-flops and represent the state of the counter, i.e. the count.

Ripple carry: This output is at logic 1 while the output state is 1111, otherwise it is at 0. In this way the counter can signal to external circuitry that the maximum count has been reached.

The variety of control inputs and outputs enable the counter to be used in many ways, two of which are described below.

6.6.3 Cascading counters

In many cases more than 16 states are required for counting, and counters like the 163 are designed to be connected together to extend the length of the count. For example, Fig. 6.38 shows three 74x163 devices **cascaded** to form a 12-bit counter, capable of sequencing through 4096 ($= 2^{12}$) states before repeating. Notice first that the clock CK is connected to all counters in parallel. However, they do not all count together. The enable inputs of counters 2 and 3 are both controlled by connection to the carry output of the previous counter.

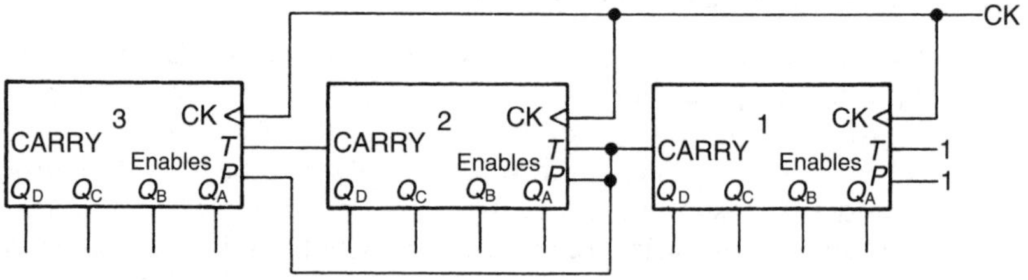

Fig. 6.38 Three 4-bit synchronous binary counters cascaded to form a 12-bit counter.

Assume that all three counters are cleared. Then, as the clock begins, only counter 1 will count: the other two are inhibited from counting by the state of the carry outputs connected to them. However, when counter 1 reaches 1111 (state 15 in Table 6.5), its carry output rises to logic 1, enabling counter 2 to respond to the next clock pulse. Hence both counters change to produce state 16 in the table. Counter 2 is now inhibited from counting again until counter 1 reaches 1111 again in state 31.

Counter 3 is first enabled when it is necessary to change from 0000 1111 1111 to 0001 0000 0000, and then again whenever counters 1 and 2 have both reached their maximum count of 1111. Hence the circuit in Fig. 6.38 behaves as

Table 6.5 Part of the 12-bit counter sequence

	Counter 3	Counter 2	Counter 1
State 14	0000	0000	1110
State 15	0000	0000	1111
State 16	0000	0001	0000
State 17	0000	0001	0001

a fully synchronous 12-bit counter. The principle can be extended to form a counter of any length from these 4-bit devices.

6.6.4 Modulo-*n* counters

The modulus of a counter is the number of states generated by the counter before it repeats itself. Thus a **modulo-*n* counter** generates *n* states before repeating. A 2-bit counter can be described as modulo-4, a 3-bit as modulo-8, and so on. The 74x163 as it stands is modulo-16. However, a common requirement is to count a number of states that is not a power of 2, and this can be achieved in two main ways with the 74x163 and similar devices.

The first method is to use the carry output to preset a number from which the count restarts. An example of a modulo-12 counter is given in Fig. 6.39. Initially, the code 0100 (= denary 4) is permanently set up on the data inputs. When a count of 1111 (= denary 15) is reached, the carry output goes high, and through the inverter the synchronous $\overline{\text{Load}}$ input is brought to a logic 0. Therefore on the next clock edge, the preset code 0100 is loaded into the counter, and on subsequent clock pulses the count proceeds from there. So the counter sequences through the states 0100, 0101, 0110, 0111, 1000, 1001, 1010, 1011, 1100, 1101, 1110, 1111, 0100, 0101, ..., etc. This involves the twelve different states from 0100 to 1111. The process of loading a starting value is often called **presetting** the counter.

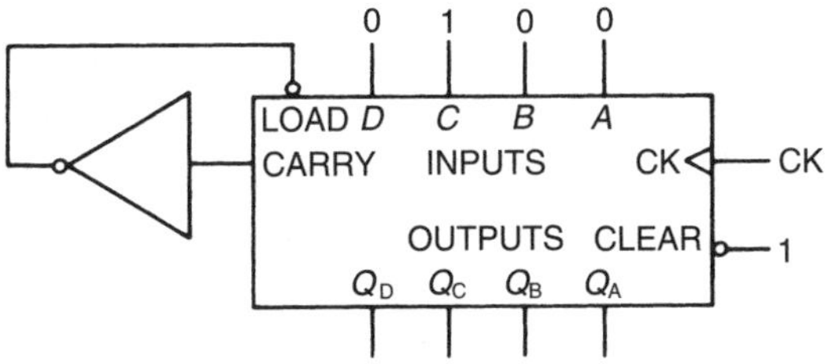

Fig. 6.39 A modulo-12 counter formed by presetting.

Some IC counters have asynchronous clear and load inputs, which respond immediately to a clear or load signal instead of making the change at the next clock edge. Techniques similar to those described here can be used to cascade or to form modulo-*n* counters. However, the resulting counter is not fully synchronous and may have brief, unwanted states between changes, as did the ripple counter in Fig. 6.29.

An alternative method for making a modulo-12 counter with the 74x163 is outlined in Fig. 6.40. The three inputs of a NAND gate are connected to counter outputs *A*, *B* and *D*. When a count of 1011 (= denary 11) is reached, the NAND gate output goes low. This is connected to the synchronous $\overline{\text{Clear}}$ input, so that on the next clock pulse the counter is reset to 0000. In this case the counter sequences through the states 0000, 0001, 0010, 0011, 0100, 0101, 0110, 0111, 1000, 1001, 1010, 1011, 0000, 0001, ..., etc. This includes the twelve different states from 0000 to 1011. Because this sequence starts at 0000, it is

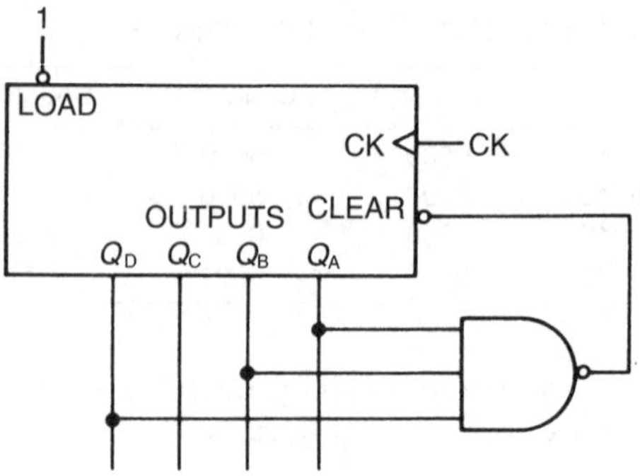

Fig. 6.40 A modulo-12 counter formed by clearing.

often more convenient to use than the twelve states obtained above by loading a preset number.

Counting to modulo-10 is such a common requirement (since denary numbers are the basis of most forms of measurement and display) that modulo-10 versions of many four-bit binary counters are available. For example, the modulo-10 74×162 is identical to 74×163 in all respects except that the counting sequence repeats after 10 states (0000, 0001…1000, 1001, 0000, 0001, …) and the carry output is high during the state 1001.

Self-assessment question

10 Why are only the three NAND gate connections sufficient to detect the count of 1011 in Fig. 6.40? (i.e. why is it not necessary to have a connection to Q_c as well?)

6.6.5 Frequency division

In this section, the emphasis is shifted to looking at the binary waveforms generated by the counter. First look back at Fig. 6.36, which shows the waveforms generated by a 4-bit binary counter.

Worked example

If the frequency of the clock is 4 kHz, what are the frequencies of the other waveforms?

Output Q_A is 2 kHz, effectively dividing the clock frequency by 2. The other outputs are 1 kHz, 500 Hz and 250 Hz, respectively, representing division by 4, 8 and 16.

Thus a binary counter offers the possibility of **frequency division** by a power of 2, depending upon which bit position is used. A familiar example of this idea is found in the digital watch. The problem here is to provide a compact, precision oscillator of high stability to produce a sequence of pulses at one-second intervals. Low-frequency oscillators are not compact (they involve the use of large capacitors) nor do they operate with the required precision (the watch may be required to be accurate to one second a month). The solution is to use a high-frequency oscillator regulated by a quartz crystal and divide the frequency to the required 1 Hz. Crystal-controlled oscillators can be designed to give good long-term stability coupled with high accuracy. In the process of dividing down, this accuracy is maintained all the way through to the final 1 Hz output.

Worked example

A crystal frequency commonly used in watches is 32.768 kHz. How many bits should be provided by the binary counter to provide the 1 Hz signal?

A 15-bit counter would be required, because $2^{15} = 32\ 768$.

However, the general problem in frequency division is to reduce the frequency of a binary waveform from f to f/n, where n is not necessarily a power of 2. The solution to this is to use a modulo-n counter. All that is required is that one of the bits changes state once and once only during the basic sequence.

Although the clock waveform may be supplied as a true squarewave (equal durations high and low), the divide-by-n output may not itself be a squarewave. If n is an odd number then the output could not possibly be a squarewave, whereas if n is an even number it may not be. You can see this by writing out the sequence of codes for the states in a modulo-10 counter which starts from 0000. In this case, the output from the MSB is 'low' for eight states and 'high' for the final two, which produces a waveform of frequency $f/10$ but which is very unsymmetrical.

Therefore, if a squarewave is required it may be necessary to modify the sequence so that the binary values of the states include one bit that changes only once and with an equal number of sequence steps as 0 or 1. For example, a modulo-10 counter could be designed so that the outputs follow the sequence:

$$0 - 1 - 2 - 3 - 4 - 8 - 9 - 10 - 11 - 12-$$

then back to 0 again. The binary values of these states are:

0000
0001
0010
0011
0100
1000
1001
1010
1011
1100

The MSB has five 0s and five 1s and therefore produces a squarewave. The only penalty is that the count no longer follows the natural binary sequence, which is often the case with divide-by-n counters. This is not always easy to achieve with integrated-circuit counters like the 74x163, but there are other methods for designing counters to give any sequence of states, and one of these is described in Sections 6.7 and 6.8.

Frequency division using modulo-n counters is used as the basis for frequency determination in 'digitally synthesized' tuners for radio reception. In this case a waveform corresponding to the required signal frequency is divided by a large value of n, and this divided value is synchronized to a fixed frequency from a stable oscillator. By altering the value of n, the 'synthesized' signal frequency can be stepped through a range of values.

6.6.6 Summary of Sections 6.4 to 6.6

Registers consist of a group of D-type latches or flip-flops which are clocked simultaneously to store a binary word. Set-up and hold times are important parameters that must be taken into consideration. Practical registers and other devices may have three-state outputs, so that several devices can be connected together to share output lines.

Shift registers allow data to be moved from one bit position to another. This enables conversion between serial and parallel forms of data to take place, and some types of arithmetic operation to be carried out.

Counting is a common requirement in sequential logic circuits. Synchronous and asynchronous counters of any bit length can be made from *JK* flip-flops and gates. Versatile integrated-circuit counters are available that can be adapted to fit many applications. In particular, the counting sequence can be made shorter by using the preset or clear inputs, or longer by cascading more than one device.

6.7 THE STATE-TRANSITION DIAGRAM

Sequential circuits can be constructed in a large number of ways from the wide variety of flip-flops, registers, counters and other devices that are available as integrated circuits. However, the design process can be extremely complex and many methods have been developed to lead more quickly and reliably to a working circuit.

This section will describe one common way to simplify the design process by representing the behaviour of any sequential circuit in a systematic, graphical form. Section 6.8 then extends this idea to a general method by which this sequential design may be implemented as a working circuit.

The discussion will be limited to designing **synchronous sequential circuits**. The definition of a synchronous sequential circuit is one which responds only to a pulse on a particular input, and where the actions in any part of the circuit are 'synchronized' with each other and with the pulse input. Many sequential circuits are asynchronous, but the complex design techniques of these are not appropriate to an introductory text such as this.

> Synchronous sequential circuits (often also known as **clocked logic**) are generally 'better' designs than asynchronous ones. Their operation does not depend on clocks of well-defined frequency, the states are well defined (without transitory effects such as those shown in Fig. 6.31), and they are less prone to noise.

Each time a clock pulse is received, a synchronous sequential circuit changes from its present state into a new state. The **state** of the circuit is represented by a combination of all the individual states of the memory elements within it. For example, each time a counter like that in Fig. 6.35 counts up, it is entering a new state represented by the condition (set or reset) of each of the *JK* flip-flops.

A **state-transition diagram** is a graphical way of representing the behaviour of a sequential system. In it, the states are represented by circles containing a symbol for the state and the value of the output associated with that state. The transition or change from one state to the next is indicated by an arrow. So a 2-bit counter with four states can be represented by the diagram in Fig. 6.41. Note that the four states **a**, **b**, **c** and **d** have the outputs 0, 1, 2 and 3 respectively.

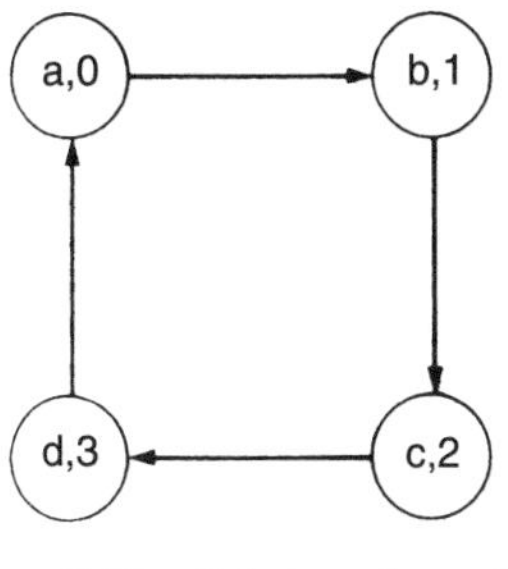

Fig. 6.41 State-transition diagram for a 2-bit binary counter.

Drawing a state-transition diagram for a synchronous binary up counter like this is a relatively simple process because we always know what the counter is going to do on the next clock pulse. A slightly more complex example is a counter which can count up or down.

Suppose that a counter is required that will count up if an input *U* is 1 and count down if the input *U* is 0. There needs to be some way of indicating on the state-transition diagram the value of the input that causes the transition, and one possible method is shown in Fig. 6.42. The numbers next to the arrows represent the input, *U*. Look at the transition from state **a**. When *U* is 1, the counter changes from state **a** to state **b**, so that the output changes from 0 to 1. When *U* is 0, the counter changes from state **a** to state **d** so that the output changes from 0 to 3. Similarly in state **b**, if *U* is 1 the counter changes to state **c** and the output changes to 2, whereas if the *U* is 0 the counter changes to state **a** and the output changes to 0.

As a convention in this book, lower case letters are used as the symbol denoting the state of a sequential circuit, and upper case italic letters denote Boolean variables.

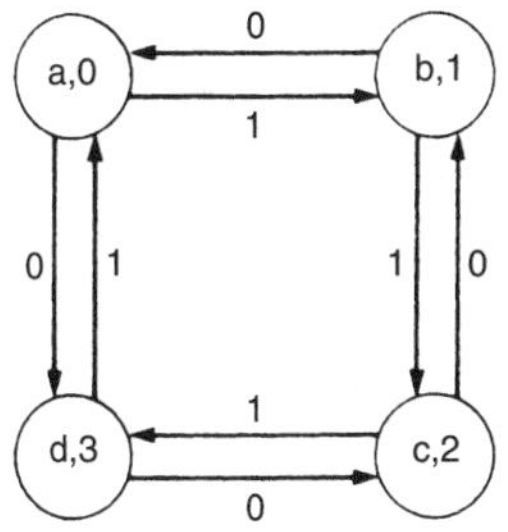

Fig. 6.42 State-transition diagram for a 2-bit up/down counter.

The effect of having an external input to the counter is that the state changes can depend on that input. In general, the state transitions of a system can be conditional on one or more inputs. Figure 6.43 shows the state-transition diagram of a system which has two inputs, so that the numbers next to the arrow represent the binary values of the inputs, namely 00, 01, 10 or 11. The transitions from state **e** show all four possible input combinations.

But what if the transition takes place regardless of the condition of one or more of the inputs? In these cases the notation used is a '-' to indicate that the transition will take place irrespective of the value of an input. Figure 6.43 shows that some of the transitions are independent of the inputs. The transition from state **a** to state **b** is independent of one of the inputs, whereas the transition from state **b** to state **c** is independent of both inputs.

Another feature of the system represented by Fig. 6.43 is that state **c** has a new type of transition when the inputs are 01, represented by the curved arrow. When this happens the transition takes place but the state remains the same. This is quite common when you want the machine to wait in a particular state until the inputs change.

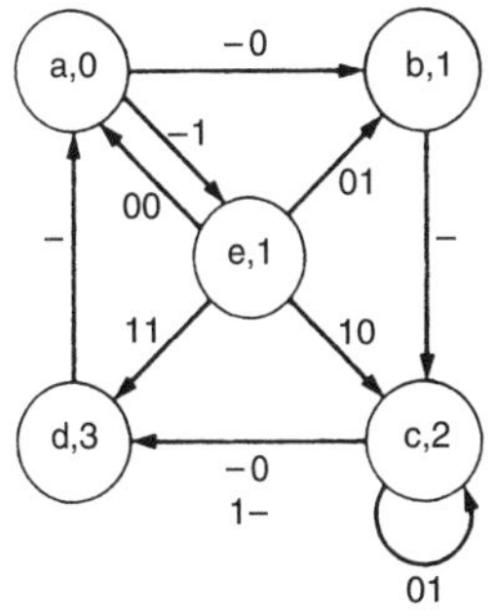

Fig. 6.43 State-transition diagram example.

A further point about the transition from state **c** is that the remaining three input combinations that cause the machine to change from **c** to **d** are 00, 10 and 11. We could write all three of these next to the arrow, but instead –0 and 1– have been used. This saves some space, and is a kind of shorthand notation: –0 is short for 00 and 10, and 1– is short for 10 and 11.

To summarize, a state-transition diagram is read as follows. A state and the output for that particular state are represented by a symbol, a comma, a number, respectively, in a circle. The arrows from one state to another state indicate possible changes of state that could take place on the arrival of the next clock pulse. The number or numbers next to the arrow show you under what input condition the machine will change to a particular state.

Self-assessment question

11 Complete the state-transition diagram in Fig. 6.44 using the following statements:

- State **a** changes to state **b** when the input is 1, or to state **c** when the input is 0.
- State **b** changes to state **c** if the input is 0, or to state **a** if the input is 1.
- State **c** changes to state **a** regardless of the input.
- The binary output is 0 except in state **c**.

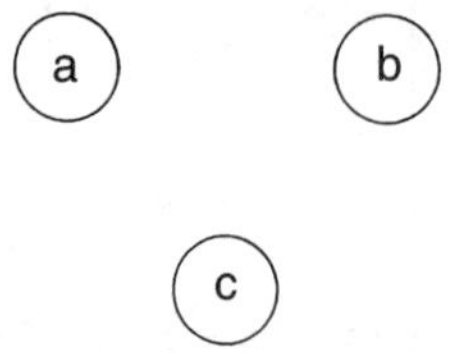

Fig. 6.44 State-transition diagram for Self-assessment question 11.

Self-assessment question

12 What will the next state and output be for the hypothetical machine represented by the state-transition diagram shown in Fig. 6.45 when:

(a) the machine is in state **e** and the input is 0?
(b) the machine is in state **d** and the input is 1?
(c) the machine is in state **a** and the input is 0?

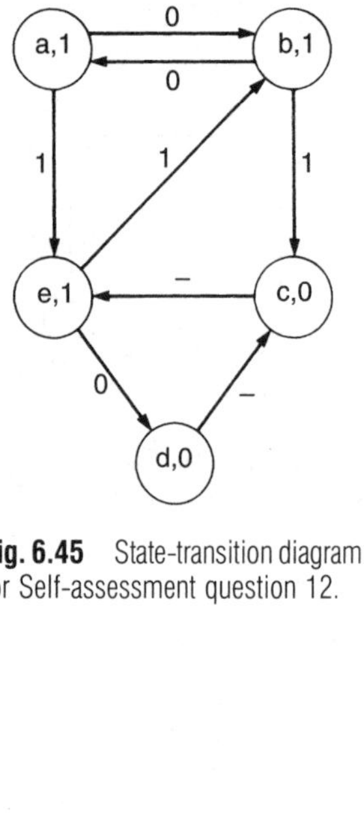

Fig. 6.45 State-transition diagram for Self-assessment question 12.

6.8 A GENERAL SEQUENTIAL MACHINE

The state-transition diagram is a clear graphical representation of all the possible states of a sequential circuit. From this data, it is possible to determine what will be the next state to be entered, depending upon what the present state is and upon the current values of the external inputs to the circuit.

A general implementation of sequential logic that is based on the present state/next state information is outlined in Fig. 6.46. This **general sequential machine** consists of two main blocks. The first is an edge-triggered register, whose outputs represent the **present state** of the circuit. The second is a block of combinational logic which generates two sets of outputs. The first is the present output required from the circuit by whatever it is connected to. The second output from the combinational logic represents the state that the circuit is to be in after the next clock pulse i.e. the **next state**. The inputs to the combinational block are the present state of the circuit and the external inputs.

The broad pathways in Fig. 6.46 are a common convention used to represent a number of parallel digital connections without drawing all the wires. The arrows show the direction of data flow.

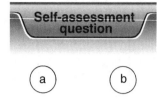

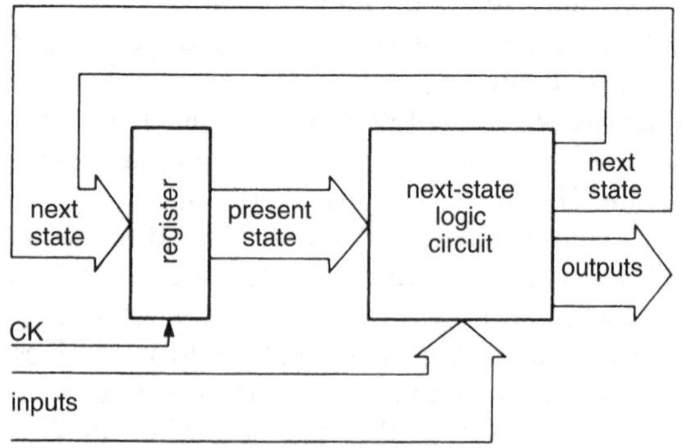

Fig. 6.46 Block diagram of a general sequential machine.

When a suitable clock pulse is applied to the register, the output changes to the next state data that is present on the register inputs. Thus what was previously the 'next state' becomes the new 'present state' of the register. The circuit has jumped from one circle in the state-transition diagram to another.

This new state has a different output, and a different choice of next state that is determined by the current inputs. In the circuit, the combinational block now has a different input supplied by the register, and so generates the new 'next state' that corresponds to these present conditions. This new 'next state' is then clocked into the register on the following clock pulse, and the process continues with each succeeding clock pulse.

In this way the circuit changes from state to state (i.e. from circle to circle on the state-transition diagram) with each clock pulse, with the combinational logic block each time defining what the next state is to be for every particular combination of present state and external inputs.

The circuit is then a general implementation of *any* sequential problem. All that is necessary is to develop a state-transition diagram for the problem, and from this to make out a comprehensive table of present state/next state information that can be implemented as the combinational block. This is illustrated by the 2-bit up/down counter in the next section.

6.8.1 The state table

The state-transition diagram is a pictorial description of a sequential problem. In order to implement it as a general sequential machine (described in the last section) we have to convert the information in the diagram into the form of a truth table to define the combinational logic block. This is done in two stages, the first of which is to construct a **state table**, which shows what state the system will change to on the arrival of the next clock pulse. At each stage, the new state of the system depends on the current state and the condition of the inputs. As already mentioned in the last section, the current state of the system is called the present state, and the state that it is going to change to is called the next state.

The state table is constructed by listing all of the states down the left-hand column, then along the top is a list of all the possible input combinations. Finally, the next state and the value of the output for the present state are written in. Table 6.6 shows the state table for the 2-bit up/down counter whose state-transition diagram was shown previously in Fig. 6.42. The table shows, for example, that when the present state is **b**, the output is 1 and if the input is 0 the next state will be **a**, whereas if the input is 1 the next state will be **c**.

Table 6.6 State table for a 2-bit up/down counter

Present state	*Next state*		*Output*
	Input = 0	*Input = 1*	
a	d	b	0
b	a	c	1
c	b	d	2
d	c	a	3

Self-assessment question

13 Construct the state table for the machine described by the state-transition diagram in Fig. 6.43.

6.8.2 The state-assignment table

Having obtained a state table, we are almost at the point where we have a complete definition of the problem in a form that can be handled using

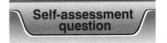

Table 6.7 State table with all binary values

Present state	*Next state*		*Output*
	Input = 0	*Input = 1*	
00	11	01	00
01	00	10	01
10	01	11	10
11	10	00	11

Table 6.8 State-assignment table for a 2-bit up/down counter

ABU	*CDPQ*
000	1100
001	0100
010	0001
011	1001
100	0110
101	1110
110	1011
111	0011

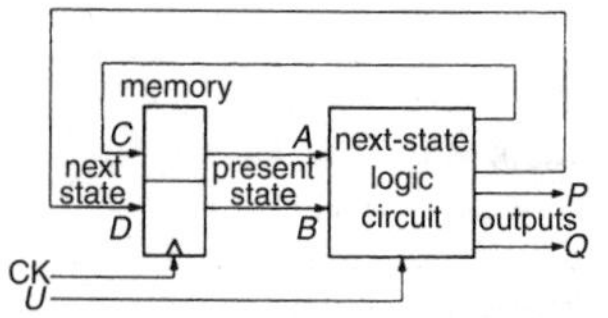

Fig. 6.47 2-bit up/down counter.

the skills that were developed for combinational circuits. One further stage is needed, which is to have all values in binary form and then to draw up a truth table. This truth table is referred to as a **state-assignment table**, but can be interpreted in exactly the same way as any ordinary truth table. Hence methods developed for handling combinational logic problems can be used to simplify and implement it.

The first step is to convert the state symbols into a binary code. The second step is to convert the denary values for the output into natural binary. Taking the counter described earlier, the state table shown in Table 6.6 can be converted as shown in Table 6.7. The choice of binary values for the states is arbitrary. One possible solution has been used above, where the state has been given the same binary value as its output on the grounds that this will probably simplify the final implementation.

Next, label the two bits in the present state A and B, and the two bits in the next state C and D. The input has already been given the label U, so all that remain are the two output bits, which can be labelled P and Q. The state table can now be re-drawn as if the present state bits, A and B, and the input U are all inputs to a combinational logic circuit. The next state bits, C and D, and the outputs, P and Q, are written as if they are the outputs of the combinational circuit. The result is shown in Table 6.8

Using a 2-bit register, the circuit for the 2-bit up/down counter can then be completed as in Fig. 6.47. The truth table for the combinational block is identical to the 1-1 state assignment table in Table 6.8, and so this logic can be

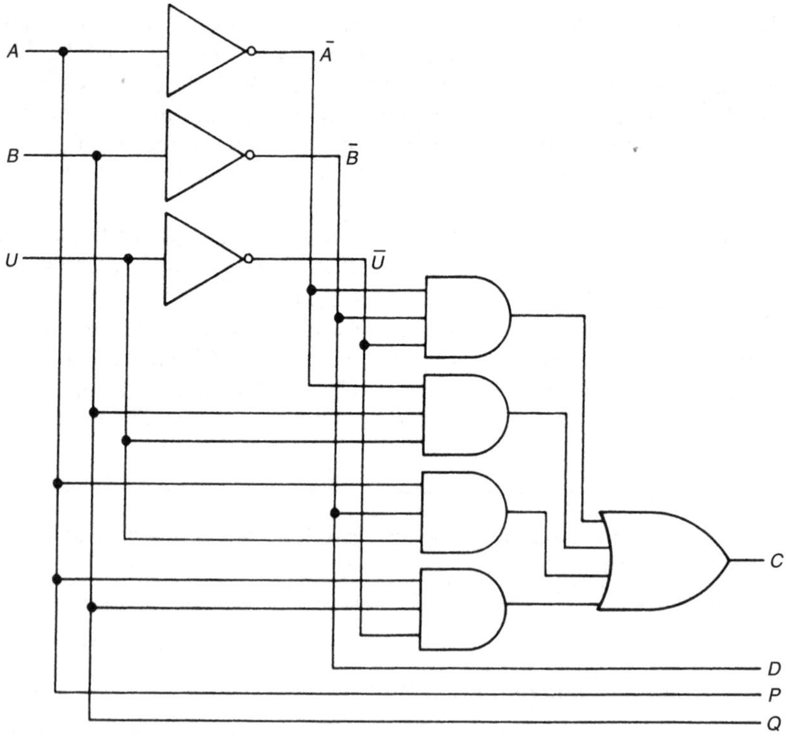

Fig. 6.48 Next-state logic for a 2-bit up/down counter.

implemented using the techniques outlined in Chapter 5. The following expressions are obtained:

$$C = \overline{A} \cdot \overline{B} \cdot \overline{U} + \overline{A} \cdot B \cdot U + A \cdot \overline{B} \cdot U + A \cdot B \cdot \overline{U}$$

$$D = \overline{B}, \qquad P = A, \qquad Q = B$$

One possible way that this circuit could be built is given in Fig. 6.48. The Boolean expressions have been directly implemented with AND gates, OR gates and inverters. If this block, the **next-state logic circuit**, is connected to a 2-bit register as in Fig. 6.47, then it will function in the required way as a 2-bit up/down counter.

Such an implementation using present-state/next-state logic can be applied to any sequential problem. It may not be the cheapest or the most appropriate way of constructing every circuit, but as long as the present-state/next-state information is correctly defined, then it will work in all cases as a fully synchronous solution.

6.8.3 Sequential PLDs

In the last few sections you were shown how any sequential circuit could be designed and implemented using the general sequential machine. The two main blocks in this system are a group of flip-flops forming a register, and a combinational circuit.

The terms PLD (programmable logic device) covers PLAs and PLSs. Some manufacturers do not distinguish between PLAs, PLSs and related devices, and use PLD for everything.

In Chapter 5 you were introduced to the PAL device as a versatile solution for combinational logic problems. It is comparatively simple to extend the architecture of a PAL device to include flip-flops on some or all of the outputs so that they are able to implement sequential logic as well.

An example of such a device is the GAL16V8, where GAL stands for *generic array logic*, originally introduced by Lattice Semiconductor. This has the overall structure similar to that of the PAL16L8 in Fig. 5.59. However, each output 'cell' has the structure shown in Fig. 6.49. The 8-input OR gate (replacing

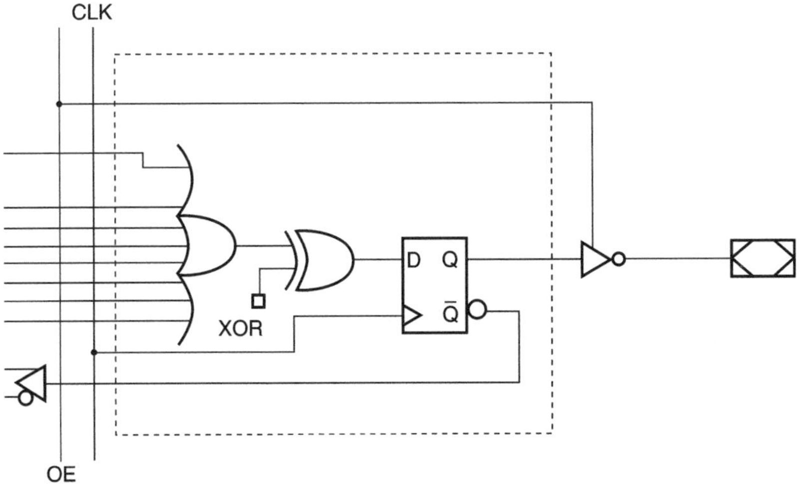

Fig. 6.49 The output cell for a GAL16V8 device in registered mode.

the 7-input OR gate of the PAL device) is followed by an exclusive-OR (XOR) gate and an edge triggered *D*-typc flip-flop. The XOR gate simply allows the output to be inverted or not, depending upon the fixed level applied to the input labelled XOR. The *D*-type flip-flop holds the present state, allowing any sequential function to be implemented from the next state generated from the AND–OR array that makes up the bulk of the chip.

Two other features of the GAL16V8 are worth noting. First, remember that the bipolar PAL16L8 in Chapter 5 was programmed by permanently blowing internal fuses to generate the required logic function. The GAL16V8 is a CMOS device, and holds the programming information in 'floating gate' field effect transistors. Hence this device can be erased and reprogrammed to change its function, which is a very useful characteristic in some situations. However, the downside to this is that the programmed data is only guaranteed to be retained for a minimum of 10 years. The second feature to note is that the output flip-flops have programmable links which allow them to be bypassed, giving a purely combinational output. Hence with their lower consumption, erasability and greater versatility these CMOS devices have largely replaced PAL devices in newer circuit designs.

There are many other programmable logic devices available, reaching increasingly high levels of size, complexity and versatility. However a discussion of these is beyond the scope of this book.

6.8.4 Summary of Sections 6.7 and 6.8

A method has been presented for the design of a general sequential circuit for any sequential logic problem. The stages involved were:

1. construct a state-transition diagram for the problem;
2. construct a state table;
3. construct a state-assignment table.

The state-assignment table is very much like a truth table and can be used to design the next-state logic circuit. This circuit, together with a register, forms the basis for the implementation of the design as a general sequential machine.

Answers to self-assessment questions

1. The function table for this circuit, called a *D* latch, is as follows:

Inputs		Output		Meaning
D	C	Q_{n+1}	$\overline{Q}_{n+1}$	
x	0	Q_n	$\overline{Q}_n$	Hold
0	1	0	1	Reset
1	1	1	0	Set

2. The complete timing diagram is shown in Fig. 6.50.

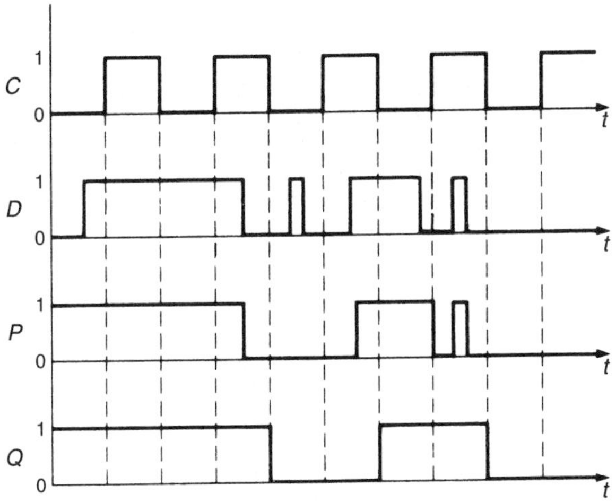

Fig. 6.50 Solution to Self-assessment question 2.

3. (a) The function table for the latch shown in Fig. 6.51 is as follows:

$\overline{S}$	$\overline{R}$	Q_{n+1}	$\overline{Q}_{n+1}$	
1	1	Q_n	$\overline{Q}_n$	
0	1	1	0	
1	0	0	1	
0	0	1	1	$(Q = \overline{Q})$

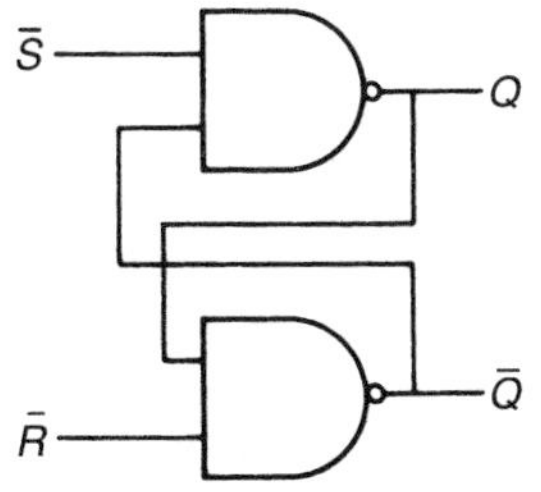

Fig. 6.51 NAND gate latch for Self-assessment question 3(a).

Returning from the 00 input condition, in which both outputs are 1, to the 11 input condition leads to an unpredictable state.

(b) The completed table is:

	C	D	E	F	G	H	Q	$\overline{Q}$
Initial state	0	0	0	1	1	1	1	0
Positive clock edge	1	0	0	1	0	1	0	1
Change D	1	1	0	1	0	1	0	1
Negative clock edge	0	1	1	1	1	0	0	1
Positive clock edge	1	1	1	0	1	0	1	0
Negative clock edge	0	1	1	1	1	0	1	0
Change D	0	0	0	1	1	1	1	0
Positive clock edge	1	0	0	1	0	1	0	1

4. The *D* flip-flop is initially cleared, $Q = 0$, so the output $\overline{Q}$ and consequently *D* will both be 1. The first pulse on the control input will transfer this input to the output *Q*, changing its value from 0 to 1. The output $\overline{Q}$ and the input *D* will also change from 1 to 0.

The second pulse on the control input will change the output *Q* from 1 to 0 again. The process repeats with every subsequent pulse on the control input. The following sequence therefore results:

Initial output $Q = 0$
1st pulse $Q = 1$
2nd pulse $Q = 0$
3rd pulse $Q = 1$
and so on.

If the flip-flop was replaced by a *D*-type latch, consider what happens when the latch is in its transparent mode. If $Q = 0$, then $\overline{Q} = 1$, and so this opposite state is immediately transmitted through the latch. $\overline{Q}$ then changes, and again the opposite state is fed back to the *D* input, is transmitted through, and the state toggles once again. In other words, while the latch was transparent it would oscillate between the values 0 and 1, the frequency of oscillation being related to the propagation delay in the latch.

5. The characteristic table for the *JK* flip-flop shows that the next output will always be a 1 when the present *J* and *K* inputs are 1 and 0, respectively, and the next output will be a 0 if the *J* and *K* inputs are 0 and 1, respectively. Therefore, to function as a *D*-type flip-flop the *JK* flip-flop requires an additional INVERTER on the *K* input, as shown in Fig. 6.52.

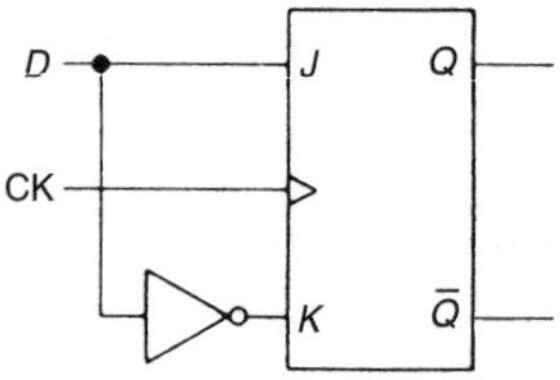

Fig. 6.52 Solution to Self-assessment question 5.

6. (a) The table is

T	C	Q_{n+1}
0	↑	Q_n
1	↑	$\overline{Q_n}$

The last line of the function table shows that when J and K are both connected to 1, the output will always change state on the receipt of a pulse on the control input.

(b) The binary waveform generated when a squarewave is applied to the control input is shown in Fig. 6.53. Since the flip-flop is a positive edge-triggered device, each change takes place on the 0 to 1 transition of the clock waveform. This circuit generates a sequence of states 0, 1, 0, 1, 0, 1, 0, . . ., etc.

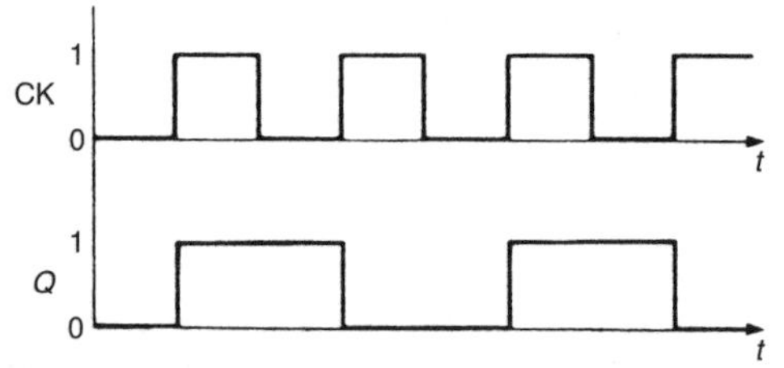

Fig. 6.53 Solution to Self-assessment question 6(b).

7. In the timing and switching tables in the data sheet, the column headed '$T_A = 25°C$' is used if the ambient temperature is stated to be 25°C, regardless of whether the register is a SN54HCT573 or SN74HCT573.

(a) This is the propagation delay t_{pd} from D changing. The table of switching characteristics at the end of the data sheet extract gives (for 4.5 V) no minimum time, a typical time of 25 ns and a maximum time of 35 ns. No minimum time is stated, but it would be unwise to assume a minimum time larger than 0 ns when designing the circuit that is to incorporate this register (this is the worst-case value). Hence the time interval could be anything from 0 to 35 ns, with a typical value of 25 ns.

(b) This is the propagation delay from LE changing. This time interval could be anything from 0 to 35 ns, with a typical value of 28 ns.

(c) This is the set-up time, and is quoted as at least 10 ns.

(d) $2Q$ is in the high-impedance state because $\overline{OE}$ is 1 (high). When $\overline{OE}$ goes to 0, $2Q$ changes after the enable time t_{en}, which is quoted as having a maximum time of 35 ns and a typical time of 26 ns. Again, no minimum time is quoted, so a worst case of 0 ns should be assumed. The time interval could be anything from 0 to 35 ns, with a typical time of 26 ns.

8. The waveforms are shown in Fig. 6.54. In the initial state of 0000, the inverted Q_D output sets D_A to be 1. Therefore the first clock pulse shifts in a 1, as do the second, third and fourth. The 1s are progressively shifted to the right. The first 1 is now in Q_D, so the next four clock pulses shift in 0s from D_A. After the eighth clock pulse, the shift register is back to its initial state of 0000, so the sequence begins again.

The repeating sequence formed by Q_A, Q_B, Q_C and Q_D is:

0000 1000 1100 1110 1111 0111 0011 0001 0000 etc.

This circuit is known as a *4-bit Johnson counter*.

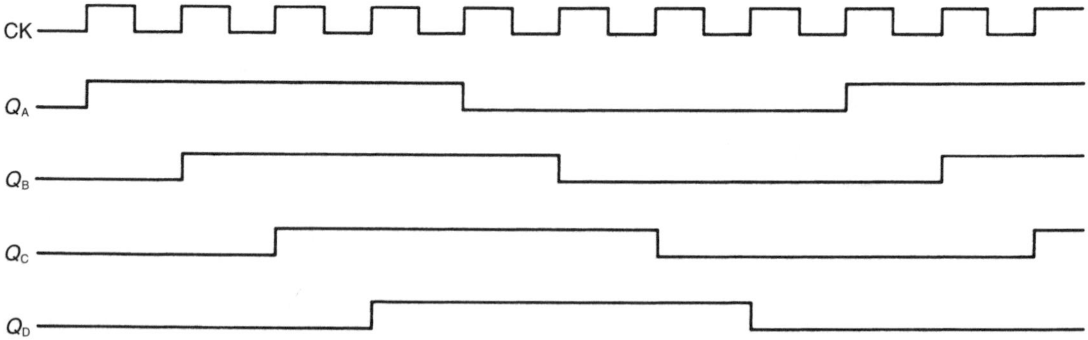

Fig. 6.54 Solution to Self-assessment question 8.

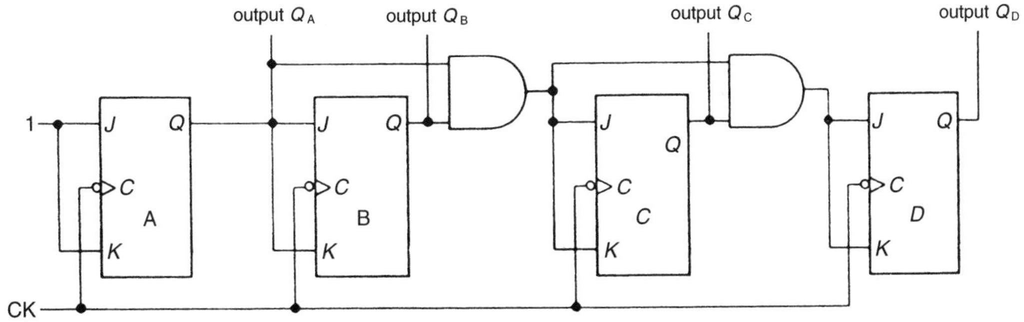

Fig. 6.55 Solution to Self-assessment question 9(b).

9. (a) Flip-flop 4 toggles on the correct falling edge of the clock when the outputs of all of the other flip-flops are 1. Therefore, if we use the AND of all the outputs of the previous flip-flops as the input to both *J* and *K* of flip-flop D it will toggle at the correct time.

(b) Rather than use a 3-input AND gate connected to Q_A, Q_B and Q_C, we can take the output of the 2-input AND gate used for flip-flop C as the input to another 2-input AND gate. The other input for D is the output of flip-flop C, as shown in Fig. 6.55.

10. The output of the NAND gate connected as in Fig. 6.40 would drop to 0 for the output 1111 as well as for output 1011. However, in a binary up counter the code 1011 is reached before 1111. Since this code resets the counter in Fig. 6.40 to 0000, the state of 1111 is never reached and the NAND gate is all that is needed to detect the required code unambiguously. The use of a NAND gate connected only to the '1s' in this way applies to the detection of any code which is to be the maximum count.

11. The completed state-transition diagram is shown in Fig. 6.56.

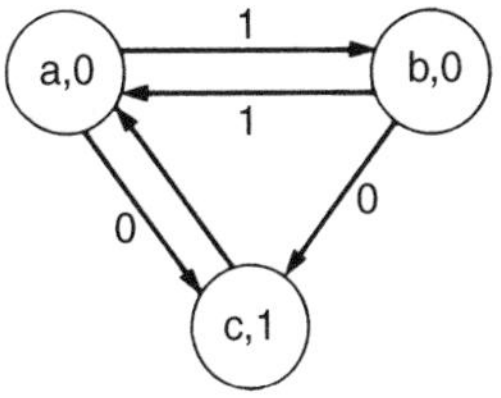

Fig. 6.56 Solution to Self-assessment question 11.

12. (a) **d**, 0; (b) **c**, 0; (c) **b**, 1.

13. The state table is

Present state	*Next state input* 00	01	10	11	*Output*
a	**b**	**e**	**b**	**e**	0
b	**c**	**c**	**c**	**c**	1
c	**d**	**c**	**d**	**d**	2
d	**a**	**a**	**a**	**a**	3
e	**a**	**b**	**c**	**d**	1

Notice that state **c** has itself as the next state when the input is 01.

7 Analogue–digital conversion

AIM

The aim of this chapter is to introduce the principles and circuits used in the design of analogue-to-digital and digital-to-analogue converters.

GENERAL OBJECTIVES

Understand the meaning of the following terms:

- acquisition time
- alias
- aliasing
- analogue to digital (A-D) converter
- anti-aliasing filter
- aperture
- aperture error
- binary-weighted resistor network
- bipolar
- comparator
- conversion rate
- conversion time
- counter-ramp converter
- digital analogue (D-A) converter
- droop-rate
- dual slope converter
- dynamic range
- feed-through
- flash converter
- integrating converter
- integrator
- multiplexer
- offset binary
- quantization error
- quantization intervsal
- quantization level
- quantization noise
- *R*-2*R* ladder
- sample-and-hold
- sampling rate
- sampling rule
- settling time
- successive approximation converter
- tracking converter
- under-sampling
- unipolar

SPECIFIC OBJECTIVES

1. To be able to calculate the output voltage of a D-A converter given the input codeword, the voltage of the source, the feedback resistance and the weighting resistances.
2. To be able to calculate the errors introduced in a D-A converter output voltage due to the non-ideal characteristics of switches.
3. To draw circuits and explain the operation of the binary weighted resistor network and the *R*-2*R* ladder network.
4. To draw the block diagram and explain the operation of the flash, counter-ramp, successive approximation and dual slope A-D converters.

5. To be able to calculate the conversion time for an A-D converter and the maximum permissible bandwidth of the input signal to that A-D converter.
6. To be able to explain the sampling rule.
7. To be able to describe the basic design of a sample-and-hold circuit and explain how it works.

7.1 INTRODUCTION

In the earlier chapters of this book analogue and digital systems were treated separately. However, there are very many systems in which both analogue and digital subsystems are combined, especially in the fields of computer, communication and entertainment technologies, and measurement and control.

Digital recording and transmission of sound and video signals are now well established through the many technologies of CD audio, DAT, DVD, digital TV, digital telephony, and the whole world of PC multimedia. These digital technologies offer much higher quality storage, transmission, filtering and other forms of processing than older analogue techniques could achieve. In addition, they provide entirely new abilities for adding data to signals, for sophisticated searching and recall of material, and for manipulating the form and content of sound and vision. Many examples of domestic and industrial equipment and systems rely heavily on computer technologies to provide cheap, effective and reliable control functions, ranging in scale from washing machines to power stations.

However, real world signals, including the inputs from microphones, video cameras, temperature sensors and other transducers, are usually analogue. Outputs to loudspeakers, displays and motors are often required in analogue form. Hence conversion between analogue and digital subsystems, particularly between analogue inputs and outputs and digital processing, is a very common requirement in electronic design. Some of the basic building blocks for interfacing between analogue and digital signals are the subject of this chapter.

7.2 DIGITAL-TO-ANALOGUE CONVERSION

7.2.1 Introduction

The first interface device to be covered is the **digital-to-analogue (D-A) converter**. Its function is to convert a digital input codeword into a signal that can be processed by analogue circuits, and so it forms the output interface from a digital to an analogue subsystem.

Figure 7.1 shows the outline of a 4-bit D-A converter. Each of the 16 possible input combinations that can be set up on the four digital inputs should lead to a particular analogue voltage generated on the output. In principle, any input code could be associated with any analogue voltage. However, most D-A

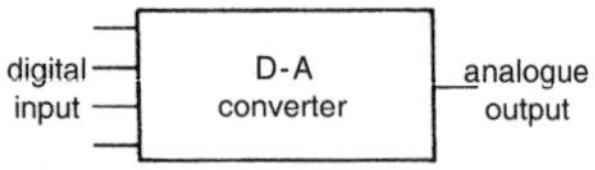

Fig. 7.1 4-bit D-A converter schematic.

converters produce an analogue voltage (or, in some cases, current) that is proportional to the natural binary value of the input code. Thus, if the device has a full scale of 15 V,

0000 represents 0 V
0001 represents 1 V
0010 represents 2 V
1000 represents 8 V
1111 represents 15 V

with other values in proportion.

Now, electronically it would be possible to detect each bit pattern to operate a switch when the pattern was present. The switch would turn on a power source giving the voltage for the number represented by each particular pattern of bits.

However, a more economical solution takes advantage of the conventional way of coding binary numbers in which each bit position represents a numerical weight of a power of 2. Hence a 4-bit pattern of binary digits represents a natural binary number in the form

$$(\text{bit } 3 \times 2^3) + (\text{bit } 2 \times 2^2) + (\text{bit } 1 \times 2^1) + (\text{bit } 0 \times 2^0)$$

Figure 7.2 shows how the conversion can be achieved in practice. If bit 3 (the MSB) is a 1, an 8 V source would be switched on; if bit 2 is 1, a 4 V source;

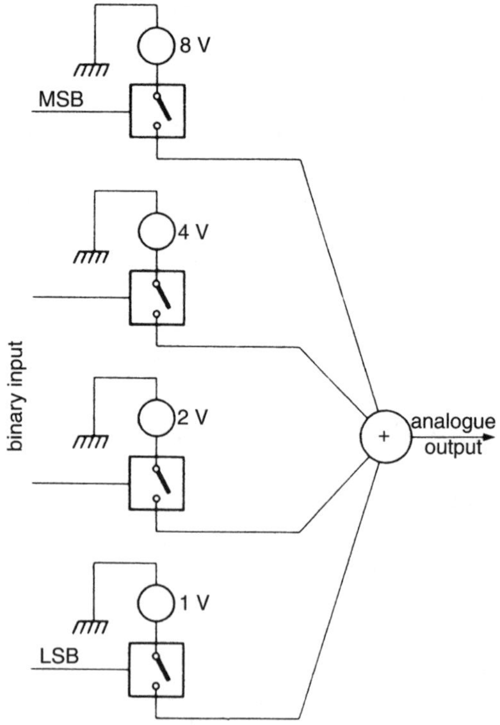

Fig. 7.2 4-bit D-A converter.

and so on. The voltages are then added by an analogue circuit. If, for example, a binary code 1001 was applied, then the output of a power source providing 8 V (representing the most significant bit) is added to the output of a power source giving 1 V (representing the least significant bit), to give the correct final output of 9 V.

For a 4-bit converter there are then four power sources of 8, 4, 2 and 1 V and a device that can perform addition. To obtain greater precision, more power sources and switches can be added. The range of voltage output is determined by the values of the signals available from the power sources.

D-A and A-D converters which do not assume the natural binary value of the digital code are available for special purposes. For example, converters which follow a power law or logarithmic characteristic are used to compress audio signals (which cover a very wide range of amplitude) into a smaller range of codes

Self-assessment question

1 Choose the voltage values to be generated by each of the five power sources of a D-A converter that converts a 5-bit binary input word to voltages in the range 0 to 7.75 V.

Sections 7.2.2 to 7.2.6 describe some examples of electronic circuits that can be used to build practical D-A converters following this natural binary interpretation. There are so many different circuits that the ones given represent only a small selection.

7.2.2 The binary-weighted resistor network

To begin with let us see how we can generate a voltage that corresponds to a binary codeword using the circuit shown in Fig. 7.3. The circuit comprises a 4-bit register which stores the binary codeword, and a resistor network. The voltage level at an output of the register will depend upon whether a 0 or a 1 is stored in that bit position. For convenience assume that the voltage levels are 0 V and 5 V for a 0 and a 1 logic level, respectively. Also assume that the values of R_1 to R_4 are $R_1 = R$, $R_2 = 2R$, $R_3 = 4R$ and $R_4 = 8R$. This is called a **binary-weighted resistor network**.

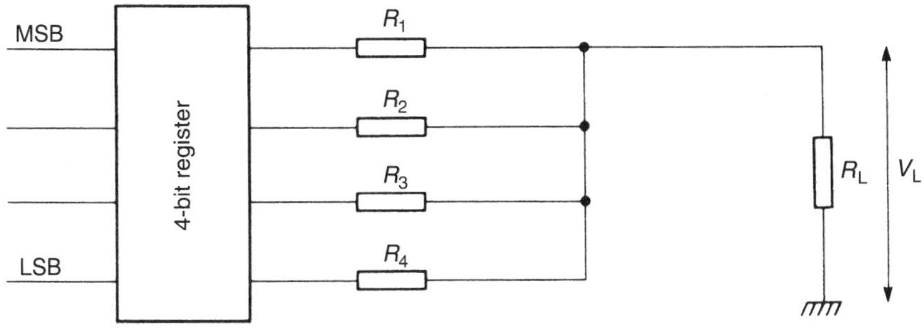

Fig. 7.3 A simple D-A converter circuit.

For an input codeword of 1000, the equivalent circuit is that shown in Fig. 7.4a, because only R_1 is connected to 5 V, while R_2, R_3 and R_4 are connected to 0 V. With the binary-weighted resistor values, the parallel combination of R_2, R_3 and R_4 equals $8R/7$ and the circuit can be redrawn as shown in Fig. 7.4b.

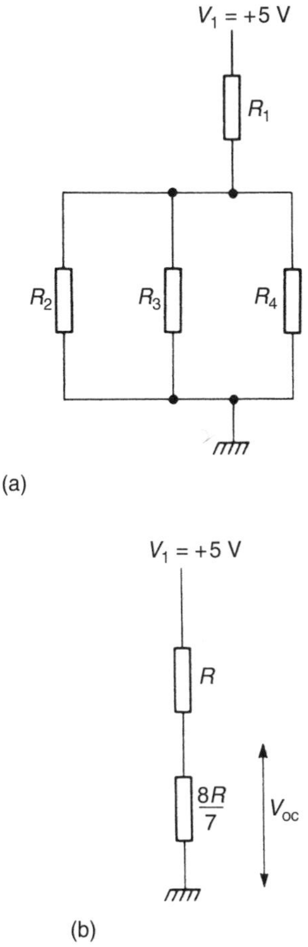

Fig. 7.4 Equivalent circuit for the binary-weighted resistor network.

Table 7.1

Codeword	Output voltage, V_{oc}	Denary equivalent
1000	$\frac{8}{15}V_1$	8
0100	$\frac{8}{60}V_1=\frac{4}{15}V_1$	4
0010	$\frac{8}{60}V_1=\frac{2}{15}V_1$	2
0001	$\frac{8}{120}V_1=\frac{1}{15}V_1$	1

The open-circuit voltage V_{oc}, for an input codeword of 1000 can be calculated from this equivalent circuit. From Fig. 7.4b we can see that

$$V_{oc}=V_1\times\frac{8R/7}{(8R/7)+\mathrm{R}}=V_1\times\frac{8}{15}$$

The output voltages for the codewords 0100, 0010 and 0001 are calculated in exactly the same way, and are shown together with the value of 1000 in Table 7.1. We can see that the output voltage for each single bit is related to the binary weighting of that bit in the codeword. For example, the least significant bit contributes a voltage of 1/15 V_1 and its denary value is $2^0=1$, and the most significant bit contributes a voltage of 8/15 V_1 and its denary value is $2^3=8$.

Self-assessment question

2 Draw up a table of the open-circuit voltages V_{oc} for all possible input codewords for the circuit in Fig. 7.3. Assume $R_1=R$, $R_2=2R$, $R_3=4R$ and $R_4=8R$, and that the register outputs are 0 V and 5 V for a logic 0 and 1 respectively. (*Hint*: assume that the system is linear and apply superposition.)

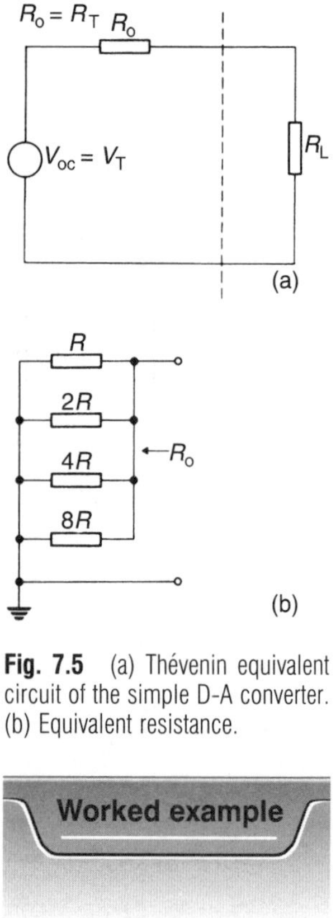

Fig. 7.5 (a) Thévenin equivalent circuit of the simple D-A converter. (b) Equivalent resistance.

A general expression for the output voltage can be obtained using the principle of superposition since the system is linear. This assumption is valid if the output resistance of the voltage source is the same when the input is 0 as it is when the input is 1. Then if b_0 to b_3 are the bits of the binary codeword, the output voltage is given by the expression

$$V_{oc}=\left(\frac{8}{15}b_3+\frac{4}{15}b_2+\frac{2}{15}b_1+\frac{1}{15}b_0\right)\times V_1$$

Hence the choice of binary-weighted resistor values for the network has led to the natural binary relationship between the input codeword and the output voltage. Such binary-weighted resistor networks are frequently used in discrete component D-A converters, because of their simplicity.

To find the voltage across the load R_L, we use the Thévenin equivalent circuit shown in Fig. 7.5a. V_L is given by the expression

$$V_L=\frac{R_L V_{oc}}{R_L+R_o}$$

Here V_{oc} is the weighted sum that we have just found and R_o is the output resistance of the weighting network, which can be calculated from Thévenin's theorem.

Worked example

How would you find the value of the output resistance?

Thévenin's theorem states that the output resistance is equal to the resistance that would be measured between the output terminals if the load were removed and all sources were replaced by their internal resistances (see Chapter 1).

If we take the circuit shown in Fig. 7.3 and replace the sources by their internal resistances (which we can assume are zero) the resulting network is shown in Fig. 7.5b. The value of the output resistance R_o is equal to the parallel combination of the four resistances which is:

$$R_o = \frac{8R}{15}$$

Hence the effect of loading can be calculated.

7.2.3 Buffering the resistor network

The potential-divider effect of the load and the weighting network is a major disadvantage of the D-A converter circuit of Fig. 7.3; any change in R_L is seen as a change in V_L.

Worked example

What would be the range of output voltages, V_L, if $R_L = R = 1$ kΩ approx, and $V_1 = 5$ V?

$$V_L = \frac{R_L V_{oc}}{R_L + R_o} = \frac{10^3}{10^3 + (8/15) \times 10^3} \times V_{oc} = \frac{15}{23} \times V_{oc}$$

When the input is 0000 the output will be 0 V, and when the input is 1111 the value of V_{oc} is 5 V, so the range of voltages is 0 to 3.26 V.

We can obviously improve matters either by increasing the value of R_L or by reducing R_o.

The best practical solution is to follow the resistor network with a buffer amplifier. As well as ensuring that the current out of the resistor network is virtually independent of the load R_L, the op amp also provides a low-impedance output stage for connection to subsequent circuits, and also allows the possibility of being able to choose the gain, to give the range of output voltages that best suit a particular application.

Figure 7.6 shows the resistor network connected to the inverting input of an op amp. This is the inverting feedback configuration described in Chapter 4. Owing to the very high impedance of the amplifier, all the current flowing out of the weighting network flows through the feedback resistor R_F. The node S is usually referred to as the summing junction, or virtual-earth point. The output voltage V_o is given by the expression:

$$V_o = -I_F \times R_F = -(I_1 + I_2 + I_3 + I_4) \times R_F$$

To calculate the output voltage, the individual currents from the resistor network are summed. For example, consider the circuit in Fig. 7.6 for an input codeword of 1010, if a logic 1 is 5 V and a logic 0 is 0 V, and $R = R_F = 1$ kΩ. The currents I_1 to I_4 are equal to the input voltage (either 5 V or 0 V) divided by the resistance in the corresponding branch. The output voltage is therefore

$$V_o = -R_F\left(\frac{5}{R} + \frac{0}{2R} + \frac{5}{4R} + \frac{0}{8R}\right)$$

Since $R_F = R$, the expression simplifies to

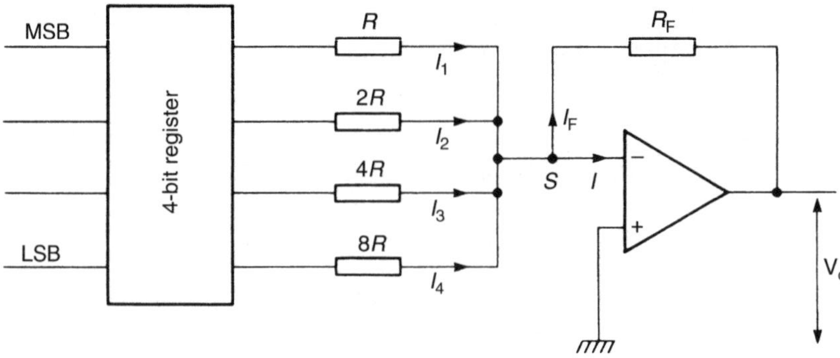

Fig. 7.6 The amplifier resistor network.

$$V_o = -\left(\frac{5}{1} + \frac{5}{4}\right) = -6.25 \text{ V}$$

To get the output voltage in the range of, say, 0 to −5 V, we have to ensure that its value when all the inputs are 1 is equal to −5 V:

$$V_o = -R_F\left(\frac{5}{R} + \frac{5}{2R} + \frac{5}{4R} + \frac{5}{8R}\right) = -\frac{5R_F}{R}\left(1 + \frac{1}{2} + \frac{1}{4} + \frac{1}{8}\right)$$

$$= -\frac{5R_F}{R}(1.875) = -\frac{9.375R_F}{R}$$

If $V_o = -5$ V and $R = 1$ kΩ, then

$$-5 = -\frac{9.375R_F}{1 \times 10^3}, \text{ so } R_F = 533\,\Omega$$

Self-assessment question

3 Use the equivalent circuit that you found earlier for the resistor network to determine the output voltage of the non-inverting amplifier circuit shown in Fig. 7.7, when $V_1 = 5$ V and $R_1 = R_2 = 1$ kΩ.

By supplying a sequence of values it is possible to use the D-A converters of Fig. 7.6 and Fig. 7.7 to generate a varying output voltage. If, for example, the registers were to be replaced by a 4-bit binary counter, the output voltage would increase in a series of steps as the counter was incremented.

However, once varying signals are generated, it becomes necessary to take into account the dynamic characteristics of the circuit. Of particular importance is the time taken for the output voltage of the op amp to settle to a steady-state value after a change in the input codeword. This time interval is known as the **settling time** and is dependent on a number of factors including the time response of the op amp to a step input.

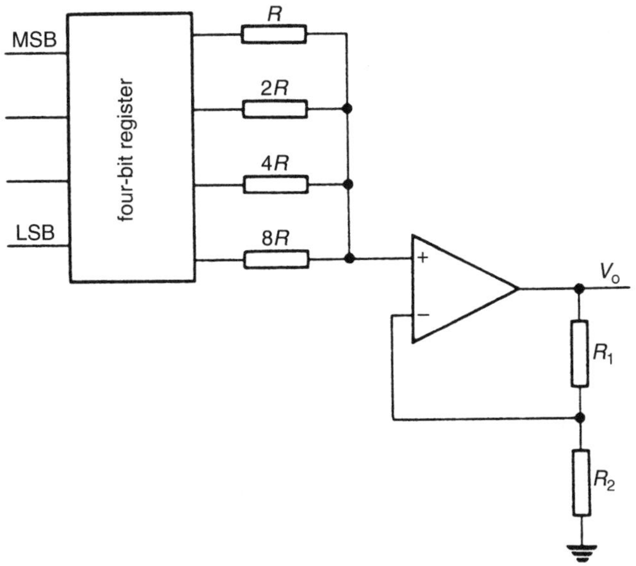

Fig. 7.7 Circuit for Self-assessment question 3.

7.2.4 D-A converter with a stabilized voltage source

Practical logic gate output voltages are specified to lie only within a certain range of voltage. We can improve the accuracy of the D-A converter by deriving the voltage levels from a stabilized voltage source which varies over a much narrower range. In such a design the logic signals are used to close switches connecting the voltage source to the resistor network as shown in Fig. 7.8. Apart from this modification, the circuit operates as before.

While the introduction of switches removes the problem associated with variable voltage sources, their use may introduce other errors. In order to determine what effect the introduction of the switches has upon the performance of the D-A converter, a switch in an 'on' state can be modelled by a resistor, R_{on}, and in an 'off' state the switch is replaced by an open circuit.

Figure 7.9 shows how this model can be used to establish the errors in the D-A converter of Fig. 7.8, when the input codeword is 1000. For this example the switch in series with resistor R is replaced by resistor R_{on} and the other three switches are effectively disconnected. The total current I_s entering the summing junction is

$$I_s = \frac{V_s}{R_{on} + R}$$

If the switches were ideal (that is, no 'on' resistance) the current I entering the summing junction would be:

$$I = \frac{V_s}{R}$$

Electronic switches are based on field-effect transistors. These are connected to provide a low-resistance path with an appropriate bias voltage, and a much higher resistance path if this voltage is removed. Thus analogue signals can be conveniently switched on and off by applying logic levels. The 'on' resistance is typically 100 Ω, while the 'off' resistance is usually at least many tens of megohms.

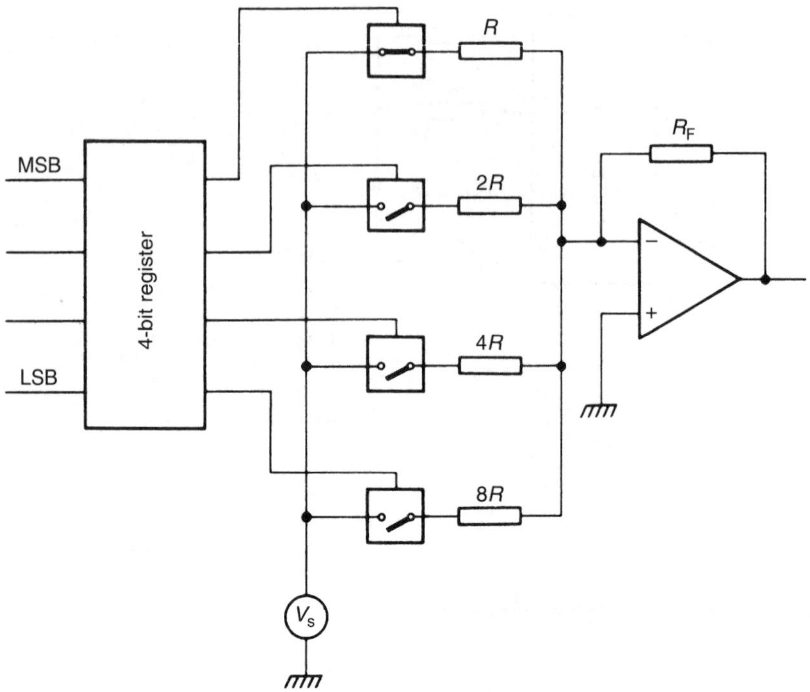

Fig. 7.8 Binary-weighted D-A converter with a stabilized reference voltage.

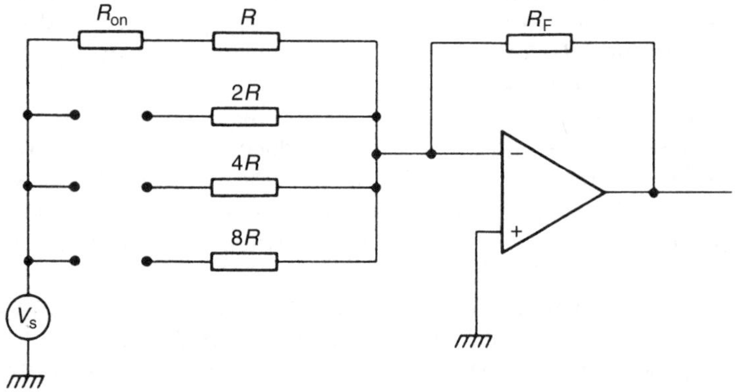

Fig. 7.9 Equivalent circuit of the D-A converter.

The difference $(I - I_s)$ times the feedback resistor equals the error in the output voltage of the op amp.

For example, suppose that a 4-bit converter uses a switch that has an 'on' resistance of 100 Ω. If the switches were ideal, the output voltage for an input codeword of 1111 would be

$$V_o = -R_F\left(\frac{V_s}{R}+\frac{V_s}{2R}+\frac{V_s}{4R}+\frac{V_s}{8R}\right)$$

With $V_1 = 5$ V and $R = R_F = 10$ kΩ, this becomes

$$V_o = -10^4\left(\frac{5}{10^4}+\frac{5}{2\times10^4}+\frac{5}{4\times10^4}+\frac{5}{8\times10^4}\right) = -9.375\,\text{V}$$

Taking account of the 'on' resistance R_{on}, the output voltage becomes

$$V_o = -R_F\left(\frac{V_s}{R+R_{on}}+\frac{V_s}{2R+R_{on}}+\frac{V_s}{4R+R_{on}}+\frac{V_s}{8R+R_{on}}\right)$$

With $R_{on} = 100$ Ω, V_o is

$$V_o = -10^4\left(\frac{5}{1.01\times10^4}+\frac{5}{2.01\times10^4}+\frac{5}{4.01\times10^4}+\frac{5}{8.01\times10^4}\right) = -9.309\,\text{V}$$

The error is the difference between these two outputs

$$-9.375 - (-9.309) = -0.066\,\text{V}$$

7.2.5 The *R*-2*R* ladder resistor network

Although the binary-weighted D-A converter has many practical applications, it is seldom used when there are more than 6 bits in the input codeword. The problem is that more bits means that a larger range of weighting resistors is required.

Worked example

What range of resistors would be required for a 16-bit D-A converter?

If the smallest resistor in the network is R then the largest in the network is $2^{n-1} \times R$, where n is the number of bits in the codeword. So, for n = 16, the largest resistor is 32 768R. The smallest resistor R will have to be accurate to within 1 part in 32 768, otherwise the current from the least significant bit will be swamped. This means that you need an accuracy of better than 0.003% which is difficult and very expensive to produce.

An alternative resistor network exists which incorporates resistance values that differ by only 2 : 1 for any number of bits in the codeword, and the principle involved is illustrated in Fig. 7.10. The current I entering node N must leave by way of the two resistors R_1 and R_2 (Kirchhoff's current rule). If these resistors are equal, $I/2$ will flow in each branch. Such a resistor network gives us a simple method for dividing the current flowing in a network. The principle can be extended indefinitely; the only requirement is that the equivalent resistance of each current path leading away from the node is the same.

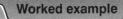

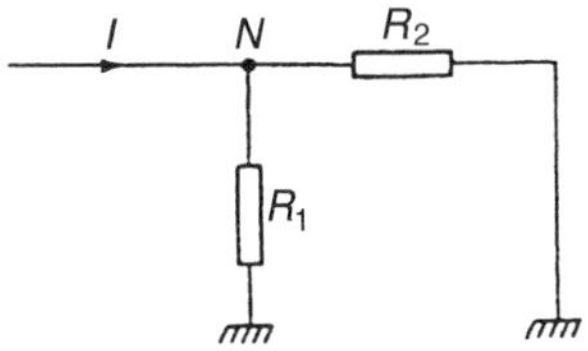

Fig. 7.10 Current division at a node.

Figure 7.11a shows a circuit for dividing the input current by 4. $I_2 = I/4$, but to check this we must first calculate I_1. This is done by calculating the equivalent resistance of the circuit to the right of line A, and then applying Kirchhoff's rule. The equivalent resistance R_{eq} is equal to R plus the value of the two $2R$ resistors in parallel, that is

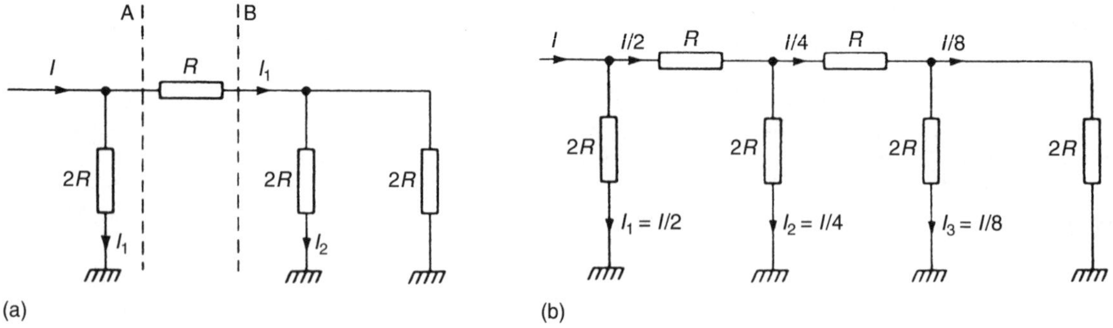

Fig. 7.11 Current-dividing networks.

$$R_{eq} = R + 2R/2 = R + R$$
$$R_{eq} = 2R$$

I_1 is therefore equal to $I/2$, and an equal current $I/2$ flows through the first $2R$ resistor. Because the resistances of the two branches to the right of line B are equal, $I_2 = I_1/2 = I/4$.

The network of three resistors to the right of line A has an equivalent resistance of $2R$, and so the right-hand $2R$ resistor can be replaced by a copy of this network without changing the current flowing into it. This is shown in Fig. 7.11b, giving a current $I_3 = I/8$, as well as $I_2 = I/4$ and $I_1 = I/2$. Notice that all the resistors are either R or $2R$, hence such networks are referred to as ***R*-2*R* ladder networks**.

Table 7.2

Bit	Current
5	I
4	$I/2$
3	$I/4$
2	$I/8$
1	$I/16$
0	$I/32$

Now let us see how we can use the R-$2R$ ladder to build a D-A converter. Our problem is to ensure that the current controlled by each bit of the input codeword is related to a bit's position in the codeword. If, for example, the most significant bit of a natural binary codeword switches in a current I, then the next most significant bit must switch in a current $I/2$, and so on. For a 6-bit D-A converter the current switched in by each bit is shown in Table 7.2.

Remember that in the case of the binary-weighted network the current switched by each bit was determined by a resistor, so the resistance was increased by a factor of 2 going from each bit to the next less significant bit. However, if we use the R-$2R$ ladder the current switched by each bit is determined by the point along the ladder at which it is tapped off.

Figure 7.12 shows a block diagram for a complete 4-bit D-A converter incorporating an R-$2R$ ladder network. As in Section 7.2.4, the outputs of the register are used to control the state of a switch, and thus do not themselves provide the reference voltages for the network. The op amp operates in exactly the same way as for the buffered binary-weighted network D-A converter, by summing the current flowing into its inverting input.

The output voltage is again proportional to the natural binary value of the input codeword:

$$V_o = -R_F(b_3I/2 + b_2I/4 + b_1I/8 + b_0I/16)$$

Notice that the switches themselves do not affect the magnitude of the currents flowing through the $2R$ resistors. In one position, the current flows directly to 0 V. In the other position, the current flows into the summing junction of the op amp, which is also effectively at 0 V (virtual earth), thus leaving the current unchanged.

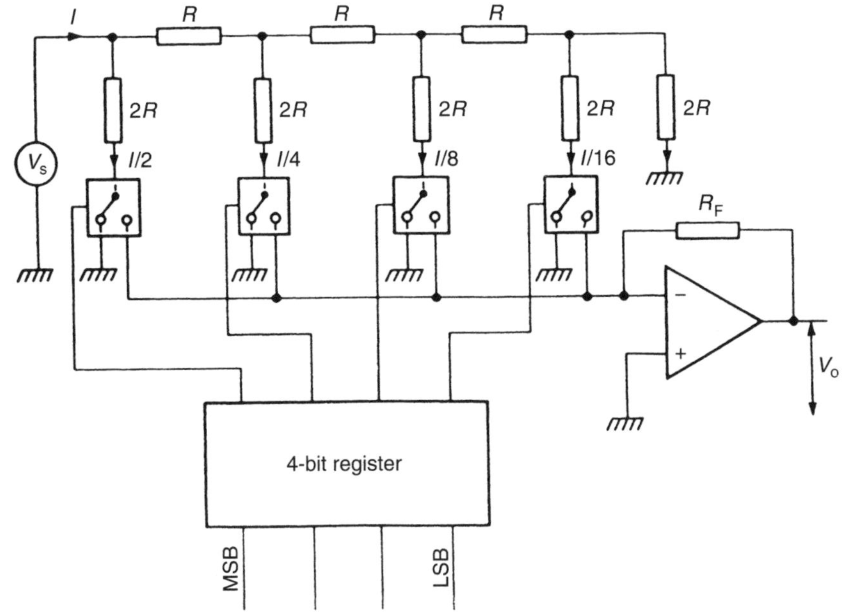

Fig. 7.12 An *R*-2*R* ladder network D-A converter.

Self-assessment question

4 What is the full-scale output voltage of an 8-bit D-A converter of the form illustrated in Fig. 7.12, designed around an *R*-2*R* ladder network and a 5 V reference source, if the feedback resistor is 10 kΩ and $R = 5$ kΩ?

7.2.6 Bipolar conversion

In most of the D-A converter designs we have examined so far the output voltage has been negative. This is because the reference voltage was chosen to be positive and, with the exception of the circuit in Self-assessment question 3, the D-A converters have employed inverting amplifiers. We could obtain a positive output voltage from the same D-A converter by using a negative reference voltage. Such single-polarity-output D-A converters are called **unipolar converters.** In many practical applications, it is necessary to design a D-A converter that will produce both positive and negative outputs; they are called **bipolar converters.**

To obtain bipolar outputs it is necessary to modify the design of the unipolar D-A converter so that it will accept binary input codewords which contain both polarity and magnitude information. Figure 7.13a shows the output voltage for a unipolar converter, where the smallest output is produced by a binary input of 0000, the largest by an input of 1111. Figure 7.13b shows the simplest

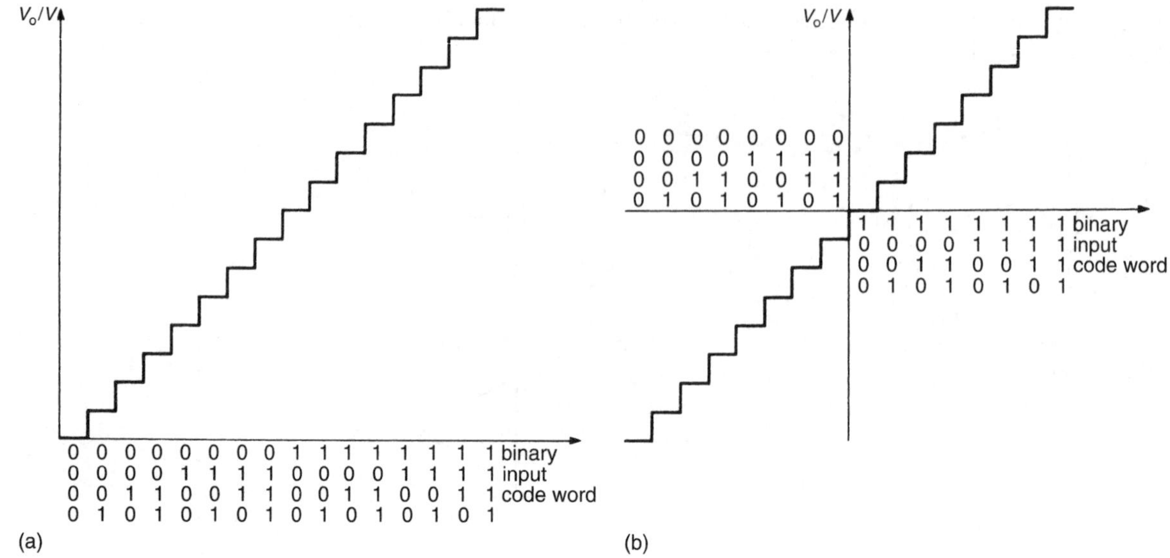

Fig. 7.13 Unipolar and bipolar converter outputs.

method of producing a bipolar output. In this case a voltage equal to half the full-scale output has been subtracted from the output of the D-A converter. For this modified converter an input of 0000 gives the maximum negative output while 1111 gives the maximum positive output. A zero output is achieved when a binary input of 1000 is applied to the converter. This type of coding is referred to as **offset binary**, simply because it is a normal binary D-A converter with the output offset by half the full-scale output.

Let us now calculate the full range of the D-A converter output voltage for the case $R_F = R$. For an input codeword of 0000 the D-A converter output is 0 V, but from this we must subtract $V_{ref}/2$, so the actual output is $-V_{ref}/2$.

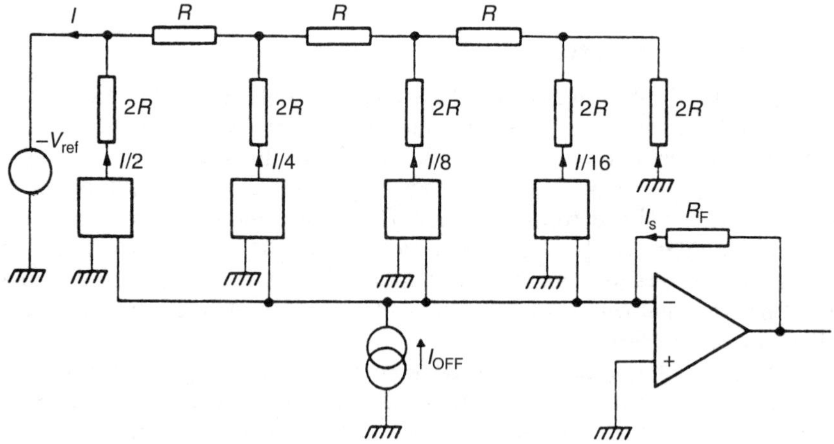

Fig. 7.14 A bipolar D-A converter.

When the input codeword is 1111, the D-A converter output is 15/16 V_{ref}, so the offset output is $+7/16\ V_{ref}$. The full-range output of the bipolar D-A converter is therefore $+7/16\ V_{ref}$ to $-V_{ref}/2$.

There are several methods for actually producing the necessary offset voltage. One method is to inject an equivalent offset current, I_{OFF}, into the summing junction of the op amp as shown in Fig. 7.14. The value of this current is adjusted to be equal in magnitude, but opposite in sign to the current injected by the ladder network when a binary input of 1000 is applied to the converter. This equal and opposite current ensures that no current flows through the feedback resistor R_F for an input codeword of 1000, resulting in zero volts output from the amplifier.

Self-assessment question

5 What will be the output voltage of the D-A converter shown in Fig. 7.14 for input codeword 1010, if $R = 5\ k\Omega$, $V_{ref} = -5$ V and $R_F = 10\ k\Omega$?

7.2.7 Quantization

You have seen how a D-A converter can be used to generate a d.c. voltage, or, when driven from a binary counter, a staircase waveform. In practice, D-A converters can be used to generate any waveform given the appropriate sequence of input codewords. However, the output waveform does not vary continuously, but can only jump between the distinct levels defined by the input codes. For example, suppose we wanted to generate the sawtooth waveform shown in Fig. 7.15a for use in the timebase of an oscilloscope. An approximation is to use the staircase waveform obtained by driving a D-A converter from a 4-bit counter, shown in Fig. 7.15b.

The 16 possible values of the D-A converter output voltage are called the **quantization levels**, and the difference in voltage between two adjacent quantization levels is termed a **quantization interval**.

The slope of the staircase waveform is determined by the rate at which the input codewords to the D-A converter change. For example, to obtain a steep slope the D-A converter input codewords must be changed quickly, while for a shallow slope they need only be changed slowly. For a uniform slope, the duration of each step is equal to the period of the ramp divided by the number of quantization levels. In our example the period of the ramp is T, so the duration of each step is $T/16$, say τ, then τ is the period of the clock pulse applied to the counter. Remember that τ must be long enough to allow the D-A converter output to settle before the arrival of the next clock pulse.

Because of quantization, the staircase waveform is only an approximation to the ramp. Figure 7.16a shows a short portion of the two waveforms of Fig. 7.15 drawn to a much larger scale. You can see that during the interval t_1 to t_2 the required ramp voltage increases from V_A to V_B, while the D-A converter output is constant at V_1. The two waveforms are therefore equal at only one instant of time during the interval t_1 to t_2, and at all other points they differ. In terms of our approximation these differences produce an error, known as the **quantization error**.

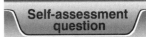

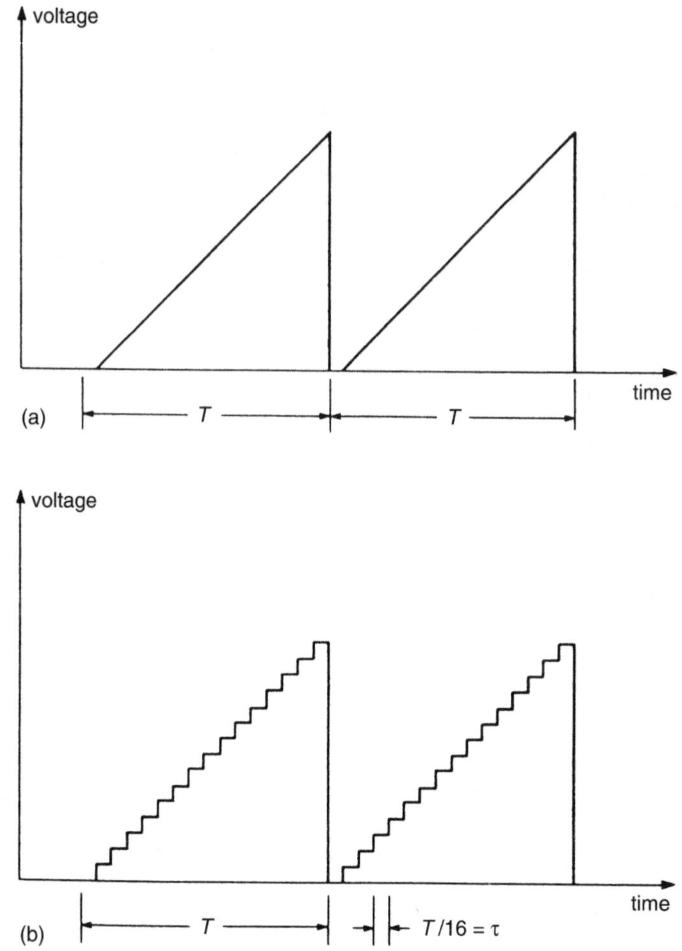

Fig. 7.15 Analogue and digital ramp waveforms.

We can see the effects of quantization error by subtracting the ideal ramp voltage from the D-A converter output voltage which approximates to it. For example, using the waveforms of Fig. 7.16a, the difference between the D-A converter output and the ramp is shown in Fig. 7.16b. You can see that the error alternates about zero volts, and has a peak-to-peak value equal to the quantization interval. So the maximum quantization error is half the quantization interval.

It is convenient to model the approximation to the ramp voltage produced by the D-A converter as a ramp with noise added to it. Because the noise is due to quantization, it is given the name **quantization noise**. To reduce the quantization error we must use a D-A converter controlled by more bits, because if the full-scale output voltage remains the same this will reduce the size of the quantization interval. For example, the quantization error for a 4-bit D-A converter is always 1 part in 30 (full-scale output is equal to 15 quantization intervals), and is 1 part in 510 for an 8-bit D-A converter, no matter what the full-scale output voltage.

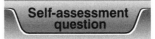

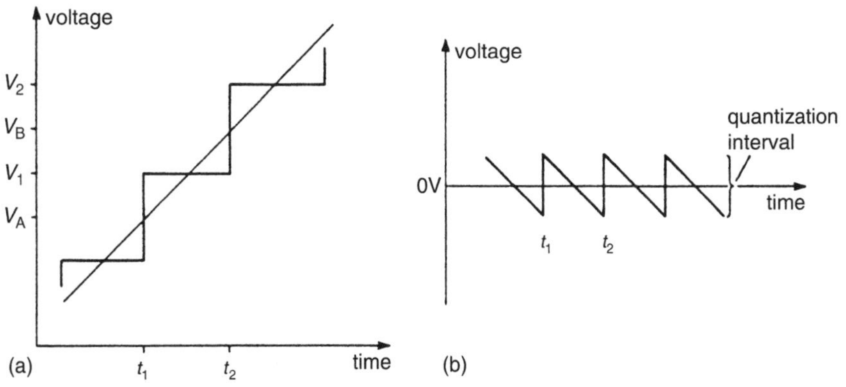

Fig. 7.16 Quantization errors.

Self-assessment question

6 What is the maximum quantization error for a 10-bit bipolar D-A converter with a full-scale output of 10 V?

An alternative way of specifying the quantization effect is in terms of the **dynamic range** of the D-A converter. The dynamic range in decibels is defined as

$$\text{Dynamic range} = 20\log_{10}\frac{\text{full-scale output}}{\text{quantization interval}}$$

and indicates the relative magnitudes of the smallest non-zero and largest output voltages that can be generated by a D-A converter.

The dynamic range can also be obtained by substituting into the above equation the actual value for the full-scale output in terms of quantization intervals. For an n-bit converter there are 2^n quantization levels, but only $2^n - 1$ quantization intervals, hence the full-scale output is $(2^n - 1) \times q$, where q is the quantization interval. The dynamic range is therefore

$$\text{Dynamic range} = 20\log_{10}\frac{(2^n - 1) \times q}{q} = 20\log_{10}(2^n - 1)$$

A simple rule of thumb to remember is that each bit of the input codeword of a D-A converter adds 6 dB to the dynamic range.

7.2.8 Summary of Section 7.2

Section 7.2 introduced you to devices that convert a digital codeword into an analogue voltage by means of resistor weighting networks, in particular the binary-weighted and the R-$2R$ ladder networks. The R-$2R$ ladder is particularly convenient for incorporation into integrated circuits as a wide range of resistor values is not required.

The addition of an amplifier minimizes loading effects upon the resistor weighting network and also provides a low-impedance output stage. High-performance D-A converters also require a stabilized voltage reference source and switches to connect this source to the appropriate resistors in the weighting network. However, the introduction of the switches can add errors associated with on-resistance.

The conversion from digital to analogue form involves the quantization process, which limits resolution and introduces quantization noise. This quantization error can only be reduced by increasing the number of bits in the converter.

7.3 ANALOGUE-TO-DIGITAL CONVERSION

This section will examine the principles that can be used to convert an analogue quantity, such as a voltage or current, into a digital codeword. Such a subsystem is called an **analogue-to-digital (A-D) converter**.

7.3.1 The comparator and flash converter

An electronic **comparator** is a device which continuously compares two signals A and B. If $A > B$ then the comparator output is in one logic state (0, say), while if $B > A$ then it is the opposite state (1). In essence a comparator is simply an op amp used without feedback, so that a tiny difference between two signals connected to the differential inputs causes the output to be close to one supply line or the other.

Consider the case of such a device illustrated in Fig. 7.17, which is to compare a reference voltage with an analogue input signal. Suppose that the comparator is connected to a voltage supply of 10 V. When the input signal is less than or equal to the reference, the output is close to 0 V, corresponding to a logical 0. When the input signal is greater than the reference, the output is close to 10 V, corresponding to a logical 1.

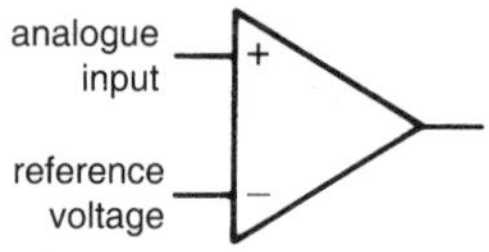

Fig. 7.17 Comparator.

Now suppose that two comparators are supplied each with a different reference, one quantization interval apart. An analogue input signal with a value between the two references will set one comparator output to 1 and the other to 0 (the signal is less than one and greater than the other). An input signal level that is less than either sets both comparator outputs to 0. An input greater than either sets both to 1. This would enable a circuit to detect when the incoming signal is within a particular quantization interval and whenever this happened it could, with suitably connected logic gates, set the output to the appropriate digital representation.

The **flash converter** is a system with a comparator and a reference for each of many discrete levels represented by the digital output, and an example for 7 levels is shown in Fig. 7.18. The references have been chosen at 1 V intervals from 1 V to 7 V. Thus a signal below 1 V leads to all the comparators having an output of 0. Between 1 V and 2 V comparator A has an output of 1 while the others are still at 0. Between 2 V and 3 V comparators A and B are 1, and so on up to >7 V when all comparators are 1. The comparator outputs for all possible input voltages are shown in Table 7.3.

Taken together the comparator outputs make up a 7-bit code that represents the magnitude of the input signal, but this is not in a very useful form. Hence the system includes an encoder, a block of combinational logic

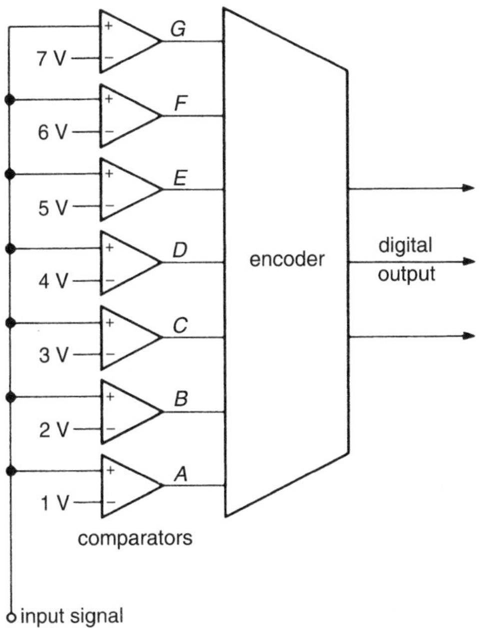

Fig. 7.18 A flash converter.

Table 7.3 Encoding table for the parallel converter

Converter input range (V)	Comparator outputs							Encoder output
	A	B	C	D	E	F	G	
≤1	0	0	0	0	0	0	0	0 0 0
1–2	1	0	0	0	0	0	0	0 0 1
2–3	1	1	0	0	0	0	0	0 1 0
3–4	1	1	1	0	0	0	0	0 1 1
4–5	1	1	1	1	0	0	0	1 0 0
5–6	1	1	1	1	1	0	0	1 0 1
6–7	1	1	1	1	1	1	0	1 1 0
>7	1	1	1	1	1	1	1	1 1 1

which converts the comparator output code to the more usual natural binary sequence. This output code, of only three bits to represent all possible ranges of input signal, is also given in Table 7.3.

The number of comparators needed in a flash converter is equal to the number of quantization intervals. Thus a 10-bit converter contains 1023 comparators, together with control and encoding circuity. Because so many components are involved, flash converters are usually found in the form of integrated circuits and have no more than 10 bits output, and are often quite expensive devices. However, their principle of operation is very simple, and because all comparisons are carried out simultaneously they are capable of much faster operation than any other type of A-D converter.

Some commercial 8-bit flash converters take less than 1 ns ($= 10^{-9}$ s) to perform a conversion.

7.3.2 The counter-ramp converter

A block diagram of another type of converter is shown in Fig. 7.19. It comprises a D-A converter, a binary counter, a clock and some control logic, and a single comparator.

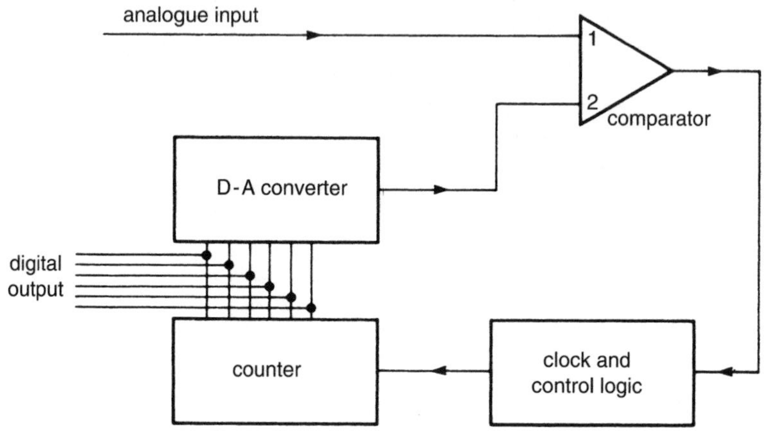

Fig. 7.19 A counter-ramp A-D converter.

The A-D converter works as follows. When a conversion is required, a signal is sent to the converter from an external digital subsystem requesting a conversion. This signal sets all the counter outputs to 0, and hence forces the D-A converter output to zero volts. Input 2 of the comparator is therefore also zero volts, so that if the analogue signal at input 1 of the comparator is non-zero the comparator output is set to logic 0. Once the D-A converter output has settled, a clock increments the counter by one count, thus increasing the D-A converter output. If this new output is greater than the input signal, the output of the comparator changes to a logic 1, and this stops the clock signal.

If, on the other hand, the D-A converter output is less than the analogue input, the counter is incremented by another clock pulse. This comparison and increment sequence is repeated until the D-A converter output exceeds the input voltage. Figure 7.20 is a timing diagram for the conversion process, and shows the interrelationship between the conversion request signal, the clock pulses, the D-A converter output and the comparator output. Notice that the counter increments on the rising clock edge, and the comparator output only changes when the D-A converter output exceeds the input voltage. This design of converter is called a **counter-ramp A-D converter**.

The conversion cycle is completed once the comparator output changes from logic 0 to logic 1. At this point the 'clock and control logic' block generates an 'end of conversion signal' to indicate to the digital subsystem

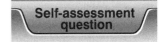

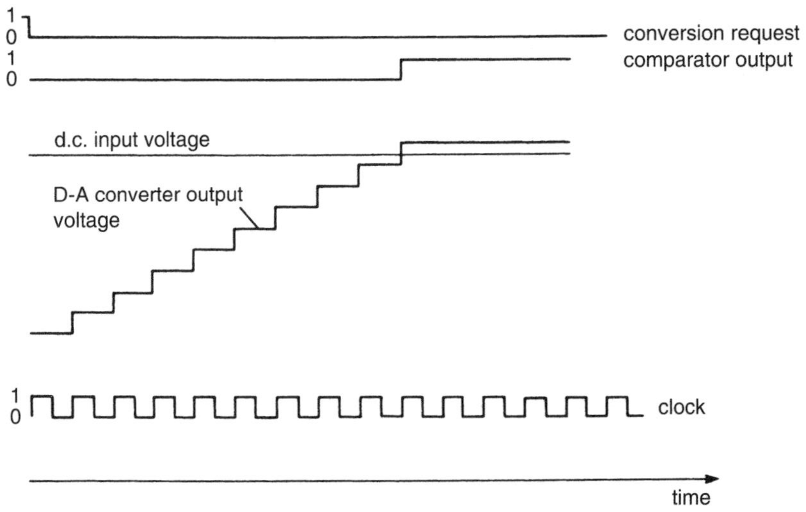

Fig. 7.20 Timing diagram for the counter-ramp A-D converter.

that the binary representation of the input voltage can be read from the counter outputs.

The time lapse between the start of a conversion and the generation of the end of conversion signal is known as the **conversion time** of the A-D converter, and the reciprocal of the conversion time is the **conversion rate**. The one major drawback of the counter-ramp A-D converter is its relatively long conversion time for large input signals, arising from the fact that the D-A converter output must ramp from zero. The conversion time can be reduced by decreasing the counter clock period, but there are practical limitations to this approach because the clock period must allow the D-A converter output to settle after each counter increment. Most commercial devices operate with maximum clock rates of 1 MHz or less.

Self-assessment question

7 A 10-bit counter-ramp A-D converter has a full-scale input of +10 V. If the clock period is 15 μs, how long will it take to convert an input of 4 V? How long for an input of 10 V?

A variation of the counter-ramp converter is the **tracking converter**. This has a very similar configuration to the counter-ramp arrangement of Fig. 7.19, but the counter has an up/down control and is not reset to all zeros before a conversion. When a conversion is required, then if the comparator indicates that the input voltage is larger than the present D-A output, the counter counts up until the comparator switches. If, however, the comparator initially indicates that the input voltage is smaller, then it counts down until the comparator switches. If the signal has not changed by much (either an increase or a decrease) since the previous conversion, then the counter will not have to count through many levels. Hence, this method works well for relatively slowly changing signals, and conversion is very much quicker than having to count from zero each time. However, like the counter-ramp, the tracking converter still suffers from the uncertainty of conversion time, and in practice a 'worst-case' value of conversion time must always be assumed which is a great limitation to speed.

7.3.3 The successive approximation A-D converter

This section will examine another A-D converter design, one which allows much faster conversion rates than the counter-ramp type. It is called a **successive approximation converter**.

Figure 7.21 is a block diagram for a 4-bit successive approximation A-D converter. The major difference from the counter-ramp A-D converter is that the counter has been replaced by a 4-bit register. The states of the output bits of this register are controlled by the 'clock and control logic' block.

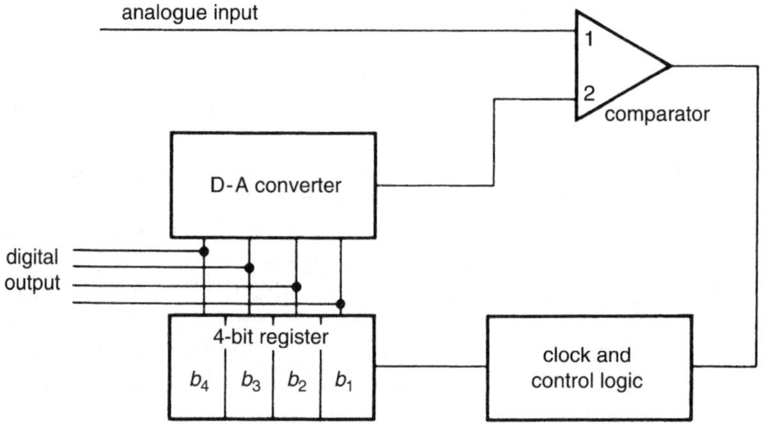

Fig. 7.21 A successive approximation A-D converter.

The converter works as follows. At the start of a conversion the first clock pulse clears all the register outputs to logic 0. On the second clock pulse the most significant bit of the register is set to logic 1, which causes the D-A converter output to change to half its maximum value. If the analogue input is more than the D-A converter output, the comparator output causes the logic control to leave the register bit set to a logic 1. If, however, the analogue input is smaller than the D-A output then the register bit is reset to 0. On the next clock pulse this process is repeated for the next most significant bit, and so on until all four bits of the register have been tested.

The total process is somewhat like a guessing game in which one person thinks of a number and another tries to guess what it is. The person guessing is told whether the number is less than or greater than each guess. Eventually the right answer can be determined.

Let us work through all the steps for the 4-bit converter, for a particular d.c. input voltage. Figure 7.22 is a timing diagram for the conversion process, and it

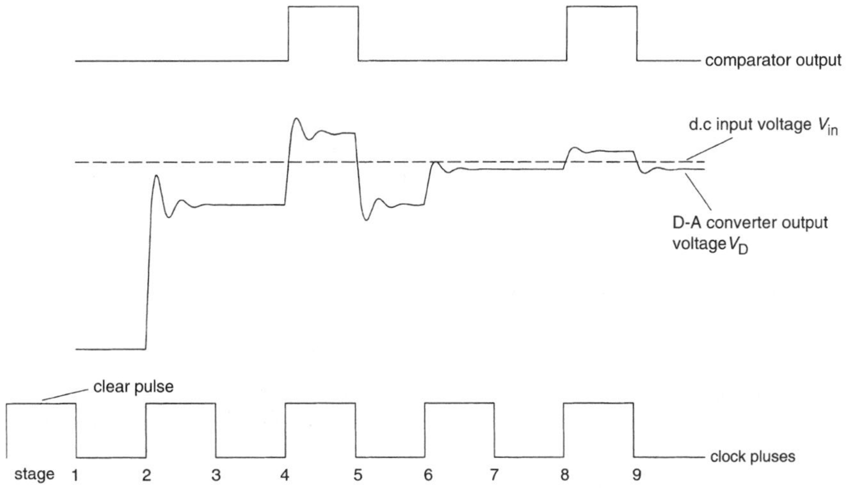

Fig. 7.22 Timing diagram for the successive approximation A-D converter.

illustrates the interrelationships between the clock pulse train, the D-A converter output and the comparator output. V_{in} is the d.c. analogue input, and V_D is the D-A converter output. There are the following nine stages to the conversion, and the register contents at each stage are listed in Table 7.4. These stages are indicated on Figure 7.22 against the clock edge which initiates them.

1. Clear all register outputs to logic 0. This forces $V_D = 0$ and the comparator goes low.
2. Set the most significant bit to logic 1. A value of V_D equivalent to 1000 is applied to the comparator.
3. Comparator output stays low, indicating that $V_D < V_{in}$. Hence b_4 must be 1 in the final answer (all codes above 1000 begin with a 1). The value of this bit is now determined, and is shown in bold in Table 7.4.
4. Set next bit b_3 to a logic 1, and V_D changes accordingly.
5. Comparator output goes high, showing that $V_D > V_{in}$. So b_3 is cleared to zero (all codes less than 1100 begin with 10××). The first two bits are now certain.
6. Set bit b_2 to logic 1. V_D changes.
7. Comparator output signals that $V_D < V_{in}$, so b_2 is left at logic 1. Three bits now determined.
8. Set bit b_1 to logic 1. V_D changes again.
9. Check comparator output. b_1 is cleared because $V_D > V_{in}$. Signal conversion completed. The final value obtained is 1010.

Table 7.4

1	0 0 0 0
2	**1** 0 0 0
3	**1** 0 0 0
4	**1** 1 0 0
5	**1 0** 0 0
6	**1 0** 1 0
7	**1 0 1** 0
8	**1 0 1** 1
9	**1 0 1 0**

Examining Fig. 7.22 more closely we can see that the setting, testing and possibly re-setting of a bit must all occur within one clock cycle. This example has assumed that the rising edge of the clock signal is used to set a bit and that the falling edge must not occur until the D-A converter output voltage has settled. The comparator output must be ignored during this settling period, because its output might change due to transient variations in the D-A converter output. The falling clock edge therefore determines when the data bit will be re-set, if necessary. The period between the falling clock edge and the next rising edge is again provided to enable the D-A converter output to settle if the tested bit was re-set.

The total conversion time is equal to $n+1$ clock cycles, one for each bit of the codeword and one to initialize the D-A converter output to zero. This is much faster than the counter-ramp A-D converter, if we ignore the few occasions when the input voltage is near zero. An additional advantage is that the conversion time is independent of input amplitude, which is useful in systems that require a constant conversion rate. However, the flash converter, taking the equivalent of a single clock pulse, is much the fastest.

Self-assessment question

8 What increase in speed can be gained by using a 12-bit successive approximation converter as opposed to a 12-bit counter-ramp design assuming a full-scale input voltage?

7.3.4 Integrating converters

An entirely different approach to A-D conversion, which is much slower in operation but capable of high precision at low cost, is widely used in digital voltmeters and similar applications. This technique is based on the circuit configuration shown in Fig. 7.23, consisting of an op amp with a capacitor as the feedback element. The circuit behaves as follows.

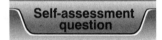

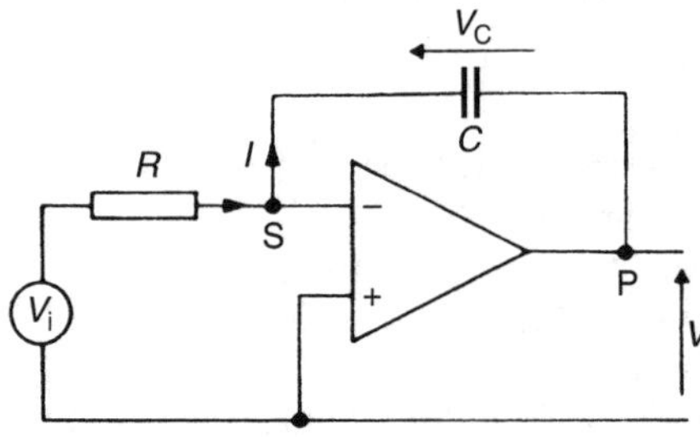

Fig. 7.23 A basic integration circuit.

Assume that a fixed analogue voltage V_i is connected to the input, so that a current I flows through R to the virtual earth point S. Since S is effectively at zero volts, Ohm's law gives

$$I = V_i/R \tag{7.1}$$

Because the op amp has a very high input impedance, all this current must flow into the capacitor. Since current = amount of charge flowing per unit time, then the total charge Q that accumulates on the capacitor during a time t (while the current I is constant) is

$$Q = It$$

Therefore the voltage across the capacitor V_C is given by

$$V_C = Q/C = It/C$$

Figure 7.23 shows that the voltage at point P is given by V_o or by $-V_c$. So

$$V_o = -V_C = -It/C$$

Substituting for I from Equation 7.1 results in

$$V_o = \frac{-V_i t}{RC} \qquad (7.2)$$

Hence if a *negative* voltage V_i is applied to this circuit, the output voltage will *increase* linearly with time as in Fig. 7.24. The rate of increase is proportional to the applied voltage. If V_i is *positive* then V_o will *decrease*. This circuit is called an **integrator** because it effectively integrates any input voltage variations with respect to time.

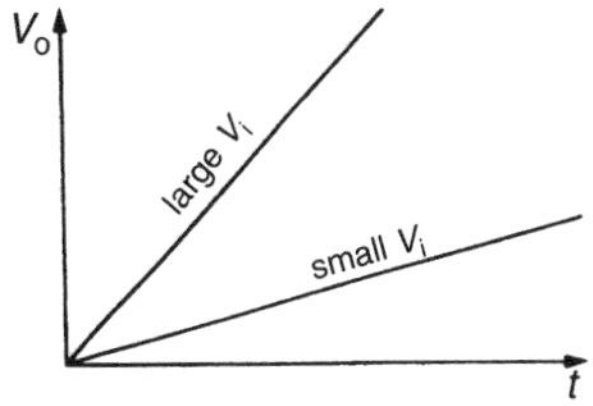

Fig. 7.24 Output voltage variation with time for the integrator.

One way in which the integrator can form the basis of an A-D converter is outlined in Fig. 7.25. The integrating circuit of Fig. 7.23 is controlled by a comparator and a block of sequential logic. The conversion is carried out in three stages which are illustrated on the waveform of V_o in Fig. 7.26.

The analysis leading to Equation 7.2 is restricted to a constant current I and hence to a constant input voltage V_i. In the general case, if V_i varies then $Q = \int I dt$ and the output voltage is

$$V_o = -\frac{1}{RC}\int V_i dt$$

Hence the circuit acts as an integrator (with respect to time) for any form of varying signal.

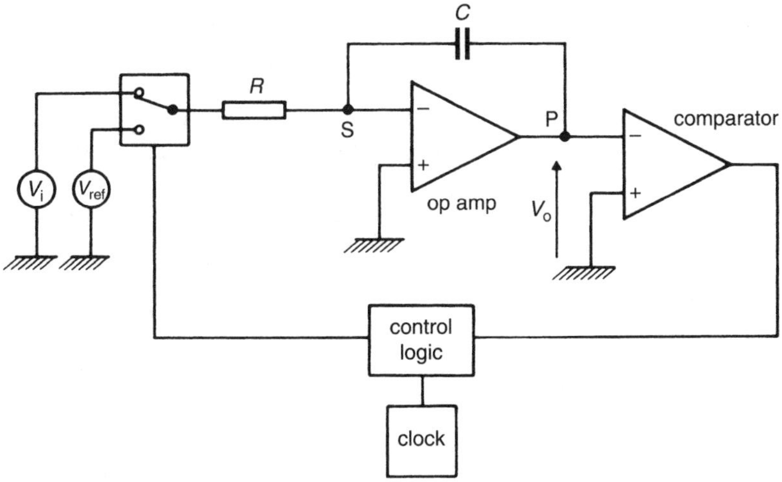

Fig. 7.25 Schematic outline of a dual-slope converter.

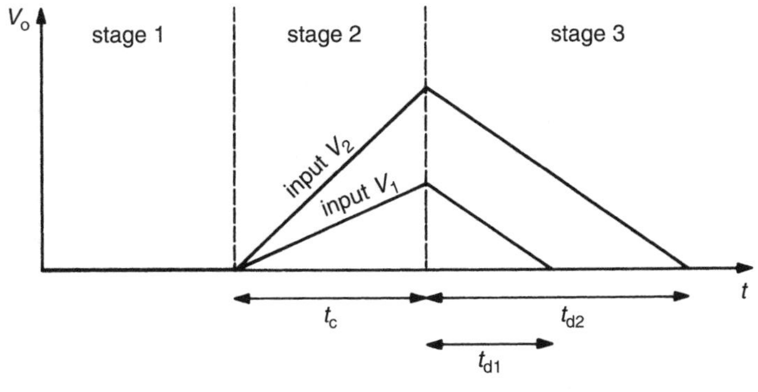

Fig. 7.26 Variation of op amp output voltage V_o during a dual-slope conversion.

In practice the offset voltage and bias currents in the integrator circuit cause the output voltage to drift unless it is periodically re-zeroed. So the first stage of the conversion process is for FET switches (not shown on the figure, to avoid clutter) to be used to discharge the capacitor C and ensure that the output voltage V_o begins at 0 V.

The second stage is to connect the input to the unknown voltage V_i for a *fixed* period of time. Assume that V_i is a negative voltage. Then during this fixed period V_o increases linearly with a slope that is proportional to V_i. Two lines are shown in Fig. 7.26, corresponding to $V_2 > V_1$. If the fixed time during which the capacitor charges is t_c then the value of V_o at end is

$$V_{oi} = \frac{-V_i t_c}{RC} \tag{7.3}$$

The third stage is to replace V_i with a positive reference voltage V_{ref}. V_o therefore decreases linearly with time, as shown in the third section of Fig. 7.26. The slope of the decrease is fixed (since V_{ref} is fixed) so that the time taken to reach an output of 0 V depends upon the value of V_{oi}, and hence upon the value of V_i. So, in Fig. 7.26,

$$\frac{t_{d2}}{t_{d1}} = \frac{V_2}{V_1}$$

Therefore a measurement of this discharge time by means of a digital counter yields a binary value that is proportional to the original input voltage V_i, and A-D conversion has been achieved.

This method is called the **dual slope** A-D conversion technique, and is the most common form of **integrating converter**. It has some very useful properties, which can be seen from a little more analysis of the circuit.

If the time taken to discharge the capacitor to zero is t_d, then from Equation 7.3,

$$V_{oi} = \frac{-V_i t_c}{RC} = \frac{V_{ref} t_d}{RC}$$

Remember that the minus sign is in Equation 7.4 because V_i is a negative voltage, and V_{ref} is positive, or vice versa.

And so

$$\frac{-V_i}{V_{ref}} = \frac{t_d}{t_c}$$

Time is measured by counting pulses from a regular clock signal of frequency f, so that

$$t_c = N_c f \quad \text{and} \quad t_d = N_d f$$

where N_c is the fixed number of pulses counted during the charging cycle and N_d is the number of pulses needed to discharge the capacitor to zero. Thus

$$\frac{-V_i}{V_{ref}} = \frac{N_d f}{N_c f} = \frac{N_d}{N_c}$$

and so

$$N_d = -V_i \frac{N_c}{V_{ref}} \tag{7.4}$$

Hence the binary code which represents the number of clock pulses counted, N_d, is a linear conversion of the analogue input voltage V_i. But it is very important to note that the value of this count depends *only* upon the number of pulses used during the charging phase (the fixed value of N_c) and the reference voltage V_{ref}. It is *independent* of the input resistance R, the capacitor C, the clock frequency f and all other effects such as amplifier offsets etc., *as long as all these parameters are constant during the two main stages of the conversion*. Thus the conversion can be made to a very great accuracy, depending mostly on the accuracy and stability of V_{ref}.

Because the method is based on timing a certain interval, the resolution of the conversion can be increased simply by raising the clock frequency, and so allowing more pulses to be counted during the appropriate time. Thus it is quite easy to achieve measurements with a resolution of 16 bits or better, without changing the basic circuit.

Conversion time for the dual slope configuration is usually rather slower than the other methods of A-D conversion described earlier, being typically a fraction of a second. However, this is perfectly suitable for measurements such as those of digital voltmeters (DVMs), in which the displayed reading needs updating only every second or so.

In fact, a long integration time can be used to advantage to cancel out some forms of interference. Consider, for example, Fig. 7.27 showing a d.c. voltage which has superimposed on it a small amount of 50 Hz interference picked up from a nearby mains cable. Suppose that the integration time for the converter is 60 ms, exactly three times the period of this interfering waveform. Then there are as many positive excursions from the mean d.c. value as there are negative ones, and over the integration time these positives and negatives will cancel out. If the integration time is not an exact multiple of 20 ms, then the interference will not cancel out and so there will be a slight error in the reading. Hence DVMs are usually designed to cancel out or **reject** interference by the selection of an appropriate integration time.

Self-assessment question

9 A DVM is required which will reject mains interference in countries which have 50 Hz mains and in those which have 60 Hz mains. What is the shortest integration time that would reject both frequencies?

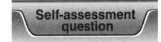

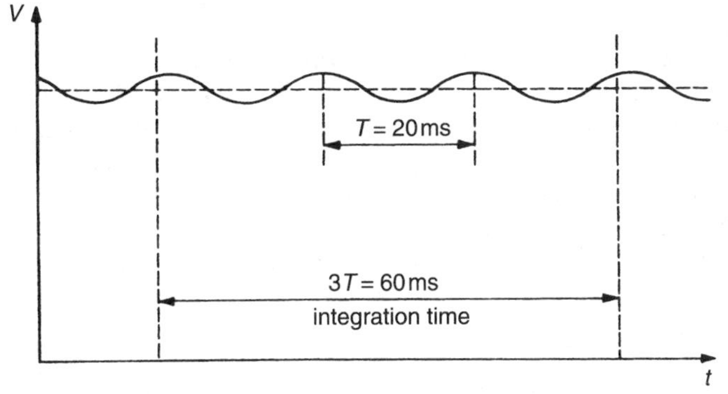

Fig. 7.27 A d.c. level containing interference of period T, integrated over time interval $3T$.

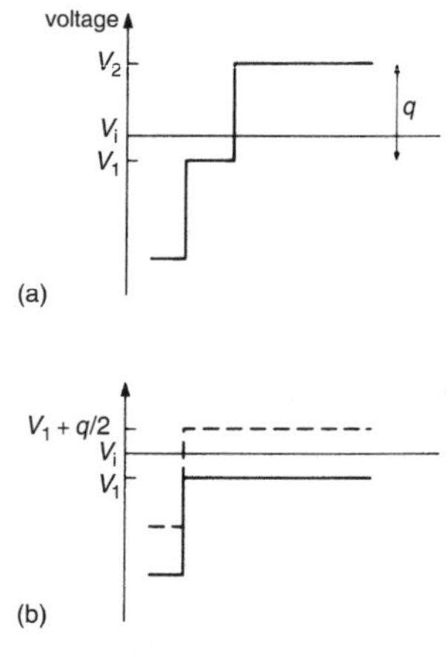

Fig. 7.28 Quantization errors in an A-D conversion.

7.3.5 Quantization

The binary output of an A-D converter can only be one of a limited number of possible codes. This means that if we interpret the output of the A-D converter as the digital equivalent of the input voltage, there will be an error which is the equivalent of quantization error and quantization noise. For example, Fig. 7.28a shows how quantization errors arise in a counter-ramp A-D converter. V_1 and V_2 are two adjacent quantization levels of the D-A converter output voltage and V_i is the d.c. analogue input voltage. Let us assume that the D-A converter output voltage has been stepped up to the level V_1. Because $V_i > V_1$, the comparator output has stayed at a logic 0. On the next clock pulse the D-A converter output rises to V_2, and because $V_2 > V_i$ the comparator output changes to a logic 1, and so stops the conversion cycle. However, the binary codeword read from the counter is an exact representation for V_2 not V_i, hence there is an error. The maximum error occurs when $V_i = V_1$ and is equal to the quantization interval q. It is possible to reduce this error to half a quantization interval by adding a voltage offset of $q/2$ to the D-A converter output. Let us see what effect this has on the conversion error.

When the D-A converter output equals V_1, the actual comparator inputs are V_i and $V_1 + q/2$. From Fig. 7.28b we can see that $V_1 + q/2 > V_i$ so the comparator output changes state and the conversion process stops. The codeword at the input of the D-A converter represents V_1 not $V_1 + q/2$, because the $q/2$ offset is added to the converter's output. Hence the digital value assigned to V_i equals V_1, and the conversion error is $|V_i - V_1|$ which is less than $q/2$.

What happens when $V_1 + q/2 < V_i < V_2$? In this case, the comparator inputs must be V_i and $V_2 + q/2$ for the conversion to stop. But the input codeword to the converter represents the voltage V_2, so the conversion error is $V_i - V_2$ which is also less than $q/2$.

Adding an offset of $q/2$ to the D-A converter output is analogous to rounding in everyday arithmetic, where, for example, we might round 101.7 upwards to 102 or round 101.2 downwards to 101.

The only way to reduce the error due to quantization is, of course, to increase the number of bits. Hence for digital voltmeters (DVMs) and other applications where a high accuracy is needed, A-D converters of 16 bits or more may be needed. However, although a measurement may be known to a *resolution* of 16 bits, unless the reference voltage which sets the range of the converter is known to the same precision, the absolute *accuracy* of the conversion will be less.

To take an example, suppose that a dual-slope converter has a range of 16 bits and so can resolve 65 536 levels, but that the value of V_{ref} is only known within a possible error band of $\pm 1\%$. Then although it is known that a count of 60 001 represents a slightly higher voltage than a count of 60 000, the *actual* voltage value that these codes represent is only known to an accuracy of $\pm 1\%$, corresponding to a range of about ± 600 steps. This can be a problem with digital voltmeters, for example, in which the displayed number of denary digits may give a misleading impression of the accuracy of measurement.

7.3.6 Multiplexers

The A-D converters described above have all been single-input devices. However, it is frequently necessary in a given system to convert several

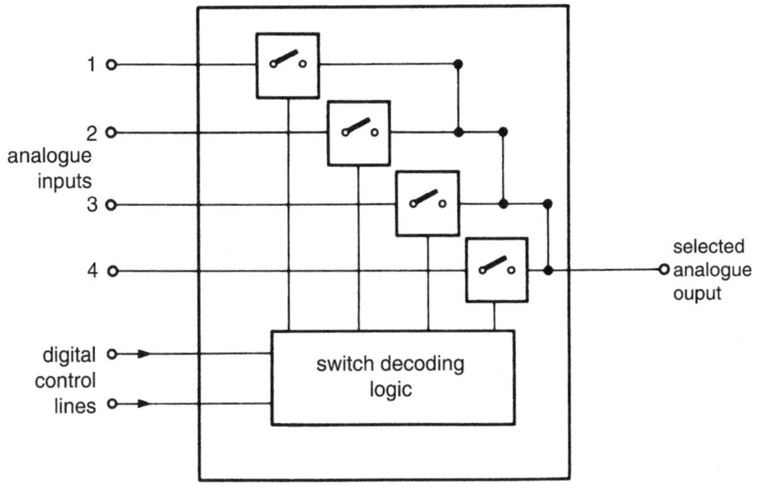

Fig. 7.29 Schematic of a 4-input multiplexer.

analogue signals to binary codewords. As an alternative to using a separate, and possibly expensive, A-D converter for each channel, integrated circuit **multiplexers** are available which can select one analogue signal at a time to connect to a single A-D converter. Multiplexers contain field-effect transistors (FETs) connected so that they provide a low-resistance analogue signal path which can be switched on or off by means of logic signals.

Figure 7.29 is a schematic diagram of a four-input multiplexer. As well as the switches, the device contains a block of decoding logic which allows the 2-bit code on the control lines to determine which of the four analogue input signals is connected through to the analogue output. Devices like this are available in a great many different configurations, and are sometimes incorporated into A-D converter ICs.

7.3.7 Summary of Section 7.3

Analogue-to digital (A-D) converters are used to obtain a binary representation of either voltage or current. The five types examined were the flash, counter-ramp, tracking, successive approximation and dual-slope converters.

Flash converters are very fast, but since they employ a comparator for each quantization step, they are comparatively complex devices and are unsuitable for high-resolution applications.

The counter-ramp is a simpler circuit, but for an n-bit converter the maximum conversion time is proportional to 2^n. A tracking converter is much faster since it only counts the difference between the present and previous samples. The successive approximation converter is in practice a much more widely used device, and the conversion time only increases at a rate proportional to n.

The dual-slope converter uses a simple integration circuit to achieve very high precision, although usually needing a long conversion time. Interference of a known frequency can be rejected by selecting an integration time which is an exact multiple of the interference period.

The only way to reduce quantization errors is to use an A-D converter with a larger number of bits. However, the absolute *accuracy* of a conversion may not be as good as the *resolution* unless care is taken with the accuracy of the reference voltage and other critical components.

A multiplexer enables one A-D converter to be switched between several signal inputs.

7.4 THE CONVERSION OF A.C. SIGNALS

The finite conversion time of an A-D converter presents no special problems when converting d.c. signals, but what about a.c. signals which will change in voltage between successive conversions?

To see how to convert a.c. signals let us examine a simple analogy, that of manually reading room temperature and plotting it against time. It is not practicable to have a continuous measurement of the temperature at every instant of time; it takes several seconds just to read a thermometer and write down the result. Instead, measurements are taken at fixed intervals of time, and the result is a sample of all the possible temperature measurements. The rate at which measurements are taken is therefore known as the **sampling rate**, and this determines the eventual agreement between the graph and the actual temperature variations.

The faster the sampling rate, the better the graph will represent the actual temperature variation, but simply sampling as fast as possible is sometimes inefficient. Figure 7.30a shows three samples of the room temperature, A_1, A_2 and A_3, taken at times t_1, t_2 and t_3, respectively. If the actual room temperature variation between times t_1 and t_3 can be approximated by a straight line, then the measurement at t_2 is superfluous, because it could have been predicted from A_1 and A_3, using linear interpolation. On the other hand, if the actual temperature variation is very rapid, as shown in Fig. 7.30b, then linear interpolation will not enable accurate prediction of temperatures between t_1 and t_3. To maintain accurate prediction many more sample points would be needed. This argument would suggest that the sampling rate should be related to the rate of change of temperature, which in turn will be related to the frequency content of the temperature waveform. Indeed there is a simple rule that relates the minimum sampling rate to the signal bandwidth, called the **sampling rule**. This states that a continuous signal with a bandwidth from d.c. to f Hz can be completely represented by, and reconstructed from, a set of equally spaced samples of instantaneous voltage which are taken at a rate which exceeds $2f$ samples per second. So for an instrumentation signal with a bandwidth of d.c. to 10 kHz, samples must be taken at a minimum rate of 20 000 samples per second.

Whenever possible the sampling rate should be chosen to be adequate for the bandwidth of the signal, in which case the sampling process is straightforward. Occasionally though, the situation arises where the available sampling rate is inadequate for the bandwidth of the signal to be measured. In this case some of the information contained in the signal must be discarded by limiting the signal bandwidth for the following reason. Figure 7.31a shows a sine wave S_1 that has been sampled at a rate below that required by the sampling rule giving the sample points shown. This is

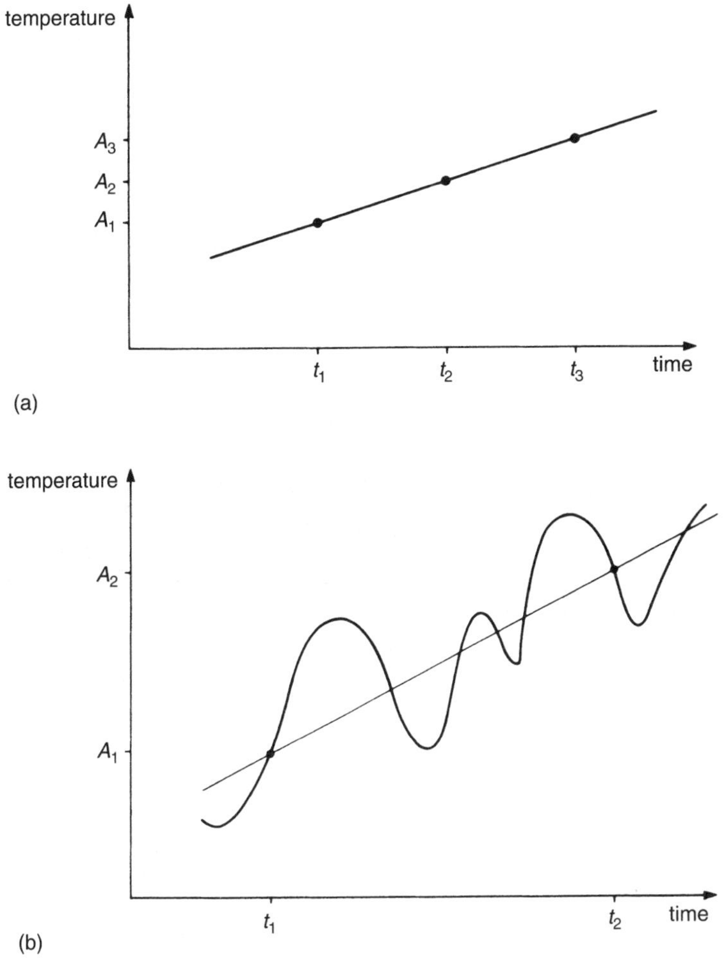

Fig. 7.30 Sampling a.c. signals.

called **undersampling**. Figure 7.31b shows that there is a second, lower-frequency sine wave S_2 that is better represented by the same set of samples. S_2 is called the **alias** of S_1. Given only the sample points, it is impossible to determine whether they are the result of sampling S_1 or its alias S_2.

Whenever we sample a signal too slowly it is possible that energy contributed by high-frequency components is being interpreted as arising from their low-frequency aliases. The errors that arise are said to be caused by **aliasing**. High-frequency noise also creates errors in the conversion process. For example, Fig. 7.32 shows that the true sample voltage at time t_1 is V_1, but owing to high-frequency noise the measured value becomes V_2. To minimize errors due to both aliasing and high-frequency noise it is essential that the A-D converter is preceded by a low-pass analogue filter with a bandwidth determined by the available sampling rate. Such filters are frequently referred to as **anti-aliasing filters**.

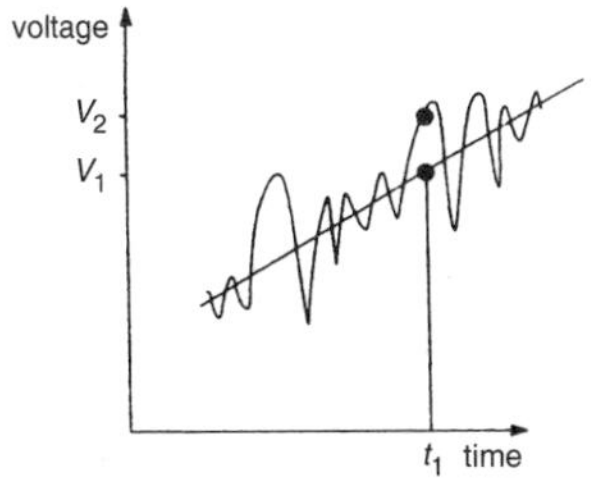

Fig. 7.32 Sampling errors caused by high-frequency noise.

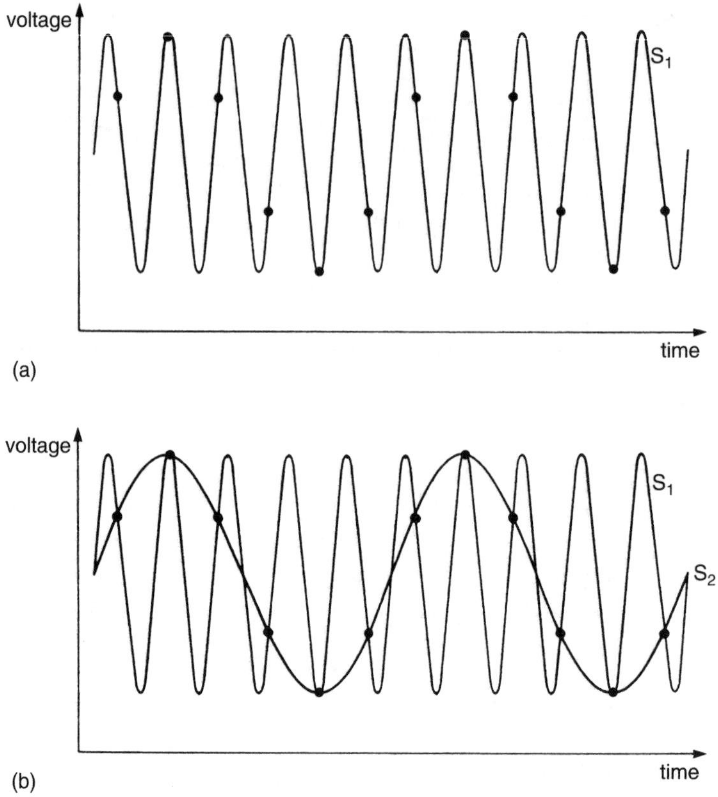

Fig. 7.31 The effects of under-sampling.

Self-assessment question

10 What is the maximum bandwidth of signal that can be converted by an A-D converter with a conversion time of 0.25 ms?

7.4.1 Sample-and-hold devices

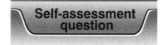

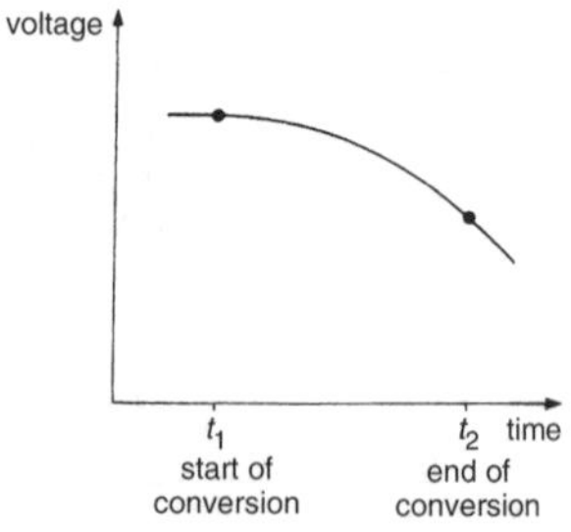

Fig. 7.33 Variation of input signal during a conversion period.

The sampling rule tells us at what rate to make conversions, but there is still another problem associated with changing signals. Figure 7.33 shows part of a signal to be converted by an A-D converter. Assume that the sampling rate is adequate and that a conversion starts at time t_1. However, because an A-D converter takes a finite time to complete a conversion, the end-of-conversion signal does not occur until time t_2. A problem arises because throughout the conversion period the input signal is changing in amplitude. In some types of counter-ramp device the conversion may never terminate, as the D-A converter tries to follow the input variations; and although the successive approximation A-D converter always terminates a conversion, it will produce an unpredictable error. Even flash converters can lose accuracy if the signal changes significantly whilst the comparators are settling.

A device is needed to precede the A-D converter that is capable of taking an instantaneous sample of the input signal and then holding the sampled value until the conversion is completed. Such a device is called a **sample-and-hold**. A simple circuit that meets these basic requirements is illustrated in Fig. 7.34. When the switch is in the closed position the voltage across the capacitor equals the input voltage. Upon command the switch is opened and the charged capacitor retains this value, isolated from subsequent variations at the input. However, a practical version of this simple circuit for use in precision measurement systems demands attention to several points.

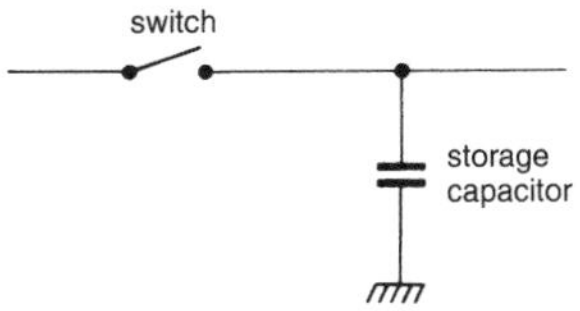

Fig. 7.34 A basic sample-and-hold circuit.

First let us consider the input of the circuit. Connecting a capacitor across the signal source, as with the switch closed, will load the circuit producing the signal. This loading can be overcome by preceding the switch with a unity-gain buffer which has a high input impedance and low output impedance. This is shown on the left-hand side of Fig. 7.35. Such a buffer amplifier also ensures that on subsequent closures of the switch the time constant for charging the capacitor is limited by the switch 'on' resistance, and not by the preceding circuit.

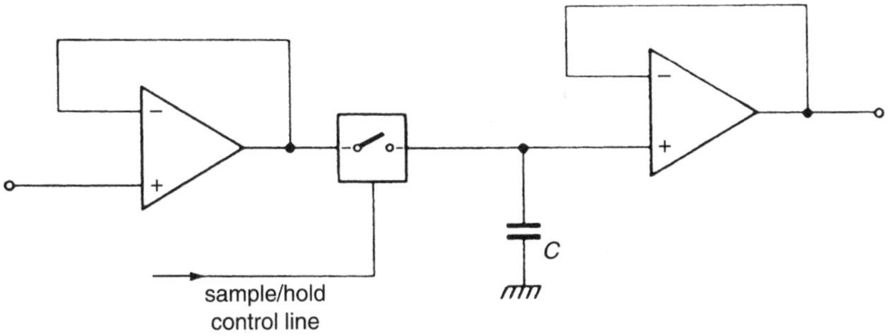

Fig. 7.35 A buffered sample-and-hold.

Now let us look at the output of the sample-and-hold, bearing in mind that it is connected to an A-D converter with a finite input impedance. Any current flowing out of the capacitor when the switch is open will reduce the stored charge level and hence the voltage across the capacitor. If this voltage change is greater than a quantization interval, errors may arise in the conversion process. Adding a second high-input-impedance unity-gain buffer amplifier after the capacitor, as in Fig. 7.35, will minimize the charge loss due to loading, by the A-D converter for example, but cannot eliminate it entirely. The total charge loss from the capacitor is specified in terms of the rate of change of voltage, dv/dt, and is called the **droop rate**.

7.4.2 Specifying the sample-and-hold

The first parameter that must be specified by the system designer is related to the speed at which the switch can change from the 'closed' to the 'open' state upon receipt of a command to hold. Real switches cannot change state instantaneously and there is always a short delay between the inception of the hold command and the time the capacitor voltage ceases to follow the input.

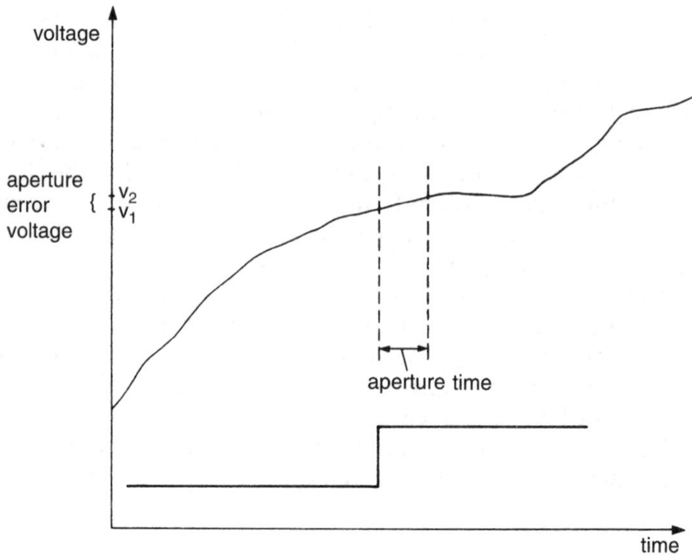

Fig. 7.36 Error due to the sample-and-hold aperture time.

This delay is called the **aperture time**, and its effect is to introduce an **aperture error** in the stored voltage level as shown in Fig. 7.36. At the inception of the hold command the input voltage is v_1, but because the capacitor continues to track the input during the aperture time, the voltage actually stored is v_2. It is up to the designer to ensure that the error $(v_2 - v_1)$ is commensurate with the overall system accuracy, such as specifying a maximum error of plus or minus half a quantization interval.

There is also a delay associated with the change in state from 'hold' to 'sample'. Upon receipt of a 'sample' command, the switch closes and the capacitor either charges or discharges until its voltage equals the input voltage. The switch 'on' resistance limits the current that can flow between the buffer amplifier and the capacitor and so limits the response time of the sample-and-hold. The response time can be specified in terms of the time constant produced by the switch resistance and the storage capacitor. Manufacturers usually quote the response in terms of the time required for the capacitor voltage to settle to within a fixed percentage of its final value.

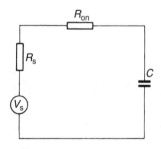

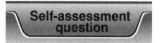

Fig. 7.37 *CR* equivalent circuit of a sample-and-hold.

We can estimate the response time with the aid of Fig. 7.37, which shows the buffer amplifier output stage, modelled as a voltage source and series output resistance, together with the switch 'on' resistance and the storage capacitor C. The time constant for charging C is $(R_s + R_{on}) \times C$, but since R_{on} is typically 100 Ω and R_s is usually less than 1 Ω, it is reasonable to approximate the time constant as $R_{on}C$. For the capacitor voltage to settle to 0.01% requires roughly 9 time constants.

Self-assessment question

11 What is the maximum size of capacitor that can be used in a sample-and-hold, if the switch resistance is 100 Ω and the maximum settling time to 0.01% is 10 μs?

The maximum settling time, and hence **acquisition time**, occurs when the capacitor voltage has to change full-scale, say from −10 V to +10 V, so this is the condition normally quoted by the manufacturers.

The final parameter to consider is **feed-through**, which tells us how much of the input signal appears at the output of the sample-and-hold when it is in the hold state. In all previous discussions of the errors associated with switches, the 'off' state switch is assumed to be modelled as an open circuit. For low-frequency signals this model is quite adequate, but it is not adequate for high-frequency signals. This apparent discrepancy can be attributed to the effects of stray capacitance within the switch. At low frequencies the stray capacitance offers a high impedance to the flow of alternating current but at high frequencies the impedance decreases and so some of the input signal appears at the output. The ratio of the input signal to the sample-and-hold appearing at the output is termed the feed-through.

7.4.3 Summary of Section 7.4

The conversion of d.c. signals is straightforward, but for a.c. signals we must sample the input. The rate at which we sample is determined by the sampling rule, which requires that a signal of bandwidth f Hz be sampled at a rate in excess of $2f$ samples per second.

If the A-D converter cannot maintain this rate of conversion it is essential that the input signal bandwidth be restricted, even though we may lose information, to prevent aliasing errors.

To prevent errors arising from variations in the input signal during the A-D conversion cycle, a sample-and-hold is used to sample the signal and present a d.c. level to the A-D converter.

The operational characteristics of a sample-and-hold are defined by the aperture and acquisition times, and the droop-rate and the feed-through.

Answers to self-assessment questions

1. A 5-bit converter has 32 possible outputs. In this case:

00000 represents 0 V
11111 represents 7.75 V
00001 must therefore represent 7.75/31 V = 0.25 V.

The voltage sources are then:

1×0.25 V = 0.25 V
2×0.25 V = 0.5 V
4×0.25 V = 1.0 V
8×0.25 V = 2.0 V
16×0.25 V = 4.0 V

You can check that the sum of these voltages, corresponding to an input code of 11111, is 7.75 V.

2. Table 7.1 gives the open-circuit output voltage for each bit in the input codeword. Because the system is linear, we can apply the principles of superposition and say that the total output voltage will be equal to the sum of the individual output voltages corresponding to each bit that is a logic 1. With $V_1 = 5$ V, the output voltages corresponding to each bit in ascending order are 0.33 V, 0.67 V, 1.33 V and 2.67 V. The following table shows the resulting output voltages for all the 16 possible input codewords.

Codeword	Output voltage	Codeword	Output voltage
0000	0	1000	2.67
0001	0.33	1001	3
0010	0.67	1010	3.33
0011	1	1011	3.67
0100	1.33	1100	4
0101	1.67	1101	4.33
0110	2	1110	4.67
0111	2.33	1111	5

3. Figure 7.38 shows the amplifier circuit again, this time with the resistor network replaced by its equivalent circuit. Chapter 4 shows that the gain of the non-inverting amplifier is given as

$$G = 1 + \frac{R_1}{R_2}$$

In this circuit, $R_1 = 1\ \mathrm{k}\Omega$ and $R_2 = 1\ \mathrm{k}\Omega$, so $G = 2$. The output voltage is therefore:

$$V_o = 2 \times V_1\left(\frac{8}{15}b_3 + \frac{4}{15}b_2 + \frac{2}{15}b_1 + \frac{1}{15}b_0\right)$$

The output voltage should therefore range between 0 V for an input codeword of 0000 and 10 V for an input codeword of 1111.

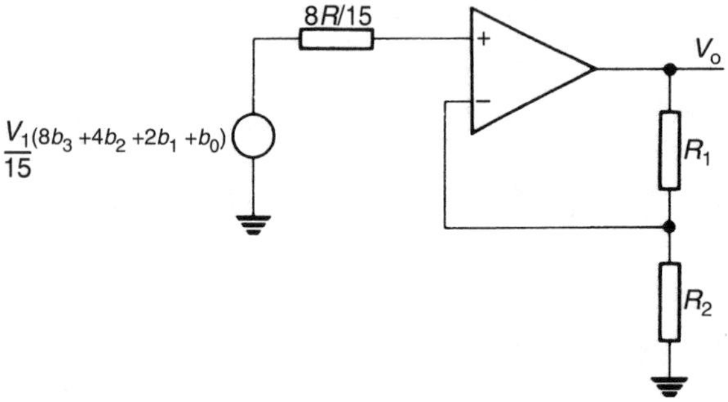

Fig. 7.38 Equivalent circuit for the non-inverting binary-weighted network.

This shows that it is possible to build a D-A converter without having to produce an output voltage which is negative. However, circuit designers nearly always prefer to use the inverting amplifier whenever signals need to be summed.

4. The output voltage of the D-A converter is given by

$$V_o = -I_F \times R_F$$

To calculate I_F we need to know the current flowing in each switch, but first we must determine the maximum current flowing into the ladder. This current is equal to the reference voltage divided by the R-$2R$ ladder input impedance. Combining all the parallel and series resistors of the ladder gives an input impedance of R, so the maximum current is 5 V/5 kΩ = 1 mA.

The maximum output voltage is obtained for an input codeword consisting entirely of logic 1s, so the current entering the feedback loop is

$$I_F = \frac{I}{2} + \frac{I}{4} + \frac{I}{8} + \frac{I}{16} + \frac{I}{32} + \frac{I}{64} + \frac{I}{128} + \frac{I}{256}$$

$$I_F = \frac{255}{256}I$$

Hence the output voltage is

$$V_o = \frac{255}{256} \times 1\ \mathrm{mA} \times 10\ \mathrm{k}\Omega = -9.96\ \mathrm{V}$$

5. When the input codeword to the D-A converter is 1010, the total current entering the summing junction of the amplifier will be

$$I_s = I/2 + I/8 - I_{off}$$

But $I_{off} = I/2$, so $I_s = I/8$, hence the output voltage of the amplifier will be

$$V_o = -I_s R_f$$

$$V_o = \frac{-I}{8}R_f$$

We can calculate I by dividing the reference voltage by the input resistance of the ladder network. Starting at the right-hand end of the network and combining the parallel and series resistors, it is quite simple to show that the ladder has an input resistance of R. The input current I is therefore

$$I = \frac{V_{ref}}{R} = \frac{-5}{5 \times 10^3} = -1\ \mathrm{mA}$$

and the output voltage V_o is

$$V_o = -\frac{I}{8} \times 10\,\mathrm{k}\Omega = \left(\frac{10^{-3}}{8} \times 10^4\right)\mathrm{V} = 1.25\ \mathrm{V}$$

6. The maximum quantization error equals half a quantization interval. Because there are 1024 states in a 10-bit converter, the quantization interval is 20 V/1023 = 19.5 mV. So the maximum quantization error is 19.5 mV/2 = 9.75 mV.

7. To determine the conversion time for the 4 V input it is necessary to calculate the quantization interval. For a 10-bit converter there are 1024 codewords, so the quantization interval is 10/1023 = 9.755 mV. The number of states through which the converter must ramp is

$$4/(9.775 \times 10^{-3}) = 409.2$$

Assuming that the comparator changes state at the following level, the conversion time is 410 × 15 μs = 6.15 ms. For a 10 V input signal, the converter must ramp through all 1024 states, taking a total of 1023 × 15 μs = 15.35 ms.

8. For a 12-bit counter-ramp A-D converter the output must ramp through 4096 states, hence 4095 clock periods are required. In the successive approximation converter only $n+1$ (that is 13) clock periods are required so the increase in speed is 4095/13 = 315 times.

9. To reject 50 Hz the integration time needs to be a multiple of 20 ms, while to reject 60 Hz it needs to be a multiple of 16.66 ms. The shortest time which is a multiple of both is

100 ms (5×20 ms and 6×16.66 ms). Thus a DVM with an integration time of 100 ms would successfully reject interference at either frequency, and this principle is often used in commercial instruments.

10. If the conversion time of the A-D converter is 0.25 ms, then the conversion rate is $1/(250 \times 10^{-6})$ Hz or 4000 Hz. The sampling rule tells us that minimum sampling rate is twice the bandwidth of the input signal, which in this case limits the bandwidth to 2 kHz.

11. Modelling the sample-and-hold as a simple *CR* network, we know that 9 time constants are required for the output to settle to 0.01%. If the total settling time is 10 μs, then $CR = (10 \times 10^{-6})/9$ s or 1.11 μs. Because $R = 100\ \Omega$, $C = (1.11 \times 10^{-6})/100$ F $= 11.1$ nF.

8 Diodes and power supplies

AIMS

1. To describe and explain the d.c. characteristics of pn junction diodes.
2. To illustrate the principles of rectification and regulation in power supplies.

GENERAL OBJECTIVES

After studying this text you should be able to:
Explain and use correctly the following terms:

- acceptor
- bridge rectifier
- d.c. to d.c. converter
- donor
- doping
- dropout voltage
- forward bias
- full-wave rectification
- half-wave rectification
- heat-sink
- holes
- linear voltage regulator
- majority carrier
- minority carrier
- n-type
- overload protection
- p-type
- regulation
- reverse bias
- ripple
- saturation current
- smoothing capacitor
- switched-mode power supply
- thermal resistance
- transition region

Explain the function of each part of a regulated d.c. power supply, and carry out calculations on output regulation and temperature rise.

SPECIFIC OBJECTIVES

1. Calculate the current through a pn junction diode.
2. Calculate the peak voltage and ripple of the output from a capacitor-smoothed d.c. power unit.
3. Determine the output characteristics of a three-terminal linear regulator circuit.
4. Carry out calculations relating to power dissipation and thermal resistance for a linear regulator or similar device.

8.1 INTRODUCTION

The most important building blocks of all circuits for amplification, logic, and conversion between the analogue and digital worlds described in Chapters 4 to 7 are diodes and transistors. Transistors are introduced later in Chapter 9. This chapter concentrates on the simplest semiconductor device—the diode—which has one basic property: it conducts current much more easily in one direction compared to the other. This depends above all on the properties of pn junctions, which are junctions between two different types of semiconductor formed within a single crystal. In this chapter the manufacture and properties of pn junction diodes will be described, as a first step towards understanding the structure and operation of more complex integrated circuits.

As well as forming a component of ICs, huge numbers of diodes are made as discrete components, and one of their biggest applications is for rectification, the conversion of a.c. into d.c. Hence Chapter 8 continues with a description of power supplies, particularly those which convert the a.c. mains into a stable d.c. voltage to power electronic circuits. Rectification is an important step in this, and is followed by smoothing and regulation. Two forms of regulation are described. Linear regulators are analogue circuits, which dissipate unwanted power, and are inefficient but easy to use. Switching power supplies, essentially digital, are much more complex, but are also much more efficient and versatile.

8.2 DIODES

The fundamental characteristic of a diode is that it passes current more easily in one direction than the other. In the early days of electronics this behaviour was achieved by means of an evacuated tube containing two electrodes with the property that electrons could be emitted from one electrode to cross the gap between them, but not from the other. Nowadays diodes are 'solid state', usually consisting of a junction between two differently doped pieces of semiconductor such as silicon. They can be either discrete devices or part of much more complex integrated circuits. This basic property of passing current well in only one direction is fundamental to the operation of many analogue and digital circuits.

8.2.1 Electrons and holes

In a metal wire, electric current consists of a flow of electrons from the more negative to the more positive end. However in silicon, as in other semiconductors, electric current appears to be carried not only by electrons but also by means of something else, called **holes**. Holes are not some new kind of sub-atomic particle: the silicon still consists of electrons, protons and neutrons and nothing else. Holes are created by missing electrons in some of the covalent bonds which hold the crystal structure together. When a potential is applied across the crystal, the movement of electrons between these incomplete bonds *gives the appearance* of conduction occurring by means of independent, positive charge carriers. So a hole behaves like a mobile positive charge in the crystal, and can be thought of in those terms.

Although the solid state physics behind the description and analysis of holes is very complex, the way they can be used is very simple. When describing how semiconductors such as silicon conduct, we can think in terms of two types of charge carrier mixed up together. A current will consist of a quantity of negatively charged electrons moving from negative to positive, and a different quantity of positively charged holes moving from positive to negative. The direction of current conventionally marked on circuits is, of course, the same as the direction of the holes.

Pure silicon has an equal number of free holes and electrons that can contribute to conduction. However the relative proportions of the two can be changed by **doping** the silicon crystal with certain impurity atoms. Doping with **acceptor** atoms such as boron produces **p-type** silicon with very many free holes and very few free electrons. In this case holes are the **majority carrier** and electrons are the **minority carrier**. Doping with **donor** atoms such as phosphorus produces **n-type** silicon with very many electrons (the majority carrier) and very few holes (the minority carrier). Most diodes consist of a junction between n-type material and p-type material formed with a continuous crystal structure.

8.2.2 The structure of pn junctions

A pn junction is a junction, within a single crystal, between p-type silicon and n-type silicon. In other words the doping changes from primarily acceptors to primarily donors at a plane within the crystal.

The usual way to produce a pn junction is as follows. A thin slice or wafer of silicon, as illustrated in Fig. 8.1a, is first cut from an ingot of silicon crystal which has been produced in a special furnace. This starting wafer is doped during the growth of the single crystal by adding either donors or acceptors to the molten silicon so that, as it crystallizes, the silicon becomes either n-type or p-type. Each wafer then forms the substrate for perhaps hundreds or thousands of pn junctions. These junctions are commonly formed by a high temperature diffusion process, in which acceptor atoms are diffused into an n-type substrate, or donor atoms into a p-type substrate, as illustrated in the cross-section diagrams of Figs. 8.1b and c.

The diffusion process is used in the fabrication of silicon pn junctions as follows. First a film of silicon dioxide is grown over the surface of the silicon by heating the silicon in an atmosphere of oxygen. Then a hole—or 'window'—is etched in the oxide layer as indicated by the gap in the oxide in Figs. 8.1b and c.

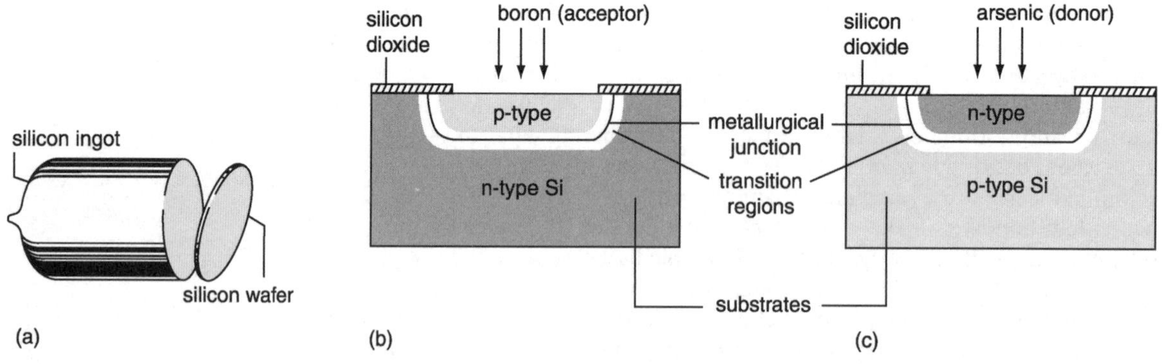

Fig. 8.1 Forming a pn junction in silicon. (a) Cutting a thin wafer from a single-crystal ingot of silicon to form the substrate for many pn junctions, (b) diffusing boron through a 'window' in the oxide layer that covers the surface of an n-type substrate to form a single pn junction, or (c) diffusing arsenic into a p-type substrate. The extent of the 'transition region' on either side of the metallurgical junction is shown by the white area.

When the silicon is heated to over 1000°C in an atmosphere of the dopant, the donor or acceptor atoms diffuse into the silicon through this 'window' in the oxide. The density of the dopant atoms decreases with distance from the surface; both depth and density can be accurately controlled by controlling the temperature and time of the process. If the starting material is n-type and the diffused material is an acceptor such as boron, as in Fig. 8.1b, the surface is converted to p-type silicon because the acceptor density there exceeds the original donor density. But deeper down the donors still dominate, so a boundary is formed between the p-region and n-region at which the dominant dopant changes from donors to acceptors. This boundary is called the metallurgical junction. Even if the change of doping from donors to acceptors is abrupt, the change of majority carriers from electrons to holes is not so abrupt, for reasons which are beyond the scope of this book. The result is that the pn junction as a whole extends a little on either side of the metallurgical junction, to form the so-called **transition region** whose properties are at the heart of pn junction and transistor action.

Figure 8.1c illustrates the complementary process of diffusing donors into a p-type substrate in order to convert the top surface to n-type. Again a pn junction is formed at the interface.

With both forms of pn junction, if wires are now attached to the two regions it will be found that it is possible to pass a much greater current through the device in one direction than in the opposite direction for the same magnitude of applied voltage.

8.2.3 The properties of pn junction diodes

A typical d.c. characteristic of a silicon pn junction diode, and the circuit for measuring it, are shown in Fig. 8.2. Figure 8.2a shows a circuit for measuring the forward current of a diode. (Forward direction is when the p-region is made more positive than the n-region. This is called **forward bias**.) The supply voltage is varied and the voltage and current are recorded. Some typical results are shown in Fig. 8.2b. For a positive value of V_D up to about 0.5 V there seems to be little current flowing, while above this the current can be seen to increase, progressively more steeply, with voltage.

The conventional symbol for the diode is also included in Fig. 8.2a, where the arrow-shaped part indicates the direction in which current can flow easily. The bar-shaped end of the symbol always corresponds to the n-type material in a pn junction.

When the voltage is reversed in polarity, not surprisingly called **reverse bias**, a very small but measurable current flows. This is called the **saturation current**, I_S, and is typically 10^{-13} A in small silicon diodes.

Worked example

What modifications should be made to the circuit of Fig. 8.2a in order to measure the diode's reverse current, other than reversing the battery and the meters?

The circuit in its present form measures the current through the diode plus the current through the voltmeter. Since the current through the reverse biased diode is so small it is likely that the voltmeter current will be comparable to or much larger than that of the diode, so the meter will give a false reading of diode current. If the current meter is put in series with the diode, so that the voltmeter measures the voltage across both diode and current meter, the error is likely to be much less.

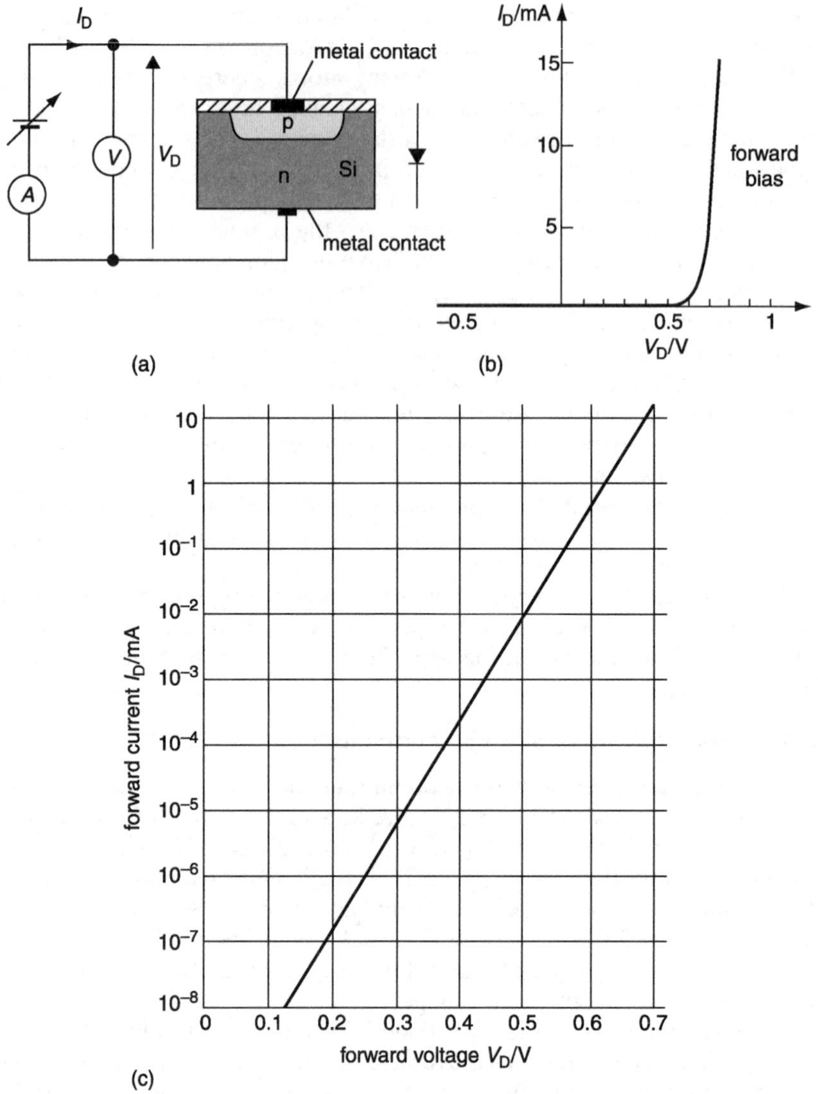

Fig. 8.2 (a) A cross-sectional diagram for a pn junction, its graphical symbol and a circuit for measuring its forward d.c. characteristic. (b) A possible forward d.c. characteristic plotted on linear axes. (c) The same data plotted on log-linear axes.

A detailed explanation of why this device behaves the way it does is beyond the scope of this book. However one simple way of looking at the phenomenon is as follows.

Under forward bias, current is carried by majority holes flowing from the p region to the n region, and majority electrons flowing from the n region into the p region. There are plenty of these carriers available, but they are faced with a 'potential barrier' in the transition region which restricts the numbers which can cross. The height of this barrier decreases as the forward bias voltage increases, more carriers can pass, and hence the diode current increases rapidly with voltage.

With reverse bias, current is carried by minority electrons flowing across the junction from the p region and minority holes from the n region. There is no potential barrier to hinder this movement, but the constraint to current flow is simply the very small supply of minority carriers ready to cross the junction. Hence the reverse current is much smaller than the forward current. The supply of minority carriers is not affected by the reverse bias voltage, and so the reverse saturation current is essentially constant over a wide range of voltage.

Further analysis shows that the forward characteristics of junction diodes can be described by this equation, often called the **diode equation**:

$$I_D = I_S(\mathrm{e}^{KV_\mathrm{D}} - 1) \qquad (8.1)$$

where I_S is the saturation current referred to above, and K is a temperature dependent parameter which varies somewhat between devices, but is often approximately equal to 35 V^{-1} for silicon diodes at room temperature.

The forward bias characteristic illustrated in Fig. 8.2b is plotted on logarithmic axes in Fig. 8.2c, the straight line showing that the diode equation is being followed. The exponential form means that each increase of a factor of ten in current through a silicon junction diode corresponds to an increase of only about 57 mV across the diode.

Self-assessment question

1 Using the diode equation with $K = 35\ \mathrm{V}^{-1}$ and $I_\mathrm{S} = 10^{-13}$ A, what forward voltage would produce a forward current of 1 mA in such a pn junction? What current would be produced by a forward voltage of 0.4 V?

What the solution to Self-assessment question 1 means in practice is that, over the typical range of currents found in low power electronic circuits, say 0.1 to 10 mA, the voltage across a diode is in a range between 0.6 to 0.7 V. This is clearly shown in the linear plot in Fig. 8.2c which covers this range of currents. This is a good rule of thumb for circuit design.

Another rule of thumb is that for a forward bias of less than about 0.4 V, or with a reverse bias, the current flowing is likely to be quite negligible compared to other currents in the circuit.

An ideal diode would be an insulator under reverse bias and have no resistance (and hence drop no voltage) under forward bias. Real diodes only approximate to this ideal, but are close enough for many practical purposes. Diodes of all shapes and sizes are used in very many applications: for example, converting a.c. to d.c. in a car alternator or a mains power supply; extracting the wanted signal from the radio frequency carrier in a radio or TV; protecting a PDA when the batteries are inserted wrongly; improving the switching speed of digital circuits. Light emitting diodes are a particular form of pn junction device made not from silicon but from compound semiconductors such as gallium arsenide phosphide (GaAsP). These emit light when a forward current is passed, and are commonly used in bicycle lamps, and various forms of indicator and display.

When considering which diode to use, there are three main electronic parameters which determine their suitability for an application.

1. The maximum forward current, I_F. For all forms of diode there is a high current level at which the device will be irreversibly damaged. Much of this is

due to heating, and diodes for high current use have to be in a package with high thermal conductivity so that the temperature rise can be minimized. The safe maximum value of I_F depends on such factors as the external temperature and the way the diode is mounted on a circuit board or heat sink. Values can range from less than 100 mA for a small signal diode to thousands of amps for a power rectifier.

2. Forward voltage drop, V_F. As shown above, the exponential form of the diode equation means that for a particular range of forward currents, the voltage across the diode junction does not vary very much. For a silicon diode in the milliamp range, a value of 0.65 V is a good rule of thumb. The forward voltage drop is greater than this with higher currents, and may be increased even further in a practical device by the voltage drop across the resistance of the semiconductor material. If a lower V_F is needed, then there are special diodes such as Schottky barrier diodes. These contain a junction between a metal and a semiconductor which has a much higher value of I_S than a silicon pn junction and hence a much lower value of V_F than ordinary silicon diodes.
3. Peak reverse voltage, V_R. When a small reverse bias voltage is applied to the diode, only a small leakage current flows. As the reverse bias increases, however, it reaches a value at which the semiconductor junction will break down, allowing a large and potentially damaging current to flow. Hence it is important to ensure that the peak voltage in a circuit is less than this critical V_R. There is, however, a particular type of diode called a Zener diode in which the reverse bias breakdown is sharply defined, reproducible and reversible. These devices are one form of voltage reference, used to set particular voltage values in a circuit.

The next section will describe power supplies, illustrating one important application of diodes in the conversion of alternating current to direct current.

8.3 POWER SUPPLIES

A regulated d.c. supply is a circuit for deriving a well-defined d.c. voltage source from the a.c. mains supply. Every piece of mains-driven electronic equipment includes such a d.c. supply, and it is often the heaviest part of the equipment because of the transformer that is normally used. Although d.c. supplies derived from the mains sometimes have to be capable of supplying many hundreds of amps of d.c. current, we are only concerned here with d.c. supplies for electronic circuits, capable of supplying only a few amps at most.

Regulated d.c. supplies have the basic structure shown in Fig. 8.3. They can be thought of as comprising (a) a d.c. power unit, which creates a d.c. source from the a.c. mains, plus (b) some regulating circuitry which gives a precisely controlled d.c. voltage.

The output from the d.c. power unit on its own may be unsatisfactory in electronic apparatus for two reasons: firstly, because it usually carries a significant amount of a.c. ripple superimposed on the d.c. voltage; secondly, because the output resistance of the d.c. power unit is usually higher than desirable, implying that the voltage output can be significantly affected by variations in the current drawn from the unit. The success with which the output voltage is held constant as the load varies is called the regulation of the d.c. power unit. The regulating circuitry added to the d.c. power unit in Fig. 8.3

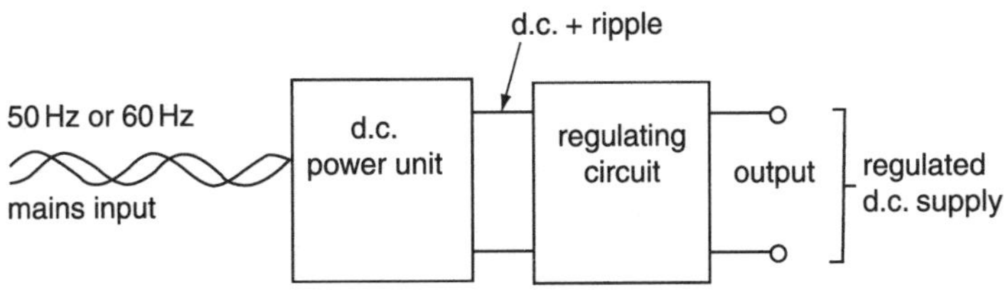

Fig. 8.3 The structure of a regulated d.c. supply.

reduces the ripple and improves the regulation. Typically such additional regulating circuitry would ensure that the d.c. output voltage would vary by less than 0.1 V (perhaps a few millivolts) as the current supplied to the load varied between the maximum and minimum specified levels, and that the a.c. ripple superimposed on the d.c. voltage would be less than 1 mV.

It is often convenient for either the d.c. power unit or the complete regulated supply to be built into the mains plug to form a power adapter. All safety issues are then confined to the adapter, and the system which is to be powered can be smaller and lighter. In the case when the adapter is unregulated, the regulation circuit may be built in to the equipment itself.

Most power supplies also include simple circuits to ensure that an accidental short circuit of the load does not cause damage to the regulated d.c. supply. These are called **overload protection circuits** and are usually designed to limit the output current to a level which will not cause damage—say 100 mA or 1 A, depending on the intended application.

D.C. power units

D.C. power units convert the a.c. mains into a d.c. source of much lower voltage than the amplitude of the mains. Their output usually has a significant ripple superimposed on it. They usually have a larger output resistance (i.e. worse regulation) than is acceptable for most electronic apparatus. Power units are therefore no more than a first step towards the design of a regulated d.c. supply. They consist of three parts, as indicated in Fig. 8.4.

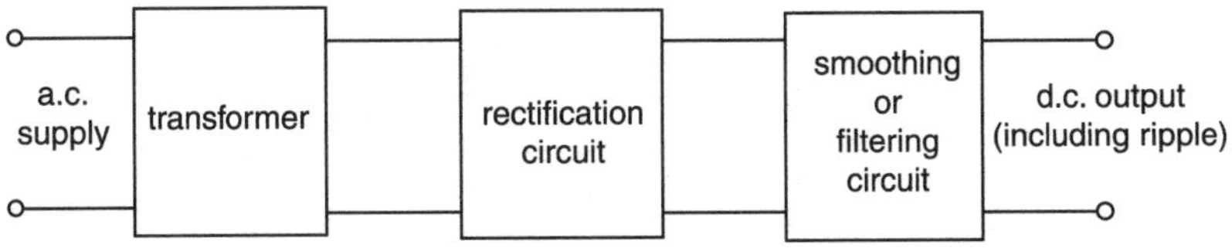

Fig. 8.4 The elements of a d.c. power unit: the transformer, the rectification circuit and the smoothing circuit.

(i) the transformer to provide a lower a.c. voltage than the mains voltage;
(ii) a rectifier circuit to convert the a.c. into d.c., but with a great deal of ripple on it;
(iii) a smoothing circuit which gets rid of most of the ripple, though by no means all of it.

These three aspects of d.c. supplies are considered in turn.

8.3.1 The transformer

Few electronic circuits operate at the sort of voltage supplied by the a.c. mains, so a transformer is usually needed. The isolation from the a.c. mains provided by a transformer (i.e. no direct connection between primary and secondary windings) is also an important safety feature.

Worked example

What is the voltage amplitude of the a.c. mains whose r.m.s. voltage is 240 V?

The r.m.s. value of a sinusoidal waveform is the amplitude divided by $\sqrt{2}$, so in this case the amplitude, or maximum positive voltage, is $240 \times \sqrt{2} = 339$ V.

To obtain a d.c. supply whose output voltage is in the range of 5 V to 15 V, a step-down transformer is needed. For reasons that will soon become clear, the amplitude of the a.c. output voltage of the transformer in a power unit must be at least a volt or two higher than the d.c. output of the required d.c. power unit, more often 4 or 5 V higher. So, for a 12 V d.c. supply, a typical transformer output might have an amplitude of about 17 V (though less than this is enough if the rest of the circuit is well designed). To give a 17 V amplitude, a turns ratio in the transformer of $339/17 = 20$ is needed. In addition the transformer must be capable of passing the peak currents which are needed to drive the smoothing circuit (see Section 8.3.4) without significant loss of output voltage due to the internal resistance of the transformer winding. The magnitude of these currents varies very much with the kind of smoothing used in the design, so this topic is taken up again after the smoothing circuits have been considered.

The transformer in a d.c. power unit uses either a single secondary winding or a centre-tapped one, depending on the design of the rectifier circuit, as described in the next section.

Transformers were considered in some detail in Chapter 2. Figure 8.5 shows the construction of a typical transformer for electronic equipment. The important conclusions as far as this chapter is concerned are:

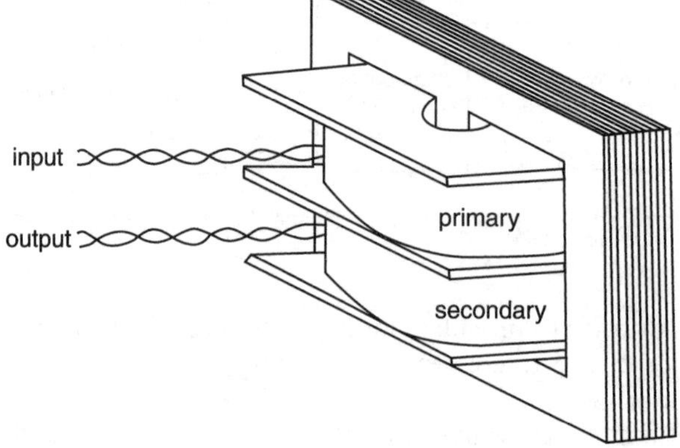

Fig. 8.5 The structure of a typical transformer. Note that the core encircles both windings, and that the two windings occupy similar amounts of space.

(i) The voltage ratio is equal to the turns ratio. That is,

$$\frac{\text{Output e.m.f.}}{\text{Input voltage}} = \frac{\text{number of turns in secondary}}{\text{number of turns in primary}}$$

or

$$\frac{e_{out}}{v_{in}} = \frac{N_S}{N_P}$$

(ii) The primary impedance should be as large as possible to keep the magnetizing current small, so the primary should have as many turns as possible and the reluctance of the magnetic circuit should be as small as possible. Both these requirements tend to lead to large transformers at 50 or 60 Hz.

(iii) The output resistance is mainly due to the resistance of the secondary winding. Therefore the wire of the secondary winding should have as large a gauge as possible in the space provided for the secondary winding.

(iv) The weight of the transformer should normally be kept as small as possible, to save cost and to achieve a lightweight design. This conflicts with requirements (ii) and (iii), so a compromise is needed which optimizes the design for the particular application—which is where good transformer designs differ from bad ones. In general it is best to ensure that the primary and secondary windings each occupy about half the space available for windings around the core, as indicated in Fig. 8.5. So in step-down transformers the secondary is wound from a thicker gauge of wire than the primary.

(v) The secondary winding may need to be centre-tapped, depending on the type of rectification used.

8.3.2 Half-wave rectification

In practice most d.c. power units use **full-wave rectification**, which is achieved either by the use of the full-wave rectifier circuit with a centre-tapped transformer winding, or else by a **bridge rectifier** with a single secondary transformer winding. Both these circuits will be explained after the simpler **half-wave rectification** circuit has been described.

Figure 8.6a shows the circuit of a d.c. power unit with half-wave rectification but no smoothing. The rectifier is a silicon pn junction diode, which readily conducts current when it is forward biased by more than about 0.65 V, and will behave like an open circuit when reverse biased or when the forward bias is less than about 0.4V (see Section 8.2.3). Figure 8.6b shows the input and output waveforms of the circuit. The transformer output waveform is the continuous line, whilst the voltage across the load resistor R_L is shown as a dashed line. The difference between them is caused by the presence of the silicon diode. When the transformer output voltage is sufficiently positive for the diode to conduct, the voltage applied to the load resistance is the a.c. voltage minus the voltage dropped across the diode. When the transformer output voltage is negative, no current flows, so there is zero voltage drop across the load, as shown. The voltage dropped across the diode depends a little on the current being supplied. At small currents it is the usual 0.65 V or 0.7 V, as shown in the figure, but at a current of an amp or so it is likely to be nearer 1 V. The diode voltage drop is one reason why the amplitude of the transformer output must be more than the final d.c. output.

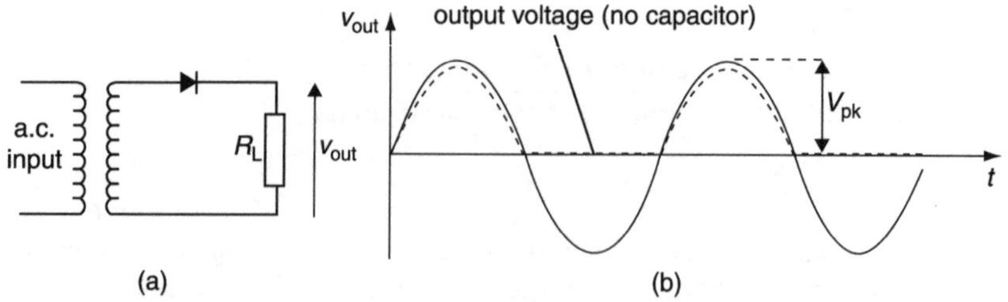

Fig. 8.6 Half-wave rectification. (a) The circuit of a power unit which contains only a transformer and a half-wave rectifier. (b) The input and output waveforms.

8.3.3 Full-wave rectification

The full-wave rectifier circuit

The full-wave rectifier circuit, shown in Fig. 8.7a, can be analysed in much the same way as the half-wave rectifier. Note that the secondary winding of the transformer is centre tapped in this circuit (that is, a connection is made to the centre of the secondary winding, dividing it into two halves in series). Relative to the centre tap, the sinusoids from each half of the secondary winding are out of phase with each other.

Each diode is connected across one half of the winding, so one diode is reverse biased during one half-cycle of the waveform and the other is reverse biased during the other half-cycle. At any particular time, therefore, only one half of the transformer is supplying current. But the diodes are arranged so that the current through the load resistance is in the same direction during each half-cycle of the a.c. waveform. The circuit is essentially two half-wave rectifiers in parallel, each using a different half of the a.c. waveform to produce the same current through the load.

The current waveforms in the two diodes are shown in Figs. 8.7b and c. These currents add together to produce the voltage across the load shown in Fig. 8.7d.

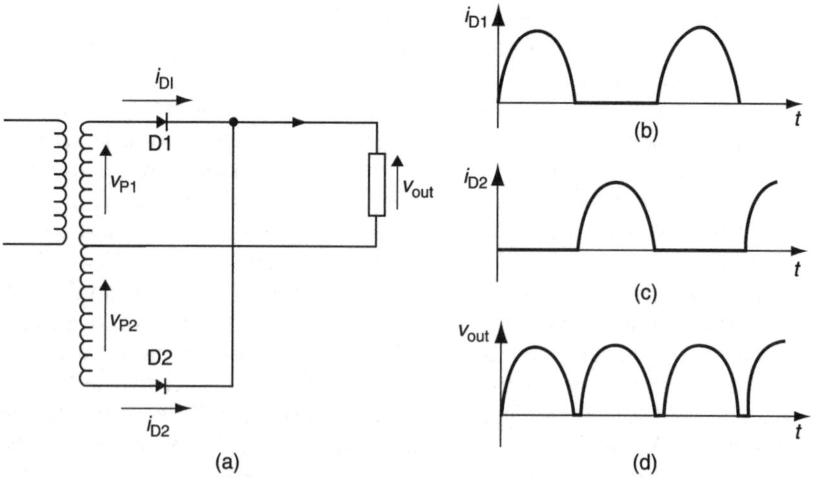

Fig. 8.7 The full-wave rectifier. (a) The circuit diagram: note the centre tap on the transformer secondary. (b) The output due to D1 only. (c) The output due to D2 only. (d) The overall output.

The bridge rectifier

An alternative way of producing full-wave rectification is shown in Fig. 8.8a. In this circuit there are always two diodes conducting, either A and C (as shown in Fig. 8.8b) or B and D (as shown in Fig. 8.8c). The circuit ensures that even though the sign of the output voltage from the transformer is reversed with each half-cycle of the a.c. supply, the voltage across the load has the same sign: positive in this case. The output voltage of this arrangement is therefore almost the same as that of the full-wave rectifier of Fig. 8.7.

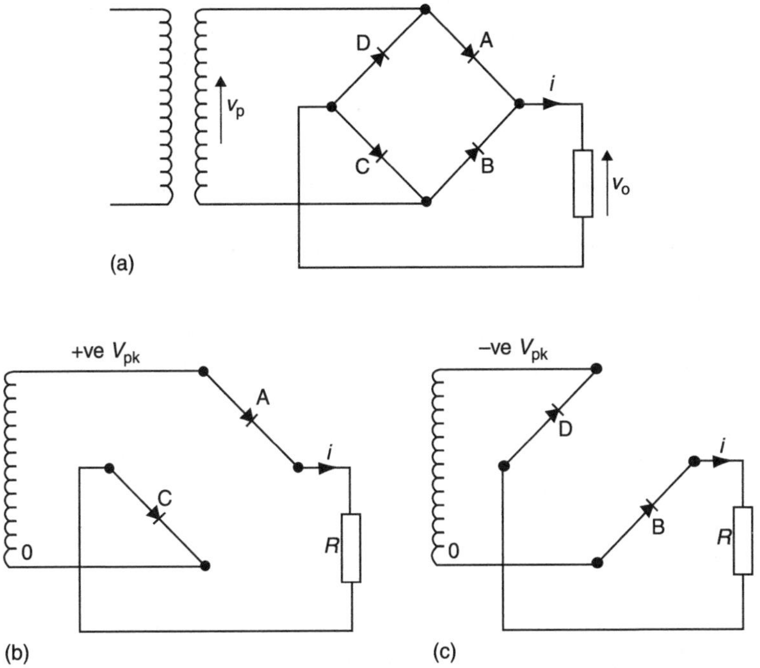

Fig. 8.8 (a) The bridge rectifier circuit. (b) The effective circuit during the positive half-cycle of the waveform. (c) The circuit during the negative half-cycle.

Self-assessment question

2 (a) What is the difference between the output voltage from a bridge rectifier and the output voltage of a full-wave rectifier as described in Fig. 8.7, assuming that the a.c. voltage applied to the diodes is the same in both cases?

(b) How would you modify the circuits of Figs. 8.7 and 8.8 in order to generate a *negative* d.c. supply of the same voltage?

8.3.4 Filtering circuits

All the waveforms produced by the rectifier circuits alone consist of a sequence of half-sine waves and are therefore quite unsuitable as a form of d.c. supply. The filtering circuits described in this section filter out, in one way or another, most of the non-zero frequency components of this waveform and so greatly reduce the amplitude of the undulations until they are no more than a 'ripple'.

Capacitive smoothing

The circuit of the simplest d.c. power unit is shown in Fig. 8.9a. It consists of a transformer, a half-wave rectifier and a capacitor; this is the same as Fig. 8.6a but with a capacitor across the load. The voltage waveforms it produces are shown in Fig. 8.9b. The dashed line is the waveform produced without the capacitor, the same as that shown dashed in Fig. 8.6b. The continuous line shows the voltage waveform across the load. Thus the output waveform follows the rectified waveform from the diode as far as its peak value, but then falls much more slowly, until the next cycle of the waveform arrives. The explanation of this output waveform is as follows.

Current only flows through the diode when it is forward biased; that is, when the transformer output voltage is 0.65 V more positive than the voltage across the capacitor and the load. When current flows it not only charges up the capacitor to the peak value of the transformer output voltage (minus the diode voltage drop) but also supplies current to the load. But when the input voltage falls again to below its peak value the diode cuts off, disconnecting the transformer from the load. The capacitor then discharges into the load, so that between peaks in the waveform the load current is supplied by the capacitor only.

The capacitance of the capacitor must be sufficiently large that it only loses a small fraction of its charge through the load between each cycle of

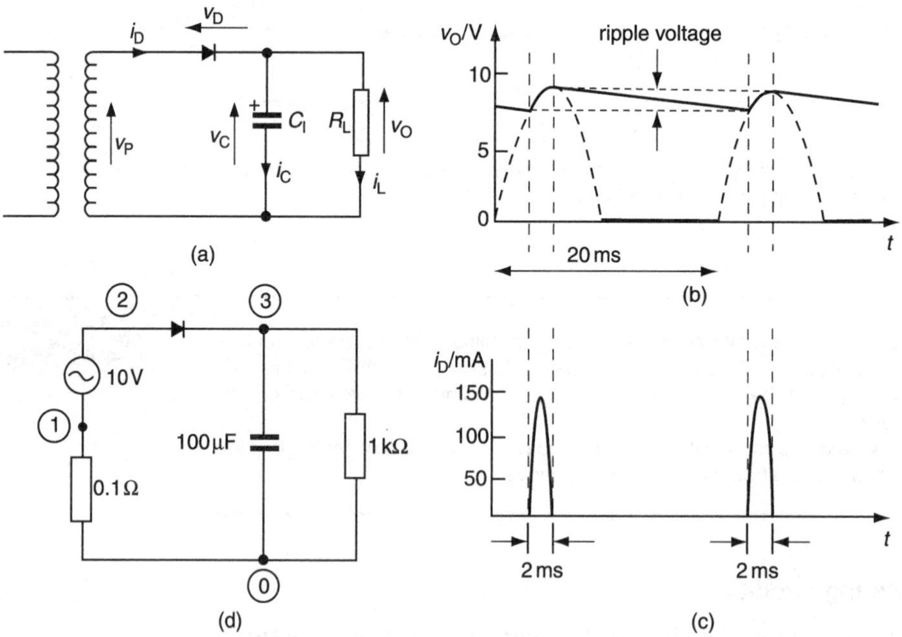

Fig. 8.9 (a) A simple d.c. power unit, with half-wave rectification and capacitive smoothing. (b) The output voltage waveform. (c) The current flow (in mA) through the diode and transformer. (d) The equivalent circuit of (a) with values suitable for computer simulation.

the waveform. The 'droop' of the output voltage between each cycle, shown in Fig. 8.9b, is the exponential decay of the capacitor voltage as it discharges through the load. The discharge is virtually linear if the droop is small. Evidently, the bigger the capacitance or the load resistance, or both, the smaller the voltage droop. The magnitude of the droop, or the peak-to-peak variation in the output, is called the ripple voltage, as indicated in the figure.

Self-assessment question

3 (a) *Estimate* (rather than calculate precisely) the maximum and minimum voltages across the capacitor, given the following data. The transformer is driven by the 240 V, 50 Hz mains supply. The step-down transformer turns-ratio is 20. The diode is a silicon pn junction diode. The capacitance of the smoothing capacitor is 1 mF and the load resistance is 1 kΩ. Assume that the time taken to recharge the capacitor is negligible and that the discharge is linear.

(b) What is the peak-to-peak variation of the output voltage (i.e. the ripple voltage)?

The capacitors involved in this kind of smoothing usually have very large capacitances—often greater than 10 mF (i.e. 10 000 μF)—so they are electrolytic capacitors and must be connected the right way round. They are likely to claim to be only within +50% and −30% of their nominal value, so allowance should be made for this wide tolerance in the smoothing circuit design. Note that it is usually a good idea to connect a resistor (e.g. an indicator lamp or light-emitting diode) across the capacitor, whose resistance is much larger than the design load resistance, so that the capacitor is automatically discharged when the power unit is on open circuit. An unexpected residual charge on such capacitors can sometimes cause damage to other equipment.

Figure 8.9c shows the current pulses that flow through the diode. In the brief time that the diode is conducting, sufficient current must flow through it to replenish the capacitor charge which flowed away through the load resistor between cycles of the a.c. mains. The surges of current through the diode are therefore much larger than the steady current flowing through the load. For example, if the capacitor is recharged in a millisecond, and discharged in 19 ms every cycle of the mains, the mean charging current must be 19 times the average load current. Hence the transformer must be capable of providing large peak currents even for continuous load currents which are quite modest.

The waveforms shown in Fig. 8.9 illustrate the point. The magnitude of the ripple voltage in Fig. 8.9b, multiplied by the capacitance of the smoothing capacitor, gives the amount of charge lost by the capacitor during its discharge. Similarly, the area of the current pulses shown in Fig. 8.9c, which have the dimensions of current and time, indicates the amount of charge flowing into the capacitor each time it is charged up. These two measures of charge should evidently be equal.

Computer work

Simulate the performance of the half-wave rectifier circuit shown in Fig. 8.9a. Include a 0.1 Ω resistor in series with the a.c. source, as indicated in Fig. 8.9d, in order to observe the current flow.

In setting up the simulation, select a sinewave generator at a frequency of 50 Hz. Display the transient response over an interval of three cycles, first at node 3 to give the output voltage waveform, and then at node 1 to display the current waveform. (Note that the current pulses you observe are inverted, as compared with Fig. 8.9c, because the current in the circuit flows from node 0 to node 1.) Check that the voltage and current waveforms are mutually consistent as was done in Self assessment question 4.

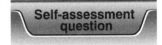

Self-assessment question

4 (a) In Fig. 8.9, note the *maximum and minimum voltages* of the output waveform, and note the *peak currents and durations* of the current pulses of the current waveform. Assuming that the current pulses are triangular (and remembering that the area of a triangle is half the base width times the perpendicular height) relate the charge they represent to the charges represented by the droop in the voltage waveform. (i.e. compare the two ways in which you can calculate the change of charge on the capacitor.) The capacitor value is 100 μF.

(b) What difference would be made to the two waveforms if the capacitance of the capacitor were doubled?

If the half-wave rectifier is replaced by a full-wave rectifier circuit, the frequency at which the capacitor is recharged will be doubled, so that both the ripple voltage and the current peaks will be approximately halved. These are both obviously improvements in performance, which is why half-wave rectification is rarely used.

Self-assessment question

5 Design a 12–0–12 V d.c. power unit (producing positive and negative voltages of 12 V maximum) which has capacitor smoothing giving a ripple of 0.5 V on the positive supply and a ripple of 1 V on the negative supply when the current from both sources is 0.5 A.

Inductive filtering

If the current pulses arising from capacitance smoothing are unacceptably high (e.g. because they cause too much I^2R heating in the transformer or too much hum), an inductor can be included in the filter circuit. The simplest inductive circuit consists of an inductor between the full-wave rectifier and the load, as shown in Fig. 8.10a. Because inductors tend to oppose any sudden changes in current (just as the capacitor in capacitive smoothing tends to oppose sudden changes in voltage across it), the circuit tends to smooth out the half-sinewaves that the rectifiers produce. Or, to put it another way, the impedance of the inductor is greater for the high-frequency components of the rectified waveform than for the low-frequency ones and so the high-frequency components are attenuated more.

Figure 8.10b shows a typical current waveform through the inductor and the load. With a resistive load this is also the output voltage waveform, v_{out}, as shown (because $v_{out} = i_{out}R_L$). The dashed line shows the rectified waveform which would be produced if there were no inductor. Figures 8.10c and d show the currents through the diodes in the rectifier circuit.

Note that the diodes do not turn on and off at clearly indicated points, as they do with capacitive smoothing. This is because the voltage across the inductor is out of phase with the current through it. The continuous line in Fig. 8.10b shows the current through the inductor or the voltage across the load, but not the voltage at the *input* to the inductor. Thus, when the current through the inductor is falling, the back-e.m.f. lowers the voltage at the input to the inductor, even making it negative at times, thus ensuring that one or other diode is conducting even when the output voltage from the transformer has dropped to zero. The current flow does not therefore switch from one diode to the other when

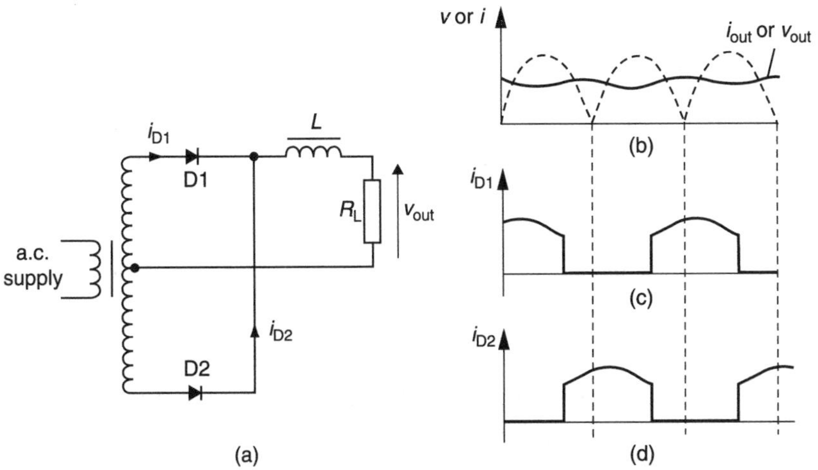

Fig. 8.10 (a) Full-wave rectification with inductive smoothing. (b) The output current and voltage waveforms. (c), (d) The currents through the diodes.

the transformer output voltage goes through zero. The precise output waveform is difficult to calculate but can readily be simulated by computer.

Note also that the mean output voltage in this case does not approach the peak voltage of the a.c. supply, as it does with capacitive smoothing; it is at the mean voltage of the rectified waveform, namely $2/\pi$ times the peak voltage.

A further reduction in the ripple produced by a d.c. power unit can be achieved by combining inductive and capacitive smoothing. However, inductors of significant impedance at 100 Hz are heavy and expensive so capacitive smoothing alone is normally used.

8.4 VOLTAGE REGULATION

As we have seen from the last section the output from the d.c. power unit on its own may be unsatisfactory in electronic apparatus for two reasons. First, because even after smoothing there is usually a significant amount of a.c. ripple superimposed on the d.c. voltage. Second, due to the internal impedance of the transformer and the rest of the circuit, the level of the output voltage can be significantly affected by variations in the current drawn from the unit. The success with which the output voltage is held constant as the load varies is called the **regulation** of the d.c. power unit. The regulating circuitry added to the d.c. power unit in Fig. 8.3 reduces the ripple and improves the regulation. But even power drawn from other sources, such as batteries or solar cells, can vary too much to allow circuits to function properly under all conditions. Hence some form of regulation is required in nearly all electronic equipment.

The structure of this additional circuitry is shown in the block diagram of Fig. 8.11. The output is 'regulated' by a feedback system which compares the final output with a fixed reference voltage and corrects any departure from the intended voltage. The actual d.c. output voltage available from the overall circuit can be adjusted by means of the 'sample unit' to be any value greater than the reference voltage by comparing a fraction of the output voltage with the reference voltage.

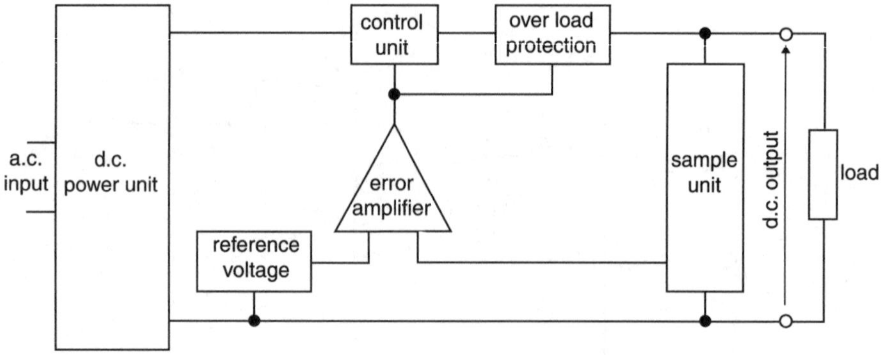

Fig. 8.11 A block diagram of a regulated d.c. supply. The regulation circuitry is between the d.c. power unit and the load.

The functions of the various blocks following the d.c. power unit may be briefly summarized as follows.

(a) The reference voltage. This provides a constant d.c. reference voltage, which should be independent of temperature, supply voltage, etc. A simple voltage reference device which is completely free of ripple is a battery. However within regulator circuits it is more common to use a Zener diode (mentioned in Section 8.2.3), or, better still, a 'bandgap' reference source, which relies on the properties of a forward biased pn junction to maintain a very stable value over a wide range of conditions. This reference voltage is connected to one input of the 'error amplifier'. The other input to the error amplifier is connected to a fraction of the output of the d.c. supply. The output of the error amplifier is therefore proportional to the difference between the reference voltage and this fraction of the regulated d.c. output voltage.

(b) The sample unit. The sample unit is just a potential divider across the output terminals. It connects a fraction of the output voltage to the inverting input of the error amplifier. Thus if the output tends to fall, the output of the amplifier rises to counteract this change.

(c) The control unit. This consists of one or more transistors, connected so that the output of the error amplifier can control the current passing to the output.

(d) The error amplifier. The output of this amplifier is proportional to any deviation of the regulated supply voltage from the intended value, and is used to increase or decrease the current supplied by the control unit so that the final output is restored to its intended voltage. This correcting action applies to ripple voltages as well as to changes in the output voltage caused by changes in the load. The improvement in the regulation and ripple of the final output depend upon the gain of this amplifier. The amplifier is powered by the unregulated d.c. supply and therefore, despite the rail-rejection properties of such amplifiers, it will create a small amount of ripple.

(e) Overload protection. This circuit provides protection of both the control unit and perhaps the load by limiting the output current to some preset value even when the load is a short circuit.

8.4.1 Linear voltage regulators

The simplest and most common method to achieve voltage regulation is by adding a device called a **linear voltage regulator**, an IC containing all the circuitry necessary to implement the control function outlined in Fig. 8.11. The word 'linear' implies that this is an analogue device which works by continuously adjusting the output current to maintain a constant output voltage, rather than by the 'switching' principle which is described in the next section. A wide variety of linear regulator ICs is available, both for fixed voltages (such as +3 V or +5 V for logic circuits, or ±12 V for op amp circuits), or variable types which use external resistors to set the output level.

A fixed voltage regulator is usually a three terminal device and is used in the way outlined in Fig. 8.12. The unregulated output from a d.c. power unit is connected between the input and common I/O terminals of the regulator. The regulated output is taken from the output and common I/O terminals. The regulator functions by maintaining the output voltage close to its nominal value over a range of input voltages and output loads. Small capacitors are also sometimes connected across the input and output to improve the response to transient fluctuations of voltage.

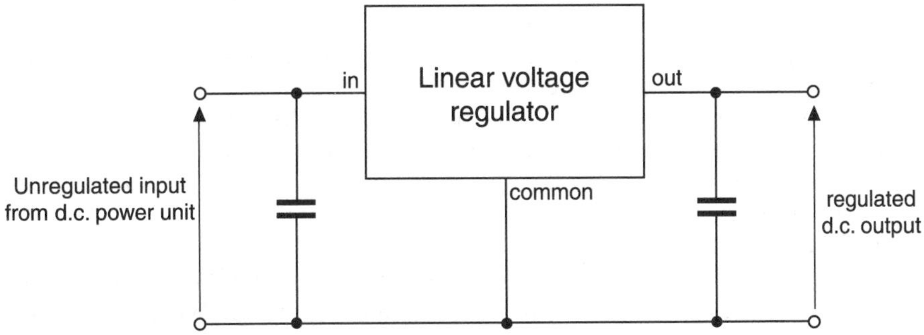

Fig. 8.12 A regulation circuit using a linear regulator.

To consider a specific example, Table 8.1 shows some values for the main performance parameters for a typical 7805 regulator, a common type of fixed +5 V regulator which is produced by a number of manufacturers.

The output voltage of the 7805 regulator in Table 8.1 is nominally 5.0 V, to match common logic families, but can vary from this by ±0.2 V. This amount of variation is within the allowable range for TTL (see Chapter 5), and so is accurate enough for most applications. In some related devices a much closer tolerance for the output voltage is achieved by trimming the internal reference of each device during manufacture.

The input voltage can be anywhere in the range from 7 V to 25 V. The minimum value is needed because there has to be more than about 2 V across the device (between the input and output terminals) for it to work correctly. This 2 V is called the **dropout voltage**. An absolute maximum value of 35 V input voltage is specified for this device, above which there is the possibility of reverse breakdown of pn junctions inside it. The recommended maximum of 25 V up to which the device is characterized gives a good margin of error and assured reliability. However, the power that can be dissipated in the device

Table 8.1 Some performance characteristics for a typical 7805 fixed voltage linear regulator in a plastic package, as specified for a temperature of 25°C. The maximum recommended output current for this device is 1.5 A, and the maximum recommended operating junction temperature is 125°C

	Min	*Typ*	*Max*	*Unit*
Output voltage	4.8	5.0	5.2	V
Input voltage	7		25	V
Input voltage regulation, $V_I = 7$ V to 25 V		5	100	mV
Output voltage regulation, $I_O = 5$ mA to 1.5 A		1.3	100	mV
Ripple rejection (120 Hz)		68		dB
Thermal resistance, junction-to-ambient		65		°C W^{-1}
Thermal resistance, junction-to-case		5		°C W^{-1}

often sets a lower practical limit to the maximum input voltage that can be used in a particular application (see below).

The input voltage regulation parameter gives the variation in output voltage when the input changes between the recommended limits of 7 V and 25 V.

The output voltage regulation demonstrates how the output voltage deviates from its nominal value as the load current is varied between a very low value of 5 mA and the maximum for the device of 1.5 A.

The ripple rejection is the amount by which a small ripple signal (measured at 120 Hz) will be attenuated.

The final two items in Table 8.1 are two forms of thermal resistance for the device, and the use of these important parameters will be covered in the next section.

In addition the 7805 contains output protection circuitry which limits the output current to a safe level in the event of a short circuit or an excessive junction temperature.

Self-assessment question

6 A 5 V regulated power supply is to be constructed using the 7805 voltage regulator specified in Table 8.1.

(a) If the peak-to-peak ripple on the input to the regulator is 100 mV, what value of ripple is to be expected on the output?

(b) If the output voltage from the d.c. power unit is 8 V and the load current is 1 A, how much power is dissipated in the load, and how much in the regulator?

Heat dissipation

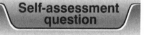

The temperature of the regulator, as with any electronic component, will rise according to the amount of power dissipated within it. Most of the temperature rise will occur at a pn junction in the pass transistor, the component which actually controls the output current. Damage can occur if the operating temperature of this junction is allowed to rise above a critical value, but problems usually arise with the packaging before this can occur. The 7805, like many other silicon devices in plastic packages, is specified with an **absolute maximum operating junction temperature** of 150°C. To allow a margin of error and to improve reliability, a **recommended maximum** operating junction temperature of 125°C is also quoted, and this lower value is normally the one to design for.

The actual junction temperature reached will depend not only upon the power dissipated, but also upon the conduction of heat through the case and the transfer of heat from the case to the surroundings. The junction temperature can be calculated if this heat loss is known, and the parameter normally provided to allow this calculation to be carried out is called **thermal resistance**. It is defined as 'the temperature rise divided by the power transferred', and has units of $°\mathrm{C\,W^{-1}}$.

The simplest situation is when the body of the device (apart from the input connections) is not in contact with anything, and is cooling in free air. In Table 8.1 the parameter which represents this situation is 'thermal resistance, junction-to-ambient', quoted as $65°\mathrm{C\,W^{-1}}$ for this particular package. The temperature of the junction is then given by

$$T_{\mathrm{J}} = T_{\mathrm{A}} + \theta_{\mathrm{JA}}P$$

where T_{J} is the junction temperature, T_{A} is the ambient temperature (i.e. of the surroundings), θ_{JA} is the junction-to-ambient thermal resistance, and P is the electric power dissipated by the device. In other words the rise in temperature above ambient is simply the power dissipated multiplied by the thermal resistance.

Worked example

If 3 W is dissipated by the regulator specified in Table 8.1, what will the junction temperature be if the ambient temperature is 20°C? Is it possible to operate the device under these conditions?

Using the above equation, $T_J = 20°\mathrm{C} + (65°\mathrm{C\,W^{-1}} \times 3\,\mathrm{W}) = 215°\mathrm{C}$. *This is well above even the quoted absolute maximum of 150 °C, so the device cannot be used in this way.*

The way that such a power level can be dealt with is by attaching the regulator to a metal **heat-sink**, which is designed to transfer heat more effectively to the surroundings. In this case the overall heat path to the surroundings involves three thermal resistances in series:

θ_{JC} is the thermal resistance from junction to the device case (package), and so depends on the package type. For the 7805 in Table 8.1 the value is specified as $5°\mathrm{C\,W^{-1}}$ for its particular form of the standard TO220 package. This is a common plastic encapsulation for transistors and ICs that need to pass currents of about an amp or so, and has a metal tab attached so that it can be easily attached to a heat-sink.

θ_{CS} is the thermal resistance of the contact between the case and the heat-sink. This depends upon the tightness of the contact, and whether or not insulating washers or heat conducting grease are used. It is usually less than $0.3°\mathrm{C\,W^{-1}}$, and may be much smaller.

θ_{SA} is the thermal resistance between the heat-sink and the surroundings. With the TO220 package, this might vary from 25°C W^{-1} for a small clip, down to less than 1°C W^{-1} for a large finned aluminium extrusion.

These values are simply added to obtain the total thermal resistance, and so the junction temperature can be calculated from the following equation:

$$T_J = T_A + (\theta_{JC} + \theta_{CS} + \theta_{SA})P \qquad (8.2)$$

Self-assessment question

7 (a) Suppose that a 7805 regulator as specified in Table 8.1 is mounted on a small aluminium heat-sink with thermal resistance of 7°C W^{-1}. The thermal resistance of the contact is 0.3°C W^{-1}. Calculate the junction temperature if the power dissipated is 3 W and the ambient temperature is 20°C. Compare this with the previous Worked example.

(b) Calculate the maximum current that the 7805 regulator can safely pass if the maximum expected input voltage is 20 V, the ambient temperature is 50°C, and the device is mounted on a large finned heat-sink of thermal resistance 1.9°C W^{-1} through a contact of thermal resistance 0.1°C W^{-1}.

8.4.2 Switched-mode power supplies

The type of regulated d.c. supply described hitherto is an analogue circuit, in essence operating by connecting the output through a variable attenuator which is continuously adjusted to keep the output constant. This is not very efficient, as the power dissipated in the d.c. power supply itself might be comparable with the power supplied to the load. When high powers are involved, this is not only wasteful but can lead to excessive heating in the power supply. The power is dissipated mostly in the regulator and in the rectifier circuit because all the output current flows through them, and the total voltage drop across them may be about the same as the output voltage.

An alternative way of controlling a supply voltage is by means of rapid 'switching' rather than 'attenuation'. An important advantage of such switched-mode power supplies is that they achieve much greater efficiency and so can deliver much more power to the load than they dissipate themselves. Typically over 90% of the input power can be transferred to the output. However, other possibilities for this technique include eliminating the expensive 50 Hz (or 60 Hz) transformer, converting positive d.c. supplies to negative and vice versa, and producing a regulated supply at a higher voltage than the d.c. input.

Switched-mode d.c. power supplies make use of the fact that very little power is dissipated in a switch. This is clearly seen with a mechanical switch: when the switch is on, the resistance is very small and so there is very little voltage across it; when the switch is off, no current flows. In either case voltage × current is very small. In electronic switching circuits a transistor (see Chapter 9) is used rather than a mechanical switch. The details are not important at this stage, but in essence a transistor can be used as a voltage (or current) controlled switch, with a small voltage drop in the 'on' state, and a small leakage current in the 'off' state. But the principle is the same: very little power is dissipated in either the on or off states.

When such a switch in the output circuit of a power supply is switched rapidly between one state and the other, essentially switching the output

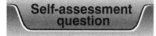

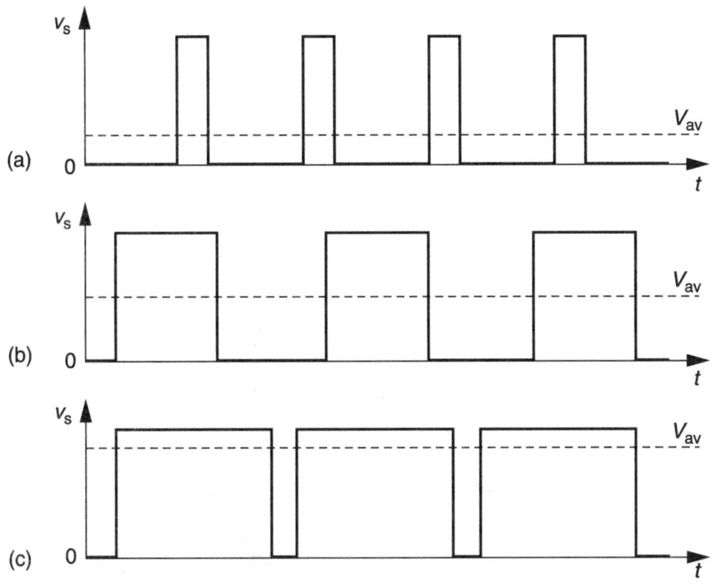

Fig. 8.13 Voltage waveforms from a switched output transistor before smoothing. The following low-pass filter delivers a d.c. voltage equal to the mean voltage of the waveform, shown dashed.

on and off, it will never dissipate much power itself, but it can still deliver a significant power to a load. The important point is that it is the relative time spent in the two switching states (the 'duty cycle') which controls the average power delivered.

Some typical voltage waveforms delivered to a load by a transistor switch are shown in Fig. 8.13. These are also the current waveforms in the circuit, so the power delivered to the load is proportional to v_S^2. The corresponding voltage waveforms across the switch are simply the inverse of these waveforms. That is, when the voltage across the load is v_S, the voltage across the switch is almost zero, and vice versa. So, at the switch, either the voltage or the current is almost zero, and the power dissipated in it is quite small in both the ON and OFF conditions.

The average delivered voltage of each waveform is shown by the dashed line in Figs. 8.13a to c. So, by varying the duty cycle, or the 'mark-space ratio' of the waveform (i.e. the ratio of the ON time to the OFF time) the average output voltage can be varied.

To deliver a smooth d.c. voltage to the load—equal to the average voltage of such a switch waveform—an efficient low-pass filter is needed.

Worked example

Why is the capacitor type of smoothing circuit described in Section 8.3.4 unsuitable for this application?

Because the capacitor type of smoothing circuit gives an output voltage approximately equal to the peak voltage, rather than the mean, of the waveform.

The filter usually used is an inductor-capacitor circuit such as that shown in Fig. 8.14a. The switch represents the transistor which is being switched on

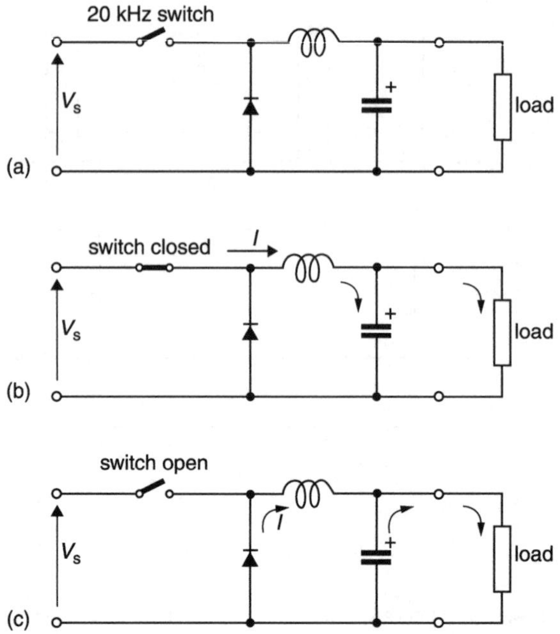

Fig. 8.14 An *LC* smoothing circuit for a switched-mode supply. The inductor and capacitor store energy when the switch is closed, and release it when the switch is open.

and off, and V_S is the (unregulated) d.c. voltage derived from a d.c. power unit. The circuit operates as follows. When the switch is closed and V_S is connected to the inductor, as in Fig. 8.14b, the diode is cut off, so current builds up in the inductor, as indicated in the figure, and flows into the load and the capacitor. Remember that current increases relatively slowly when a voltage is applied to an inductor and does not die away immediately when there is no longer a voltage applied to it. So the longer the voltage V_S is applied to the inductor, the larger the inductor current becomes. When the switch is opened, the current can continue to flow through the inductor as indicated in Fig. 8.14c because the diode becomes forward biased. Any 'sag' or 'droop' in the inductor current is largely made up in the load by the discharge of the capacitor. So although the input voltage changes abruptly between V_S and zero, the load current tends to flow steadily with only a small ripple superimposed on it.

It is probably best to think of the inductor and capacitor as storing the energy supplied by the source when the switch is closed and releasing it steadily again when the switch is open. The greater the proportion of the time that V_S is connected to the load by the switch, the greater the final smoothed d.c. output voltage.

Evidently, the circuit must include an oscillator which rapidly switches the control transistor on and off, usually at about 20 kHz, but sometimes considerably higher. The oscillator must be such that its duty cycle can be controlled by an applied input voltage. Given such an oscillator, Fig. 8.15 shows how a regulated, switched d.c. supply can be produced. A fraction of the d.c. output voltage is compared with a reference voltage, just as in the 'analogue' d.c. supply circuit previously described, but this time any difference between

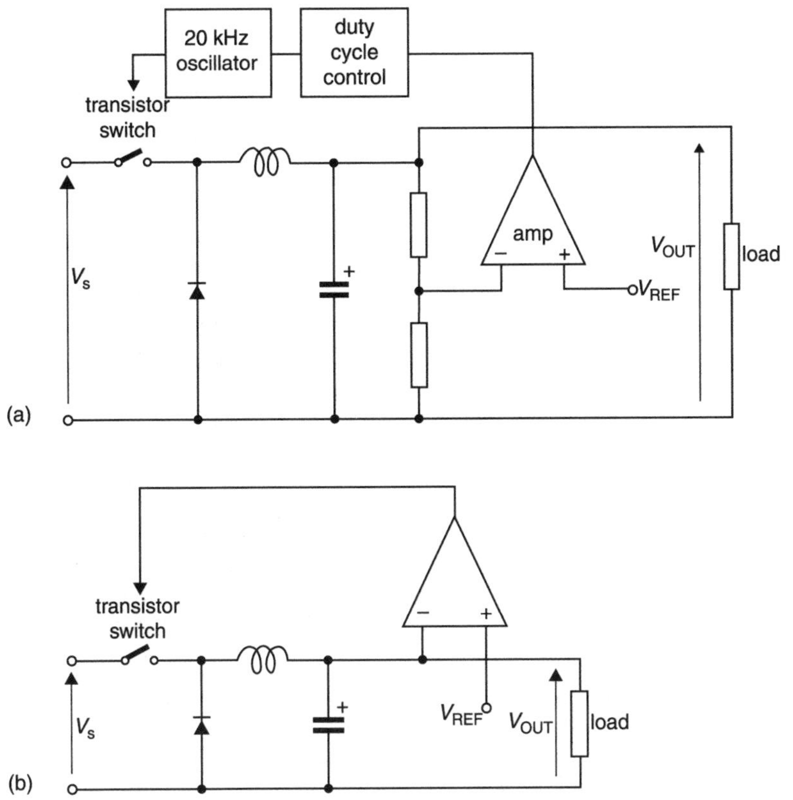

Fig. 8.15 Switched-mode d.c. supply circuits. (a) The standard form of the circuit. (b) A simpler switched-mode circuit which does not need a separate oscillator.

them is amplified and used to control the duty cycle of the oscillator. If the output voltage drops, the duty cycle is changed in the direction of the waveform of Fig. 8.13c, and so increases the current supplied, thus tending to restore the intended output voltage. A high switching rate of 20 kHz or more means that the inductance of the inductor does not have to be very big to achieve excellent smoothing. Such switching speeds also have the advantage that they are above most people's audible range and so cannot be heard.

The circuit just described is the standard design of switched d.c. supply, but simpler circuits are also used. Figure 8.15b shows a switched-mode power supply which does not need an oscillator; it generates its own switching waveform. The amplifier is again essentially an op amp, but here it is driven repeatedly from fully-on to fully-off as the input changes slightly. Because the amplifier has a high gain, only a few millivolts difference between the output voltage and the reference voltage is sufficient to drive the output of the amplifier from fully-on to fully-off. Thus a small drop in output voltage below the reference voltage causes the transistor switch to be turned on, and a small rise above the reference voltage turns the transistor off again. The arrangement is rather like a central heating system, in which the boiler is switched on and off in response to small changes of room temperature, and in which the gross changes in heat supplied are smoothed out by the thermal inertia of the building. Here, however, the power is switched on and off thousands of times a second instead of every few minutes. This circuit differs from that of Fig. 8.15a in that the frequency of switching is not predetermined. The frequency at which the transistor is switched depends on how rapidly the output voltage drops; so, as the load increases, both the duty cycle and the frequency of switching increase. This has the fortuitous effect of improving the filtering as the demands on the d.c. supply increase.

As mentioned earlier, a great advantage of these switched-mode circuits is that they do not need a mains transformer. If, as indicated in Fig. 8.16, the a.c. mains supply of 240 V r.m.s. (or 110 V r.m.s.) is rectified and smoothed, a d.c. supply, with ripple included, of about 330 V (or 150 V) is produced. If this is then chopped at a high frequency, the resulting waveform can be transformed down to an appropriate a.c. voltage by means of a much smaller and cheaper transformer, as shown in the figure.

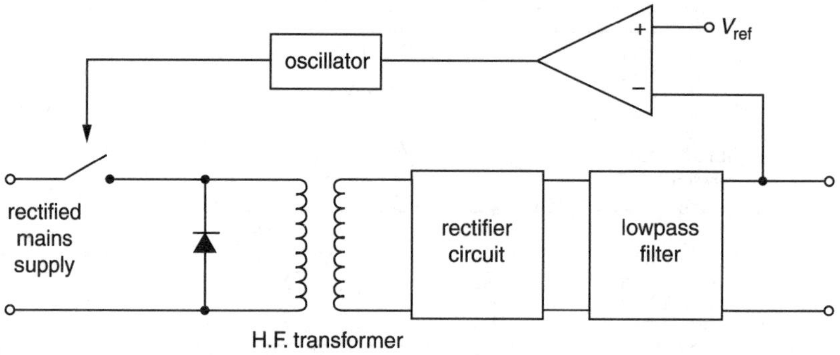

Fig. 8.16 A switched-mode regulated d.c. supply in which the usual low-frequency transformer is replaced by a high-frequency one (to save weight and cost).

Worked example

What is the function of the diode placed across the primary of the transformer in Fig. 8.16?

To allow current to continue to flow in the primary transformer winding when the switch has opened.

The lower output voltage from the transformer can then be rectified and filtered as before, with the feedback controlling the duty cycle (rather than the frequency) of the oscillator.

One of the reasons for using a mains transformer in a linear power supply is to achieve isolation between the mains and the circuit that is being powered, for safety. Such isolation can be achieved in the circuit of Fig. 8.16 by introducing an opto-isolator into the feedback path between the output and input. The opto-isolator is a small plastic package containing a light emitting diode (which emits light when a forward current is passed) close to a photodiode (through which a large reverse current can flow depending upon the amount of light falling on it). Hence a signal can be passed from the output to the switch of this device with no electrical connection between them, but using light as the pathway.

The subject of power supplies that operate using switching principles is very complex, and further discussion is beyond the scope of this book. However it should be apparent from the switching circuits described so far that this is a very versatile technique. With suitable circuit design, the regulated output can be higher or lower in voltage, or of opposite polarity to the input. Hence a common method of producing several different voltage supplies in a complex electronic system is to have a single d.c. supply which feeds a number of switching **d.c. to d.c. converters**. It is often convenient for this single d.c. supply to be built into the mains plug to form a power adapter, which may be regulated or unregulated. The d.c. to d.c. converters would then be included in the powered system itself. The high efficiency of switching circuits means that this is usually the most efficient way of producing multiple supply voltages.

Self-assessment question

8 A 5 V regulated supply of 1 A current is to be produced from a 12 V d.c. power unit.

(a) If a linear voltage regulator is to be used, what is the input current and hence what is the overall efficiency of the regulation (i.e. output power/input power)?

(b) If a switching regulator with a power efficiency of 93% is to be used, what is the input current to the regulator?

8.5 SUMMARY

1. A d.c. power unit converts the a.c. mains into one or more smoothed but unregulated d.c. sources. These sources still usually possess a significant ripple voltage superimposed on the d.c. output voltage.

   D.C. power units contain the following elements:

   - A step-down mains transformer. Each d.c. source uses a separate secondary winding of the transformer. For a 'full-wave rectifier' a centre-tapped secondary winding is needed. Full-wave rectification can also be achieved using a bridge rectifier.
   - A rectifying circuit involving 1, 2 or 4 rectifiers depending on the design.
   - Filtering circuitry. Capacitive smoothing is cheapest and lightest but requires large current pulses from the transformer; the output approaches the peak voltage. Inductive smoothing avoids current pulses but, at 50 or 60 Hz, requires a large, heavy inductor.

2. A simple three-terminal IC regulator includes:
   - A d.c. voltage reference
   - A sampling unit or potential divider to feed back a fraction of the output voltage to the error amplifier.
   - An error amplifier, which supplies a current to the output transistor in proportion to the difference between the reference voltage and the sampled output voltage.
   - A control unit, which controls output current according to the output of the error amplifier. It is usually mounted on a heat-sink to minimize temperature rise.

- Protection circuit which limits the output if the output current or junction temperature exceed specified limits.

3. Power dissipation is a major design issue with linear regulators (as with other types of 'power' device), and appropriate heat-sinks and other forms of heat transfer must be used to maintain the junction temperature below specified limits.
4. Switched-mode power supplies make use of the fact that a transistor switch dissipates negligible power both when it is 'on' and when it is 'off'. By switching the output transistor rapidly between these two states at a high frequency (e.g. 20 kHz), and filtering out the large high-frequency 'ripple' thereby produced, a d.c. output can be obtained.
   - In regulated switched-mode d.c. supplies the error amplifier varies the mark-space ratio of the oscillator in the circuit.
   - The low-frequency transformer is usually replaced by a high frequency one operating at the oscillator frequency.
   - Switched-mode regulators can produce voltages higher, lower, or inverted with respect to the d.c. input.
   - Switched-mode regulation efficiency is usually more than 90%.

Answers to self-assessment questions

1. Substituting the data into Equation 8.1 gives 1 mA = 10^{-13} A × [exp(35 V_D)−1]. The (−1) in the bracket is negligible compared with the exponential, so it can be ignored. Therefore,

$10^{-3}/10^{-13} = 10^{10} = \exp(35\ V_D)$

$35\ V_D = \log_e 10^{10} = 23.02$

so $V_D = 23.02/35 = 0.658$ V

If the forward voltage is 0.4 V, then Equation 8.1 becomes

$$I_D = 10^{-13}[\exp(35 \times 0.4) - 1]$$
$$= 10^{-13} \times e^{14}$$
$$= 10^{-13} \times 1.2 \times 10^{6}$$
$$= 1.2 \times 10^{-7} = 0.12\ \mu\text{A}$$

2. (a) With the bridge rectifier there are two diodes between the transformer and the load, so the peak output voltage of the rectifier circuit is less than the peak transformer output by about 1.4 V. This is 0.7 V less than the peak output voltage of the full-wave rectifier circuit.

(b) In both circuits the end of the load resistor which is regarded as 0 V is arbitrary, so to produce a negative d.c. source in either case it is only necessary to regard the other end of the load as 0 V, and to connect it to earth. However with the full-wave rectifier it is usual to regard the centre tap as the

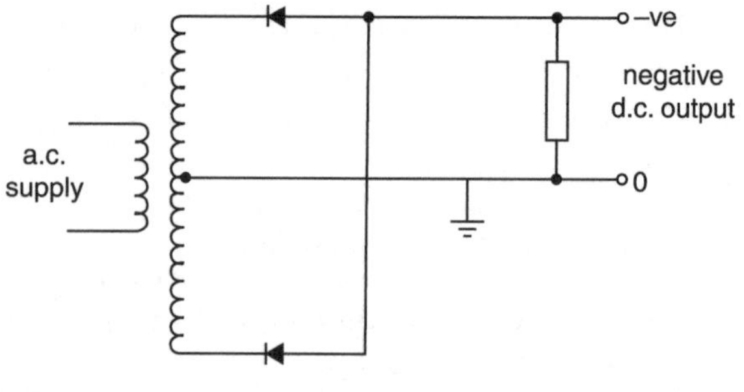

Fig. 8.17 Producing a negative d.c. voltage. See Self-assessment question 2.

0 V connection, so to reverse the polarity of the output it is necessary to reverse both diodes as shown in Fig. 8.17. Reversing the diodes in the bridge rectifier is just another way of saying that the other end of the load is 0 V!

3. (a) There are several steps to this calculation.

(i) Since the transformer has a turns ratio of 20 its output voltage is 12 V r.m.s., which corresponds to an amplitude of $12 \times \sqrt{2} = 17$ V.

(ii) Since the capacitor is charged to the peak amplitude minus the voltage drop across the diode, it is charged to about 16.3 V.

(iii) Since the time taken to charge the capacitor is assumed to be negligible, the time during which the capacitor discharges through the load is very nearly 1/50 s = 20 ms. Furthermore, since the drop in voltage during discharge is much smaller than 16 V, it can be assumed that the discharge current is almost constant at about 16.3 V/1 kΩ = 16.3 mA.

(iv) If this current flows for 20 ms, the total charge leaving the capacitor is 20 ms × 16.3 mA = 326 μC.

(v) Taking this amount of charge from a 1 mF capacitor will produce a voltage change of Q/C = 326 μC/1 mF = 326 mV.

(b) The peak-to-peak variation of the voltage across the capacitor is therefore 326 mV which is also the peak-to-peak variation of the output voltage.

4. (a) There are two measures of charge flow each time the 100 μF capacitor is charged. The area of the current pulse is one measure, the change of voltage across the capacitor, times its capacitance, is another measure. They should give the same result.

The max. and min. voltages of the voltage waveform are 9.3 V and 7.7 V. The *change* of voltage across the capacitor is 1.6 V, so the charge delivered to it during each cycle of the a.c. supply is 1.6 V × 100 μF = 160 μC.

The peak value of each current pulse is about 150 mA and its duration is 2 ms. The area of each current pulse is (150 mA × 2 ms/2) = 150 μC.

This is as good an agreement as can be expected between these two estimates of charge supplied to the capacitor.

(b) The voltage swing of the voltage waveform will be halved. The current pulses will be shorter and higher.

5. The circuit is shown in Fig. 8.18. Assuming that the time between each pulse of current recharging the capacitors is 10 ms, the charge taken from the capacitors is

$$\Delta Q = 10\,\text{ms} \times 0.5\,\text{A} = 5\,\text{mC}$$

This must discharge the capacitor on the positive side by no more than 0.5 V. So, since

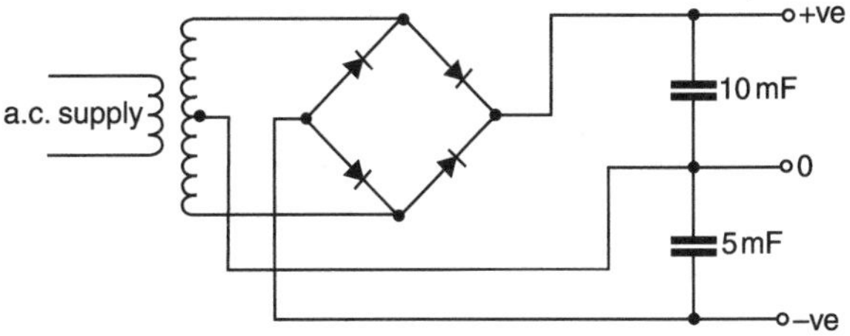

Fig. 8.18 See Self-assessment question 5.

$$\Delta Q = C \times \Delta V$$

$$C = \frac{5 \times 10^{-3}\ \text{C}}{0.5\ \text{V}} = 10^{-2}\ \text{F} = 10\ \text{mF}$$

Similarly, the capacitor on the negative supply should have a capacitance of 5 mF.

6. (a) The parameter to use here is the ripple rejection, with a typical value quoted in Table 8.1 of 68 dB. This corresponds to a voltage ratio of $10^{68/20}$=2512. Hence the ripple on the output is expected to be 100 mV / 2512 = 39.8 μV.

(b) With an input of 8 V and an output of 5 V, 3 V is dropped across the regulator. Hence the power dissipated in the regulator is 3 V × 1 A = 3 W. The power dissipated in the load is 5 V × 1 A = 5 W. Hence 3/8 (=37.5%) of the total input power is dissipated in the regulator, which is clearly not a very efficient means of control. If the input voltage is higher, the efficiency is even less.

7. (a) T_J=20°C + (5°C W^{-1} + 0.3°C W^{-1} + 7°C W^{-1}) × 3 W = 56.9°C.

With this arrangement, the junction temperature is well within the permitted operating conditions for the device. It is much lower than the example of free cooling calculated in the Worked example.

(b) The maximum recommended junction temperature is 125°C, and the total thermal resistance is 5°C W^{-1} + 0.1°C W^{-1} + 1.9°C W^{-1}=7°C W^{-1}.

If the maximum power that can be dissipated is P_{max} then

125°C = 50°C + 7°C W^{-1} × P_{max}

and so

P_{max} = 75°C / 7°C W^{-1} = 10.71 W.

The maximum voltage across the device is then 20 V − 5 V = 15 V.

Hence the maximum current is 10.71 W / 15 V = 0.71 A.

This calculation demonstrates that worst case power dissipation can limit the maximum current for a regulator to below its nominal maximum value, and must always be taken into account when designing systems in which comparatively large currents are flowing.

8. (a) A linear regulator uses little current to power itself, so that we can assume that the input current is equal to the output current, 1 A. Hence the power delivered to the regulator is 12 V × 1 A = 12 W. The power delivered to the load is 5 V × 1 A = 5 W. Hence the efficiency of the regulation is 5/12 = 0.4167 or 41.67%.

(b) For the switching regulator with an efficiency of 93%, the input power must be 5 W / 0.93= 5.376 W. Hence the input current to the regulator is 5.376 W / 12 V = 0.448 A. Hence a d.c. power unit with a much lower current capability can be used compared with the example of a linear regulator.

Basic transistor circuits

9

AIMS

1. To describe and explain the d.c. characteristics of bipolar junction transistors (BJTs) and field-effect transistors (FETs).
2. To describe and explain the a.c. characteristics and equivalent circuits of bipolars and FETs.
3. To introduce and explain the operation of some basic analogue and digital circuits using both types of transistor.

GENERAL OBJECTIVES

On completing this chapter you should understand, and be able to use correctly, the following terms:

As regards bipolar transistors

- common-base current gain α, common-emitter current gain β
- d.c. characteristics
- early voltage
- emitter, base and collector regions and terminals
- hybrid-π equivalent circuit
- input resistance, output resistance, emitter resistance
- mutual conductance
- npn, pnp
- planar process

As regards bipolar transistor circuits

- bias resistor(s), emitter resistor
- common-emitter amplifier
- differential gain, common-mode gain, common-mode rejection ratio (CMRR)
- emitter bypass capacitor
- emitter-follower, double emitter-follower output stage
- long-tailed pair (differential pair)]
- Schottky diode
- speed-up capacitor
- switching (digital) circuits

As regards field-effect transistors

- channel length modulation factor, output resistance
- depletion mode and enhancement mode MOSFETs
- drain, gate, source and substrate
- junction type (JFET), metal-oxide-silicon type (MOSFET)
- n-channel and p-channel MOSFETs
- pinch-off point, pinch-off voltage of a JFET
- threshold voltage of a MOSFET

As regards field-effect transistor circuits

- CMOS digital circuits
- common-source amplifier
- long-tailed pair (differential pair)
- source-follower, double source-follower output stage

SPECIFIC OBJECTIVES

1. Calculate the mutual conductance g_m of a bipolar junction transistor, given its collector operating current.
2. Calculate the output resistance r_o of a bipolar junction transistor in the common-emitter configuration, given its collector operating current and its Early voltage.
3. Calculate the input resistance r_i of a bipolar junction transistor in the common-emitter configuration, given its collector operating current and its common-emitter current gain β.
4. Calculate the voltage gain of a common-emitter amplifier, with and without an emitter-bypass capacitor.
5. Calculate the differential voltage gain of a BJT long-tailed pair.
6. Calculate the CMRR of a BJT long-tailed pair.
7. Calculate the input and output resistances of an emitter-follower, given its load and source resistances.
8. Calculate the output resistance r_o of an FET in the common-source configuration, given its drain operating current and its channel length modulation factor λ.
9. By analogy with the common-emitter amplifier, calculate the voltage gain of a common-source amplifier, with and without a source-bypass capacitor, given the value of g_m.
10. By analogy with the BJT long-tailed pair, calculate the differential voltage gain of an FET long-tailed pair.
11. By analogy with the BJT long-tailed pair, calculate the CMRR of an FET long-tailed pair.

9.1 INTRODUCTION

Transistors are the 'active' devices of electronics. They are capable of controlling power from a d.c. power supply, in response to a low-power input signal, and providing a much greater output power. So they can provide power gain. This is the basic function of an amplifier. An analogue amplifier should do this with minimum distortion of its input waveform. A digital switch or gate is also required to provide power gain, but should convert any ill-defined input voltage level (within limits) into a 'clean' well-defined voltage level at its output.

There are two principal types of transistor, the bipolar junction transistor (BJT) and the field-effect transistor (FET). Both types are described in this chapter, together with their a.c. equivalent circuits and the circuits of some simple amplifiers and switches.

9.2 BIPOLAR TRANSISTORS

Silicon bipolar transistors are three-terminal devices consisting of two pn junctions formed back to back in a single crystal of silicon. They are made in npn and pnp configurations as shown diagrammatically in Figs. 9.1a and b. The three regions of the transistor are called **emitter**, **base** and **collector** as shown. The graphical symbols for the two versions of the device are shown too. In these symbols the arrow is always on the emitter, and points in the direction of conventional current flow.

A cross-section of a typical npn transistor is shown in Fig. 9.1c. Such a device is produced in much the same way as a pn junction diode except that there is one additional diffusion. Beginning with an n-type substrate, a p-type region is

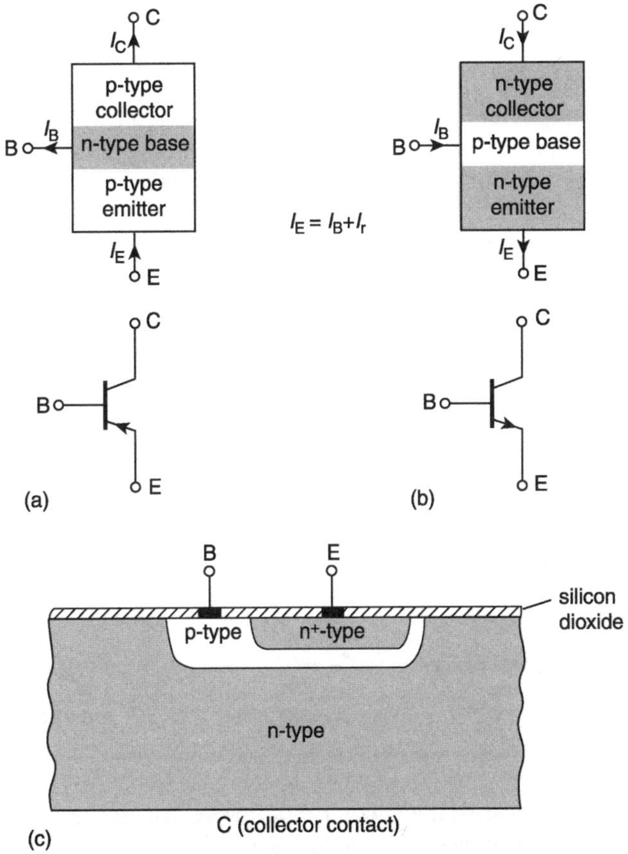

Fig. 9.1 Bipolar transistors. (a) A diagrammatic representation of a pnp transistor, plus the standard graphical symbol for a pnp transistor. (b) The same for an npn transistor. (c) A typical structure of a discrete npn transistor.

diffused into it through a window in the protective layer, as for a diode. But then a new oxide layer is grown over the surface, a new and smaller window is etched in it and some *donor* material is diffused into the p-type region, converting its surface layer back into n-type material. This final diffusion produces the emitter region which is much more heavily doped than the base region, and is therefore labelled n^+. This method of making transistors is called the **planar process.**

9.2.1 Operation of bipolar transistors

The essence of transistor action is that a small voltage applied between the *base* and *emitter* terminals causes a current to flow in the *collector* circuit, as shown in Fig. 9.2. The thicker arrow of Fig. 9.2(b) indicates the path of most of the current. As V_{BE} is varied, producing a small input current I_B, then the much larger currents I_E and I_C vary in response. The reason that the transistor can be used as an effective amplifier or switch is that a large output power generated from I_E or I_C can be controlled by a much smaller input power from I_B.

In an npn transistor, the emitter is n-type and the base is p-type. The input voltage V_{BE} is of the polarity which forward biases the base-emitter junction. Hence the way that I_B varies with V_{BE} is as shown in Fig. 9.3a. This is of the same exponential form as the characteristic of a forward biased junction diode shown in Fig. 8.2.

The corresponding variation of collector current I_C for this configuration is shown in Fig. 9.3b. The graph is very similar to Fig. 9.3a, but the magnitude of the current is about 100 times larger. The ratio of collector current to base current is an important transistor parameter, normally referred to as the **common-emitter current gain** β, where $\beta = I_C/I_B$. (This parameter is sometimes called h_{fe}.) In the example in Figs. 9.3a and b, $\beta = 100$.

You can see from Fig. 9.2 that the emitter current is the sum of the base and collector currents, that is, $I_E = I_C + I_B$. The ratio I_C/I_E is called α, the **common-base current gain**, with a value slightly less than one.

Worked example

Show that $\alpha = \beta/(1+\beta)$ and that $\beta = \alpha/(1-\alpha)$. If the range of values of β is 100 to 500 what is the range of α?

$\alpha = I_C/I_E$ by definition, and $I_E = I_C + I_B$, so $\alpha = I_C/(I_C + I_B)$. Dividing numerator and denominator by I_B, and using $\beta = I_C/I_B$ gives

$$\alpha = \beta/(1+\beta)$$

This equation can be rewritten as $\alpha\,(1+\beta) = \beta$. Rearranging gives

$$\alpha = \beta - \alpha\beta = \beta(1-\alpha), \text{ or } \beta = \alpha/(1-\alpha)$$

If $\beta = 100$ then $\alpha = 100/101 \approx 0.99$. If $\beta = 500$ then $\alpha = 500/501 \approx 0.998$.

In this npn device, most of the current that flows through the forward-biased emitter-base junction consists of electrons passing from the heavily doped emitter region to the more lightly doped base region. (Remember that electrons, because of their negative charge, actually travel in the direction that is opposite to the conventional current flow shown in Fig. 9.2.) There is

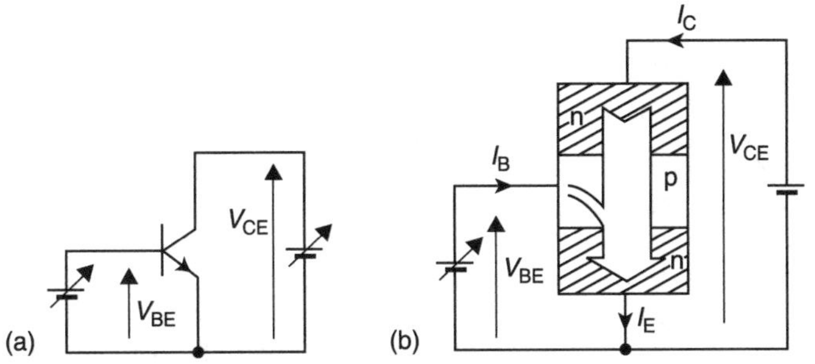

Fig. 9.2 Currents in an npn transistor.

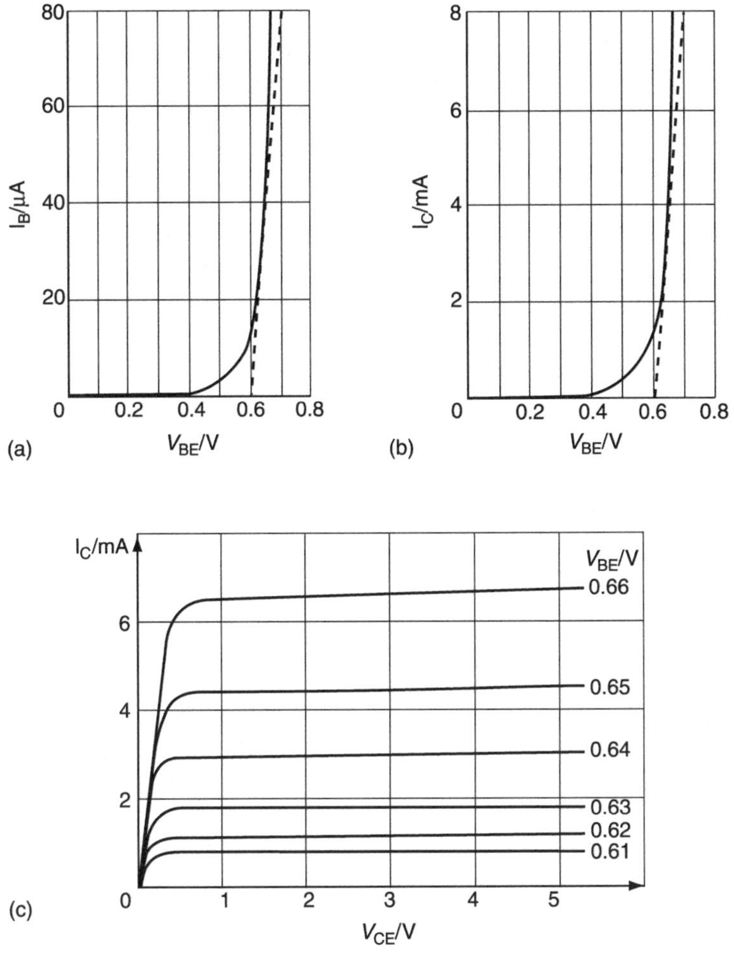

Fig. 9.3 The d.c. characteristics of a bipolar transistor in the common-emitter configuration. (a) The input characteristic. (b) The forward transfer characteristic. (c) The output characteristics for different constant values of V_{BE}.

a much smaller hole current flowing from the base region to the emitter. But notice from Fig. 9.2 that the polarity of V_{CE} is such that the collector-base junction is reverse-biased. This might seem surprising—after all, with a simple pn junction diode, reverse bias means a very small current. However bipolar junction transistors are not that simple! The large emitter current consists of electrons that flow through the forward-biased emitter-base junction, cross the very thin base region, and pass through the reverse-biased collector-base junction. This current flow is controlled not simply by the bias voltages on the junctions, but by the distribution of both types of carrier (electrons and holes) within the base region. The much smaller base current consists mainly of holes flowing from the base to the emitter.

A more complete explanation is beyond the scope of this book, and for now you will just have to accept this description of 'transistor action' as it is. The important point to remember is that small changes to the forward bias voltage on the base-emitter junction cause large changes in the current flowing through the emitter and collector terminals.

Another graph to note is the variation of collector current with V_{CE}, the voltage between collector and emitter, shown in Fig. 9.3c. Each curve corresponds to a different value of input voltage V_{BE}. The curves are not equally spaced, even though the voltage increments which they represent are all the same at 10 mV. This is a consequence of the exponential relationship between V_{BE} and the currents.

For a value of V_{CE} greater than about 0.6 V, the collector current varies only slightly with V_{CE}. We will take account of this slight variation when calculating the small-signal parameters in a later section, but for now we consider the variation to be negligible.

Small signal amplification

Figure 9.3b shows the relationship between the collector current I_C and the input voltage V_{BE}. Suppose that a small change is made to a particular V_{BE}. The corresponding change in I_C is then given by the slope of the curve at that point. For example, the dashed line on the graph is the tangent drawn at a value of $I_C = 2$ mA, and its gradient shows how T_C would change for a small increase or decrease in V_{BE}. If the input change is an a.c. signal that varies V_{BE} by a small amount above and below its d.c. value, the gradient of the line allows us to see what is the subsequent a.c. variation in I_C. It is clear, from the shape of the curve, that the higher the d.c. value of I_C, the bigger the change is in I_C for a given small-signal input voltage. In other words, the ratio of small-signal output to small-signal input, represented by dI_C/dV_{BE}, varies with the d.c. value of current.

This shows that the small-signal characteristics depend on the d.c. **operating point**. We can be more precise about this. Under normal conditions of forward bias, the d.c. equation for a pn junction (Equation 8.1) also applies to the emitter-base junction so, neglecting the –1 under forward bias conditions, and noting that $I_C \approx I_E$,

$$I_C \approx I_E = I_{SE} e^{KV_{BE}} \tag{9.1}$$

where K for a bipolar junction transistor is about 40 V^{-1} at room temperature. So, when the collector voltage is held constant at, say, 5 V, I_C varies with V_{BE} according to the usual pn junction characteristic.

The small-signal value of the increment in I_C for an increment in V_{BE} (at a fixed value of V_{CE}) is defined as the **mutual conductance g_m**:

$$\begin{aligned} g_m &= \frac{dI_C}{dV_{BE}} \\ &= \frac{d(I_{SE}e^{KV_{BE}})}{dV_{BE}} \\ &= KI_{SE}e^{KV_{BE}} \\ &= KI_C \end{aligned} \qquad (9.2)$$

$$\text{Thus } g_m = (40\,\text{V}^{-1}) \times I_C \qquad (9.3)$$

For instance, at a collector current of 1 mA, $g_m = 40\ \text{V}^{-1} \times 1\ \text{mA} = 40\ \text{mS}$, sometimes referred to as 40 mA/V.

Substituting the values of V_{BE} shown in Fig. 9.3c into Equation 9.2 explains why the 10 mV increments in V_{BE} correspond to increases in I_E and I_C by a factor of about 1.5. (That is, $\exp(40\ \text{V}^{-1} \times 0.01\ \text{V}) = e^{0.4} \approx 1.5$.)

Self-assessment question

1 Consider the point $V_{CE} = 5$ V and $V_{BE} = 0.65$ V in Fig. 9.3c. For this point, calculate the value of g_m from the adjacent curves. Compare this with the value obtained from the formula.

9.2.2 A simple amplifier and the hybrid-π equivalent circuit

The d.c. conditions

Figure 9.4 shows a simple common-emitter amplifier circuit using an npn bipolar transistor. Circuit (a) shows just the d.c. components which, together with the transistor, decide the bias voltages and currents or d.c. **operating point**. The resistor R_B provides a path for the base current. The resistor R_C is chosen to drop about half the collector supply voltage V_{CC} at the chosen collector current, so that the collector operating voltage is about $V_{CC}/2$. This

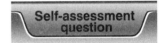

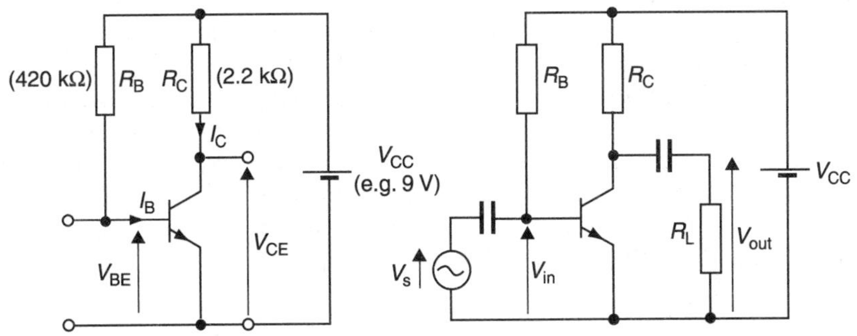

Fig. 9.4 A simple common-emitter amplifier circuit with components appropriate for a transistor whose β is 100, operating at a collector current of 2 mA. (a) The d.c. circuit arrangement. (b) The d.c. circuit coupled via capacitors to the source and load. (The capacitors are to ensure that the average d.c. operating point is not disturbed by the input waveform.)

allows the maximum collector voltage excursion, or 'swing', when an input signal is applied at the base.

In this example, with a 9 V supply, the collector voltage is to be 4.5 V. A collector current of 2 mA is chosen. So the collector resistor must have the value $R_C = 4.5\,\text{V}/2\,\text{mA} \approx 2.2\,\text{k}\Omega$.

The base resistor must drop a voltage $V_{CC} - V_{BE}$ at a base current of $I_B = I_C/\beta$. Clearly, this depends on the value of β. Say we assume a value of 100. Then $I_B = 2\,\text{mA}/100 = 20\,\mu\text{A}$. Assuming $V_{BE} \simeq 0.65$ V, the base resistor is

$$\begin{aligned} R_B &= \frac{V_{CC} - V_{BE}}{I_B} \\ &= \frac{(9 - 0.65)\,\text{V}}{20\,\mu\text{A}} \\ &\approx 420\,\text{k}\Omega \end{aligned}$$

Note that the range of possible values (the 'spread') of β is very wide, even for a given type of transistor. Typically β may be anything from 100 to 400 or more, so this is not a very practical circuit. We use it at this stage because it is relatively easy to analyse.

Note that, as it stands, this simple circuit passes a collector current through the load which is independent of the value of the collector voltage V_{CE} (still assuming negligible slope of the curves in Fig. 9.3c). So I_C does not change if R_C is changed. The limit on this is when $I_C R_C = V_{CC}$, and all supply voltage is 'used up' across R_C, but for resistors of lower value than this, the collector current has a very nearly constant value, determined by the value of V_{BE}. So the transistor acts as a base-voltage-controlled current source. In fact, this circuit is also the basis of circuits designed for the sole purpose of providing a constant current supply.

The a.c. performance

Circuit (b) of Fig. 9.4 is the d.c. circuit coupled at input and output via capacitors. These are chosen so that they have negligible impedance at signal frequencies. The a.c. input signal voltage, represented by V_s, is superimposed on the base bias voltage so that the base-emitter voltage becomes $v_{BE} = V_{BE} + V_s$. So the base-voltage-controlled current source has an a.c. component added. Increases in V_{BE} cause increases in I_C, and hence increases in the voltage drop across R_C and *decreases* in the collector voltage. The collector output voltage becomes $v_{CE} = V_{CC}/2 + AV_s$, where A is the voltage gain of the circuit (which is negative). The output signal waveform is an inverted, amplified, version of the input waveform, superimposed on $V_{CC}/2$.

To calculate the voltage gain, first recall that $g_m = \text{d}I_C/\text{d}V_{BE}$. To a first approximation, ignoring the slopes of the characteristic curves, the collector current I_C is independent of the collector-emitter voltage V_{CE}, and is determined only by the base-emitter voltage V_{BE}. So

$$\text{d}I_C = g_m \text{d}V_{BE}$$

The collector *signal* voltage is

$$\begin{aligned} \text{d}V_{CE} &= -\text{d}I_C R'_L \\ &= -g_m R'_L \text{d}V_{BE} \end{aligned}$$

where R'_L is the effective load resistance to a.c. signals. So the voltage gain is

$$A = \frac{dV_{CE}}{dV_{BE}} = -g_m R'_L \tag{9.4}$$

For example, if $I_C = 2$ mA and the effective load is 1 kΩ then

$$\begin{aligned} g_m &= (40\,\mathrm{V}^{-1}) \times I_C \\ &= (40\,\mathrm{V}^{-1}) \times (2\,\mathrm{mA}) \\ &= 80\,\mathrm{mS} \end{aligned}$$

and the voltage gain is

$$\begin{aligned} A &= -g_m R'_L \\ &= -(80\,\mathrm{mS}) \times (1\,\mathrm{k\Omega}) \\ &= -80 \end{aligned}$$

Note that, because g_m is a linear function of I_C, this *a.c. signal gain* is a linear function of the *d.c. collector operating current* I_C.

Equivalent circuits

Figure 9.5 shows small-signal a.c. equivalent circuits. These represent the action of the transistor and its circuit components to the *signal components* of the voltages and currents in the amplifier. These are shown as a.c. (phasor) small-signal quantities, replacing the incremental quantities used before.

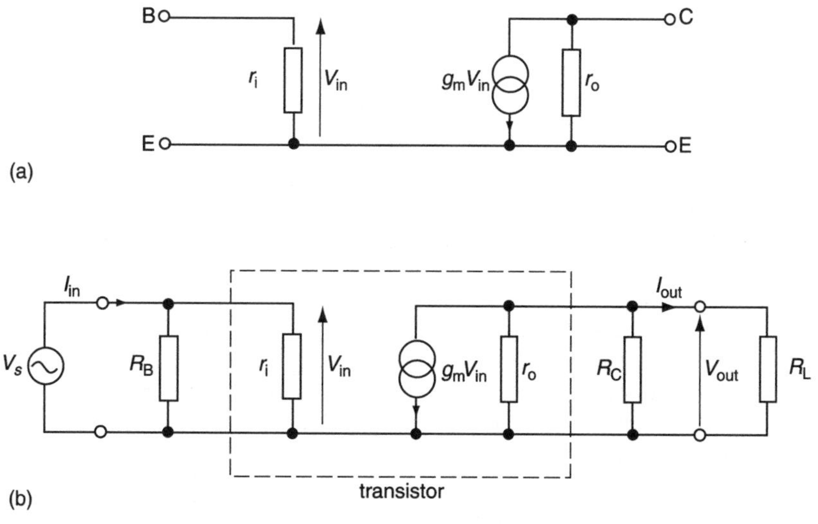

Fig. 9.5 (a) The small-signal equivalent circuit of the transistor in the common-emitter amplifier. (b) The equivalent circuit of the whole amplifier.

Hybrid-π equivalent of the transistor

Figure 9.5a is a small-signal a.c. equivalent circuit of the transistor alone. This is the low-frequency version of the **hybrid-π equivalent**. We will look in turn at the three components:

(i) In the circuit of Fig. 9.4b, the d.c. voltage-controlled current source considered before has an a.c. component added to its input voltage, resulting in an a.c. voltage-controlled component with constant amplitude in its output. In the a.c. equivalent circuit of the transistor (Fig. 9.5a), the d.c. source is ignored, and the current generator $g_m V_{in}$ represents the a.c. component of collector signal current $I_c = g_m V_{in}$, where the a.c. values I_c and V_{in} (with lower-case subscripts) are substituted for dI_C and dV_{BE}. In this case,

$$g_m = (40\,\text{V}^{-1}) \times I_C = 40\,\text{V}^{-1} \times 2\,\text{mA} = 80\,\text{mS}.$$

(ii) The input resistance r_i is

$$\begin{aligned} r_i &= \frac{dV_{BE}}{dI_B} \\ &= \frac{dV_{BE}}{dI_C/\beta} \\ &= \frac{\beta dV_{BE}}{dI_C} \\ &= \frac{\beta}{g_m} \approx \beta r_e \end{aligned} \tag{9.5}$$

where r_e is called the **emitter resistance** and has the value

$$r_e \approx 1/g_m \approx 1/\left(40\,\text{V}^{-1} \times I_C\right) \approx 25\,\text{mV}/I_C \quad \text{at room temperature} \tag{9.6}$$

In this circuit $r_e \approx 25\,\text{mV}/2\,\text{mA} = 12.5\,\Omega$ and

$$r_i \approx \beta r_e = 100 \times 12.5\,\Omega = 1.25\,\text{k}\Omega$$

Self-assessment question

2 Calculate the values of r_e and r_i from the graphs of Figs. 9.3a and b, and compare with the two values above. (Fig. 9.3a is drawn on the assumption that $\beta = 100$.)

(iii) Now consider the resistance r_o. This is the output resistance of the transistor. It represents the effect of the increasing collector current as the collector voltage is increased, shown by the slope of the characteristic curves of Fig. 9.3c. The slope is caused by the **Early effect** (named after J M Early of the Bell Laboratories who first drew attention to it). Figure 9.6 shows curves drawn with an exaggerated slope to show that, when the straighter parts are extended 'backwards', they tend to converge at a point labelled $-VA$ on the voltage axis. VA is the **Early voltage** and is typically between 50 and 200 V. You can see from this figure that the slope of any chosen curve is

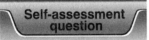

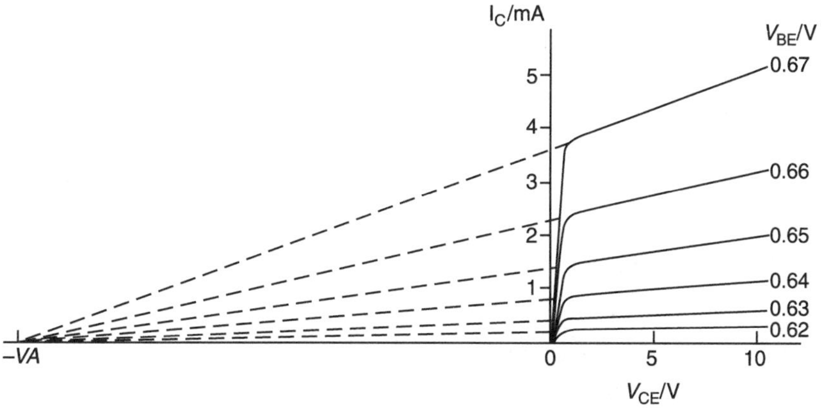

Fig. 9.6 The common-emitter d.c. characteristics drawn with an exaggerated slope to show how, for $V_{CE} > 1$ V, they tend to converge on a particular voltage, $-VA$, where VA is called the Early voltage.

$$g_o = \frac{I_C}{VA + V_{CE}}$$

where $g_o = 1/r_o$ is the output conductance. Thus the output resistance is

$$r_o = \frac{VA + V_{CE}}{I_C} \qquad (9.7)$$

Suppose $VA = 100$ V; then at the collector operating point 4.5 V, 2 mA, the output resistance is

$$r_o = \frac{104.5\,\text{V}}{2\,\text{mA}} \approx 52\,\text{k}\Omega$$

Equivalent circuit and performance of the complete amplifier

Figure 9.5b is the small-signal a.c. equivalent circuit of the complete amplifier of Fig. 9.4b. Fixed d.c. voltages, such as the supply V_{CC}, have zero a.c. voltage. They are at the same a.c. voltage as the emitter connection (commonly taken as earth, ground or 0 V rail), so components connected to them go to the 0 V rail in the equivalent circuit.

At the input, the equivalent circuit includes the signal source and the resistor R_B. In the real circuit, the 'top' end of R_B goes to the supply voltage V_{CC}. This is a fixed voltage, and its a.c. component is zero. So, in the a.c. equivalent circuit, this point is connected to the 0 V rail, and R_B appears in parallel with r_i. The signal source is connected via a capacitor of negligible reactance at signal frequencies, so it too appears in parallel with r_i.

Similarly, the 'top' end of the collector resistor R_C is connected to V_{CC}, so R_C appears in parallel with the current source and r_o in the equivalent circuit. The load resistor is connected via a capacitor of negligible reactance at signal frequencies, so it too appears in parallel with r_o.

We can now find the input resistance and voltage gain of the complete circuit.

The input resistance is simply R_B in parallel with r_i.

In this circuit, R_B = 420 kΩ and r_i = 1.25 kΩ, so the input resistance is still about 1.25 kΩ.

The total load on the output current source is the effective load resistance R'_L, the resistance of r_o, R_C and R_L all in parallel. The corresponding conductance is $g'_L = g_o + g_C + g_L$.

In this circuit, say R_L = 10 kΩ. Then

$$\begin{aligned} g'_L &= g_o + g_C + g_L \\ &= 1/52\,\text{k}\Omega + 1/2.2\,\text{k}\Omega + 1/10\,\text{k}\Omega \\ &= 19\,\mu\text{S} + 454\,\mu\text{S} + 100\,\mu\text{S} \\ &\approx 570\,\mu\text{S} \end{aligned}$$

So $R'_L = 1/570\,\mu\text{S} \approx 1.75\,\text{k}\Omega$, and the voltage gain is

$$\begin{aligned} A &= -g_m R'_L \quad \text{(Equation 9.4)} \\ &\approx -80\,\text{ms} \times 1.75\,\text{k}\Omega \approx -140 \end{aligned}$$

Note that, with collector resistors and load resistors of a few kilohms, the effect of the transistor output resistance (or conductance) is quite small, and can be ignored for a good approximation to the gain.

Self-assessment question

3 Suppose that the load resistance in Fig. 9.4b is 8 kΩ, and that the input is driven by a source whose open-circuit voltage amplitude V_s is 5 mV and whose internal resistance is 1 kΩ. Assume that the coupling capacitors have negligible reactances. Draw the small-signal equivalent circuit of this arrangement and calculate the amplitude of the output signal across the load. (First calculate V_{in} by regarding the input circuit as a potential divider driven by V_s. Then calculate the voltage gain V_{out}/V_{in}.)

9.2.3 A practical, discrete common-emitter amplifier

In this subsection the performance of a common-emitter circuit with a better d.c. design than the circuit of Fig. 9.4a is considered.

Self-assessment question

4 Explain why the circuit of Fig. 9.4a is in general unsatisfactory. (Consider the operating point of the circuit when $\beta = 200$ or more, instead of 100.)

Figure 9.7 shows the usual way of reducing the variation of operating point, due to differences in the value of β, in a discrete common-emitter amplifier.

The base resistor R_B of Fig. 9.4a is replaced by resistors R_1 and R_2, which form a potential divider and set the *voltage* level of the base terminal. The emitter current is then determined mainly by the emitter resistor R_E.

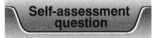

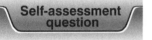

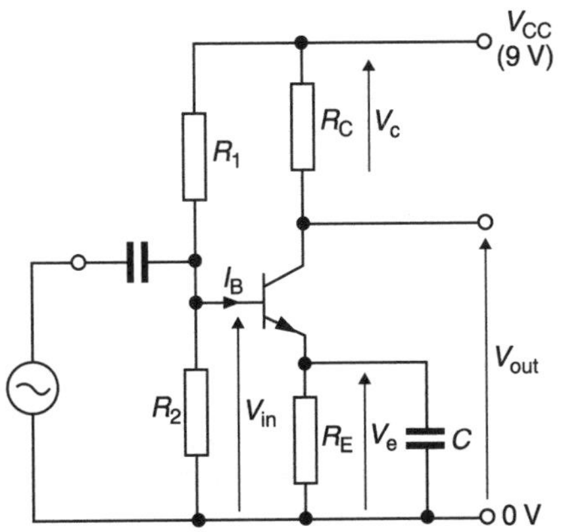

Fig. 9.7 A common-emitter amplifier in which the operating point of the transistor is well defined despite variations of β. The base *voltage* is defined by R_1 and R_2 so that the emitter current is set primarily by the resistor R_E. The capacitor in parallel with R_E restores the small-signal gain without upsetting the d.c. operating point.

For example, if you want to design a circuit in which the emitter operating current is 2 mA, you can choose R_E to be 500 Ω to produce a voltage drop across it of 1 V. This is much greater than the likely variations in V_{BE}, so it fixes the emitter current quite accurately for a given value of base voltage. Since $V_{BE} \approx 0.65$ V, it follows that the base terminal should be held at a voltage of about 1.65 V. With $V_{CC} = 9$ V, and if I_B is small compared with the current flowing in R_1 and R_2 for all likely values of β, this can be achieved with the potential divider by making the ratio $R_2/R_1 = 1.65/7.35$ (e.g. $R_1 = 1.65$ kΩ and $R_2 = 7.35$ kΩ or $R_1 = 16.5$ kΩ and $R_2 = 73.5$ kΩ). Then I_E and I_C will not deviate much from 2 mA, even for extreme β values of 100 or 400.

The gain of the circuit of Fig. 9.7

If the emitter capacitor C is omitted from the circuit, we can estimate the gain approximately as follows, assuming that g_o of the transistor is negligible compared with g_c. The signal currents through R_C and R_E are very nearly the same ($I_c = \alpha I_e$), so the voltage drops, V_{out} and V_e, across them are in proportion to their resistances. That is $V_{out} / R_C = -V_e / R_E$ (the minus sign indicating the 180° phase difference). And since there is relatively little *signal* voltage across the base-emitter junction, $V_e \approx V_{in}$. Therefore

$$\frac{V_{out}}{R_C} \approx \frac{-V_{in}}{R_E} \text{ or } A_V \approx -\frac{R_C}{R_E} \text{ without the emitter capacitor} \qquad (9.8)$$

Worked example

If $I_C = 2$ mA and if R_C and R_E are respectively 2.2 kΩ and 500 Ω, what is the voltage gain of the circuit?

From the above equation, $A_v \approx -2.2\ \text{k}\Omega/500\ \Omega = -4.4$.

A more accurate analysis (which we do not have space for) shows that

$$A_V \approx -\frac{R_C}{R_E + r_e} \quad \text{without the emitter capacitor}$$

Evidently the inclusion of R_E in the circuit (and omitting capacitor C) reduces the voltage gain considerably. But it also defines the gain quite accurately, which may be a more useful factor than a high gain in many circumstances.

Self-assessment question

5 That the gain of the circuit of Fig. 9.7 is the ratio of two resistances, and that variations of β have little effect on the gain, suggest that the circuit is a negative feedback one. Explain the operation of this circuit from a feedback point of view. Is it voltage- or current-derived, and is the feedback shunt- or series-connected?

The emitter capacitor C in this circuit is called a **bypass capacitor**. Its purpose is to bypass, or short-circuit, the emitter resistor at signal frequencies. If its reactance is much lower than R_E, R_E has little effect. But its reactance must be lower than r_e too, at the lowest signal frequency. When this is done, the emitter terminal is simply connected to the common rail in the a.c. equivalent circuit, and the voltage gain becomes

$$\begin{aligned} A_V &\approx -\frac{R_C}{r_e} \\ &\approx -g_m R_C \quad \text{with the emitter capacitor connected.} \end{aligned} \tag{9.9}$$

9.2.4 The long-tailed pair (differential pair)

The circuit of a long-tailed pair is shown in Fig. 9.8. This is a simple form of the type of bipolar differential amplifier commonly used at the input of operational amplifiers. The long-tailed pair is one of the most effective ways of establishing the operating point of an amplifier. A constant current source establishes the d.c. current for the two transistors, whose emitters are connected together.

Note that the circuit uses a three-rail supply. The negative rail is included so that the circuit will work correctly when the input signal voltage swings both negative and positive about 0 V.

The signal voltage is applied between the base terminals of the two transistors, and is therefore a *differential* input voltage as in an op amp. When the differential input voltage is zero, the two transistors each take half of the constant current I_S which is established by the current source. But when a small differential voltage v_{in} is applied across the input terminals, the current through one transistor increases and the current through the other decreases, leaving the total current unaltered. The change in current through T2 causes a change in the voltage drop across R_{C2}, thus producing an output signal voltage V_{out}, and if V_{out} is greater than V_{in}, the circuit is a voltage amplifier.

The circuit has a number of advantages over the common-emitter amplifier. Firstly, since it has a differential input, it is not essential for one side of the signal source to be connected to the common (zero voltage) line. Secondly, the d.c. voltage levels of the two inputs can vary *together*, without affecting

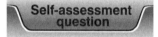

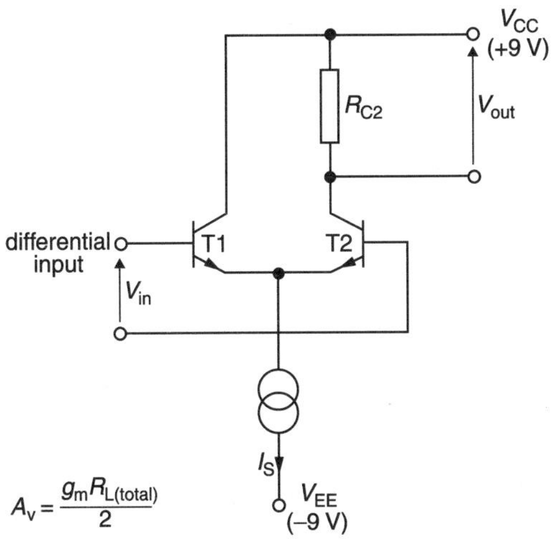

Fig. 9.8 The long-tailed pair: the basic circuit with the emitters of two transistors connected to a single current source.

the output very much. That is, a coupling capacitor is not needed to isolate changing or different *common* d.c. levels associated with the input signal, as in the common-emitter amplifier. Thirdly, variations in temperature do not affect the circuit much because they affect both transistors equally.

The d.c. design

The d.c. design is very straightforward. The operating current of each transistor is half the current established by the current source, so the collector voltage of T2 is clearly $(V_{CC} - I_S R_{C2}/2)$. To take advantage of the fact that the common d.c. voltage level of the input terminals can vary a good deal without affecting the circuit performance very much, the collector operating voltages must be chosen to allow for such variations.

Worked example

If the voltage drop across R_{C2} in Fig. 9.8 is designed to be 2 V, and if the current source requires at least 1 V across it to work properly, what is the possible range of d.c. levels that the inputs can be when $V_{in} = 0$?

Transistor T2 will still work satisfactorily with small signals even if $V_{CB} = 0$; so the input voltage can rise to +7 V (9 V minus the voltage drop across R_{C2}).
For 1 V across the current source, the d.c. input voltage must be about 0.65 V more positive than this in order to provide the base-emitter voltage for T1 and T2. So the lowest possible input voltage is about –7.3 V.
If a 1 V signal amplitude is to be allowed for, the range of allowable common d.c. input voltage levels is about ±6 V.

The differential gain

With a differential input, each transistor receives half of the input voltage V_{in} because, for small signals, the voltage of the two emitters remains midway between the voltages of the two input terminals. So each transistor acts as a common-emitter amplifier, but with half the input voltage. Thus the differential gain is

$$A_{v(diff)} = \frac{V_{out}}{V_{in}} \approx \frac{g_m R_{L(total)}}{2}$$

where $R_{L(total)}$ is the net load on T2 and includes its own output resistance. Note that the output of T2 is not inverted with respect to the input.

Common-mode

An input voltage applied to both inputs simultaneously is called a *common-mode input*. Such an input can occur if, for instance, the two wires carrying the differential signal pick up interference from a stray magnetic field, and the same voltage is induced in both wires. Ideally, a differential amplifier should reject such common-mode signals, and amplify only differential inputs.

Figure 9.9 shows a simpler long-tailed pair, using a resistor R_E in place of the current source, which exhibits poor **common-mode rejection**. If a common-mode signal is fed to the inputs, the circuit acts as two common-emitter amplifiers in parallel, but with their collector signal currents half of the total emitter signal current. The voltage gain of a single common-emitter circuit with an emitter resistor is $A_V \approx -R_C/R_E$ (Equation 9.8). So, in this long-tailed pair, we have a **common-mode gain**:

$$A_{v(CM)} \approx -\frac{R_{C2}}{2R_E}$$

The **common-mode rejection ratio** is

$$\text{CMRR} = \frac{|\text{differential gain}|}{|\text{common-mode gain}|} \approx \frac{g_m R_{C2}}{2} \Big/ \frac{R_{C2}}{2R_E} = g_m R_E \approx \frac{R_E}{r_e}$$

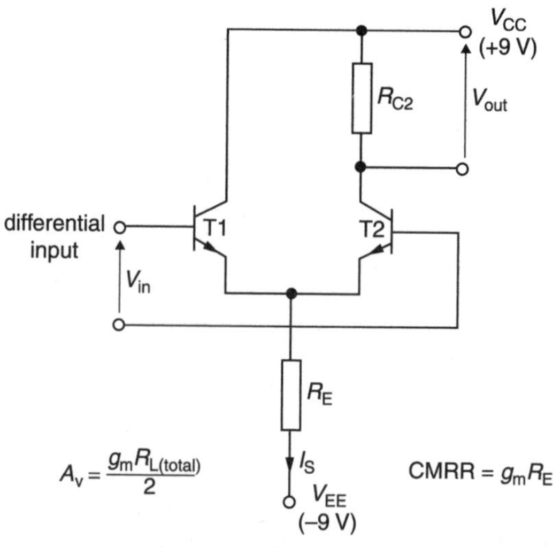

Fig. 9.9 The original form of the long-tailed pair with a resistor R_E in place of the current source.

The use of a current source in place of R_E, as in Fig. 9.8, greatly improves the CMRR because the output resistance of the current source (typically about the same value as r_o of T2) is much greater than R_E. We then have

$$\text{CMRR} = g_m \times (r_{out} \text{ of current source}) \approx \frac{r_{out} \text{ of current source}}{r_e}$$

A good CMRR also diminishes the effect of temperature changes. The base-emitter voltage at a given current decreases by about 2 mV per degree Celsius (or Kelvin). Thus a 1°C temperature rise of T1 and T2 is equivalent to a 2 mV *common-mode* input, and it is desirable that this should not affect the output substantially. The CMRR is a measure of how well the amplifier achieves this rejection of common-mode signals. A CMRR of better than 10^5, or 100 dB, is typical for an op amp.

9.2.5 The emitter-follower

The basic emitter-follower circuit is shown in Fig. 9.10a. Figure 9.10b shows it coupled by capacitors to a voltage source and to a load R_L. The output is taken across the emitter resistor. The circuit has a voltage gain of very nearly unity ($A_v \approx 1$) but it has a large input resistance and a small output resistance so it

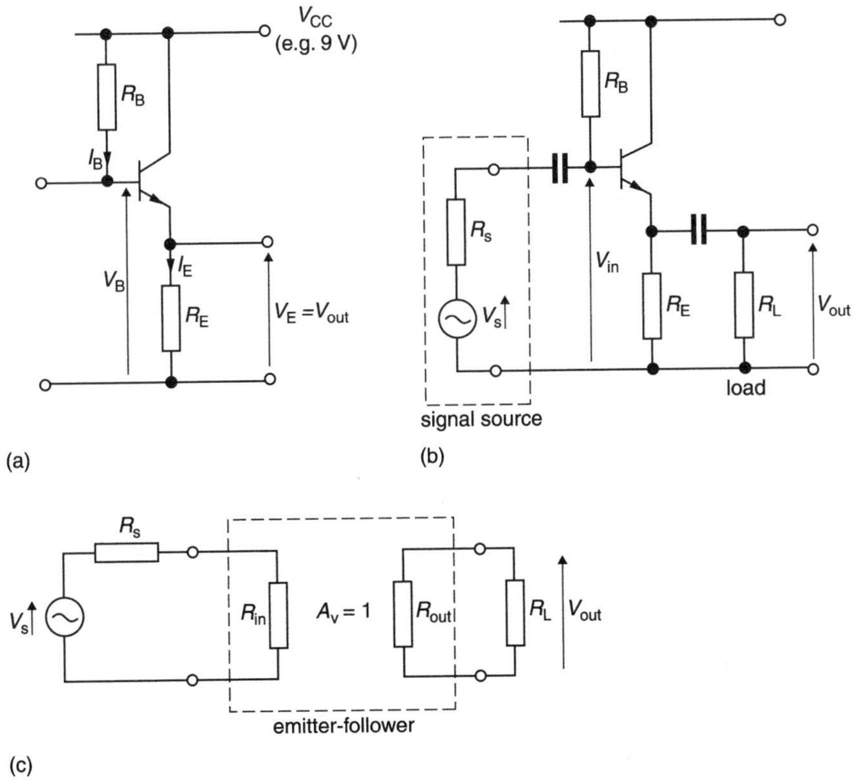

Fig. 9.10 The emitter-follower. (a) The basic circuit. (b) An a.c. capacitor-coupled circuit. (c) A system diagram of a 'buffer'.

can be used as a 'buffer', so-called because it buffers a high output-resistance source against the demands of a low-resistance load. Figure 9.10c shows the equivalent circuit of a buffer. Because it has a high input resistance and a small output resistance, and $A_v \approx 1$, the emitter-follower buffer can transfer almost the whole of V_s to a load even when the output resistance of the source is much bigger than the load resistance.

D.C. conditions

In the circuit of Fig. 9.10a the base current needed to establish the operating point is supplied through R_B, and is dependent on the value of β. A better-defined operating point is obtained by using a potential divider to set the base voltage, as in the common-emitter circuit of Fig. 9.7.

Worked example

Given that R_E in Fig. 9.10a is 1 kΩ and that the typical value of β of the transistor used is 200, what is a suitable value of R_B to give a d.c. emitter voltage V_{OUT} of about half the supply voltage?

If the d.c. emitter voltage is to be 4 V it follows that I_E must be 4 mA, and the d.c. voltage of the base terminal is about 4.65 V (since $V_{BE} = 0.65$ V). Therefore the voltage drop across R_B is 4.35 V. Now I_B must be

$$I_B \approx \frac{I_E}{\beta} = \frac{4\,mA}{200} = 20\,\mu A$$

so $R_B = 4.35\,V/20\,\mu A \approx 220\,k\Omega$

The single base resistor can be satisfactory in this circuit, because of the d.c. negative feedback from the emitter to the base. For example, a higher β value than 200 will produce a larger emitter current and therefore a larger voltage drop across R_E, thus reducing I_B.

Input and output resistances

We can find close approximations to the a.c. input and output resistances of the emitter-follower (R_{in} and R_{out} of Fig. 9.10c) as follows.

First, we find the Thévenin output resistance. When driven from a source with zero source resistance, the circuit has an a.c. open-circuit output voltage of $V_{oc} = V_{in}$. The short-circuit output current is found by shorting the signal current at the emitter terminal to 0 V. The a.c. emitter current is then $I_{sc} \approx g_m V_{in}$. Thus

$$R_{out} = V_{oc}/I_{sc} \approx 1/g_m = r_e$$

A more exact analysis shows that, with a finite input source resistance R_s and with base resistor R_B, the output resistance becomes

$$R_{out} \approx r_e + (R_s /\!/ R_B)/\beta$$

To find the input resistance, recall that the input resistance of a common-emitter amplifier is $r_i \approx \beta r_e$. In the emitter-follower, the emitter resistor and the load are in parallel at signal frequencies, between the emitter and 0 V, with an effective resistance $R_{L(eff)}$. This resistance is effectively in series with r_e, so the input resistance becomes

$$R_{in} \approx \beta(r_e + R_{L(eff)})$$

Clearly, the load can considerably affect the input resistance, just as the source can considerably affect the circuit's output resistance.

9.3 FIELD-EFFECT TRANSISTORS (FETs)

Field-effect transistors are three-terminal devices with terminals called **source, drain** and **gate**. There are two types, the **junction FET**, or **JFET**, and the **metal-oxide-silicon FET**, or **MOSFET**. MOSFETs are sometimes referred to as 'insulated-gate field-effect transistors' or **IGFETs**.

Both types of FET can be made in either p-channel form or n-channel form. In **p-channel FETs** the current is carried by holes, whilst in **n-channel FETs** it is carried by electrons.

An n-channel JFET is illustrated in Fig. 9.11a, and an n-channel MOSFET in Fig. 9.11b.

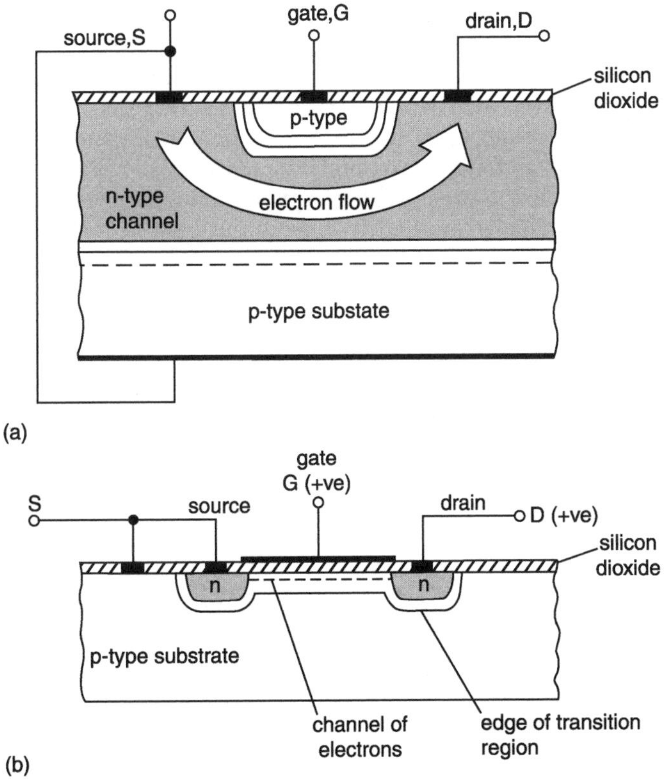

Fig. 9.11 Cross-sections of field-effect transistors. (a) An n-channel JFET. (b) An n-channel MOSFET in which the channel consists of electrons induced in the transition region which is formed just under the gate.

9.3.1 The operation of JFETs

Figure 9.11a shows the cross-section of a JFET. The pn junction transition region is shown in white, just below the p-type material under the **gate** terminal. The electron current flows through a channel of silicon whose cross-sectional area is controlled by the width of the pn junction transition region, which intrudes into the channel, as illustrated in the figure. The two ends of the channel are called **source** and **drain**. The application of a reverse bias between gate and source causes the transition region of the gate to widen, and so reduce the width of the channel through which the current flows. In this way the applied gate-source voltage V_{GS} can be used to control the source-drain current I_D. Figure 9.12 shows a family of characteristic output curves, each plotted for a fixed value of gate-source voltage V_{GS}.

The gate of a JFET is a pn junction which must be reverse-biased. Thus the gate-source voltage V_{GS} of the n-channel type must be held at zero or a negative voltage. Suppose, for example, V_{GS} is held at –5 V. This does not close the channel when the drain-source voltage V_{DS} is low but, as V_{DS} increases, the voltage V_{DG} across the gate pn junction increases, causing the transition region of the gate pn junction to widen at the drain end of the channel. The gate-source voltage which completely closes the channel is called the **pinch-off voltage** V_p, at which no drain current flows whatever the value of the drain-source voltage V_{DS}. But, if V_{DS} is insufficient to cut-off I_D, then I_D increases as V_{DS} is increased, as shown by the $V_{GS} = -5$ V line in Fig. 9.12. The region of operation up to pinch-off, where the V_{DS} versus I_D characteristics are curved, is called the 'linear region'. Beyond this is the 'saturation region', in which the current increases much more slowly than in the linear region. Here the channel gets shorter as the gate transition region, at the drain end, becomes wider due to the increased reverse voltage across it. The current is determined by the channel dimensions and the electric field.

As with the bipolar transistor, the mutual conductance is the change in output current per change in input voltage; in this case

$$g_m = \frac{dI_D}{dV_{GS}}$$

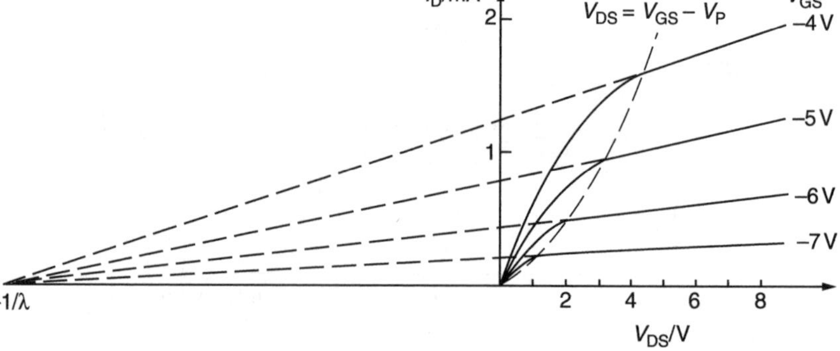

Fig. 9.12 A family of n-channel JFET characteristics, showing their tendency to converge at a voltage of $-1/\lambda$ on the V_{DS} axis ($V_P = -8$ V).

In the case of the FET, however, this turns out to be proportional to the *square root* of the drain operating current, and depends too on the structure of the device.

The output conductance is the slope of the output curve, or

$$g_o = \frac{dI_D}{dV_{DS}}$$

The saturation-region curves of Fig. 9.12 are projected back onto the V_{DS} axis, and meet at a voltage of $-1/\lambda$, where λ is known as the *channel length modulation factor.* The voltage $1/\lambda$ is analogous to the Early voltage (VA) of the bipolar; if the value of $1/\lambda$ is known, the output resistance can be calculated from an analogous expression:

$$r_o = \frac{V_{DS} + 1/\lambda}{I_D}$$

9.3.2 The operation of MOSFETs

n-channel MOSFETs

In a MOSFET the drain and source are pn junctions formed side by side in the surface of a silicon substrate, as illustrated in Fig. 9.11b. This time the gate is a conductor, originally a metal film (hence the name of the device), but nowadays it is usually a layer of well-doped silicon. This gate electrode is separated from the silicon substrate by a film of oxide, thus forming an input capacitance. The application of sufficient voltage between gate and source induces carriers in the silicon under the gate, as indicated in the diagram. The amount of charge induced in the channel is dependent on the gate voltage. When a gate-source voltage is applied, these induced carriers flow between source and drain; the larger the induced charge the greater the drain current I_D. Thus again, but owing to a quite different kind of interaction, the gate voltage V_{GS} can be used to control the source-drain current I_D. Note that in JFETs the silicon material through which the drain current flows is called the channel, but in MOSFETs it is the induced carriers, not the material in which the carriers are induced, which constitute the channel. As in the case of JFETs, MOSFETs are made in both n-channel and p-channel versions.

Figure 9.11b is of an n-channel type, where the drain current is carried by electrons. The current I_D does not simply start to flow as soon as any positive voltage is applied to the gate, but the gate voltage must exceed the threshold voltage V_T of the device before current can flow. The threshold voltage can be adjusted during manufacture to any value between a few volts negative to a few volts positive, as we shall explain.

- If the threshold voltage is positive (in an n-channel device) the device is called an **enhancement-mode MOSFET**. No drain current flows until V_{GS} exceeds the threshold voltage, so $I_D = 0$ when $V_{GS} < V_T$.
- If the threshold voltage is negative (in an n-channel device) the device is called a **depletion-mode MOSFET**, and source-drain current can flow when $V_{GS} = 0$. The gate voltage must be made more negative than V_T to stop the current flowing.

Enhancement-mode n-channel MOSFETs

We begin by discussing enhancement-mode n-channel MOSFETs with a positive threshold voltage of a volt or two, with the substrate connected to the source, as in Fig. 9.11b. This gives a zero-voltage bias to the source-substrate pn junction, so that no current normally flows through it. Also, an input V_{GS} between gate and source appears too between gate and substrate.

Since the substrate is p-type material, the carriers initially present under the gate are mainly holes. A positive gate voltage creates an electric field in the substrate which can either repel holes or attract electrons if there are any available. At gate voltages less than V_T, the electric field is only sufficient to repel holes; a region depleted of carriers is formed, and no source-drain current flows. If V_{GS} is increased to exceed V_T, the layer just under the gate becomes completely depleted of holes, and electrons can be drawn in from the source to form the channel. With a positive drain-source voltage, electron current flows through the channel to the drain.

Figure 9.13b shows typical output characteristics of an enhancement-type n-channel MOSFET.

As V_{DS} is increased from low values, the current increases too. However, as V_{DS} is increased further, the voltage between drain and gate *decreases*, so the electron density at the drain end of the channel decreases. Thus, as V_{DS} is increased, the drain current increases, but does not rise in proportion to V_{DS}, and tends to level out.

The dashed line joins the points on the curves where they start to level out, called the **pinch-off** points. This occurs when the voltage difference between the gate and drain equals the threshold voltage, so that at the drain end of the channel there are no induced electrons. That is, $V_{GD} = V_T$, or $V_{GS} - V_{DS} = V_T$, so $V_{DS} = V_{GS} - V_T$.

As V_{DS} is increased beyond pinch-off the channel gets slightly shorter, due to the widening of the drain transition region, leaving the voltage drop along the channel at the value $V_{GS} - V_T$. But, because the channel shortens as V_{DS} increases, the drain current increases slightly. This is called the **saturation region** of operation.

As in the JFET case, these essentially straight parts of the curves can be extended back to the V_{DS} axis to the point $(-1/\lambda)$, where λ is the length modulation factor. The slope of each line is the output conductance for that value of V_{GS}:

$$g_o = \frac{I_D}{V_{DS} + 1/\lambda} \quad \text{(saturation region)}$$

Figure 9.13 shows the circuit symbols and typical characteristics of all six types of FET: p-channel and n-channel JFETs; and p-channel and n-channel MOSFETs of enhancement and depletion types. The types not covered so far are described below.

Depletion-type n-channel MOSFETs

In a depletion-type n-channel MOSFET, the threshold voltage is made negative by implanting a very thin layer of donors in the surface of the p-type substrate just under the gate. The implanted donor density exceeds the density

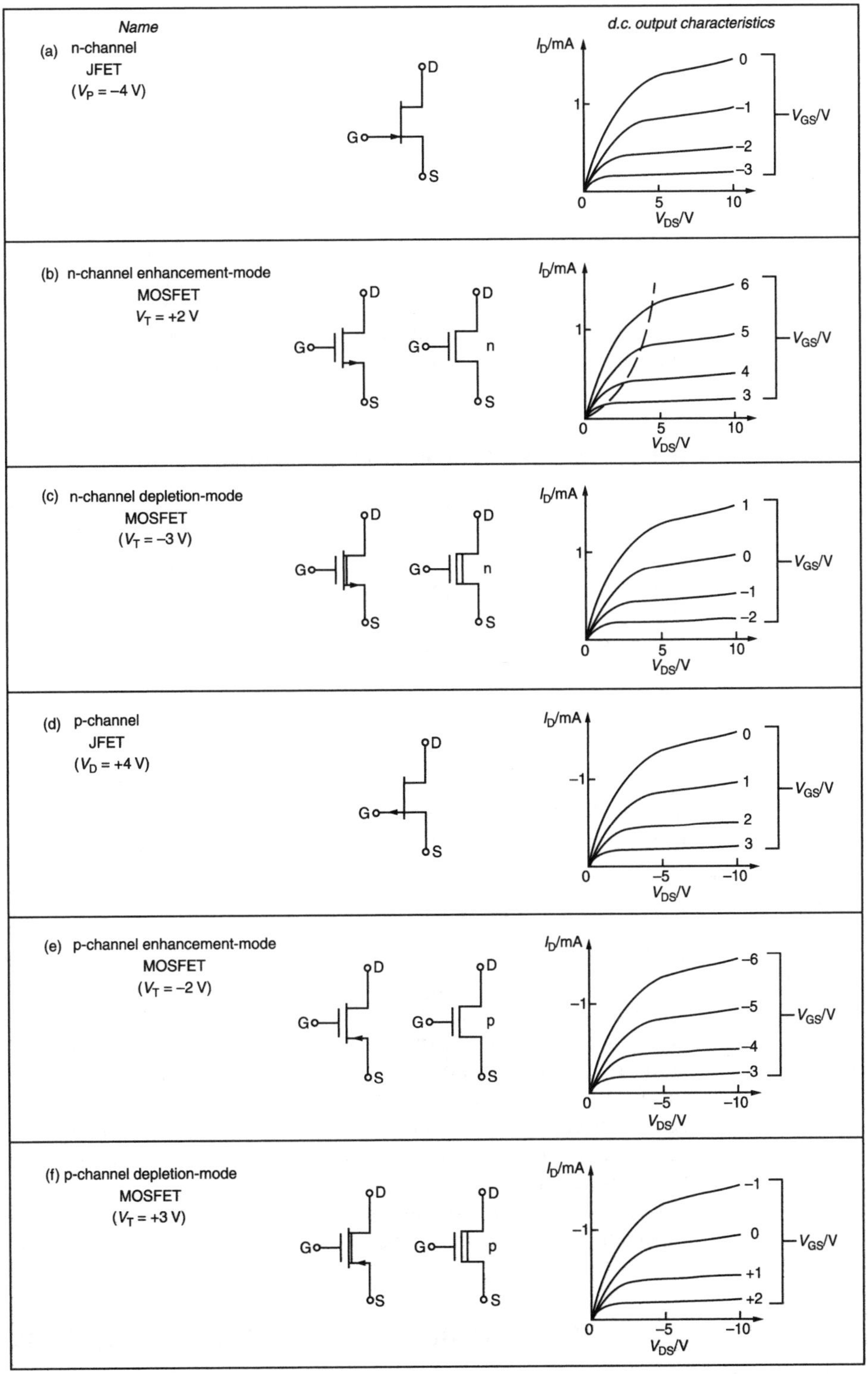

Fig. 9.13 A summary of FET characteristics, showing, for each type, (i) its name and pinch-off or threshold voltage, (ii) its graphical symbol or icon, and (iii) the form of a typical set of d.c. characteristics.

of holes already there, so that a channel of electrons is formed even when $V_{GS} = 0$, and drain current will flow as soon as V_{DS} is applied. A negative gate voltage (e.g. $V_{GS} < -3$ V) has to be applied to the gate to drive these electrons away and reduce I_D to zero.

p-channel MOSFETs

The operation of p-channel MOSFETs is identical to that of n-channel MOSFETs except that all region types, carrier types, voltages and currents are reversed. For example, the gate is made increasingly *negative* to cause increasing drain current, and the threshold voltage of an enhancement-mode p-channel device is negative. P-channel MOSFETs are inherently slower than n-channel ones because holes move through silicon more slowly than electrons do.

9.3.3 MOSFETs compared with bipolars

The main electrical advantage of MOSFETs over bipolars is their zero low-frequency gate current, because of the insulating oxide layer between gate electrode and substrate. Thus the low-frequency input power can be very low. However, at higher frequencies considerable input current flows into the gate-substrate *capacitance*, and this current has to be supplied from the previous stages, increasing the overall power dissipation in, for example, high-speed digital circuits.

The advantage of MOSFETs from a production point of view is that, in general, they are smaller and cheaper to manufacture than bipolars.

Their main disadvantage is that the control of output current by the input current is less effective, so that the mutual conductance and, hence, the voltage gain available from MOSFET analogue amplifiers is less than that from bipolar transistor amplifiers operating under similar conditions.

9.3.4 FET circuits

FETs, like bipolars, are used for a wide variety of analogue and digital circuits. FET analogue circuits have close similarity with bipolar ones, differing mainly in their gate bias arrangements. Their small-signal hybrid-π equivalent circuit is very similar to that of the bipolar transistor, with of course the parameters associated with the bipolar's emitter, base and collector replaced by those for the FET's source, gate and drain. One simplification is that the FET's input (gate-source) resistance is normally so high that it can be ignored, although its input *capacitance* is similar to that of the BJT, and is significant in high-frequency circuits.

By far the most popular use of FETs must be as MOSFETs in digital circuits, especially of computers, and some of these circuits are described in Section 9.5.

9.4 BIPOLAR AND FET PUSH-PULL OUTPUT STAGES

The output stages of analogue amplifiers are commonly designed to feed loads of relatively low resistance, such as 8 Ω loudspeakers for audio amplifiers and 75 Ω cables for video distribution amplifiers. The output stages of logic gates are designed to feed the inputs of several other logic gates, which present a significant capacitive load at high switching speeds. In both these cases, output stages are needed which have low output resistance (and impedance) compared with the load impedance, and are capable of providing relatively high voltages and/or output currents. The common-emitter or common-source type of amplifier (such as that of Fig. 9.4b) is quite unsuitable for this purpose, and two devices in a push-pull configuration are usually used.

9.4.1 A double emitter-follower

The emitter-follower of Fig. 9.10 has the required low output resistance, but cannot meet the current requirements. On large positive signal excursions, the base voltage rises, the emitter voltage follows it, and increased emitter current flows into the load. However, large negative signals tend to reverse-bias the base-emitter junction while the output coupling capacitor is discharging though R_E and R_L. In other words, this circuit is suitable only for small signals and relatively large load impedances.

The solution is to use two emitter-followers, as in Fig. 9.14. This circuit dispenses with the output coupling capacitor and uses two supply rails.

On positive signal excursions, the npn transistor's emitter voltage follows its base, and drives ('sources') current from the positive supply into the load. Meanwhile the pnp transistor is cut-off by the positive voltage across its base-emitter junction.

On negative excursions, the pnp transistor's emitter follows its base and pulls ('sinks') current to the negative supply from the load. Meanwhile the npn transistor is cut-off by the negative voltage across its base-emitter junction.

This type of circuit configuration with two devices, alternately 'pushing' current into the load and 'pulling' it out, is called *push-pull.*

The simple circuit of Fig. 9.14 has one serious drawback if used as an analogue power amplifier, as the output waveform shows. Because of the base-emitter bias voltage needed for conduction in a bipolar transistor, the input voltage has to reach ± 0.65 V or so before any output current flows. This problem is solved by biasing the bases, commonly to the extent that a small operating or 'quiescent' current flows through the two transistors under no-signal conditions.

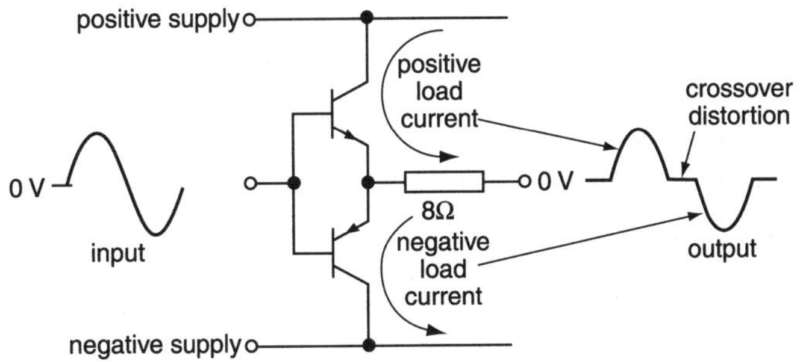

Fig. 9.14 A double emitter-folllower: two emitter-followers in push-pull.

Figure 9.15 shows one way of doing this. Here the resistors R_A and R_B and the diodes form a potential divider which sets the base voltages. The diodes are used to provide temperature compensation. As the output transistors heat up under load, the temperature rise causes a fall in their base-emitter voltages needed for the required quiescent current. This potentially leads to increased collector currents and power dissipation, and so-called 'thermal run-away'. However, as the temperature rises, the diodes' junction voltages fall, by a similar amount to that of the base-emitter voltages, and the *applied* base-emitter voltages are reduced to the value needed to keep the quiescent current at the right value. For this to work correctly, the diodes should be in good thermal contact with the transistors' heat sink, so that their temperatures track those of the transistors.

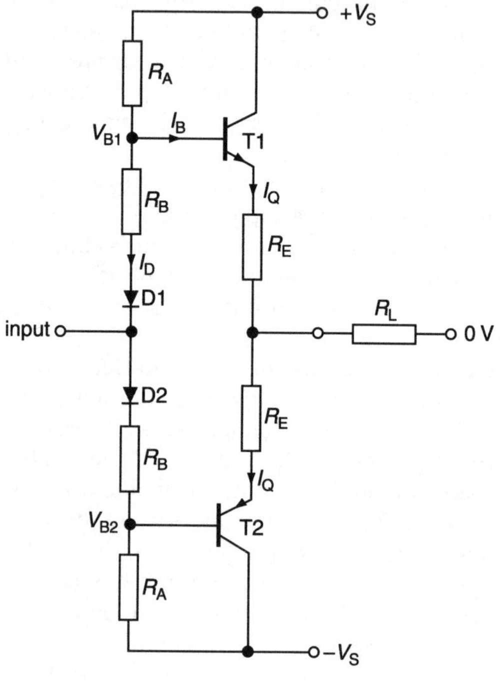

Fig. 9.15 The biassed double emitter-follower.

The emitter resistors are included to help compensate for production spreads in the values of the diode and base-emitter junction values, typically ± 50 mV. If the spreads tend to cause too much quiescent current, increased voltage drop across these resistors reduces the applied V_{BE} and 'throttles back' the quiescent current. These resistors also have an effect on the distortion of the circuit; there is an optimum value for a given value of quiescent current. Analysis of this is beyond the scope of this book.

9.4.2 A double source follower

Figure 9.16 shows the circuit of a MOSFET double source follower, using enhancement-mode devices, for analogue power output. The majority of power FETs are enhancement-mode MOSFETs, available in both n-channel and p-channel configurations.

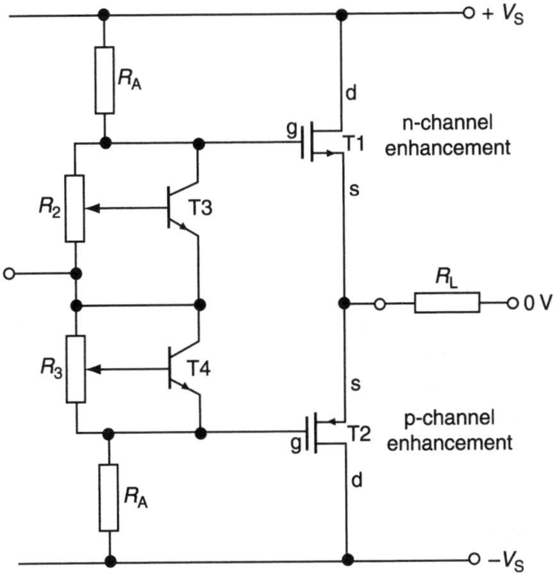

Fig. 9.16 A complementary-symmetry MOSFET double source-follower.

The principle of operation is just the same as that of the bipolar circuit: the n-channel transistor sources current to the load on positive signal excursions and the p-channel transistor sinks current from the load on negative signal excursions.

The bias arrangement is different, mainly because the enhancement-mode transistors need 3 to 4 V bias, rather than about 0.65 V. The two bipolars T3 and T4, together with their associated resistors R_2 and R_3, form 'V_{BE} multipliers' and play a similar role to that of the diodes in the bipolar circuit. They *act* as diodes with a junction voltage of a few volts, which can be set by adjusting R_2 and R_3. Their negative voltage–temperature coefficient compensates for the negative voltage–temperature coefficient of the MOSFETs of about 0.2% per °C (or K).

The V_{BE} multiplier

Figure 9.17 shows the circuit of a V_{BE} multiplier. The adjustment 'pot' is represented by the two resistors R_a and R_b. The values of these are chosen so that the transistor base current I_B is much less than the current I_R through R_a. So the current through R_b is approximately I_R also. The potential divider gives

$$V_{BE}/V_{CE} = R_b/(R_a + R_b)$$

or

$$V_{CE} = V_{BE}(R_a + R_b)/R_b$$

Thus the voltage across the circuit is V_{BE} multiplied by the factor $(R_a + R_b)/R_b$. We know that V_{BE} is about 650 mV, whatever the value of the transistor current $I_C \approx I_E$, so the V_{BE} multiplier acts like a forward-biased diode

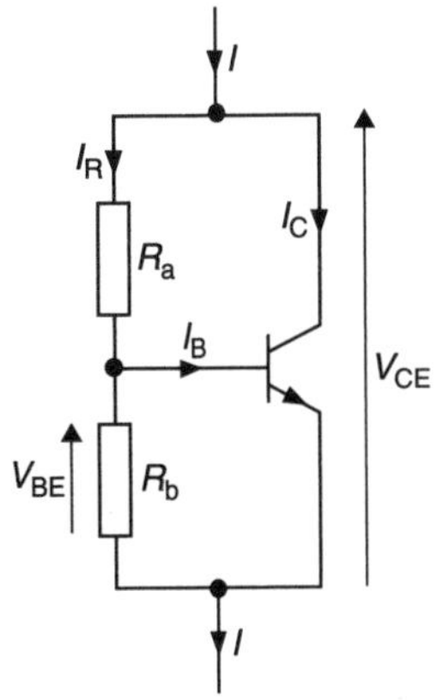

Fig. 9.17 A V_{BE} multiplier.

of the double emitter-follower circuit, maintaining a nearly constant voltage drop.

As in the case of the diodes of the double emitter-follower circuit, the V_{BE} multiplier transistors should be mounted in close thermal contact with the output transistors' heat sink, to ensure that their temperatures track. The base-emitter junction voltage V_{BE} is temperature dependent. It falls by about 2 mV per °C or K. This represents a fractional change of $(-2\text{ mV}/650\text{ mV})\text{K}^{-1}$, or $-0.3\%\text{ K}^{-1}$. In the V_{BE} multiplier, V_{CE} is a linear multiple of V_{BE}, so it too will change by $-0.3\%\text{ K}^{-1}$.

Thus the V_{BE} multipliers will overcompensate slightly for the $-0.2\%\text{ K}^{-1}$ temperature coefficients of the MOSFET gate-source voltages. This may be an advantage: the junction temperature of the V_{BE} multiplier transistors will not rise quite as much as the temperature of the MOSFETs as the MOSFETs heat up due to power dissipation, even when they are all mounted on the same heat sink.

CMOS logic gates

The output stages of CMOS logic circuits also use two MOSFETs in a push-pull configuration, known as 'complementary symmetry', with an n-channel device and a p-channel device. This is explained in the next section.

9.5 BASIC SWITCHING AND DIGITAL CIRCUITS

9.5.1 The transistor as a switch

In addition to being used as an amplifier, a transistor can also be used as a switch. An electrical switch, such as a domestic light switch, has two possible states; it is either *off*, and passes no current (for all practical purposes) or *on*, and drops negligible voltage. The transistor cannot do quite so well, but can be operated so that in its *off* state it takes a finite but negligible current, and when *on* it drops a small but negligible voltage.

In many applications, these qualities are essential in order to reduce the power dissipation in the transistor used as a switch. In the *off* state, high voltage across the device multiplied by negligible current equates to low power. In the *on* state, high current times negligible voltage also leads to low power.

The transistor has the advantage over an ordinary switch in that it can be operated by an input signal. So can an electromagnetic relay of course, but the fastest relay cannot operate any faster than about 100 times per second, whereas transistor switching speeds are into the gigahertz region in computers. Moreover, transistors are usually much smaller and cheaper than relays.

Transistors are sometimes used simply as switching devices in, for example, switched-mode power supplies (see Chapter 8), where they switch on and off the currents fed to other parts of the circuit. In other cases, they form the principal elements of digital circuits, where their function is to switch between the two voltage levels representing logical 0 and 1. In this section we look at some of the ways bipolars and FETs are used for these purposes.

9.5.2 Switching circuits using bipolar transistors

Figure 9.18 shows the basic circuit of a switch using a bipolar transistor in the common-emitter configuration. Its purpose is to switch on and off the current passing from the supply, through the load resistor, and down to 0 V.

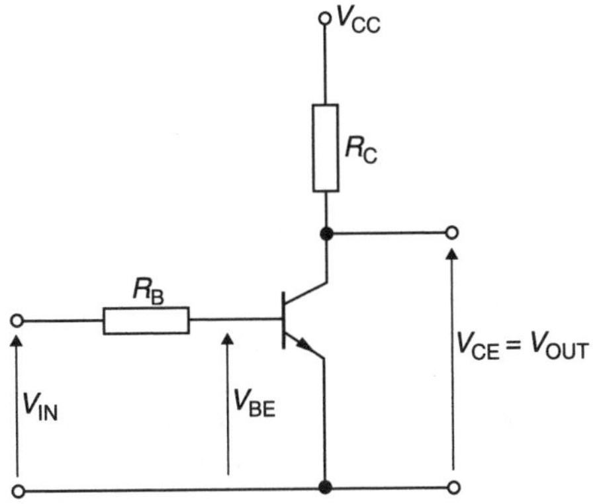

Fig. 9.18 The basic, bipolar transistor switch.

To switch the current off, the base-emitter junction must be reverse-biased, or at least its forward bias must be reduced below that needed for appreciable current to flow. This can be achieved by lowering the input voltage to about 0.5 V or less (for a silicon transistor). At this voltage, negligible base current flows, and the collector current is reduced to little more than the collector-emitter leakage current I_{CE0}. This is defined as the collector current, for normal reverse-bias on the collector, with the base-emitter junction reverse-biased.

To switch the current on, the transistor should, ideally, drop no voltage. If this were the case, then $V_{CE}=0$ and $I_C=V_{CC}/R_C$. To achieve this, we need a base current of $I_B=I_C/\beta=V_{CC}/(\beta R_C)$. Because of the wide spread in the value of β for any chosen transistor type, we have to allow for the worst case, that is the minimum guaranteed value of β, say β_{min}. So the required base current becomes $I_B=V_{CC}/(\beta_{min}R_C)$. The base resistor is then chosen to pass this current when the input voltage is at the level specified to switch the circuit on.

When a bipolar transistor is driven on in this way, the typical voltage drop across it (V_{CE}), in a low-power circuit with a current of a few milliamps, is about 0.1 or 0.2 V. In higher-power applications, such as switched-mode supplies or switched-mode motor control, the drop may be of the order of 0.5 to 1 V at currents of tens of amps. Here the power dissipation in the transistor would be tens of watts, and substantial heat sinks are needed.

Self-assessment question

6 Suppose the 'chopper' transistor in series with the rectified mains supply in a switched-mode power supply has an *on-off* duty ratio of 1:9, that its *off*-state leakage current is 100 μA and that its *on*-state current is 20 A with a 1 V drop across the transistor. Assume the smoothed rectified supply has an average voltage of 340 V.

Calculate the average power dissipation in the transistor.

Unfortunately, the circuit of Fig. 9.18 will usually be supplied with more base current than is necessary to just 'bottom' the collector voltage near 0 V, because of the need to allow for the worst-case value of β. This excess base current leads to an accumulation of charge, called the *saturation charge*, in the base-emitter junction, and this charge has to be discharged, through the base resistor, by the input drive circuit before the transistor will switch off. This slows down the switching time, and makes the circuit unsuitable for high-speed operation.

Speed-up capacitor

One solution to this problem is to shunt a *speed-up* capacitor across the base resistor, as in Fig. 9.19. Insertion of a capacitor in this way is not always possible, especially in integrated circuits where capacitors take up so much surface area, and they load the driving circuit severely. However, if a good drive source is used, the speed-up capacitor ensures a much faster response.

When the input voltage step rises, the current to supply the base charge is not limited by the base resistor but is supplied by the speed-up capacitor. This ensures a rapid switch-on. When the input falls again, the speed-up capacitor removes first the saturation charge and then the base charge and ensures a rapid switch-off.

Schottky diode

The saturation base charge can be reduced considerably by connecting a Schottky diode (see Chapter 8) across the collector-base junction as shown in Fig. 9.20. Without the diode, the collector voltage in saturation is, say, 0.15 V and the base voltage about 0.65 V, so the collector-base junction is *forward-biased* by about 0.5 V. The Schottky diode has a forward voltage drop of only about 0.35 V. So, when the input voltage of Fig. 9.20 is high and the transistor is *on*, some of the input current flows through the diode (in the direction of the arrow) to the collector, and the diode's forward voltage drop limits the forward bias of the collector-base junction to about 0.35 V. Because of the exponential relationship between the diode's current and voltage, this reduction in voltage

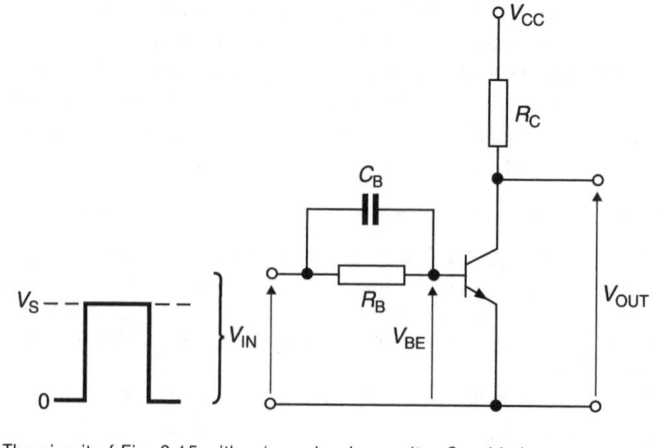

Fig. 9.19 The circuit of Fig. 9.15 with a 'speed-up' capacitor C_B added.

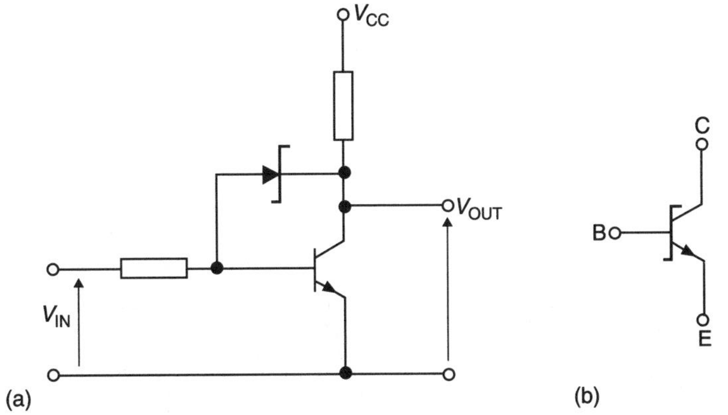

Fig. 9.20 (a) A bipolar transistor switch with a Schottky diode connected across the transistor's collector junction to prevent excessive saturation. (b) The usual symbol for this combination of transistor and Schottky diode.

by 0.15 V causes a reduction in the saturation base charge by a factor of about 400.

This is a much more effective way to speed-up the switch, and the Schottky diode is easy to incorporate in integrated circuits. As a result, most ranges of bipolar digital circuits use this technique.

Bipolar integrated-circuit logic families

There is such a wide range of bipolar logic families that it is impossible to describe their circuit details in the space available. The characteristics of several types are covered in Chapter 5.

9.5.3 Switching circuits using FETs

Figure 9.21 shows a MOSFET basic common-source circuit and its characteristics. You can see from the characteristics that, compared with a bipolar transistor, the MOSFET has a much higher output voltage in the *on* state. In this example, even with 6 V input to the gate, the output voltage exceeds 1 V unless the load is such that less than 1 mA drain current flows. This implies more power dissipation and a less well-defined *on* state output voltage. Partly for this reason, but mainly because resistors are more expensive to make in integrated form than transistors, the circuit of Fig. 9.21a is not used in integrated digital circuits. More practical MOSFET circuits of the CMOS type are used.

CMOS digital logic circuits

Figure 9.22 shows a basic CMOS logic circuit, using a p-channel FET (at the top) and an n-channel FET, both enhancement types, giving a kind of 'push-pull' output. This arrangement is called *complementary symmetry*, and digital circuits using it are called *complementary-symmetry MOSFET* circuits, or **CMOS** for short.

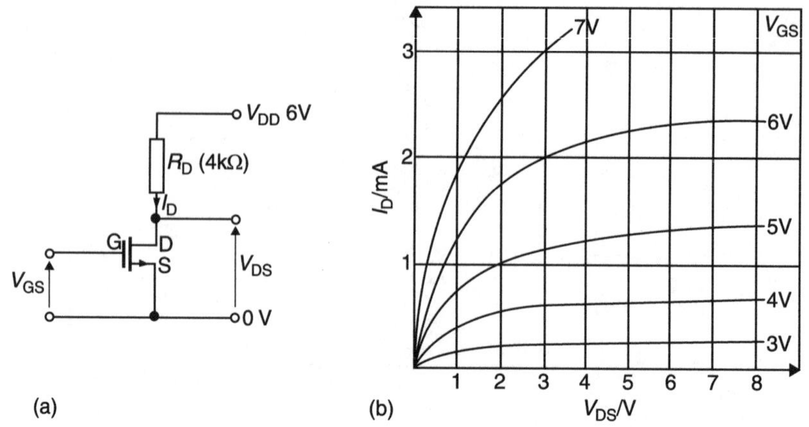

Fig. 9.21 (a) The basic common-source circuit of a MOSFET. (b) The d.c. characteristics of an enhancement-mode, n-channel MOSFET.

Note that, in contrast to the push-pull complementary-symmetry source-follower of the previous section, this push-pull stage is a *common-source* amplifier. Moreover, since it is not expected to cope with both positive and negative input and output signals, it needs only a single supply rail.

If the input is high (about 5 V in Fig. 9.22) the n-channel device (at the bottom) is driven hard into the conducting state, but the p-channel device is cut off. Since the two transistors are in series, $I_D \approx 0$ in both devices. That is, for the n-channel device $V_{GS} = 5$ V and $I_D \approx 0$ which causes V_{DS}, the output voltage, to be very small. (See Fig. 9.21b again.) Similarly, when the input voltage is low (about 0 V) the p-channel device conducts but the n-channel one is cut off, so the output voltage is close to the supply voltage: about 5 V in Fig. 9.22.

Since the output voltage is low when the input voltage is high, and vice versa, this basic CMOS circuit is an inverter.

In its quiescent state, with no load connected to the output, its power consumption is tiny, because $I_D \approx 0$ in both logic states. When the output is

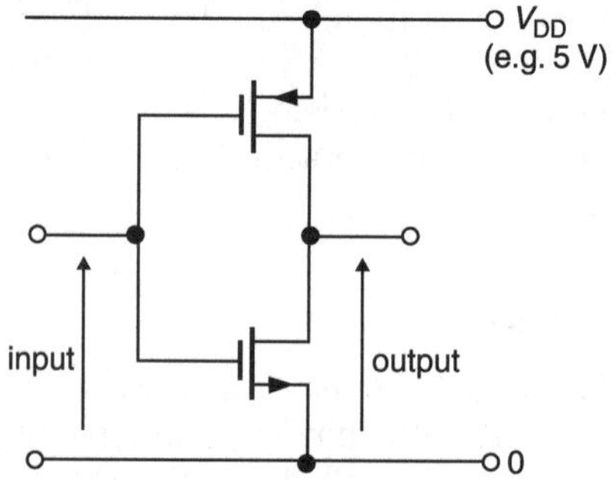

Fig. 9.22 The basic CMOS circuit in which a p-channel enhancement-mode MOSFET and a similar n-channel one are connected together so that they have a common input and output.

connected to the inputs of other CMOS circuits (invertors or gates), the quiescent load resistance is extremely high, because all the inputs are gate terminals of MOSFETS, with extremely high input resistance. So the power consumption remains tiny in the quiescent state, even with other CMOS logic circuits connected to the circuit's output.

However, when the circuit's input switches from high to low, appreciable current flows from the p-channel transistor into the *input capacitance* of the load as the output is switching from low to high. Similarly, appreciable current flows *from* the load capacitance and into the n-channel device when the output is switching from high to low. This means that *the average power consumption of CMOS is proportional to the number of times per second that the switch operates.* It is this near-zero current consumption, except when switching, which makes CMOS so suitable for digital watches, calculators and other low-powered devices in which the gates switch relatively infrequently and are powered by batteries. On the other hand, it is the increased consumption at high switching speeds which makes the CMOS VLSI chips in computers dissipate so much power that they have to be cooled by heat sinks and fans.

CMOS NAND gate

This is shown in Fig. 9.23a. Note that the two p-channel devices have their outputs in parallel, and the two n-channel devices are connected in series. If both inputs are low, T1 and T2 are conducting whilst T3 and T4 are cut-off, so the output is high. Equally, if both inputs are high the output will be low.

Worked example

What happens if only one input is high in the circuit of Fig. 9.23a?

Driving input A high enables T3 to conduct but, since T4 is still cut-off and in series with T3, no current will flow and the output remains high. The fact that T1 becomes cut-off when A goes high simply means that it can be ignored. It is in parallel with T2, which remains conducting. So the output remains high if only one input goes high.

Thus the output is low only when both inputs are high, which is the NAND function.

The number of inputs (the fan-in) can be increased by adding pairs of transistors in a manner which should be clear from this two-input example.

Because of the very low input currents involved, there would be hardly any limit on the fan-out capability if speed did not matter. However, the more inputs of the following circuits there are to be driven, the greater the total load capacitance and the longer the charging time for a given current drive. So the rise and fall times of the output increase almost in proportion to the fan-out.

CMOS NOR gate

This is shown in Fig. 9.23b. If both inputs are low, then both n-channel transistors are cut-off and both p-channel transistors are switched on, so the output goes high. If input *A* goes high (whilst *B* is low), its associated n-channel

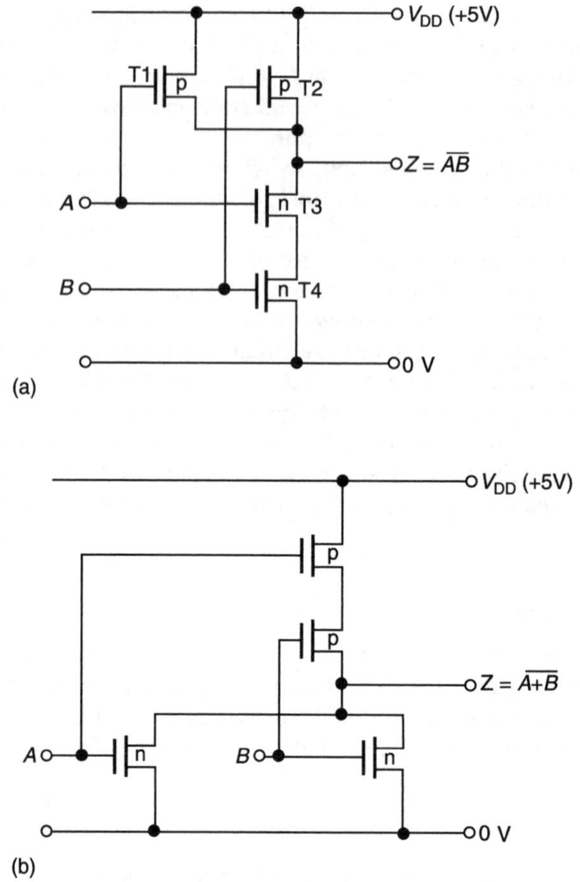

Fig. 9.23 (a) CMOS NAND gate. (b) CMOS NOR gate.

transistor switches on, and its associated p-channel transistor switches off, bringing the output low. Similarly, if B goes high with A low, the output goes low. Both inputs high also cause the output to go low.

Thus the output is low when either or both of the inputs are high, which is the NOR function.

As in the case of the NAND gate, the fan-in of the NOR gate can be increased by adding pairs of transistors in a manner which should be clear from this two-input example.

Again, the output rise and fall times are increased by the fan-out.

9.6 SUMMARY

1. Bipolar transistors have emitter, base and collector regions formed by diffusion of donor or acceptor atoms into intrinsic semiconductors such as silicon. The pnp type has a layer of n-type material sandwiched between

two p-type layers. The npn type has a layer of p-type material sandwiched between two n-type layers.

2. In an npn common-emitter amplifier, the base-emitter voltage is normally positive, making the junction forward-biased. The collector-emitter voltage is normally positive too, making the collector-base junction reverse-biased. However, electrons from the emitter, crossing the emitter-base junction, are swept across the base region and into the collector region. This constitutes a conventional-current flow from the collector to the emitter. The essence of transistor action is that a small change in the base-emitter voltage causes a relatively large change in the collector current, and substantial voltage gain across the collector load resistor.

3. The ratio of the increment in collector current to the increment in base-emitter voltage is called the mutual conductance, and has the value $g_m = 40\ V^{-1}I_C$.

4. A simple npn common-emitter amplifier has its base biased by current through a resistor connected to the positive supply. An alternative, to avoid dependence of the operating point on the d.c. value of the current gain β (h_{fe}), is a potential divider to set the base voltage. An emitter resistor then determines the emitter and collector currents. The input signal is fed in via a coupling capacitor to the base. The output signal is developed across a resistor connected to the positive supply, and is fed out via a coupling capacitor from the collector. The voltage gain of a common-emitter amplifier is $A = -g_m R'_L$, where R'_L is the effective load resistance.

5. The hybrid-π a.c. equivalent circuit of a bipolar transistor has:

   a voltage-controlled current source with value $g_m V_{be}$;
   an input resistance $r_i = \beta r_e$;

   an output resistance $r_o = VA + V_C/I_C$

   where β (or h_{fe}) is the current gain, $r_e \approx 25\,\text{mV}/I_C$ and VA is the Early voltage.

6. The long-tailed pair (differential pair) using two npn bipolar transistors has the two emitters joined, and connected via a constant-current source to the negative rail. The differential input signal is connected between the bases. The collector of one of the transistors is connected to the positive rail, and the output signal is taken from the collector of the other transistor, which is connected to the positive rail via a collector resistor. The differential gain is $A_{v(diff)} = V_o/V_{in} = g_m R'_L/2$.
   The common-mode rejection ratio is

$$\text{CMRR} = \frac{\text{differential gain}}{\text{common-mode gain}} = g_m R_E \approx \frac{R_E}{r_e}$$

7. The npn emitter-follower has its collector connected directly to the positive rail and its emitter connected to 0 V, or a negative rail, via an emitter resistor. Base bias can be provided by either of the methods used for the common-emitter amplifier. The emitter operating point is about 0.65 V

below the base voltage. The input signal is fed in via a coupling capacitor and the emitter voltage 'follows' the input, maintaining the 0.65 V difference. The output can be fed out via another coupling capacitor. The circuit has a voltage gain of nearly one, but 'buffers' a high-impedance source from a low-impedance load.

8. Field-effect transistors have source, gate and drain regions formed by diffusion of donor or acceptor atoms into intrinsic semiconductors such as silicon. The two types are the junction type, or JFET, and the insulated-gate type, IGFET or MOSFET.

9. The n-channel JFET has a p-type region diffused into an n-type channel. The source and drain terminals are at the ends of the channel, and the gate terminal connects to the p-type region. The drain is usually biased positively with respect to the source. The gate-channel junction, which is a pn junction, must be biased negatively. This reverse bias causes the transition region of the gate to widen, so reducing the width of the channel through which the current flows from drain to source. Thus the gate-source voltage controls the source-drain current. The gate-source voltage which completely closes the channel is called the pinch-off voltage.

10. In a MOSFET the source and drain are pn junctions formed side by side in the surface of a silicon substrate, with the gate between them. The gate is a conductor on the surface, insulated from the substrate by a film of oxide. Appropriate gate-source voltage induces carriers in the substrate under the gate. These carriers flow from source to drain when the appropriate drain bias is applied. Thus in the MOSFET too, the gate-source voltage controls the source-drain current. In the n-channel type, the drain current comprises electrons. The gate-source voltage must exceed the threshold voltage before current will flow.

    In an enhancement-mode MOSFET, the threshold is positive (in an n-channel device), and no current flows until V_{GS} exceeds the threshold.

    In a depletion-mode MOSFET, the threshold is negative (in an n-channel device), and current flows when $V_{GS} = 0$. V_{GS} must be more negative than the threshold to stop current flow.

11. The a.c. equivalent circuit of the FET is similar to the bipolar's hybrid-π, but with the appropriate FET quantities substituted for the bipolar ones:

$$g_{\mathrm{m}} = \frac{\mathrm{d}I_D}{\mathrm{d}V_{\mathrm{DS}}}.$$

This is proportional to the square root of the drain operating current; r_{i} is very large, although the capacitance at the gate of both types of FET is significant at high frequencies.

$$r_{\mathrm{o}} = \frac{1/\lambda + V_{\mathrm{DS}}}{I_{\mathrm{D}}}$$

where λ is the channel length modulation factor. $1/\lambda$ is analogous to the Early voltage of the bipolar.

12. Common-source and source-follower FET circuits are analogous to the bipolar common-emitter and emitter-follower circuits, differing mainly in their biasing arrangements.

13. Double emitter-follower push-pull output stages have low output impedance and are capable of 'sourcing' and 'sinking' substantial currents into low-impedance loads. Use with a 3-rail supply enables the circuits to provide both positive and negative signals about 0 V. Special base biasing is necessary to avoid crossover distortion.
    Double source-followers are also used as push-pull output stages. The enhancement-mode types usually used need a few volts gate bias, so V_{BE} multipliers are used to provide this. These provide temperature compensation for the output transistors.

14. Simple bipolar common-emitter switching circuits have a low output voltage in the 'on' state and take little current in the 'off' state. However, they suffer from poor switching speed due to saturation charge in the base-emitter junction. A 'speed-up' capacitor connected across the series base resistor at the input can make a dramatic improvement, by removing the charge rapidly, but loads the driving circuit. The bulk of the capacitor precludes its use for integrated-circuit logic gates.

15. A Schottky diode connected between collector and base reduces the 'on' state forward bias of the collector-base junction to about 0.35 V, reducing the saturation base charge by a factor of about 400. Schottky diodes are used in some of the integrated-circuit bipolar logic families.

16. Simple FET common-source switching circuits have one or two volts output in the 'on' state, which is not well defined, and the voltage drop leads to increased power dissipation. These circuits are not used in integrated circuits.

17. The basic CMOS logic circuit uses complementary symmetry, in which two MOSFETs are connected as a push-pull stage. A p-channel enhancement-mode MOSFET is connected as a common-source amplifier, with its source connected to the positive rail. Similarly, an n-channel enhancement-mode MOSFET has its source connected to the 0 V rail. The two gates are connected to the input terminal, and the two drains are connected to the output. A high input turns off the p-channel device and turns on the n-channel device, bringing the output low. A low input has the opposite effect, lifting the output high. So this is a logical inverter.

18. The CMOS NAND gate has two p-channel enhancement-mode MOSFETs connected in parallel in place of the single p-channel device of the inverter, but with one gate connected to input A, and the other to input B. It also has two n-channel enhancement-mode MOSFETs connected in series in place of the single n-channel device of the inverter, with one gate connected to input A, and the other to input B. When both inputs are high, both the p-channel devices turn off, and the two n-channel devices turn on, bringing the output low. If one or both of the inputs is low, one or both of the n-channel devices turns off, and the output is lifted high by one or both of the p-channel devices. This is the NAND function. The fan-in can be increased by adding further complementary pairs of MOSFETs.

19. The CMOS NOR gate has two p-channel enhancement-mode MOSFETs connected in series in place of the single p-channel device of the inverter, but with one gate connected to input A, and the other to input B. It also has two n-channel enhancement-mode MOSFETs connected in parallel in place of the single n-channel device of the inverter, with one gate connected to input A, and the other to input B. When both inputs are low, both the n-channel devices turn off, and the two p-channel devices turn on, bringing the output high. If one or both of the inputs is high, one or both of the p-channel devices turns off, and the output is brought low by one or both of the n-channel devices. This is the NOR function. The fan-in can be increased by adding further complementary pairs of MOSFETs.

Answers to self-assessment questions

1. From the graph, the collector current at this point is about 4.5 mA. The curve for $V_{BE} = 0.66$ V and $V_{CE} = 5$ V shows I_C at about 6.8 mA. So

$$g_m \approx \frac{\Delta I_C}{\Delta V_{BE}} \approx \frac{2.3\,\text{mA}}{10\,\text{mV}} \approx 230\,\text{mS}$$

The curve for $V_{BE} = 0.64$ V and $V_{CE} = 5$ V shows I_C at about 3.0 mA. So

$$g_m \approx \frac{\Delta I_C}{\Delta V_{BE}} \approx \frac{-1.5\,\text{mA}}{-10\,\text{mV}} \approx 150\,\text{mS}$$

The difference between these two values is, of course, because g_m increases with I_C.

From the formula, the small-signal value at the point $V_{CE} = 5$ V and $V_{BE} = 0.65$ V is

$$\begin{aligned} g_m &= (40\,\text{V}^{-1}) \times I_C \\ &= (40\,\text{V}^{-1}) \times 4.5\,\text{mA} \\ &= 180\,\text{mS} \end{aligned}$$

As you might expect, this value lies between the two values found from the finite increments in V_{BE}.

2. Figure 9.3b has a slope, at the operating point $I_C = 2$ mA, of (8 mA)/(0.1 V), so

$$r_e = 100\,\text{mV}/8\,\text{mA} = 12.5\,\Omega$$

This is predicted correctly from Equation 9.6.

Figure 9.3a has a slope, at the operating point $I_C = 2$ mA, of (80 μA)/(0.1 V), so

$$r_i = 100\,\text{mV}/80\,\mu\text{A} = 1.25\,\text{k}\Omega$$

Since Fig. 9.3a is drawn for $\beta = 100$, we expect it to show $r_i = 100\, r_e$, which is the case.

3. The small-signal equivalent circuit is shown in Fig. 9.24. It is the same as Fig. 9.5b except that the source resistance is added. V_{in} is less than 5 mV because of the generator's internal resistance: that is

$$V_{in} = \frac{V_s \times R_{in}}{R_{in} + R_s}$$

$R_{in} \approx r_i = 1.25$ kΩ. Therefore

$$V_{in} = \frac{5\,\text{mV} \times 1.25\,\text{k}\Omega}{1.25\,\text{k}\Omega + 1\,\text{k}\Omega} = 2.8\,\text{mV}$$

The net load resistance is the parallel combination of the resistances R_C, r_o and R_L:

$$\frac{1}{R'_L} = \frac{1}{2.2\,\text{k}\Omega} + \frac{1}{52\,\text{k}\Omega} + \frac{1}{8\,\text{k}\Omega} = 599\,\mu\text{S}$$

$$R'_L \approx 1.67\,\text{k}\Omega$$

At an operating current of 2 mA, $g_m = 80$ mA V^{-1}, so

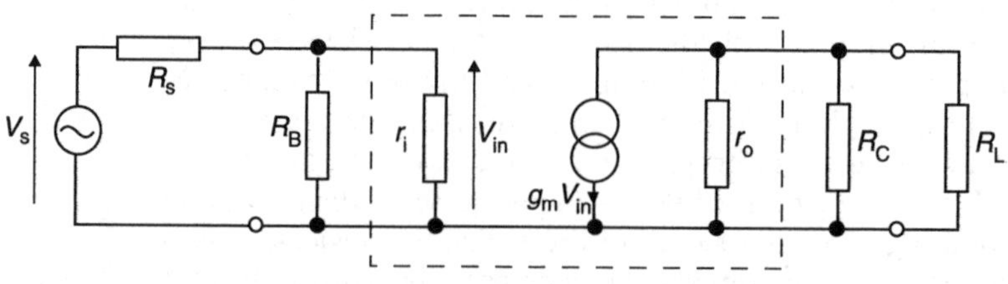

Fig. 9.24 The equivalent circuit of a common-emitter amplifier.

$$\begin{aligned} A &= -g_m R'_L \\ &\approx -80\,\text{mA V}^{-1} \times 1.67\,\text{k}\Omega \\ &\approx -134 \end{aligned}$$

Therefore $V_{out} \approx -2.8$ mV $\times 134 \approx -374$ mV.

4. In Fig. 9.4a, when $\beta = 100$, $I_C = 2$ mA and $V_{CE} = 4.4$ V. When, however, $\beta = 200$, $I_C = 4$ mA and the voltage drop across R_C is 8.8 V. So V_{CE} is 9 V − 8.8 V = 0.2 V. The transistor is therefore driven into the very curved part of the output characteristics which is called the *saturation region* of operation. See Fig. 9.3c at $V_{CE} =$ 0.2 V. The output waveform will be very distorted due to clipping. If $\beta > 200$ the transistor is driven even further into saturation and clipping is even more severe.

5. The fact that the output current from the transistor flows through R_E and produces a voltage drop at the input means that the feedback is current-derived. The fact that it subtracts directly from the input voltage means that it is series-connected to the input. It therefore increases the input resistance of the circuit.

6. The *off*-state dissipation is $P_{off} = 340\ \text{V} \times 100\ \mu\text{A}$

$$= 34\ \text{mW}.$$

The *on*-state dissipation is $P_{on} = 1\ \text{V} \times 20\ \text{A} = 20$ W. The transistor is *off* for 90% of the time and *on* for 10% of the time, so the average power dissipation is $P_{ave} = 0.9 \times 34\ \text{mW} + 0.1 \times 20\ \text{W} \approx 2$ W.

Index